QUANTUM NANOSYSTEMS

Structure, Properties, and Interactions

QUANTUM NANOSYSTEMS

Structure, Properties, and Interactions

Edited by
Mihai V. Putz, PhD

Apple Academic Press
TORONTO NEW JERSEY

Apple Academic Press Inc.
3333 Mistwell Crescent
Oakville, ON L6L 0A2
Canada

Apple Academic Press Inc.
9 Spinnaker Way
Waretown, NJ 08758
USA

Exclusive worldwide distribution by CRC Press, a member of Taylor & Francis Group

Printed in the United States of America on acid-free paper

International Standard Book Number-13: 978-1-926895-90-1 (Hardcover)

Library of Congress Control Number: 2014939543

Library and Archives Canada Cataloguing in Publication

Quantum nanosystems: structure, properties, and interactions/edited by Mihai V. Putz, PhD.

Includes bibliographical references and index.
ISBN 978-1-926895-90-1 (bound)

1. Nanostructures. I. Putz, Mihai V., author, editor

QC176.8.N35R42 2014 620'.5 C2014-902999-3

Apple Academic Press also publishes its books in a variety of electronic formats. Some content that appears in print may not be available in electronic format. For information about Apple Academic Press products, visit our website at **www.appleacademicpress.com** and the CRC Press website at **www.crcpress.com**

ABOUT THE EDITOR

Mihai V. Putz, PhD

Mihai V. Putz is a laureate in physics (1997), with an MS degree in spectroscopy (1999), and PhD degree in chemistry (2002), with many post doctorate stages: in chemistry (2002-2003) and in physics (2004, 2010, 2011) at the University of Calabria, Italy, and the Free University of Berlin, Germany, respectively. He is currently Associate Professor of theoretical and computational physical chemistry at West University of Timisoara, Romania. He has made valuable contributions in computational, quantum, and physical chemistry through seminal works that appeared in many international journals. He has actively promoted new method of defining electronegativity, DFE (density functional electronegativity) among new enzyme kinetics (logistic enzyme kinetics), a new formalization of the structure-activity relationship (SPECTRAL-SAR) model and recently the bondonic quasi-particle theory of the chemical bonding first application on nanosystems as graphene, all seeking for a unitary quantum approach of the chemical structure and reactivity targeting the bio-, pharmaco- and ecological analytical description.

He is Editor-in-Chief of the *International Journal of Chemical Modeling* and the *International Journal of Environmental Sciences*. He is member of many professional societies and has received several national and international awards from the Romanian National Authority of Scientific Research (2008), the German Academic Exchange Service DAAD (2000, 2004, 2011), and the Center of International Cooperation of, Free University Berlin (2010). He is the leader of the Laboratory of Computational and Structural Physical Chemistry for Nanosciences and QSAR at the Biology-Chemistry Department of West University of Timisoara, Romania, where he conducts research in the fundamental and applicative fields of quantum physical-chemistry and QSAR. In 2010 Mihai V. Putz was declared through a national competition the Best Researcher of Romania; in 2013 he was recognized among the first Dr. Habil in Chemistry in Romania, and from 2014 he became full member of International Academy of Mathematical Chemistry (IAMC).

CONTENTS

LIST OF CONTRIBUTORS

Neha Aggarwal
Department of Physics and Astrophysics, University of Delhi, Delhi – 110007, India

Tommaso Avellini
Dipartimento di Chimica "Giacomo Ciamician", Università degli Studi di Bologna, Via Selmi, 2 - 40126, Bologna

Massimo Baroncini
Dipartimento di Chimica "Giacomo Ciamician", Università degli Studi di Bologna, Via Selmi, 2 - 40126, Bologna

Aranya B. Bhattacherjee
Department of Physics, ARSD College, University of Delhi (South Campus), New Delhi-110021, India

Mihaela Birdeanu
National Institute for Research and Development in Electrochemistry and Condensed Matter, 1 Plautius Andronescu Street, 300224 Timisoara, Romania

Jan C. A. Boeyens
Unit for Advanced Scholarship, University of Pretoria South Africa

Fanica Cimpoesu
Institute for Physical Chemistry, Splaiul Independentei 202, Bucharest 060021, Romania

Eugenia Fagadar-Cosma
Institute of Chemistry Timisoara of Romanian Academy, 24 M Viteazu Ave., 300223-Timisoara, Romania

Mircea V. Diudea
Department of Chemistry, Babes-Bolyai University, Arany Janos 11, 400028, Cluj, Romania

Marilena Ferbinteanu
University of Bucharest, Faculty of Chemistry, Inorganic Chemistry Department, Dumbrava Rosie 23, Bucharest 020462, Romania

Gianluca Giovannetti
CNR-IOM-Democritos National Simulation Centre and International School for Advanced Studies (SISSA), Via Bonomea 265, I-34136, Trieste, Italy

Volodymyr Krasnoholovets
Indra Scientific, Square du Solbosch 26, B-1050 Brussels, Belgium

Sonam Mahajan
Department of Physics and Astrophysics, University of Delhi, Delhi – 110007, India

Dušan P. Malenov
Department of Chemistry, University of Belgrade, Studentski trg 12-16, Belgrade, Serbia

Enrico Marchi
Dipartimento di Chimica "Giacomo Ciamician", Università degli Studi di Bologna, Via Selmi, 2 - 40126 Bologna

J. Manuel Perez-Mato
Dept. Fisica de la MateriaCondensada, Fac.Ciencia y Tecnologia, Universidad del Pais Vasco, UPV/ EHU 48080 Bilbao, Spain

Silvia Picozzi
CNR-SPIN, L'Aquila, Italy

Mihai V. Putz
Laboratory of Computational and Structural Physical Chemistry for Nanosciences and QSAR, Biology-Chemistry Department, Faculty of Chemistry, Biology, Geography, West University of Timişoara, Pestalozzi Street No.16, Timişoara, RO-300115, Romania

Domenico Di Sante
University of L'Aquila, Physics Department, Via Vetoio, L'Aquila, Italy

Monica Semeraro
Dipartimento di Chimica "Giacomo Ciamician", Università degli Studi di Bologna, Via Selmi, 2 - 40126 Bologna

Alessandro Stroppa
CNR-SPIN, L'Aquila, Italy

Emre S. Tasci
Physics Dept. Middle East Technical University, METU 06800 Ankara, Turkey

Margherita Venturi
Dipartimento di Chimica "Giacomo Ciamician", Università degli Studi di Bologna, Via Selmi, 2 - 40126 Bologna

Alina D. Zamfir
Faculty of Physics, West University of Timisoara, Romania, Plautius Andronescu nr. 1, 300224, Timisoara, Romania

Snežana D. Zarić
Department of Chemistry, University of Belgrade, Studentski trg 12-16, Belgrade, Serbia

LIST OF ABBREVIATIONS

AFM	Atomic Force Microscopy
AU	Astronomic Unit
BEC	Bose Einstein Condensate
CASSCF	Complete Active Space Self Consistent Field
CE	Capillary Electrophoresis
CID	Collision-Induced Dissociation
CNS	Central Nervous System
CRM	Charge Residue Model
CSD	Cambridge Structural Database
DFT	Density Functional Theory
DNP	Dioxynaphthalene
DRIE	Deep Reactive Ion Etching
ECD	Electron Capture Dissociation
EOF	Electroosmotic Flow
ETD	Electron Transfer Dissociation
GS	GroundState
GUT	Grand Unification Theory
HCT	High Capacity Ion Trap
HG	Hydrodynamic Laws of Gravity
HPLC	High Performance Liquid Chromatography
IR	Infrared
IRMPD	Infrared Laser Multiphoton
LB	Langmuir–Blodgett
LIF	Laser Induced Fluorescence
MCP	Microchannel Plate
MLCT	Metal-to-Ligand (BPY) Charge-Transfer
MRI	Magnetic Resonance Imaging
MS	Mass Spectrometry
NMS	Normal Mode Splitting
OMC	Optomechanical Crystal
PYI	Pyrene Ligand
Q.C.D.	Quantum Chromodynamics
QD	Quantum Dots (Nanocrystals)

SCM	Single Chain Magnets
SEM	Scanning Electron Microscopy
SIM	Single Ion Magnets
SMM	Single Molecule Magnet
SORI	Sustained Off-Resonance Ion
SPM	Scanning Probe Microscopy
STM	Scanning Tunneling Microscopy
TDC	Time to Digital Converter
TGR	Theory of General Relativity
TIC	Total Ion Current
TLC	Thin-Layer Chromatographic
TOE	Theory of Everything
TOF	Time-of-Flight
TOPO	Trioctyl Phosphine Oxide
TTF	Tetrathiafulvalene
UHV	Ultra-High Vacuum
WMAP	Wilkinson Microwave Anisotropy Probe
XIC	Extracted Ion Chromatogram
ZFS	Zero Field Splitting

LIST OF SYMBOLS

Λ	amplitude of the inerton cloud of the NTH node
$\delta r_{\mathbf{n}}$	deviation of this node from its equilibrium position
dS	element of space-time
$d\Psi(w)$	extension of 3D
υ_0	initial velocity of the particle
g	lattice vector
μ_0	mass of the excited cloud of inertons
$m_{\mathbf{n}}$	mass of the n-th node
T_υ	neutrinos temperature
$\vec{B}$	outer field
$p_0 = m\,\upsilon_0$	particle's initial momentum
m	particle's mass
$\overline{T}$	period of collision
$\kappa = \Gamma/2$	phonon damping rate
ℓ_{P}	Planck length
χ	position of the centre mass of the (quantum) cloud
$\hat{S}_A$	spin operators
$\dot{r}_{\tan}$	tangential velocity of a test mass
$V^{\text{deg. cell}}$	typical average volume of a cell
$\mu_{\mathbf{n}}$	variations of mass of n-th node
V^{part}	volume of the kernel cell of the particle
η	viscosity
m_B	Bohr magneton
ω_c	unperturbed cyclotron frequency
μ_e	electrophoretic mobility
μ_{eof}	mobility of the electroosmotic flow
$\vec{r}, \vec{R}$	collection of the electronic and ionic degrees of freedom
$\hat{n}(\vec{r})$	density operator
"||"	parallel orientation
"…"	denotes hydrogen bonding
$1/T$	frequency of collisions
A_{in1}	input noise operators for the two optical modes

c	speed of light
D	effective parameters
D_{02}	initial inner diameter
e	Bouligand exponent
E	energy of the moving particle
F_{x}	x component of atomic spin
Ψ	GS (Ground State) of the system
h	Planck's constant
J_{AB}	coupling constants
K	number of repeating units
K_{B}	Boltzmann constant
M	metal atom
N_a	unpaired electrons
P	pentagon
Q	ionic charge
R	ionic radius
S_{Z}	collective spin operators of the BEC
t	time
U_{ab}	inter-center Coulomb repulsion term
V_{ee}	electron-electron interaction
X	scaling factor
Y_{lm}	Spherical Harmonics
Z	combination of spherical harmonics
Γ	force constant
$\Delta \mathrm{E}_{\mathrm{HOH/HOH}}$	energy of interaction between water molecules
Ξ	noise operator arising as the mechanical mode
γ	atomic damping rates
γ_1	angles
Ω_{r}	mechanical frequency of the cantilever mode

PREFACE

With the dawn of the 21st century, the need for new sources of energy and the increased demand for economical and multifunctional materials in the fields of mechanics, electronics, medicine, and ecology becomes more and more acute as natural, physical, and chemical resources (natural energy and fuels) reveal their limits, their difficulties, and their increased costs in storage, transport, and conversion. Therefore, knowledge of the universe and its materials (along with their self-assembly, self-regeneration, storage, and directional properties) are on the forefront of fundamental and technical research worldwide, in both academia and industry. In response to these challenges, the scientific community has developed the required synergy among involved fields, and the nanosciences emerged, aiming to combine and transfer knowledge (models, tools, practical support) from physics to chemistry to biology and related fields, producing a unitary view of how nano-phenomena trigger micro and macro cosmos, observable or hidden, and designed actions and materials to be implemented toward optimizing everyday life and secure its future.

As a consequence, new phenomena and exotic materials and properties are being explored in current research horizons. These include the quantum universe from the nano- to large-scale understanding in general, with applied subdomains at molecular levels, such as, for example:

- topological effects in nanosystems
- nanostructured solar cells
- noncovalent interactions in organometallic chemistry
- spin crossover compounds
- hysteresis in molecular magnets
- structure anistropy and related properties
- organic ferroelectrics
- coordinative chemistry of actinide ions
- modeling of bioinorganic materials
- nanoscience of carbon's allotropes (fullerenes and nanotori, self-assembly structures, new diamonds, and spongy-carbon; new theorems, tessellation, and aromaticity modeling, chemical reactivity of fullerenes)

These new research horizons also include chemistry and nanoscience and chemical nanostructures from structure to properties such as, among many others:

- oxa- and oxaaza-coronands
- nano-electronics and nano-devices
- nanobiomaterials for dentistry (such as new highly fluorescent materials, chemical captors, chemosensors)
- nanoscience of textiles, food, and ecotoxicology
- optical properties control and design of multifunctional nanomaterials

These topics viewed together assure a consistent picture of matter from quantum particles to atoms and molecules and to solid state, bonded at nanoscale, while manifesting meso to macro effects.

This book aims to collect and order the physical and chemical perspectives on nanosystems triggering macrosystems from a wide perspective of cosmic phenomenas to special nano-devices and phenomena in a succession of chapters narrating "the story of observable and induced cosmos through the eyes of nano-particles and devices."

Chapter 1 presents the so-called Fermi calculations (smart calculations), the fundamental relations of environmental large-scale physics by considering the space-time-energy properties of light and of its quantum photon; from the black-body radiation and light pressure, one may make predictions associated with the quantification of dark mass and energy in the universe, balancing between Jeans' contraction of galaxies and Hubble's expansion.

Chapter 2 describes the universe of particles and their motion as based on a mathematical space constructed as a tessellation lattice of primary balls, while the motion of a particle generates a cloud of excitations around the particle, which were named *inertons,* with remarkable consequences in treating the quantum ψ-function as a mapping of the real system "particle + its cloud of inertons." The basic feature of inertons, that is, is both the particles' inert and gravitational properties, which assures their observability in diluted cold gases and clusters of electrons, or by the Casimir effect; whereas through the associated electromagnetic field, it is able to give a deeper insight into the fundamental nature of things, playing an important role in quantum physics, chemical physics, biochemistry, biophysics, and condensed matter.

Chapter 3 surveys the limits of classical quantum mechanics because the inherent embedded linearity in the basic theory; in addition it is argued that nonlinearity plays a major role in the universe in general and in the nanomaterials and scales in particular, such as, for example, the major types of nanoparticles, including nanogold, graphene, fullerene, boron nitride, semiconductors, and their derivatives. This observation, along with the need for a description of a 4D universe, which implies descriptive geometry alike, proposes a critical understanding and modeling of the nonclassical properties of nanomaterials through number theory and universal self-similarity.

Chapter 4, instead, takes the topological route to describe the nanoworld structures: as constructive units, small cages are used to design complex structures of rotational or translational symmetry, whereas the design of multishell cages recalls the Platonic solids and their transforms, obtained by using simple map operations. The celebrated quasicrystals structures are in this way recovered by self-arrangements and multishell combinations, whereas genus and Omega polynomial calculations enrich the structure characterization of these exotic structures under the spongy manifestation.

Chapter 5 continues the observations of the previous chapter in the self-arranging feature of nanostructures by describing through atomic force microscopy (AFM) of the self-assembling porphyrins by weak Van der Waals forces or hydrophobic effects, hydrogen bond and π–π stacking interactions that finally generates a large variety of their geometries. In this way AFM evidences how the J-aggregates (edge-to-edge stacked) and H-bonding interactions are both responsible for generating different architectures, which sometimes coexist together, such as triangle type sheets, concave-convex surfaces, columns, pyramids, roofs, and even rings.

Chapter 6 takes further the use of nanosystems into optomechanical effects: apart from discussing interesting quantum phase transition in a hybrid optomechanical as a selective tool for transfer energy between a single mechanical mode and two optical modes, an advanced idea of producing laser modes of phonons by the aid of Bose-Einstein condensates is presented and documented, with switching channels controlled by adding atoms into condensate, by magnetically coupling between mechanical oscillations of a nanoscale magnetic cantilever to an ultra-cold atomic cloud in close analogy to a two-level optical laser system.

Chapter 7 presents a survey on today's hot topic of molecular magnetism related with magnetic anisotropy—a phenomena appearing by the interplay of spin and orbital magnetic moments so that obtaining systems behave as magnets at molecular and nano-scale systems. In this framework it presents the pioneering tool for revealing the poles of molecular magnets known as state-specific magnetization surfaces, with specific applications on the f-transition metal ions and d-f complexes, since revealing non-*aufbau* configuration and weak interaction with ligands and neighbor magnetic sites.

Chapter 8 is dedicated to supramolecular photochemistry or more specifically to the design and construction of nanoscale devices and machines capable of performing useful light-induced functions aiming for, in the short run, the transportation of nanoobjects, mechanical gating of molecular-level channels, and nanorobotics, while targeting the construction of chemical computers nano-scale research in the long run. From energy (solar) conversion to sensoring and catalysis, other viable supramolecular photochemistry possibilities include deposition on surfaces, incorporation into polymers, organization at interfaces, or immobilization into membranes or porous materials, providing an open field for still unpredictable challenging nano-functionalizations of materials in interaction with light.

Chapter 9 addresses new systematic insight on aromatic rings interacting with substantial energy at a very large parallel displacement (offsets), out of the ring, even beyond the C-H bond region, and also when the two rings are almost non-overlapping. Accordingly, the noncovalent interactions were explored and their consequences on the nanosystems, aromatic here, are exposed.

Chapter 10 reveals the many facets of chip systems not only in the life sciences but especially on biomedical research. Great relevance is attributed to the research on several micro- and nanofluidics analyzed by using electrospray ionization mass spectrometry (MS). In this regard the mass spectrometry technique and the allied methods (e.g., quadruple time-of-flight (QTOF), Fourier transform ion cyclotron resonance (FTICR), and high capacity ion trap (HCT) are introduced in glycomics for biomedical and clinical frontier applications, such as from screening, sequencing, and the structural analysis of *O*-glycopeptides expressed in the urine of patients suffering from Schindler's disease versus age-matched healthy controls leading to de novo design and validation of useful biomarkers.

Chapter 11 completes the book with a didactic-research exposé on how computational crystallography may use the celebrated density functional theory in studying the ferroelectric properties of materials.

As such, this book is both multidisciplinary and transdisciplinary: it combines methods from various physicochemical fields and presents novel paradigms from one discipline to be applied or transferred to other disciplines, thus offering new insights that apply to various physicochemical, biophysical, biochemical, etc., fields at nano- and mesoscales.

With chapters written by leading international scientists, the book balances both experimental results and theoretical modeling, and the new cutting research trends discussed will be equally valuable to both academia and industry.

Altogether, *Quantum Nanosystems: Structure, Properties, and Interactions* benefits from 21st-century knowledge in the nanosciences, making it pivotal in helping to open new frontiers and research horizons for understanding, synthesizing, and using matter at nanoscale (atoms and molecules) to advance the self-adapted multifunctional use and know-how of nanosystems for new energy, new materials, new technology, and new ecology.

Written by leading and open-minded scientists with international reputations, the book should be in every library of nanoscience literature. This book is enriched by the contributions from leaders from various fields, ages, and countries, all with international research experience and highly rated publications to their credit. It covers a wide and advanced variety of issues and open-issues in nanosciences. The Editor, knowing personally almost all of the contributors, has a vivid interest in connecting their fields of discipline. True thanks are offered to all the contributing authors for their dedication and high-level expertise and for helping to advance 21st-century science beyond the customary frontiers of physics, chemistry, and engineering. Equally, the Editor gratefully acknowledges the research and editing facilities provided by the Romanian Education and Research Ministry within the project CNCS-UEFISCDI-TE-16/2010–2013. Last but not least, the Editor, also on behalf of the book's contributors, thanks the Apple Academic Press team and, in particular, Ashish Kumar, president and publisher, for professionally supervising the production of the multidisciplinary scientific series in general and of this volume in particular.

— Mihai V. Putz, PhD
(Assoc. Prof. Dr. Habil)
West University of Timişoara, Romania
April 2014

FOREWORD

The latest generation of optomechanical quantum systems based on Bose Einstein condensates properties, phonon lasers, light-controlled molecular machines—the building blocks for the next-decade supramolecular photochemistry, and the most recent developments of ab-initio computational crystallography are reported here in association with more "imaginative" chapters as well, devoted to Coxeter-like atlas of novel hyperstructures or to the intriguing description of *inerton field* and the way it influences the basic properties of the matter. Outcomes of several important research studies are presented in these well-written chapters. Detailed descriptions of magnetic anisotropy phenomena and porphyrin derivatives characterized by atomic force microscopy imaging are enriched by deep insights on the emerging role of non-covalent interactions in determining the nanostructure of aromatic rings stacking.

Prof. Putz keeps a balanced eye on nanoscale on applicative and fundamental subjects, prompting graduates students and professional researchers on the latest theoretical results. Fibonacci decorations on the surface of Ag/silica particles disclose a profound theoretical re-examination of the growing mechanisms of "nanoparticles," from nanogold to semiconductors, evidencing how much quantum systems are conditioned by their inherent *four-dimensional nature*. The instrumental service of quantum theories in shaping micro- and macro- cosmos is testified the editor's original investigation on the spatial-temporal evolution of the universe resulting in an appropriate estimate of the dark matter vs. dark energy observed distribution, which captures the reader's attention on the inextricable, scale-invariant efficiency of the physical laws in describing the world.

The outstanding scientific level of the authors contributing to this book simply makes these "horizons" a must-read book on nanomaterials.

This volume constitutes an authoritative guide to nanoworld and will surely support frontier research in chemical physics, biochemistry, biophysics, condensed matter, and modern quantum theory as well.

Ottorino Ori
Actinium Chemical Research
Rome, September 9, 2013

CHAPTER 1

NANOUNIVERSE EXPANDING MACROUNIVERSE: FROM ELEMENTARY PARTICLES TO DARK MATTER AND ENERGY

MIHAI V. PUTZ

CONTENTS

ABSTRACT

"Fermi calculations" (smart calculations) for the combination of the fundamental relations of the environmental physics at large scale are considered in determining of the space-time-energy quantities of practical interest broadening the general knowledge of macro-cosmos based on the micro/nano matter-properties; to this aim, the existence of the photon and the associated (black body) radiation is explicitly or implicitly involved, according to the occurrence of the measures directly observed or through the estimated effect produced by Universe after the first milliseconds (in the era of nucleo-synthesis, nuclei, atoms and galaxies). This way, the predictions associated to the quantification of dark mass and energy are made for a Universe associated to the Jeans contraction of galaxies and to the Hubble expansion, respectively.

1.1 INTRODUCTION

In order to offer classical-quantum unitary view of the phenomena of macro-micro-cosmos, this chapter combines three fundamental themes of Physics, such as, Classical Mechanics with applications to the gravitational Newtonian mechanics; Relativistic Mechanics with the Einstein kinematic forms and consequences, with application in description of the Universe in expansion through Hubble's law, the argument for the Big-Bang Theory; statistical Physics with Bose-Einstein distribution, with applications and consequences of the radiation of black body at the level of the Universe and the elementary particles.

These themes and the related applications have been selected (from the author's experience by teaching topics of Environmental Physics) as being the most accessible and useful to the students who are in direct contact with the Modern Physics, with applications and universal models, related to the forces and fundamental particles, stars and planets, nuclear reactions and radioactive decay, while accumulating imperative notions to future approaches of modeling the Structure of the Atoms and Molecules, Structural Physical-Chemistry, Quantum Chemistry and Environmental Chemistry of course.

It was purposely omitted the presentation of other universally theory which naturally arising from the presented arguments—*The General Theory of Relativity*—which, although very important for a complete description of Cosmology (at least in terms of gravity, or precisely in terms of the principle of equivalence inertial mass-gravitational mass, sustained by the expansion of physical laws beyond the inertial systems to the accelerated systems) it would enlarge to much of this chapter, which only contains the essential cosmological ideas and determinations: Bing Bang modeling through the straight after zeroth moment—Planck Universe, Expansion of Universe, calculations of Expansion time while the Universe become cold, Radiation of Cosmic Background simulation, the quantum-gravity equivalence of photon, etc. As for the General Relativity themes, it was decided to approach it with further occasion, designed precisely for Advanced Environmental Physics.

Therefore, this chapter is made to be a basic, orientate application of the essential principles from Mechanical Physics (classical and quantum), Nuclear and Statistical (classical and quantum), for a better understanding of Universe. Actually, the Environmental Physics can be considered as being not a specific field of Physics, but the fundamental synthesis of all of the Physics chapters, having as a starting point the combination of micro- and macro-phenomena in characterization of global and local observed effects. This "art" to combine micro- and macro-cosmos is actually one of the conceptual, didactic and scientific challenge (attribution) of Environmental Physics at a classical-quantum-relativistic combined level.

The fascination and sometimes the difficulty of this subject appear when the quantities of the micro universal constants (light speed in vacuum and Planck constant) are combined with the macro ones (as the universal gravitational constant or Boltzmann constant for a statistical ensemble to the macroscopic balance type) in order to generate physical sizes (length, time, temperature, etc.) with universal value; but definitely they give the unifying key in the right knowledge, prediction and even the control of the evolution and interactions of Nature's objects.

Thus, this chapter is dedicated especially to those from Physics related specializations and interdisciplinary applications in environmental problems; yet, this chapter can be also read and used as a model-guide in combining the fundamental notions of Physics by the graduate students. Furthermore, it can be used in the Scientific Cycles and the extracurricular activities of preacademic intermediate, but also as a starting point in writing a graduation paper, in Physics and Chemistry, regarding the Cosmology and considering the Universe as an open physical system.

1.2 BACKGROUND THEORIES OF UNIVERSE

1.2.1 ON GRAVITATION AND GALACTIC CONDENSATION

One starts by observing the gravitational attraction law can be written under gradient formulation

$$\vec{F} = m \cdot \vec{g} = -m \cdot gradU \tag{1}$$

with gravitational potential given by the simple expression

$$U = -G\frac{M}{r} \tag{2}$$

Worth noting that the Eq. (2), through which the dynamic (inertial) form is equalized with the gravitational form of Newton's law, represents the premise of the so-called principle of equivalence (between the inertial mass and the gravitational mass), fundamental for the general relativity theory of Einstein! In these conditions, one can apply the Gauss's law/theory which connects the flow of a vector $\vec{g}$, through a closed surface Σ, to the divergence integral.

$$\oint_{\Sigma} \vec{g}d\vec{S} = \int_{V} div\vec{g}dV \tag{3}$$

If we consider the vector in question as being bound to the gravitational source of Newton's attractive force,

$$|\vec{g}| = |grad_r U| = \left|\nabla_r \left(G\frac{M}{r}\right)\right| = G\frac{M}{r^2} \tag{4}$$

then, the left side of Gauss's law term in Eq. (3) is consecutively written

$$\oint_{\Sigma} \vec{g}d\vec{S} = \oint_{\Sigma} |\vec{g}| \cos(\vec{g},\vec{u}_n) dS = -\oint_{\Sigma} |\vec{g}| dS_n = -\int |\vec{g}| r^2 d\Omega$$

$$= -GM\int d\Omega = -4\pi GM = -4\pi G\int_{V} \rho dV \tag{5}$$

where the natural geometry from Fig. 1.1 have been considered for the scalar product of gravitational acceleration with the unit vector (normal at the integration surface).

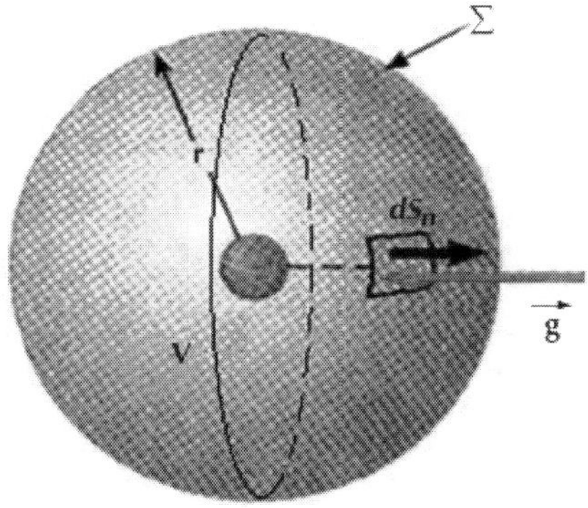

FIGURE 1.1 Geometry of gravitational action for Gauss's law application for gravitational acceleration; credit: Putz, M.V. Environmental Physics and Universe (in Romanian as "Fizica Mediului şi Universul") West University of Timişoara Press: Timişoara, 2010 (adapted after HyperPhysics 2010; http://hyperphysics.phy-astr.gsu.edu/hbase/hph.html).

By equalizing the last result with the right side of Gauss's law (3) the relation of gravitational acceleration divergence results

$$div\vec{g} = -4\pi G\rho \tag{6}$$

In terms of gravitational potential gradient, the last relation is written first by a direct replacement

$$div\left(-gradU\right) = -4\pi G\rho \tag{7}$$

which is immediately generated by recognizing the differential identity involving the Laplacin of the gravitational potential

$$div\left(gradU\right) = \vec{\nabla}\left(\vec{\nabla}U\right) = \nabla^2 U = \Delta U\ , \tag{8}$$

and the Poisson's theorem for propagation of gravitational potential based on the gravitational source represented by the mass density ρ

$$\Delta U = 4\pi G\rho \tag{9}$$

equivalent with the Newton's law of gravitational action at a distance! Moreover Poisson's gravity equation usually represents the benchmark and the classical nonrelativistic specialization of Einstein's equations of Theory of General Relativity (TGR).

Notice that, because the gravitational field is not a closed field inside the producing source volume, the gravitational acceleration corresponds to a non-rotational field, which, in terms of field equation is rewritten by canceling the associated rotor

$$rot\vec{g} = \nabla \times \vec{g} = 0 \tag{10}$$

As an application to the Universe structure, the law of gravitational acceleration divergence will be further used, in order to elaborate the classical model (nonrelativistic and nonquantum) of the cosmic systems, galaxies, stars and planets formation.

We consider the gravitational field as a fluid, which follows the hydrodynamic laws. Considering the gravitational source as being represented by the mass density ρ, the gravitational field is moving with a velocity $\vec{v}$, with the gravitational acceleration $\vec{g}$ oriented to the center of the mass source and developing a pressure $\vec{P}$ reactive to the gravitational effect (i.e., contrary to this); it satisfy three forms of hydrodynamic laws of gravity (HG):

The continuity of gravitational charge equation, which varies in space and time in a given region

$$\text{HG1:}\ \frac{\partial \rho}{\partial t} + \vec{\nabla} \cdot \underbrace{(\rho \vec{v})}_{\substack{GRAVIFIC\\ CHARGE\\ CURRENT}} = 0 \tag{11}$$

The convective hydrodynamic equation HG2 of total gravitational acceleration equation written as the (absolute) convective derivative

$$\frac{D\vec{v}}{Dt} = \frac{d}{dt}\vec{v}\left(t, \vec{x}(t)\right) = \frac{d\vec{v}}{dt} + \frac{d\vec{v}}{dx}\frac{d\vec{x}}{dt} = \frac{d\vec{v}}{dt} + \left(\vec{v} \cdot \vec{\nabla}\right)\vec{v} \tag{12}$$

resulted from the competition between the gravitational acceleration of the source and the one generated from the reactive pressure of the gravitational field (which can be of quantum nature if it's referring to the photons pressure radiated by a star, or purely of kinetic nature if it's referring to galactic or cosmic gas dust which condenses)

$$\text{HG2:}\ \frac{d\vec{v}}{dt} + \left(\vec{v} \cdot \vec{\nabla}\right)\vec{v} = -\frac{\vec{\nabla} P}{\rho} + \vec{g} \tag{13}$$

The equation of gravitational flux, connected to the gravitational acceleration divergence (and by default to the Poisson form of Newton gravitational law):

$$\text{HG3:}\ \vec{\nabla} \cdot \vec{g} = -4\pi G \rho \tag{14}$$

Then, one considers as an initial state (unperturbed) of the galactic/stellar/planetary structure the intergalactic space (-stellar or -planetary), where the gravity forces and masses are absent, i.e. the mass density and pressure being constant quantities

$$\begin{matrix} INITIAL \\ GALACTIC \\ STATE \end{matrix} \begin{cases} \vec{g} = \vec{g}_0 = 0 \\ \vec{v} = \vec{v}_0 = 0 \\ \rho = \rho_0 = ct. \\ P = P_0 = ct. \end{cases} \tag{15}$$

The formation of galactic nebulae, stellar or planetary, is thus considered to be associated with the bringing of gravitational forces in a perturbative way, where the previous conditions become

$$\begin{matrix} GRAVIFIC \\ GALACTIC \\ STATE \end{matrix} \begin{cases} \vec{g} = \vec{g}_1 \\ \vec{v} = \vec{v}_1 \\ \rho = \rho_0 + \rho_1 \ \ldots \ \rho_1 / \rho_0 \in [-1,+1] \\ P = P_0 + P_1 \ \ldots \ P_1 / P_0 \in [-1,+1] \end{cases} \tag{16}$$

which lead to the specialization of the hydrodynamic equations of gravity. Thus, HG1 becomes *galactic*-HG1 (GHG1) equation

$$\underbrace{\frac{\partial \rho_0}{\partial t}}_{=0} + \frac{\partial \rho_1}{\partial t} + \vec{\nabla} \cdot \left[(\rho_0 + \rho_1) \cdot \vec{v}_1 \right] = 0$$

$$\Leftrightarrow \frac{\partial \rho_1}{\partial t} + \rho_0 \vec{\nabla} \cdot \vec{v}_1 + \underbrace{\rho_1}_{\cong 0} \vec{\nabla} \cdot \vec{v}_1 + \vec{v}_1 \vec{\nabla} \cdot \underbrace{\rho_1}_{\cong 0} = 0$$

$$\Rightarrow \text{GHG1: } \frac{\partial \rho_1}{\partial t} + \rho_0 \vec{\nabla} \cdot \vec{v}_1 = 0 \tag{17}$$

In same manner, HG2 becomes GHG2:

$$\frac{d\vec{v}_1}{dt} + \left(\vec{v}_1 \cdot \vec{\nabla} \right) \underbrace{\vec{v}_1}_{\cong 0} = -\frac{\vec{\nabla} \cdot (P_0 + P_1)}{\rho_0 + \rho_1} + \vec{g}_1$$

$$\Leftrightarrow \frac{d\vec{v}_1}{dt} \cong -\frac{\vec{\nabla} P_1}{\rho_0} + \vec{g}_1 \tag{18}$$

If the size is introduced by taking the pressure divergence,

$$\vec{\nabla} P_1 = \frac{\partial P_1}{\partial \rho_1} \frac{\partial \rho_1}{\partial x} = v_S^2 \vec{\nabla} \rho_1 \tag{19}$$

connected with the sound speed in gravitational matter

$$v_S = \sqrt{\frac{\partial P_1}{\partial \rho_1}} \tag{20}$$

we obtain the following expression for the GHG2 law

$$\frac{d\vec{v}_1}{dt} \cong -\frac{v_S^2}{\rho_0}\vec{\nabla}\rho_1 + \vec{g}_1 \tag{21}$$

Next, by calculating the derivative of equation GHG1 in relation with the time and then using the expression GHG2 followed by the integration of equation HG3 one successively obtains

$$0 = \frac{\partial}{\partial t}\left(\frac{\partial \rho_1}{\partial t} + \rho_0 \vec{\nabla}\cdot\vec{v}_1\right)$$

$$= \frac{\partial^2 \rho_1}{\partial t^2} + \rho_0 \vec{\nabla}\cdot \underbrace{\frac{\partial \vec{v}_1}{\partial t}}_{HGH2}$$

$$= \frac{\partial^2 \rho_1}{\partial t^2} + \rho_0 \vec{\nabla}\cdot\left[-\frac{v_S^2}{\rho_0}\vec{\nabla}\rho_1 + \vec{g}_1\right]$$

$$= \frac{\partial^2 \rho_1}{\partial t^2} - v_S^2 \nabla^2 \rho_1 + \rho_0 \underbrace{\vec{\nabla}\cdot\vec{g}_1}_{HG3}$$

$$= \frac{\partial^2 \rho_1}{\partial t^2} - v_S^2 \nabla^2 \rho_1 - 4\pi G \rho_0 \rho_1 \tag{22}$$

generating a single hydrodynamic gravific condensation (HGC) equation

$$\text{HGC:} \quad \frac{\partial^2 \rho_1}{\partial t^2} = v_S^2 \nabla^2 \rho_1 + 4\pi G \rho_0 \rho_1 \tag{23}$$

which emphasizes the time variation relating the space modification for the galactic gravitational condensation through mass density of initial galactic state, together with the sound speed in the gravitational mass involved in condensation and with the contribution of universal gravitational constant G.

Finding a "monochromatic" solution of condensation gravitational density, such as, solves the equation,

$$\rho_1 \propto \exp\left[i\left(kx - \omega t\right)\right] \tag{24}$$

By formation of the directly relations

$$\begin{cases} \dfrac{\partial^2 \rho_1}{\partial t^2} = -\omega^2 \rho_1 \\ \nabla^2 \rho_1 = \dfrac{\partial^2 \rho_1}{\partial x^2} = -k^2 \rho_1 \end{cases} \tag{25}$$

the HGC equation (23) is reduced to the relationship of gravitational dispersion of condensation

$$\begin{aligned} \omega^2 &= v_S^2 k^2 - 4\pi G \rho_0 \\ &= v_S^2 \left(k^2 - \frac{4\pi G \rho_0}{v_S^2} \right) \\ &= v_S^2 \left(k^2 - k_J^2 \right) \end{aligned} \tag{26}$$

where the Jeans' wave vector has been introduced

$$k_J = \sqrt{\frac{4\pi G \rho_0}{v_S^2}} \tag{27}$$

This way, we reached the classical interpretation of the galactic condensation phenomena under the action of gravitational force: when the body mass exceeds the mass contained in Jeans' sphere, of radius

$$R_J = \frac{2\pi}{k_J} = \sqrt{\frac{\pi v_S^2}{G \rho_0}} \tag{28}$$

beyond the critical Jeans' mass

$$M > M_J = \left(\frac{4}{3} \pi R_J^3 \right) \rho_0 = \frac{4}{3} \pi \rho_0 \left(\frac{\pi v_S^2}{G \rho_0} \right)^{3/2} \tag{29}$$

it leaves with the condition

$$k < k_J \tag{30}$$

with the appearance of imaginary frequencies as the direct consequence

$$\omega_J = \pm i v_S \sqrt{k^2 - k_J^2} = \pm i \operatorname{Im} \omega \tag{31}$$

and respectively with the instability in the solution of mass density

$$\begin{aligned} \rho_1 &\propto \exp\left[-i\omega_J t\right] = \exp\left[\pm\left|\operatorname{Im}\omega\right| t\right] \\ &= \begin{cases} \exp\left[+\left|\operatorname{Im}\omega\right| t\right] \ldots \textit{contraction} \\ \exp\left[-\left|\operatorname{Im}\omega\right| t\right] \ldots \textit{aeration} \end{cases} \end{aligned} \tag{32}$$

However, the present theory gives interesting information: the universal gravitational constant G might be seen, by eliminating the mass density in the initial galactic state between the mass and the Jeans' radius, such as

$$G = \frac{4\pi^2}{3} v_S^2 \frac{R_J}{M_J}. \tag{33}$$

This form has in "its structure" the critical sizes of condensation and the contribution of speed sound by the condensed substance!

The values of this approach can be proved by applying the expression associated to sound speed (obtained in this case by eliminating the Jeans' mass from the last relation with the Jeans' radius and the initial density of condensation)

$$v_S = R_J \sqrt{\frac{G\rho_0}{\pi}} \tag{34}$$

to the Earth data (referred to the Jeans-critical-stable condition) with radius $R = 6378$ (km) and mass $M = 5.975 \times 10^{24}$ (kg) we have $\rho_0 = 3M/(4\pi R^2) = 5.5$ g/cm^3 for which the reasonable value is obtained

$$v_s \cong 2.17 \text{km/s} \tag{35}$$

Nevertheless, the gravitational constant has a decisively contribution to the galaxies, stars and planets formation, and it is essentially for systematically cosmological theory, at any level: classical, relativistic, quantum and combined [1–12].

1.2.2 HUBBLE LAW: EXPANDING OBSERVABLE UNIVERSE

A very important kinematic consequence of the special relativity theory is connected to the temporal Lorentz transformation for the time measured in a system (t'), which uniformly moves with a velocity (v) respecting a system of reference ($0x$) where the time is measured (t)

$$t' = \frac{t - vx/c^2}{\sqrt{1 - v^2/c^2}} \tag{36}$$

with c-speed of light in vacuum.

Therefore, we can immediately write the chain of equivalent temporal transformations

$$t' = \frac{t - \frac{v}{c^2}x}{\sqrt{1 - \frac{v^2}{c^2}}} = \frac{t\left(1 - \frac{v}{c^2}\frac{x}{t}\right)}{\sqrt{1 - \frac{v^2}{c^2}}} = \frac{t\left(1 - \frac{v}{c^2}c\right)}{\sqrt{1 - \frac{v^2}{c^2}}} = \frac{t\left(1 - \frac{v}{c}\right)}{\sqrt{1 - \frac{v^2}{c^2}}} \tag{37}$$

where, in frequencies $\upsilon = 1/t$, reverses as

$$\upsilon'_{SOURCE} = \upsilon \frac{\sqrt{1 - \frac{v^2}{c^2}}}{\left(1 - \frac{v}{c}\right)} = \upsilon_{OBS} \sqrt{\frac{1 + v/c}{1 - v/c}} \tag{38}$$

This relation is valid also at the wavelengths level ($\lambda = c/\upsilon$) between the source system that emits a radiation (the mobile system) and the observed system which observe it (the stationary system)

$$\lambda_{OBS} = \lambda_{SOURCE} \sqrt{\frac{1 + v/c}{1 - v/c}}. \tag{39}$$

The law allows the calculation of displacing velocities of the massive stars (even of the galaxies) when the displacing in spectrum recorded for the Hydrogen atom, the most abundant element of Universe (99.99%), is known. For example, if the green ($n_1 = 2$) – blue ($n_2 = 4$) transition line of Hydrogen is displaced inside the spectrum to the red region at 700 nm when observing a galaxy then we have

$$\lambda_{SOURCE} = \frac{c}{\nu_{SOURCE}} = \frac{ch}{h\nu_{SOURCE}} = \frac{ch}{13.6eV\left(1/2^2 - 1/4^2\right)} = 486\,\text{nm} \tag{40}$$

where the *h*-Planck's constant have been considered

$$h = 6.626068 \times 10^{-34}\ (\text{m}^2\ \text{kg/s}) \tag{41}$$

along the Bohr law transition in the spectrum of the Hydrogen atom

$$h\upsilon = E_2 - E_1 = R_H\left(\frac{1}{n_1^2} - \frac{1}{n_2^2}\right) \tag{42}$$

with Rydberg's constant

$$R_H = 13.6eV \tag{43}$$

Turning back, form the relativistic Doppler relation (39) we obtain the equation

$$\left(\frac{\lambda_{OBS}}{\lambda_{SOURCE}}\right)^2 = \left(\frac{700}{486}\right)^2 = 2.075 = \frac{1 + v/c}{1 - v/c} \tag{44}$$

with the velocity result

$$v = -0.35c \tag{45}$$

which indicates the fact that the galaxy is drawing away (the minus sign indicates the back velocity orientation -*v*) from the observer (whereupon the light signal came with the light speed *c*), which it also justifies the name of recessional velocity. Erwin Hubble, while studying the movement of galaxies, observed that all of them are drawing away from the observer, and among each other, *see* Fig. 1.2.

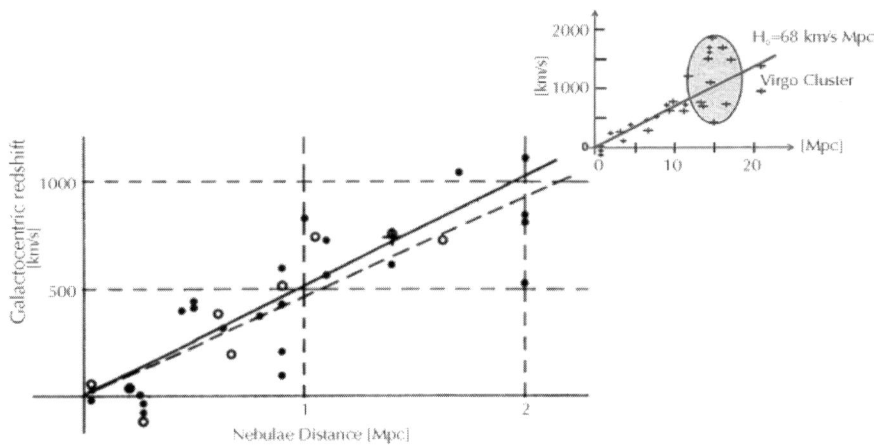

FIGURE 1.2 Left: interpolation of recessional velocities for Hubble's law determination (according to William C. Keel, The Road to Galaxy Formation, Berlin, Springer, 2007, published in association with Praxis Pub., Chichester, UK. ISBN 3540725342); credit: Wikipedia commons http://en.wikipedia.org/wiki/File:Hubble_constant.JPG continuing the excerpt of the original figure of interpolation of article of Hubble.[1]

Those observations caused the advancing of the Universe expansion hypothesis. Considering a cosmic scale factor $a(t)$ which represents the relative expansion of the Universe, sometimes called Robertson-Walker scale factor. With its aid the distance D_t can be connected at the moment t after the moment characterized by the distance D_0

$$D_0 = D_t a(t) \tag{46}$$

Worth noting that, because the relation (46) assumes as an effect the space expansion at the dimension D_t towards the earlier one, $D_0 < D_t$, the cosmic factor scale has actually the role of contracting the space-time metric while the expansion, in other words, when applied, it will be considered with minus sign in variation:

$$a(t) \xrightarrow{\partial_t} -\partial_t a(t) \tag{47}$$

Under these circumstances, a galaxy is considered at a certain point at the distance D_t from the observer (found in the present moment), the distance modification in time generates the recessional velocity equation (Hubble's law)

[1]Edwin Hubble, A Relation Between Distance and Radial Velocity Among Extra-Galactic Nebulae, *Proceedings of the National Academy of Sciences*, **1929**, vol. 15 no. 3, pp. 168–173.

$$v = \partial_t D_t = \partial_t \left(\frac{D_0}{a(t)} \right) = -D_0 \frac{\left[-\partial_t a(t) \right]}{\left[a(t) \right]^2} = \underbrace{\frac{D_0}{a(t)}}_{D_t} \frac{\partial_t a(t)}{a(t)} = \frac{\partial_t a(t)}{a(t)} D_t = HD_t \tag{48}$$

where Hubble's parameter has been introduced

$$H(t) = \frac{\partial_t a(t)}{a(t)} \tag{49}$$

it is experimentally estimated as being merely a constant, in fact by the interpolations' slope such as those presented in Fig.1.2 with the values established in Table 1.1.

TABLE 1.1 Values Measured for Hubble's Constant

Year	Value H_0 (km/s)/Mpc	Observer
2009	74.2±3.6	Hubble Space Telescope
2006	77±11.55	Chandra X-ray Observatory
2001	72±8	Hubble Space Telescope

Considering the astronomic distances in the international units:

- astronomic unit (AU), **1 AU = 149,598,871 km**, representing the medium distance between Sun and Earth along a complete Earth orbit;
- one light year—the distance traveled by light (with the light speed) in a year, **1 ly ≈ 63,241 AU**;
- the parsec—the distance from where an astronomic unit subtends a one arcsecond, **1 pc ≈ 206,265 AU**.

one may observe that, since the Hubble constant has actually the inverse temporal dimension, by considering its observed average value expressed in $\sec^{-1}$

$$H_0 \sim 2.29 \times 10^{-18} \sec^{-1} \tag{50}$$

it can be concluded that its inverse represents the actual age of the Universe

$$t_{Universe} \cong \frac{1}{H_0} = 4.35 \times 10^{17} s = 13.8[Billion\ \ Years] \tag{51}$$

If one considers the Hubble constant as a real constant, the Universe expansion equation will be simply written such as:

$$\frac{dD}{dt} = DH_0 \Rightarrow D = d_0 \exp\left(H_0 t\right) \tag{52}$$

with d_0 the Universe size at a certain moment. Note that this equation has validity on not so large temporal intervals, when complicated phenomena predicted by the generalized relativity, like dark energy, inflation, etc., do occur. For times relatively closed with the observational one (the current time) can establish, with the aid of Hubble's law, the direct relation between the displacement to the red and the recessional speed.

Therefore, if we reconsider (generalize) the relationship between the distance-observer source and the cosmologic scale factor respective for each state, so having the form

$$\underbrace{D_0 a(t_0)}_{SOURCE} = \underbrace{D_t a(t)}_{OBSERVER} \tag{53}$$

then, the condition of red-shift can be successively written

$$z = \frac{D_t - D_0}{D_0} = \frac{a(t_0)}{a(t)} - 1 \cong \frac{a(t_0)}{a(t_0) + (t - t_0)\left[-\partial_t a(t)\right]_{t=t_0}} - 1$$

$$= \frac{a(t_0)}{a(t_0)\left\{1 - \dfrac{\left[\partial_t a(t)\right]_{t=t_0}}{a(t_0)}(t - t_0)\right\}} - 1 = \frac{a(t_0)}{a(t_0)\left[1 - (t - t_0)H(t_0)\right]} - 1$$

$$= \frac{1}{1 - (t - t_0)H(t_0)} - 1 = \frac{(t - t_0)H(t_0)}{1 - (t - t_0)H(t_0)} \cong (t - t_0)H(t_0) \cong \frac{D_t}{c} H(t_0) \tag{54}$$

from where one yields the relationship

$$z \cong \frac{v}{c} \tag{55}$$

once the Hubble's law is applied in the last relation of (54).

Worth to mention the fact that the same dependence between the red-shift and the recessional speed can be also established directly by processing the relativistic Doppler relation

$$z = \frac{\Delta\lambda}{\lambda_{SOURCE}} = \frac{\lambda_{OBS} - \lambda_{SOURCE}}{\lambda_{SOURCE}} = \sqrt{\frac{1+v/c}{1-v/c}} - 1 = \frac{\sqrt{1+v/c} - \sqrt{1-v/c}}{\sqrt{1-v/c}}$$

$$= \frac{2v/c}{\sqrt{1-v/c}\left(\sqrt{1+v/c}+\sqrt{1-v/c}\right)} = \frac{2v/c}{\sqrt{1-v^2/c^2}+1-v/c} \cong \frac{2v/c}{1-\frac{1}{2}\frac{v^2}{c^2}+1-v/c}$$

$$= \frac{v/c}{1-\frac{1}{4}\frac{v^2}{c^2}-\frac{1}{2}\frac{v}{c}} \cong \frac{v}{c}\left(1+\frac{1}{4}\frac{v^2}{c^2}+\frac{1}{2}\frac{v}{c}\right) \cong \frac{v}{c} \tag{56}$$

where the Taylor development has been considered in the first order

$$(1+x)^a \cong 1+ax \tag{57}$$

while the higher powers of the ratio *v/c* have been neglected.

Through these two demonstrations (actually equivalent) the intimate connection between the Hubble's law and the Doppler's relativistic effect, as a kinematic consequence of the special relativity theory is confirmed, a fact proved by the experimental observations performed by the light receptions from galaxies and clusters of galaxies, Fig. 1.3. Of great philosophical importance, the red-shift and the Hubble law advance the idea that not even our own galaxy (Milky Way), Fig. 1.4, is not the "Middle of the Universe," but it is situated in an area of the Universe, along with other galaxies and groups of galaxies, forming the so-called Island Universe! This discovery is similar with, or generalizes, the Copernicus' discovery that the Earth is not in the middle of the World, but orbits around the Sun!

Moreover, by corroborating the conclusions related to the Universe expansion, correlated to the red-shift and the age of the Universe as the inverse of Hubble's constant (parameter), Fig. 1.5, one immediately reaches the idea for the existence of a start time moment of the Universe, associated with the Big-Bang state, when all the matter was contracted, wherefrom, due to the universal explosion the Universe is uniformly and isotropically extended (large scale), which is confirmed through the red-shift systematically recorded relatively to all of the astral objects observed [13–30].

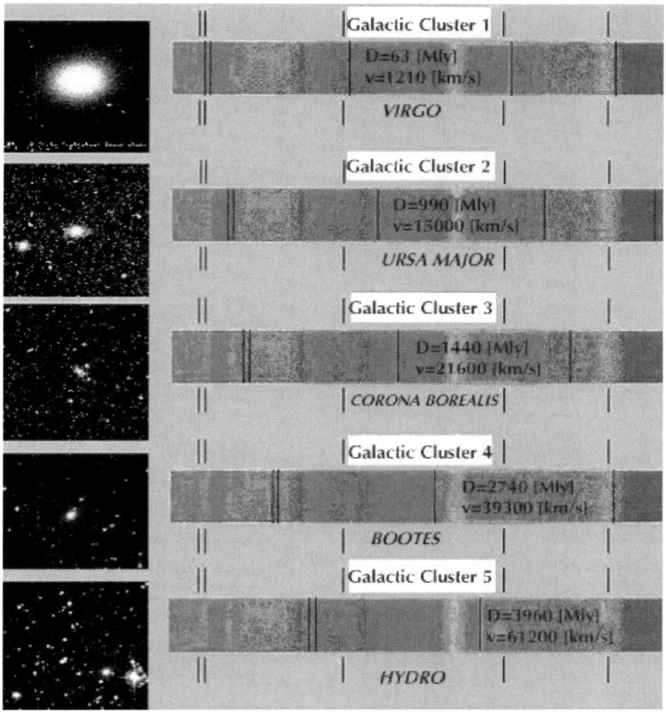

FIGURE 1.3 Illustration of red-shift of the farther and farther galactic clusters (the distance is in Mega-"light years," Mly) from the observer (situated on the Earth); credit: http://astronomy.nmsu.edu/astr110/ (July 2013).

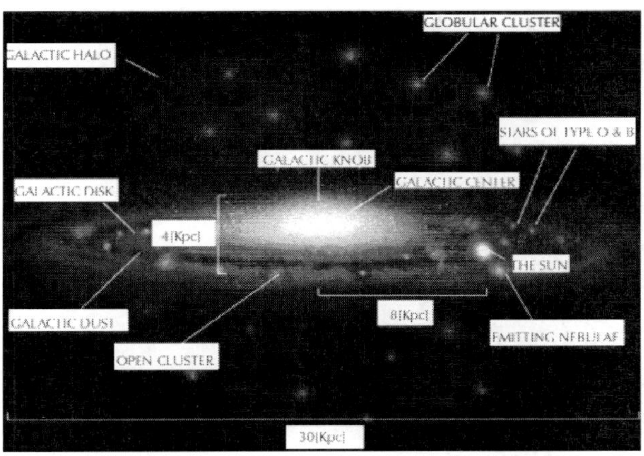

FIGURE 1.4 Insular structure of our galaxy, Milky Way; credit: http://www.euhou.net/.

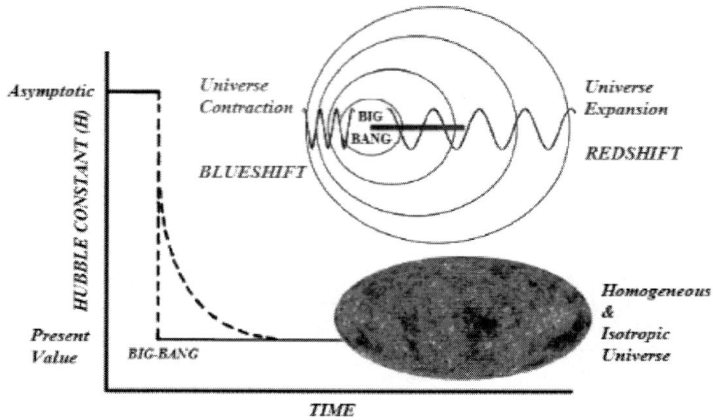

FIGURE 1.5 Homogeneous and isotropic expansion of the Universe predicted from the Big-Bang theory, in agreement with the Hubble's Law of red-shift predicted by the relativistic Doppler effect. The picture of the right-down is taken from the research spatial Wilkinson Microwave Anisotropy Probe (WMAP) and represents the "baby" picture of the Universe at about 3000K of Universe (with warm-red and cold-blue spots); adapted after HyperPhysics 2010 (http://hyperphysics.phy-astr.gsu.edu/hbase/hph.html).

1.2.3 ELEMENTARY PARTICLES AND THE PLANCK UNIVERSE

A subtle level of unification of quantum distribution is referring to: (i) the quality of fermions (semi-integer spin) to characterize the substance aka elementary particles of matter; (ii) bosons (integer spin) are attributed to the particles associated with fundamental fields (implicit to forces) of matter which intercede the interactions between elementary particles. In order to clarify, we further expose the characteristic elementary particles to the fundamental forces of Nature.

1.2.3.1 ELEMENTARY FERMIONS

1.2.3.1.1 SLIGHT FERMIONS: LEPTONS

Three fundamental families/generations of leptons list as:

$$\begin{matrix} Q=-1 \\ Q=0 \end{matrix} : \begin{pmatrix} e^- \\ \upsilon_e \end{pmatrix}, \begin{pmatrix} \mu^- \\ \upsilon_\mu \end{pmatrix}, \begin{pmatrix} \tau^- \\ \upsilon_\tau \end{pmatrix} \tag{58}$$

each generation being a combination of charged lepton connected to the corresponding neutrino lepton (neutral).

e^-, *the electron,* is the oldest known lepton, discovered by J.J. Thomson in 1897, with *electronic neutrino* υ_e anticipated by Ettore Majorana (student of E. Fermi, missing in 1938), theorized by Pauli in 1930 and experimentally discovered in 1956 by Reines and Peierls.

μ^-, *the muon,* was discovered in 1937 by Anderson (also discoverer of positron) with *muonic neutrino* υ_μ revealed in 1962.

τ^-, *triton (*from tritos = the third, in Greek), or *thaon,* is a very high mass lepton (of baryons order- see below) considered a heavy lepton, discovered in 1977 by Peierls with *tritonic (thaonic) neutrino* experimentally identified in 1990.

1.2.3.1.2 HEAVY FERMIONS: QUARKS

Three fundamental families/generations of quarks list as:

$$\begin{matrix} Q=+2/3 \\ Q=-1/3 \end{matrix} : \underbrace{\begin{pmatrix} u \\ d \end{pmatrix}}_{\begin{pmatrix} up \\ down \end{pmatrix}}, \underbrace{\begin{pmatrix} c \\ s \end{pmatrix}}_{\begin{pmatrix} charm \\ strange \end{pmatrix}}, \underbrace{\begin{pmatrix} t \\ b \end{pmatrix}}_{\begin{pmatrix} top/truth \\ botton/beauty \end{pmatrix}} \quad (59)$$

Quark concept have been introduced by M. Gell-Mann in 1964–65 at CALTECH (although a preprint with similar ideas already existed at CERN from Georg Zweig, who called them as *the asis*), by taking the term from the opera *Finnegans Wake* of James Joyce, quoting: “Three quarks for Muster Mark!”.

In 1964 were about 200 particles called elementary, without being systematically explained, until the naive theory (with 3 quarks *q=u, d, s*) of Gell-Mann; for example, the proton appears as the combination *p=3q: uud*, while the neutron has the structure *n=3q=udd.*

The strangeness of the name came to emphasize the strangeness of these particles, which have a factionary charge, in fact a subelectronic one!

The combination of *3 quarks qqq* forms the bonded states called baryons (nucleons, hyperons, etc.).

The combination of 2 quarks (one quark and one antiquark) $q\tilde{q}$ generates the bonded states of mesons (for example the mesons π, k, or the *charmonium* particle $J/\psi = c\tilde{c}$).

From leptons and quarks the substance of the matter can be formed, whatever complex it may be.

Nevertheless, recently also appear that the quarks may have, in turn, internal structure!

1.2.3.2 FUNDAMENTAL FORCES

1.2.3.2.1 STRONG INTERACTION

Strong interaction occurs between the baryonic quarks.

It is interceded by the g particle-GLUON (from glue= which bind, in English).

- With s(G)=1 spin.
- With zero rest mass *m*(G)=0.

The interaction through the gluons is the central subject of quantum chromodynamics (Q.C.D); the *cromo* attribute came from the fact that the gluons are characterized as well by the property called "color," a specific number of quantification: red (*r*), green (*g*), blue (b), turquoise (antired $\overline{r}$), (antigreen $\overline{g}$), yellow (antiblue $\overline{b}$), with the associated combinations;

The specific potential is written as

$$V_t = -k_1 \frac{1}{r} + k_2 r = \begin{cases} \infty ... r \to 0 \\ \infty ... r \to \infty \end{cases} \tag{60}$$

1.2.3.2.2 ELECTRO-MAGNETIC INTERACTION

It occurs between bodies with electric charge, being intermediated by PHOTON-γ.

- With s(γ)=1 spin.
- With zero rest mass $m_0(\gamma)=0$.

It is the subject of quantum electrodynamics (Q.E.D.).

The specific potential is written as

$$V_{EM} = -\frac{1}{4\pi\varepsilon_0}\frac{Q_1 Q_2}{r} \tag{61}$$

1.2.3.2.3 WEAK INTERACTION

It models the transformations between the nucleons, being interceded by WEAKONS (from weak, in English), both charged $W^{\pm}$ or the neutrons Z^0.

It has unspecific potential.

Charged weakons interfere in the β proton-neutron transformation reactions at the level of nucleus, such as,

$$\begin{array}{lllllll} \beta^- : & n & \rightarrow & p & + & W^- & \\ & & & & & \downarrow & \\ & & & & & e^- & + \quad \tilde{\upsilon}_e \end{array} \tag{62}$$

$$\begin{array}{lllllll} \beta^+ : & p & \rightarrow & n & + & W^+ & \\ & & & & & \downarrow & \\ & & & & & e^+ & + \quad \upsilon_e \end{array} \tag{63}$$

With s($W^{\pm}$,Z^0)=1 spin.

With masses $m_0(W^{\pm})$~84 GeV (or protonic masses m_p), $m_0(Z^0)$~94 GeV. Discovered at CERN in 1983 (in January $W^{\pm}$, in June Z^0) by the team of Carlo Rubia and Simon Van der Meer (awarded with Nobel Prize in 1984).

1.2.3.2.4 GRAVITATIONAL INTERACTION

It occurs between the bodies with mass, being mediated by GRAVITON-Γ.

- With s(Γ)=2 spin.
- With zero rest mass m_0(Γ)=0.
- It moves with the light speed.
- Graviton is carried by the gravitational waves.
- Undetected yet.

The specific potential is written as:

$$V_G = -G\frac{M_1 M_2}{r} \tag{64}$$

By unification of fundamental forces, respectively the first three interactions types of the Nature generate the so-called GUT (Grand Unification Theory) model, which extended to the fourth interaction- the Gravitational one, actually provides the weakest form the TOE (Theory of Everything) picture, with the schematically representation in Fig. 1.6.

Phenomenologically, the final unification which occurs at the so-called Planck scale era level, in Fig. 1.7 can be analytically explained in a direct manner, effectively considering the macro-micro cosmos unification, both at gravitational and quantum level, using the fundamental formulas of gravity, theory of relativity, along with the Planck quantification of energy:

$$E = h\upsilon = \frac{h}{t} \tag{65}$$

$$ma = G\frac{mM}{r^2} \tag{66}$$

This way, one obtains the universal constant of gravity relationship

$$G = \frac{ar^2}{M} \tag{67}$$

which can be eventually equivalently written in "Planck manner" involving only universal constants (supposed unified or at least inter-exchangeable at the level of the grand unification of forces, particles and energies immediately after Big-Bang)

$$G = \frac{L_P c^2}{M_P} \tag{68}$$

where everything that is not an universal constant becomes a Planck quantity, specific to the birth of Universe!

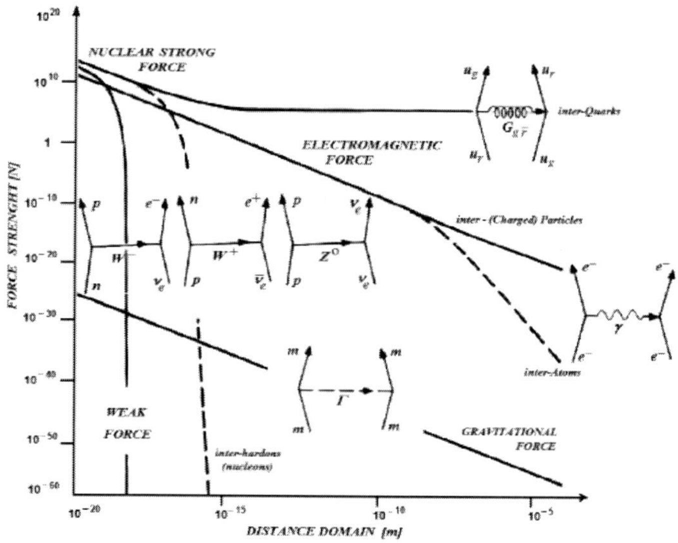

FIGURE 1.6 Representation in relativistic space-time cones' paradigm (the so-called "Feynman diagrams") of fundamental interactions, quantized by specific particles which intermediate the typical elementary particles, see the text, superimposed to the scale of fundamental forces in relation with their reciprocal strength and the domain of distance where it is applicable (adaptation after Stierstadt K., Physik der Materie, VCH-Wiley, Weinheim, 1989, p. 42).

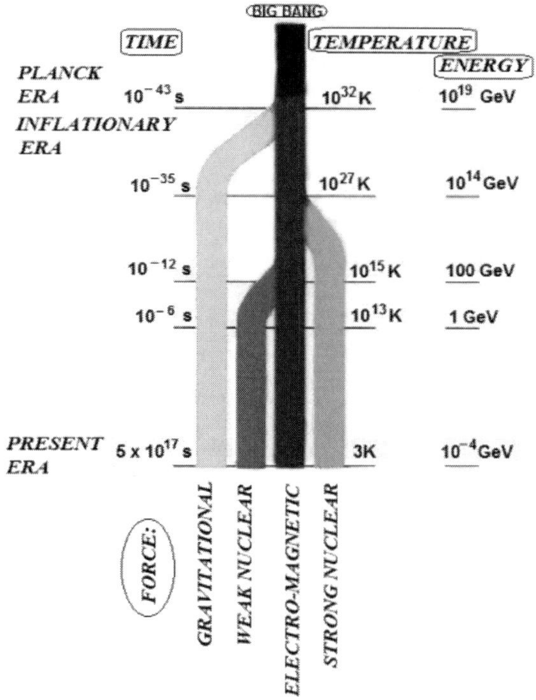

FIGURE 1.7 Unification paradigm of fundamental forces of Nature in relation with time, temperature and energy of dominant particles in the specific age along to the evolution of Universe; adapted after HyperPhysics 2010 (http://hyperphysics.phy-astr.gsu.edu/hbase/hph.html).

On the other side, using the Broglie's relation

$$p\lambda = h \tag{69}$$

rewritten at the Planck Universe level, a second working relation is obtained

$$\left(M_P c\right) L_P = h \tag{70}$$

which combined with the first from above Eq. (68), it firstly allows the elimination of Planck length, for example, by dividing the two relations, resulting in the Planck mass

$$M_P = \sqrt{\frac{ch}{G}} \cong 5 \cdot 10^{-8} [kg] \tag{71}$$

which further allows the length of early Universe (Planck) to be found

$$L_P = \sqrt{\frac{Gh}{c^3}} \cong 4 \cdot 10^{-35}[m] \tag{72}$$

generating a set of cascade determinations, as the Planck time

$$t_P = \frac{L_P}{c} = \sqrt{\frac{Gh}{c^5}} \cong 10^{-43}[s] \tag{73}$$

driving the *Planck energy* calculation

$$E_P = \frac{h}{t_P} = \sqrt{\frac{hc^5}{G}} \cong 5 \cdot 10^9[J] \cong 3 \cdot 10^{19}[GeV] \tag{74}$$

and implying the *Planck's temperature*

$$T_P = \frac{E_P}{k_B} = \sqrt{\frac{hc^5}{Gk_b^2}} \cong 3 \cdot 10^{32}[K] \tag{75}$$

while being accompanied by a colossal *density* of the early Planck Universe

$$\rho_P = \frac{M_P}{L_P^3} = \frac{c^5}{hG^2} \cong 10^{96}[kg/m^3] \tag{76}$$

This is how by unifying the quantum, gravitational and relativistic concepts one can obtain in fact a physical characterization of Big-Bang moment until the Planck horizon, by space-temporal parameters, in terms of mass, thermal and energetic expressions only, and by universal constants (Planck constant, Boltzmann constant, light speed in vacuum, universal gravity constant) as all of these would be unified in a single entity [31–70].

1.2.4 BLACK BODY RADIATION

For modeling photon radiation, the starting point is represented by the law of Bose-Einstein statistic

$$N_i = \frac{g_i}{\exp\left(\frac{\varepsilon_i - \mu}{k_B T}\right) - 1} \tag{77}$$

which it will be properly adapted, under the following peculiarities.

Considering the fact that for photons, at equilibrium, the condition of constancy of the total number of particles $N \neq ct$ is not operating, wherefrom results the appropriate Lagrange multiplier cannot be applied, and therefore the associated chemical potential is zero

$$\mu = 0 \tag{78}$$

The degeneration of multiplicities can not be applied regarding the spin, but because there are two types of electromagnetic polarization ($g = 2$), the transition from *discrete* statistic to continuous one is done through the *small space of phases*, quantum normalized (in accordance with Heisenberg localization/delocalization)

$$N_i \to dN = \frac{1}{\exp\left(\frac{\varepsilon}{k_B T}\right) - 1}\left(\frac{g}{a} d\gamma\right) \tag{79}$$

where, for the photon particles (with three degrees of freedom) we have

$$a = (2\pi\hbar)^3 = h^3 \tag{80}$$

with the elementary volume of phases

$$d\gamma = (dxdp_x)(dydp_y)(dzdp_z) = d\mathrm{r}d\mathrm{p} = dV\left(4\pi p^2 dp\right) \tag{81}$$

rewritten by adapting the energy-momenta particle relationship for the photon

$$\varepsilon = c \cdot p \tag{82}$$

expressed as

$$d\gamma_p = 4\pi V \frac{\varepsilon^2}{c^2}\frac{d\varepsilon}{c}. \tag{83}$$

With this, the photon statistics becomes

$$dN = \frac{8\pi}{ac^3} V \frac{\varepsilon^2 d\varepsilon}{\exp\left(\frac{\varepsilon}{k_B T}\right) - 1}. \tag{84}$$

Based on these statistics, the total energy of photon radiation is obtained, by passing from the discrete definition to the continuous one

$$U = \sum_i \varepsilon_i N_i \to \int \varepsilon dN \tag{85}$$

and is successively calculated

$$U = \frac{8\pi}{ac^3} V \int_0^\infty \frac{\varepsilon^3 d\varepsilon}{\exp\left(\frac{\varepsilon}{k_B T}\right) - 1}$$

$$\overset{\frac{\varepsilon}{k_B T}=x}{=} \frac{8\pi V}{ac^3} (k_B T)^4 \underbrace{\int_0^\infty \frac{x^3 dx}{\exp(x) - 1}}_{\pi^4/15}$$

$$= \frac{8\pi V}{ac^3} (k_B T)^4 \frac{\pi^4}{15} \tag{86}$$

where the value of Riemann-zeta function has been used, under its third order form

$$\int_0^\infty \frac{x^3 dx}{\exp(x) - 1} = \frac{\pi^4}{15} \tag{87}$$

If one considers the density of energy

$$u = \frac{U}{V} = \frac{8\pi^5 k_B^4}{15 h^3 c^3} T^4 \tag{88}$$

this can be used to define the so-called radiance or energy flux density or radiant flux (energy emitted by a blackbody per unit area per unit time, or the radiant power per unit area)

$$L = \int dL = \frac{P}{A} = \frac{U}{A \cdot t} = \frac{x}{t} \frac{U}{A \cdot x} = \frac{c}{4} u \tag{89}$$

which further takes the expression of the so-called Stefan-Boltzmann law

$$L = \sigma T^4 \tag{90}$$

where the Stefan constant has been introduced

$$\sigma = \frac{2\pi^5 k_B^4}{15h^3c^3} = 5.670400 \times 10^{-8} [J \cdot s^{-1} \cdot m^{-2} \cdot K^{-4}] \tag{91}$$

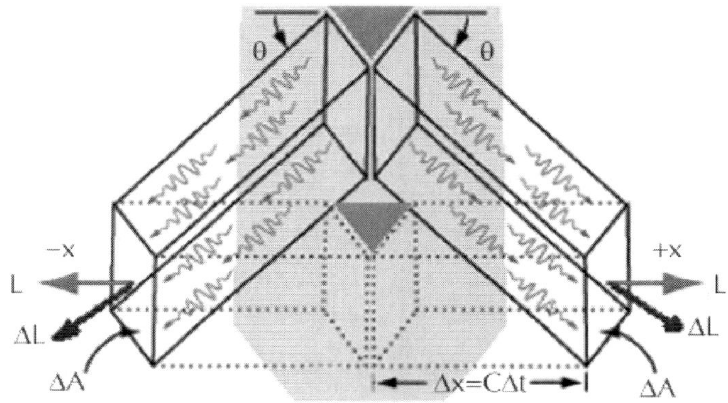

FIGURE 1.8 The geometrical configuration, which express the photonic radiance; adapted from HyperPhysics 2010 (http://hyperphysics.phy-astr.gsu.edu/Hbase/hframe.html).

Note that the factor ¼, from the radiation definition (89), comes from the geometrically condition of integration upon the values of possible radiation angles, toward a given source, as shown in Fig. 1.8:

$$\Delta U = L \cdot t \cdot \Delta A$$

$$= \underset{\pm x}{\underbrace{2}} \frac{\Delta L}{\cos\theta} \cdot \frac{\Delta t}{\cos\theta} \cdot \Delta A$$

$$= 2 \frac{1}{\cos^2\theta} \cdot \frac{\Delta x}{c} \cdot \Delta L \Delta A \tag{92}$$

wherefrom results the customary expression

$$L = \int_{\theta} \Delta L d\theta = \frac{\Delta U}{\Delta V} \frac{c}{2} \overbrace{\underbrace{\frac{1}{\pi} \int_{-\pi/2}^{\pi/2} \cos^2\theta d\theta}_{1/2}}^{\langle \cos^2\theta \rangle} = \frac{c}{4} u \tag{93}$$

Moving forward, if we rewrite the density of total radiant energy in quantification of Planck radiation $\varepsilon = h\upsilon$, then, the integral is formed

$$u = \frac{U}{V} = \frac{1}{h^3}\int_0^\infty \frac{8\pi}{c^3}\frac{(h\upsilon)^3 d(h\upsilon)}{\exp\left(\frac{h\upsilon}{k_B T}\right)-1}$$

$$= \int_0^\infty \frac{8\pi h}{c^3}\frac{\upsilon^3 d\upsilon}{\exp\left(\frac{h\upsilon}{k_B T}\right)-1}$$

$$\equiv \int_0^\infty \rho(\upsilon,T)\,d\upsilon \tag{94}$$

wherefrom the energy spectral density (in frequency) is identified

$$\rho(\upsilon,T) = \frac{8\pi h}{c^3}\frac{\upsilon^3}{\exp\left(\frac{h\upsilon}{k_B T}\right)-1} \tag{95}$$

which corresponds to the Planck formula (distribution) of the blackbody radiation! Note that Planck used his formula under the spectral intensity form as the quantity of emitted energy by a blackbody at temperature T per unit area, per unit time, per unit solid angle, in the $(\upsilon, \upsilon + d\upsilon)$ interval – that is, the unit of frequency:

$$I(\upsilon,T) = \frac{1}{\Omega}\frac{1}{t}\frac{1}{A}\frac{dU}{d\upsilon}$$

$$= \frac{1}{4\pi}\underbrace{\frac{1}{t}\underbrace{\frac{V}{A}}_{x}}_{c}\frac{8\pi h}{c^3}\frac{\upsilon^3}{\exp\left(\frac{h\upsilon}{k_B T}\right)-1}$$

$$= \frac{2h\upsilon^3}{c^2}\frac{1}{\exp\left(\frac{h\upsilon}{k_B T}\right)-1}$$

$$= \frac{c}{4\pi}\rho(\upsilon,T) \tag{96}$$

Another important quantity related to the energy density is referring to the photon radiation pressure, phenomenological deduced from the general relations

$$P_\gamma = \frac{Force}{Area} = \frac{1}{A}\frac{dp}{dt} \overset{\varepsilon=c\cdot p}{=} \frac{1}{c}\underbrace{\frac{1}{A}\frac{d\varepsilon}{dt}}_{\sim c\cdot u} \propto \frac{1}{c}(c\cdot u) = u \tag{97}$$

and which, by taking into consideration the statistical geometric conditions, in the framework of equipartition of the three considered coordinate, we will have the working formula for the radiation pressure

$$P_\gamma = \frac{1}{3}u = \frac{1}{3}\sigma^* T^4 \tag{98}$$

It is worth noting that in astrophysics, there is often used the alternative form of the Stefan-Boltzmann law, written at the energy density level

$$u = \sigma^* T^4 \tag{99}$$

with the new Stefan constant (Stefan-star)

$$\sigma^* = \frac{4\sigma}{c} = \frac{8\pi^5 k_B^4}{15h^3c^3} = 7.565767\times10^{-16}[J\cdot m^{-3}\cdot K^{-4}]$$

These are the theoretical premises that will allow us to characterize the various bodies of the universe (the Sun, the Earth, the Stars and even the Cosmos as an ensemble) based on electromagnetic radiation emitted or stored by them [71–84].

1.3 FERMI-WEINBERG TYPE ESTIMATIONS OF UNIVERSE'S DARK MATTER AND ENERGY

1.3.1 ELEMENTARY PARTICLES IN UNIVERSE AND PHYSICAL CONSEQUENCES

One starts with the Table 1.2, which resumes the physical properties of elementary particles; it can be observed that the photons and the neutrinos are the only particles assumed to have null rest mass, implicitly with the possibility of creation (from quantum or subquantum vacuum) at any nonnull temperature – thus forming the required elements for the "mysterious" lack of energy (having the expansion effect) and the dark matter (with effect of antiexpansion). Moreover, due to the fact that the photon is boson (intermediates the electromagnetic interactions) and the neutrino is fermion (is part of the structure of the matter), their existence cover, in principle, a very large part of the Universe phenomenology, the early and the current one (see the background variation). As a consequence, this section is dedicated to the main analytical explorations of the existence of these two particles in Universe, with the associated spatial-temporal consequences.

TABLE 1.2 Physical Properties of the Main Elementary Particles

	Particle	Symbol	Energy of repose (MeV)	Threshold Temperature (10^{9} K)[2]	Effective number of species[3]	Average life time (s)
	Photon	γ	0	0	$1 \times 2 \times 1 = 2$	Stable
	Electronic neutrinos	$\upsilon_e, \overline{\upsilon}_e$	0	0	$2 \times 1 \times 7/8 = 7/4$	Stable
	Muonic neutrinos	$\upsilon_\mu, \overline{\upsilon}_\mu$	0	0	$2 \times 1 \times 7/8 = 7/4$	Stable
LEPTONS	*Electron, positron*	e^-, e^+	0.5110	5.930	$2 \times 2 \times 7/8 = 7/2$	Stable
LEPTONS	*Muon, anti-muon*	μ^-, μ^+	105.66	1226.2	$2 \times 2 \times 7/8 = 7/2$	2.197×10^{-6}
	Neutral Pi meson	π^0	134.96	1566.2	$1 \times 1 \times 1 = 1$	0.8×10^{-16}
	Charged Pi mesons	π^-, π^+	139.57	1619.7	$2 \times 1 \times 1 = 2$	2.60×10^{-8}
HADRONS	*Proton, anti-proton*	$p, \overline{p}$	938.26	10888	$2 \times 2 \times 7/8 = 7/2$	Stable
HADRONS	*Neutron, anti-neutron*	$n, \overline{n}$	939.55	10903	$2 \times 2 \times 7/8 = 7/2$	920

(according to S. Weinberg, "The first three minutes of the Universe," Political Publishing House, Bucharest, 1984, p. 169)

[2]From the relation $k_B T = m_0 c^2$ results the (threshold) temperature over which a particle can be freely created from the thermal radiation.

[3]Is the (relative) contribution of each species of particle at the density of cosmic matter (energy, pressure, entropy) much beyond the specific threshold temperature; it is calculated by the product: Distinct Existence of Antiparticle (2 If Exist, 1 If Not) × Number of Spin Orientation × Degree of Freedom of Species with Spin ($2^3/2^3=1$ for Bosons and $[2^3–1]/2^3 = 7/8$ for Fermions).

From the presumptive neutrality condition of Universe, that is, the equality between the number of electrons and the numbers of protons, it can be considered that the searched sum of the number of electrons and protons in Universe (in fact the number of existent atoms of Hydrogen)

$$N_e + N_p = ? \tag{100}$$

can be equivalently expressed by the quadratic ratio of electrostatic and gravitational self-interactions, assuming the cosmic competition between these two forces at micro and macro levels of substance. Thus, there are considered the Coulombic charges

$$|Q_e| = Q_p = \frac{e}{\sqrt{4\pi\varepsilon_0}}$$

$$= \frac{1.602176487\times10^{-19}[C]}{\sqrt{4\pi\times8.854188\times10^{-12}[C^2kg^{-1}m^{-3}s^2]}} = 0.151891\times10^{-13}[kg^{1/2}m^{3/2}s^{-1}] \tag{101}$$

and respectively the *gravitational charges*

$$G_e = m_e\sqrt{G}$$

$$= 9.1093897\times10^{-31}[kg]\cdot\sqrt{6.67259\times10^{-11}[m^3kg^{-1}s^{-2}]}$$

$$= 23.5308\times10^{-73/2}[kg^{1/2}m^{3/2}s^{-1}], \tag{102}$$

$$G_p = m_p\sqrt{G} = 1836.15G_e \tag{103}$$

With them one can form an Eddington type relationship (also called as the conjecture of large numbers)

$$\frac{N_e + N_p}{2} \cong \underbrace{2}_{\substack{PHOTON\\SPECIES}} \times\left(\frac{Q_e}{G_e}\right)^2\left(\frac{Q_p}{G_p}\right)^2 \tag{104}$$

where the number of species of photons that intermediate the self-interactions was also considered, see Table 1.2; it results that the number of particles (atoms of Hydrogen) from the Universe has the size order

$$N_e + N_p \cong \frac{4}{1836.15^2}\left(\frac{0.151891\times10^{-13}}{23.5308\times10^{-73/2}}\right)^2\left(\frac{0.151891\times10^{-13}}{23.5308\times10^{-73/2}}\right)^2$$

$$= \frac{4}{(1836.15)^2}\frac{(0.151891)^4\times10^{-52}}{(23.5308)^4\times10^{-146}}$$

$$= 2.0598\cdot10^{-15}\times10^{94};$$

$$N_{UNIVERSE} \cong 2.0598\times10^{79} \text{ [particles in Universe]} \tag{105}$$

which is also very close in the absolute value of original appraisal of Eddington (2.3622×10^{79}) based on the rotation group theory combined with quantum theory[4].

Based on the number of particles in Universe, it can be also calculated its mass, from repose (creative) masses of electrons and protons contributions, related to the number of total stable species of the current Universe (of Table 1.2).

$$\begin{aligned} &\textit{STABLE} \\ &\textit{SPECIES}\ \text{electrons} + \text{protons} + \text{photons} + \text{neutrinos } (e+\mu) \\ &\qquad = \frac{7}{2}+\frac{7}{2}+2+2\frac{7}{4}=\frac{25}{2} \end{aligned} \tag{106}$$

since in estimation of the total number of particles (atoms of Hydrogen) of Universe we combined the electrons, protons, photons information, see above. Therefore, we estimate the Universe mass

$$M_{UNIVERSE} \cong \frac{2}{25}\left(N_e + N_p\right)\left(m_e + m_p\right) = \frac{2}{25}1837.15 m_e\left(N_e + N_p\right)$$

$$\begin{aligned} &= \frac{2}{25}1837.15\times9.10938188\times10^{-31}\times2.0598\times10^{79}[kg] \\ &= 2.75771\times10^{51}[kg] \end{aligned} \tag{107}$$

[4]Sir Arthur S. Eddington, *The Expanding Universe*, Cambridge University Press, 1933.

If eventually one also wanted an approximation of the Universe radius from the Universe mass, there must be considered the relationship density-volume for the spherical approximation of the Universe

$$R_{UNIVERSE} = \left(\frac{3}{4\pi}V_{UNIVERSE}\right)^{1/3} = \left(\frac{3}{4\pi}\frac{M_{UNIVERSE}}{\rho_{UNIVERSE}}\right)^{1/3} \quad (108)$$

where, for an approximate evaluation, the Universe density the density obtained from of the electromagnetic (photonic) radiation overlapped to the neutrino density at the current moment (at the temperature of background radiation T=2.74 K) can be considered:

$$\rho_{UNIVERSE}(T) = 12.5\times\left[\rho_{\gamma}(T)+\rho_{\upsilon}(T)\right] \quad (109)$$

based on the fact that the interactions between the stable species in the current Universe in bound systems (the atoms), contribute the transformations (through collisions or disintegrations) so creating the photons and neutrinos. In order to establish the photonic density the photons density energy expression of the astrophysics Stefan-Boltzmann law (deducted from the Section 1.2) is used

$$u = \sigma^{*}T^{4} = 7.565767\times10^{-16}\left[T(K)\right]^{4}[J\cdot m^{-3}\cdot K^{-4}] \quad (110)$$

Regarding the contribution to the neutrinos' density, we have to consider the photon-neutrino relationship in the early Universe, and even more specific by assuming the constant of entropy variation per unit volume at a given temperature,

$$\left(\frac{\partial S}{\partial V}\right)_{\substack{UNIVERSE\\PRIOR\\ANNIHILATION}} = \left(\frac{\partial S}{\partial V}\right)_{\substack{UNIVERSE\\POST\\ANNIHILATIONS}} \quad (111)$$

as a conservation law between the Universe era with radiation (photons) and antimatter (electrons and positrons) – prior annihilation matter with antimatter Universe

$$\underset{PRIOR-ANNIHILATION\ UNIVERSE}{SPECIES\ OF}\ \text{electrons/positrons} + \text{photons} = 7/2+2 = 11/2 \quad (112)$$

and postannihilation Universe

$$\underset{POST-ANNIHILATION\ UNIVERSE}{SPECIES\ OF}\ \text{photons} = 2 \quad (113)$$

Now, recognizing from the second principle of thermodynamics according which the entropic variations from above Eq. (111) corresponds with the pressure (considered as the astrophysical one)

$$\left(\frac{\partial S}{\partial V}\right)_{UNIVERSE} = \frac{P_{radiation}}{T} = \frac{1}{3}\sigma^* T^3 \tag{114}$$

and considering the Universe in any era in spherical approximation (with radius *R*), namely under the (quasi)isotropic expansion, then at a certain moment, we will obtain the working formula for the Universe entropy

$$S_{UNIVERSE} = \frac{4\pi}{9}\sigma^* (TR)^3 \tag{115}$$

By molding this formula with the number of species in the considered era of Universe, we have the conservation law of entropy is equivalent with

$$\frac{11}{2}(TR)^3_{\substack{PRIOR\\ANNIHILATION}} = 2(TR)^3_{\substack{POST\\ANNIHILATION}} \tag{116}$$

wherefrom the relationship

$$\frac{(TR)_{\substack{POST\\ANNIHILATION}}}{(TR)_{\substack{PRIOR\\ANNIHILATION}}} = \left(\frac{11}{4}\right)^{1/3} \tag{117}$$

Now, if we write a similar conservation relation for neutrinos (existing before and after the annihilation of the matter with the antimatter) we have

$$(T_\upsilon R)_{\substack{PRIOR\\ANNIHILATION}} = (T_\upsilon R)_{\substack{POST\\ANNIHILATION}} \tag{118}$$

where we noted with T_υ the neutrinos' temperature. Moreover, based on the fact that the neutrino has the same temperature as the photon in the PRIOR-ANNIHILATION stage, the last relation is thus extended as

$$(TR)_{\substack{PRIOR\\ANNIHILATION}} = (T_\upsilon R)_{\substack{PRIOR\\ANNIHILATION}} = (T_\upsilon R)_{\substack{POST\\ANNIHILATION}} \tag{119}$$

which allows the rewriting of the POST-PRIOR ANNIHILATION relation as following

$$\frac{(TR)_{\substack{POST\\ANNIHILATION}}}{(T_\upsilon R)_{\substack{POST\\ANNIHILATION}}} = \left(\frac{11}{4}\right)^{1/3} \tag{120}$$

wherefrom (considering the same radius of the Universe in the POST-ANNIHILATION stage) we have the relation between the photons temperature and that for the neutrinos in the epoch

$$\left(\frac{T}{T_{\upsilon}}\right)_{\substack{POST\\ANNIHILATION}} = \left(\frac{11}{4}\right)^{1/3} = 1.401 \tag{121}$$

Moving forward to the density energy approach, besides its relation with the fourth power of the epoch temperature, according to the Stefan-Boltzmann dependence, from Table 1.2 we also take into consideration the relation

$$\begin{matrix} SPECIES \\ OF \\ NULL\ MASS \end{matrix} \quad (\text{neutrino})/(photons) = (7/2)/2 = 7/4 \tag{122}$$

and obtain the relationship

$$\frac{u_{\upsilon}}{u_{\gamma}} = \frac{7}{4}\left(\frac{T_{\upsilon}}{T}\right)^{4}_{\substack{POST\\ANNIHILATION}} = \frac{7}{4}\left(\frac{4}{11}\right)^{4/3} = 0.4542 \tag{123}$$

Under these conditions, the density of photonic + neutron radiation is immediately obtained

$$u_{\gamma\&\upsilon}(T) = u_{\gamma}(T) + u_{\upsilon}(T) = 1.4542 u_{\gamma}(T)$$

$$= 11.0021\times10^{-16}\,[T(K)]^4\,[J\cdot m^{-3}\cdot K^{-4}] \tag{124}$$

Returning to the determination of mass density of the Universe, the last result is further corrected (modulated) with the presence of the stable species of particles existing in the era, that is, by 25/2 in the last expression (124), as motivated above, which by dividing to the square speed of light generates the density

$$\rho_{UNIVERSE}(T) = \frac{u_{(e\&p)(\gamma\&\upsilon)}}{c^2}$$

$$= 12.5\frac{11.0021\times10^{-16}[kg\cdot s^{-2}\cdot m^{-1}\cdot K^{-4}]}{(2.99792458)^2\times10^{16}[m^2 s^{-2}]}[T(K)]^4;$$

$$\rho_{UNIVERSE}(T) = 15.3019 \times 10^{-32} \left[T(K)\right]^4 [kg \cdot m^{-3} \cdot K^{-4}] \tag{125}$$

and which for the current state of the Universe becomes

$$\rho_{UNIVERSE}(T = 2.74K) = 862.475 \times 10^{-32}[kg / m^3] \tag{126}$$

By combining now the density with the Universe mass, a realistic appraisal must take into consideration the existence of neutrinos, as the hardest unstable particle of those considered in Table 1.2, but also with a major role in nuclear-synthesis and stellar evolution, and whose presence (in atomic nuclei) influences (through gravity) the Universe expansion and decreases its radius proportional with the number of existing species, 7/2 of Table 1.2, toward the value

$$R_{UNIVERSE}(2.74K) \cong \frac{2}{7}\left(\frac{3}{4\pi}\frac{2.75771 \times 10^{51}[kg]}{862.475 \times 10^{-32}[kg / m^3]}\right)^{1/3}$$

$$= 0.285714\left(76.3332 \times 10^{78}[m^3]\right)^{1/3} = 1.212 \times 10^{23}[km]$$

$$= 1.212 \times 10^{23} \times \underbrace{\frac{1}{(206265)(149598871)}}_{3.24075 \times 10^{-14}}[\text{parsecs}]$$

$$= 3927.79[Mpc] \tag{127}$$

For validation, we use this radius of the Universe in order to evaluate the Hubble constant, from the established relation

$$H_{UNIVERSE}(T = 2.74K) = \frac{c}{R_{UNIVERSE}}$$

$$= \frac{299792.458[km / s]}{3927.79[Mpc]}$$

$$= 76.326[km \cdot s^{-1} \cdot Mpc^{-1}] \tag{128}$$

in great accordance with the 2009 observed values, as exposed in Table 1.1. For this value, the existence age of the Universe is calculated with the transformation

$$t_{UNIVERSE}[YEARS] = \frac{1}{H\left[\frac{km}{s\cdot Mpc}\right]}\times 10^{6}\frac{pc}{Mpc}\times 3.086\cdot 10^{13}\frac{km}{pc}\times\left(\frac{1\ YEAR}{3.154\cdot 10^{7}s}\right)$$

$$= 12.8\times 10^{9}[YEARS] \cong 13[BILLION\ \ YEARS] \tag{129}$$

Worth to be noted that the direct determination of Hubble constant from quantum grounds and based on the combination of quantum properties (viz. the number of species of Table 1.2 is based on the quantum spin nature of considered particles) allow us to promote the idea of treating the Universe as an integer, as being governed by a type of quantum bosonic field (background cosmic radiation, electromagnetic, photonic radiation) that exists and circumscribes the Universe expansion. Thus, from the energy of (Einstein) relativistic existence equality with the one of quantum existence of the Universe in expansion (let's call it Hubble energy)

$$m_0 c^2 = \hbar H \tag{130}$$

results into the rest mass (existence) for bosonic (photonic) field responsible for (or covering) the Universe expansion (with Hubble constant), with an extremely small value

$$m_0 = H\frac{\hbar}{c^2} = \frac{1}{R_{UNIVERSE}}\frac{\hbar}{c}$$

$$= \frac{1}{1.212\times 10^{26}[m]}\frac{1}{2\pi}\frac{6.626068\times 10^{-34}[m^2\cdot kg\cdot s^{-1}]}{2.99792458\times 10^{8}[m\cdot s^{-1}]}$$

$$= 0.290237\times 10^{-68}[kg] \tag{131}$$

but a nonnull value!

The idea is not new or surprising, because revives the existence of the "ether" at a cosmic level, being in agreement with the prediction of the Broglie theory, according which any particle has a rest mass different by zero, no matter how small, an idea invigorated in the second part of twentieth century by the Vigier papers (a very close collaborator of de Broglie)[5]! [71–84].

[5]Mariano Moles, Jean-Pierre Vigier, *Interpretation possible de déplacements anormaux vers rouge au voisinage du soleil à l'aide de la Mécanique ondulatoire du photon*, C.R. Acad. Sc. "Paris," 24 May 1974, t. 278, Série B, pp. 969–972.

1.3.2 DARK MATTER IN UNIVERSE AS AN EFFECT OF JEANS CONDENSATION

Having at the hand the Planck's radiation of the black body, respectively the condition of galactic formation (so that a ball of matter will give birth to a gravitational system) now we can formulate in a direct analytical manner the condition for the gravitational energy developed to overcome the internal energy developed by the existent radiation

$$U_G >> U_{RADIATION} \tag{132}$$

Meaning by the gravitational energy the self-gravity effect associated to the pregalactic nebula, say of mass M and radius R

$$U_G = -G\frac{M^2}{R} \tag{133}$$

and by the radiation energy the thermodynamic relationship

$$P = -\left(\frac{\partial U_{RAD}}{\partial V}\right)_{S(entropy)=ct} \tag{134}$$

we have the working expression

$$U_{RAD} = -\left(\frac{4\pi}{3}R^3\right)P(T) \tag{135}$$

Therefore, until this moment we have the galactic formation condition

$$G\frac{M^2}{R} >> \frac{4\pi}{3}R^3 P(T) \tag{136}$$

If one further considers the mass of proto-galaxy as being related with the density of era (and volume) through the density-volume relationship in the spherical approximation, the radius of proto-galaxy can be considered from it

$$R = \left(\frac{3}{4\pi}\frac{M}{\rho}\right)^{1/3} \tag{137}$$

Rewriting the last inequality (136) such as

$$GM^2 >> \frac{4\pi}{3}\left(\frac{3}{4\pi}\frac{M}{\rho}\right)^{4/3} P(T) \tag{138}$$

it can be equivalently rearranged as

$$M^{2/3} >> \left(\frac{3}{4\pi}\right)^{1/3} \frac{P(T)}{G\rho(T)^{4/3}} \tag{139}$$

where from there results the Jeans' mass condition

$$M >> M_J \tag{140}$$

with *Jeans' mass* of galactic formation, given by the expression

$$M_J(T) = \left(\frac{3}{4\pi}\right)^{1/2} \frac{P(T)^{3/2}}{G^{3/2}\rho(T)^2} \tag{141}$$

which corrects (only through the numerical factor) the expression obtained by the classical hydrodynamic bases of Section 1.2.1. In order to make lucrative this indicator expression of the critical mass from where the proto-galaxies can be organized in Galaxies, the pressure is replaced by its radiation expression (of black-body) and the Jeans mass is rewritten as

$$\begin{aligned} M_J(T) &= \left(\frac{3}{4\pi}\right)^{1/2} \left(\frac{1}{3}\right)^{3/2} \frac{u(T)^{3/2}}{G^{3/2}\rho(T)^2} \\ &= \frac{1}{3\sqrt{4\pi}} \frac{c^3\rho(T)^{3/2}}{G^{3/2}\rho(T)^2} \\ &= \frac{1}{3\sqrt{4\pi}} \frac{c^3}{G^{3/2}} \frac{1}{\rho(T)^{1/2}} \end{aligned} \tag{142}$$

where the density of radiated energy was converted (with Einstein relation) in mass density,

$$u(T) = c^2 \rho_{EPOCH}(T)\cdot \tag{143}$$

For numerical applications, there is considered the Universe density provided above, here reloaded

$$\rho_{UNIVERSE}(T) = 15.3019 \times 10^{-32} \left[T(K)\right]^4 [kg \cdot m^{-3} \cdot K^{-4}], \tag{144}$$

as based on the essential particles specificities involved in macrocosm, and we have the working relationship for Eq. (142)

$$M_J(T) = \frac{1}{3\sqrt{4\pi}} \frac{1}{(6.67259)^{3/2} \times 10^{-33/2}[m^{9/2}kg^{-3/2}s^{-3}]}$$

$$\times \frac{(2.99792458)^3 \times 10^{24}[m^3 \cdot s^{-3}]}{\sqrt{15.3019} \times 10^{-16}[T(K)]^2 [kg^{1/2} \cdot m^{-3/2} \cdot K^{-2}]};$$

$$M_J(T) = \frac{118829 \times 10^{50}}{[T(K)]^2}[kg \cdot K^2] \tag{145}$$

or in Solar masses ($M_{Sun} = 1.987 \times 10^{30}[kg]$)

$$M_J(T) = \frac{59803.2 \times 10^{20}}{[T(K)]^2}[M_{Sun} \cdot K^2] \tag{146}$$

The last relation actually (approximately) also expresses the number of stars in a galaxy, at a given temperature, if it is considered that the major part of the stars in a galaxy are of the Sun type, as a reasonable hypothesis, supported by the placement of the most observed stars in the main sequence (where Sun is also present) of the Hertzsprung-Russel type diagrams.

Thus, the galaxies formation starts from the moment when the Universe was cooled at a temperature about T=3000 [°K], for which the Jeans mass required for the galactic condensation accordingly becomes from Eq. (146)

$$M_J(3000K) = \frac{59803.2 \times 10^{20}}{9 \times 10^6}[M_{Sun}] = 6.645 \times 10^{17}[M_{Sun}] \tag{147}$$

compare with $10^{11}[M_{Sun}]$, the mass evaluated for our galaxy (Milky Way)! While the Universe has been cooled, until the current value of temperature T=2.74 [°K] the Jeans mass required for the galactic formation increases, respectively:

$$M_J(2.74K) \cong \frac{59803.2 \times 10^{20}}{9}[M_{Sun}] = 6.645 \times 10^{23}[M_{Sun}] \tag{148}$$

which in fact told us (despite the Universe expansion) that: or the galaxies are no longer formed at the current moment due to the too high Jeans mass required, or their formation is made due to the existence of "dark matter." This can be now revealed through the calculation of the difference between the Jeans mass (in the current conditions of Universe temperature) and the mass Universe, provided in Eq. (107).

$$M_{\substack{DARK \\ MATTER}} = M_J(2.74K) - M_{\substack{UNIVERSE \\ (OBSERVED)}}$$

$$= 6.645 \times 10^{23}[M_{Sun}] - \frac{2.75771 \times 10^{51}}{1.987 \times 10^{30}}[M_{Sun}]$$

$$= 6.645 \times 10^{23}[M_{Sun}] - 1.38788 \times 10^{21}[M_{Sun}]$$

$$= 664.5 \times 10^{21}[M_{Sun}] - 1.38788 \times 10^{21}[M_{Sun}];$$

$$M_{\substack{DARK \\ MATTER}} = 663.112 \times 10^{21}[M_{Sun}] \tag{149}$$

wherefrom indeed there results that the "dark matter" mass is with two degree of size higher than the one observed (calculated) from Universe:

$$\frac{M_{\substack{DARK \\ MATTER}}}{M_{\substack{UNIVERSE \\ (OBSERVED)}}} = 447.788 \tag{150}$$

Of course, this value remain to be experimentally determined (directly or indirectly); in any case it deeper or greater clarifies (by referential value, subsequently to be confirmed) the mystery of galaxies formation in an expansive Universe and certainly require further theoretical approaches [85–146].

1.3.3 DARK ENERGY IN UNIVERSE AS AND EFFECT OF HUBBLE EXPANSION

One finally looks on the topic of establishment of the line of the Universe time, in its various ages, depending on the density of Universe era. Noted that the situation is different comparing with the way in which the previous cosmic quantities were determined, especially Hubble constant, because in this case we have to specify that we will consider a subsystem of the Universe where, at a certain expansion distance R_{EXPAND}, the interaction between a Galaxy (or a sphere of galaxies or a proto-galactic nebula) with the mass M_{GALAX} and the rest of the mass M_{EXPAND} contained by the volume with the ray R_{EXPAND} is considered:

$$M_{EXPAND} = \frac{4\pi}{3} R_{EXPAND}^3 \rho_{UNIVERSE}(T) \tag{151}$$

where the density will still be kept as being the one of Universe found at a certain epoch temperature T, previously deduced, see Eq. (144).

Further, in order to capture the image of expansion in expressing the Hubble constant-epoch density, we will have to use the condition that the total energy of the era in expansion E_{EXPAND} of the (proto)galaxy concerned, it has to remain constant in relation with its dilatation radius, namely

$$\frac{dE_{EXPAND}}{dR_{EXPAND}} = 0 \tag{152}$$

Taking into consideration that the total energy of (proto)galaxy in expansion is consisted by the competition between the term of gravitational energy due to the sphere mass on which expansion front (with radius R_{EXPAND}) is calculated:

$$U_{G-EXAPAND} = -G\frac{M_{GALAX}M_{EXPAND}}{R_{EXPAND}} = -4\pi G\frac{M_{GALAX}}{3}R_{EXPAND}^{2}\rho_{UNIVERSE}(T) \tag{153}$$

and the term of kinetic energy, expressed according to the running speed of (proto)galaxy, expressed itself by the Hubble era law:

$$v_{EXPAND} = H_{EPOCH}R_{EXPAND} \tag{154}$$

That is,

$$T_{EXAPAND} = \frac{M_{GALAX}v_{EXPAND}^{2}}{2} = \frac{1}{2}M_{GALAX}H_{EPOCH}^{2}R_{EXPAND}^{2} \tag{155}$$

Together, there results the total energy of (proto)galaxy, on the sphere of Universe in expansion in gravitational interaction with the sphere inside

$$E_{EXPAND} = U_{G-EXAPAND} + T_{EXPAND}$$

$$= M_{GALAX}R_{EXPAND}^{2}\left[\frac{1}{2}H_{EPOCH}^{2} - \frac{4}{3}\pi G\rho_{UNIVERSE}(T)\right] \tag{156}$$

wherefrom, through the condition of (constant) equilibrium in expansion the relationship is yielded

$$\frac{1}{2}H_{EPOCH}^{2} = \frac{4}{3}\pi G\rho_{UNIVERSE}(T) \tag{157}$$

This condition prevents the sphere of the Universe to be accelerated to the infinite, both with negative and positive energy, so that, in expansion to have a constant velocity, unaccelerated. In any case, the relation eases the expression of Hubble era constant, according to the specific density of Universe, and we retain

$$H_{EPOCH} = \sqrt{\frac{8}{3}\pi G \rho_{UNIVERSE}(T)} \sim \sqrt{\rho_{UNIVERSE}(T)} \tag{158}$$

The difference respecting the Hubble constant of Universe previously derived in (49) will be next discussed, for the moment being obvious that only the kinematic conditions restricted the expansion movement of a proto-galaxy not extended at the level of the entire horizon of (light) Universe

$$R_{UNIVERSE} = c \cdot t_{UNIVERSE} = \frac{c}{H_{UNIVERSE}} \tag{159}$$

In these conditions, in order to obtain the equation of the epoch expansion time from the History of Universe, we have to consider the two big ages in which it was split. *The era dominated by matter, until the annihilation of matter with antimatter, that is, in the first millisecond of the Universe, when in principle existed the conservation of the total mass of the Universe*

$$\frac{4\pi}{3} R^3_{EPOCH}(t) \rho_{\substack{MATTER\\EPOCH}}(t) = CONST. \tag{160}$$

wherefrom results the proportionality

$$\rho_{\substack{MATTER\\EPOCH}}(t) \propto \frac{1}{R^3_{EPOCH}(t)} \tag{161}$$

Here one renews the indices of the sizes at "epoch" because the searching for the instantaneous equation of Universe expansion, from where the expression of specific time to the specific era results; moreover, for the framed purpose the relation of proportionality is adequate (see below in this section).

The era dominated by radiation, after the annihilation of matter with antimatter (the electrons with the positrons, for example, wherefrom resulted the primordial sea/soup of photons etc.); this era/epoch is proportional with the fourth power of the temperature and is inversely proportional with the radius (a fact which is obvious from the relations of entropy conservation at the frontier PRIOR-POST ANNIHILATION, as explained in Section 1.3.1 by Eq. (118)).

$$\rho_{\substack{RADIATION\\EPOCH}}(t) \propto T^4 \propto \frac{1}{R^4_{EPOCH}(t)} \tag{162}$$

By unifying the two dependencies (the density-radius of epochs of Universe) in the two main ages (of matter and of radiation) we obtain the generalized form

$$\rho_{\substack{UNIVERSE\\EPOCH}}(t) \propto \frac{1}{R^{n}_{EPOCH}(t)} \tag{163}$$

where the power index introduced makes the era difference of the Universe

$$n = \begin{cases} 3...MATTER \;\; ERA \\ 4...RADIATION \;\; ERA \end{cases} \tag{164}$$

Now we are able to formulate the kinematic equation of Universe expansion, by using the era Hubble law

$$\begin{aligned} v_{EPOCH}(t) &= H_{EPOCH}(t)R_{EPOCH}(t) \\ &\propto \left[\rho_{EPOCH}(t)\right]^{1/2} R_{EPOCH}(t) \propto \frac{R_{EPOCH}(t)}{R^{n/2}_{EPOCH}(t)} = R^{1-n/2}_{EPOCH}(t) \end{aligned} \tag{165}$$

which effectively generates the kinematic working equation

$$v_{EPOCH}(t) = \frac{dR_{EPOCH}(t)}{dt} = \chi R^{1-n/2}_{EPOCH}(t) \tag{166}$$

where it has been formally introduced $\chi = ct.$ as proportionality constant. This equation is the simplest differential equation type

$$v = \frac{dx}{dt} = \chi x^{a} \tag{167}$$

whose integration for the temporal solution

$$\frac{1}{\chi}\int_{1}^{2} \frac{1}{x^{a}} dx = \int_{1}^{2} dt \tag{168}$$

immediately leads to the equality

$$\frac{1}{-a+1}\left[\frac{x}{\chi x^{a}}\right]_{1}^{2} = t_2 - t_1 \tag{169}$$

rearranged as

$$\frac{1}{-a+1}\left[\frac{x}{v}\right]_{1}^{2} = t_2 - t_1 \tag{170}$$

where it can be observed the disappearance of the direct dependency of the proportionality factor χ. Adapting the result to the kinematic equation (of expansion) of the Universe era, we have $a = 1 - n/2$ and therefore the temporal solution that we are looking for

$$t_2 - t_1 = \frac{2}{n}\left[\frac{R_{EPOCH}(t_2)}{v_{EPOCH}(t_2)} - \frac{R_{EPOCH}(t_1)}{v_{EPOCH}(t_1)}\right] \tag{171}$$

which can be immediately rewritten with the era Hubble law (or constant)

$$t_2 - t_1 = \frac{2}{n}\left[\frac{1}{H_{EPOCH}(t_2)} - \frac{1}{H_{EPOCH}(t_1)}\right] \tag{172}$$

and finally with the aid of the era density (in relation with the era Hubble constant, found above) we get:

$$t_2 - t_1 = \frac{2}{n}\sqrt{\frac{3}{8\pi G}}\left[\frac{1}{\sqrt{\rho_{EPOCH}(t_2)}} - \frac{1}{\sqrt{\rho_{EPOCH}(t_1)}}\right] \tag{173}$$

If it is still wanted an instantaneous temperature density relation of (epoch) Universe, and not through the difference, there can be used the idea according to which in performing the difference of times, the difference from an era with a higher density to one with a smaller one is made, thus neglecting the last term (the initial term is approximate to zero due to the inverse of radical of the high density in relation with the final term of temporal evaluation). Thus, the working expression for the time era looks like

$$t_{EPOCH} = \frac{2}{n}\sqrt{\frac{3}{8\pi G \rho_{EPOCH}(T)}} = \frac{2}{n}\frac{1}{H_{EPOCH}} \tag{174}$$

At this point, we are able to customize both the Hubble constant and the epoch time at different relevant periods for the history of Universe. To comparing and discussing, see Table 1.3 where for the conversion from the time era to Hubble constant one may use the relation

$$H\left[\frac{km}{s \cdot Mpc}\right] = \frac{1}{t[s]} \times 10^6 \frac{pc}{Mpc} \times 3.086 \cdot 10^{13} \frac{km}{pc} \tag{175}$$

Note that for Planck era (in the era dominated by matter) the temperature and the density previously calculated in Section 1.2.3 were used, and for the density era (in the era dominated by radiation) the density previously calculated, see Eqs. (125) and (144), and frequently used from the Stefan-Boltzmann modeling of black body of the epoch, $\rho_{UNIVERSE}(T) = 15.3019\times10^{-32}\left[T(K)\right]^4 [kg\cdot m^{-3}\cdot K^{-4}]$, corrected/molded with spin properties (and number of species) specific to the Universe POST-ANNIHILATION (of the matter with the antimatter, which marks the passing form the era of the matter to the one of radiation), *see* Section 1.2.4.

TABLE 1.3 The Values of Epoch's Temperature, Radius, time and Hubble's Constant for the Essential Moments of Evolving Universe

No	EPOCH	T $[K]$	ρ_{EPOCH} [kg/m^3]	t_{EPOCH} [s, Min, Years]	H_{EPOCH} $[km\cdot s^{-1}\cdot Mpc^{-1}]$
3	Planck	10^{32}	10^{96}	0.28×10^{-43}[s]	72.84×10^{61}
4	Nucleosynthesis	10^8	15.3019	90.1 [Min]	28.5×10^{14}
4	Thermalization of Radiation	10^5	15.3019×10^{-12}	171.43 [Years]	28.5×10^8
4	Radiation Decoupling (of Cosmic Background)	3000	12.4×10^{-18}	190.477 [Years]	25.65×10^5
4	Predicted Present	2.74	86.25×10^{-31}	228.341 [10^9 Years]	2.14

The difference between the actual age of the Universe and the Hubble constant of expansion toward the values reordered earlier of this chapter (Section 1.3.1) can be "solved" by considering, in adaptive manner- of Fermi type, also of the other adjustment constants, namely:

- The number of stable species (protons, electrons, photons and neutrinos) in Universe, 25/2;
- The number of the least unstable species (neutrons) in Universe: 7/2; An eventual correction with/dropping the 2/4=1/2 factor only for the current epoch when, by comparing to the previous ages, the times of evolutions are wider.

Thus, these information will combine into the factor

$$\frac{25}{2}\times\frac{7}{2}\times\frac{1}{2}=\frac{175}{8}=21.875 \tag{176}$$

and will multiply with the Hubble constant prediction at this moment in Table 1.3, generating the Hubble size order corrected constant:

$$H_{2.74}^{PREDICTED} = (21.875)(2.17) = 47.47\left[km \cdot s^{-1} \cdot Mpc^{-1}\right] \quad (177)$$

For the age of the Universe the prediction corresponds to

$$t_{UNIVERSE}^{PRESENT} = \frac{1}{43.15}(228.341) = 10.44[BILLION\ \ YEARS] \quad (178)$$

As long as the discrepancy is still keept toward the current accepted values, even through the corrections performed, it can be concluded that the model of Universe description, for the ages immediate after the Big-Bang, they are based on the condition of constant running speed (Hubble constant) on era/epoch – and implicitly on the establishment of the temporal scalar equations of the Universe.

However, for molding the current (present) image of the Universe there seems that the model of neutrality of electric charge-gravitational charge (presented at the beginning of the chapter and suggested by Sir Eddington) is better answering to the values measured/observed in current conditions.

Nevertheless, this discrepancy (especially toward the values of Table 1.3 for the present time) gives the opportunity of dark energy evaluation, responsible for the Universe expansion, beyond the current Universe (wherefrom the appellative "in shadow"), eventually until the time limit – with the Hubble's constant associated from the last row of Table 1.3. Thus, we can evaluate the energy density which should correspond to the actual age of the Universe (see Section 1.3.1) of about 12.8 Billions of years, through employing (174):

$$\underbrace{\left(12.8 \cdot 10^{9}\right)}_{\substack{AGE\ OF\\ UNIVERSE\\ (IN\ YEARS)}}\underbrace{\left(3.1536 \cdot 10^{7}\right)}_{\substack{SECONDS\\ IN\ A\ YEAR}} = \frac{1}{2}\sqrt{\frac{3}{8\pi\left(6.67259 \cdot 10^{-11}\right)\rho_{\substack{EXPECTED\\ PRESENT}}(2.74)}}$$

$$\Rightarrow \rho_{\substack{EXPECTED\\ PRESENT}}(T = 2.74K) = 274469 \times 10^{-32}[kg/m^{3}] \quad (179)$$

It appears now the explicit and huge difference between the expected density of matter and the one predicted by the black body radiation applied to the entire Universe: thus it can be considered numerically equal with the density of dark energy found in the Universe:

$$u_{\substack{DARK\\ ENERGY}} = c^{2}\left[\rho_{\substack{EXPECTED\\ PRESENT}}(T = 2.74K) - \rho_{\substack{OBSERVED\\ PRESENT}}(T = 2.74K)\right]$$

$$\begin{aligned} &= \left[274469 - 862.475\right] c^2 \times 10^{-32} [kg / m^3] \\ u_{\substack{DARK \\ ENERGY}} &= 273607 [c^2 \times 10^{-32} kg / m^3] \end{aligned} \tag{180}$$

Under these conditions, the ratio of the dark energy density of the Universe to the density of observed energy (both for the moment of Universe cooling to temperature 2.74°K) will give the ratio of the masses of dark energy of the observed Universe (in the same volume of the Universe), namely

$$\frac{M_{\substack{DARK \\ ENERGY}}}{M_{\substack{UNIVERSE \\ (OBSERVED)}}} = \frac{u_{\substack{DARK \\ ENERGY}}}{u_{\substack{UNIVERSE \\ (OBSERVED)}}} = 317.234 \tag{181}$$

In conclusion, corroborating the results related to the dark matter (responsible for galactic condensations in Universe) with those related to the dark energy (responsible for Universe expansion) we obtain in a first instance the total mass from the Universe (the observed one + the one "in shadow/dark")

$$\begin{aligned} M_{\substack{UNIVERSE \\ (TOTAL)}} &= M_{\substack{UNIVERSE \\ (OBSERVED)}} + M_{\substack{DARK \\ MATTER}} + M_{\substack{DARK \\ ENERGY}} \\ &= M_{\substack{UNIVERSE \\ (OBSERVED)}} \left(1 + 447.788 + 317.234\right) \\ M_{\substack{UNIVERSE \\ (TOTAL)}} &= 766.022 M_{\substack{UNIVERSE \\ (OBSERVED)}} \end{aligned} \tag{182}$$

Very interesting, in percentages, from the total mass of the Universe, we have:

- percentange of Observed Universe

$$\%\, \substack{UNIVERSE \\ OBSERVED} = \frac{M_{\substack{UNIVERSE \\ (OBSERVED)}}}{M_{\substack{UNIVERSE \\ (TOTAL)}}} \times 100 = \frac{1}{766.022} \times 100 = 0.13\% < 1\% \tag{183}$$

- percentage of Dark Matter Universe

$$\%\, \substack{DARK \\ MATTER} = \frac{M_{\substack{DARK \\ MATTER}}}{M_{\substack{UNIVERSE \\ (TOTAL)}}} \times 100 = \frac{447.788}{766.022} \times 100 = 58.46\% \tag{184}$$

- percentage of Dark Energy Universe

$$\%\frac{DARK}{ENERGY}=\frac{M_{\substack{DARK\\ENERGY}}}{M_{\substack{UNIVERSE\\(TOTAL)}}}\times 100=\frac{317.234}{766.022}\times 100=41.41\% \tag{185}$$

These numbers also represent the percentage of relative (observational) knowledge of Universe and estimates a relative equilibrium between the dark energy (antigravitational) and the dark (gravitational) matter, with a slight advantage for the latter one (at the current moment about 2.74°K), which explains the continuous formation of galaxies, solar systems etc., despite the global expansion of the Universe [85–146].

1.4 CONCLUSION

Reaching a relative global-comprehensive view of the micro- and macro-Universe (unless the notions of the Generalized Relativity) we presented an active resume (through new calculations and suggestive prospects of the spatial-temporal evolution of the Universe) involving the main notions and the "actors" of the actual picture of the wonderful cosmos (*coelestium aegregium*)—this being considered as the "best world from all the possible worlds," according to the Leibnitz dictum.

Regarding the observed Universe, its knowledge is estimated, through the exposed Fermi-Weinberg calculations, much under a percentage, which somehow corrects the "optimistic" view of the 4% of knowledge at this moment.

Note that the computational demarche (at the level of the global estimates regarding the Universe), along the model and the approximations chosen are always determinant for the final evaluations. Other considerations combining the quantum fundamental principles (of black body, species of particles, of spin and Dirac spinor- see the Dirac sea that can introduce further corrections) with those gravitational (here in a classical Newtonian manner, also with the possibility of being corrected by the percepts of the generalized relativity theory) can produce different results regarding the percent distribution of the Observed Universe vs. Dark Matter vs. Dark Energy, but ideational with the same meaning.

ACKNOWLEDGMENTS

The support by CNCS-UEFISCDI, project number PN II-RU TE16/2010–2013 is kindly thanked. This work resumes for the first time in this form and for the first time in English international edition the main topics linking the nano/micro-to-macro Universe from the author's original lecture textbook "Environmental Physics and Universe" appeared on 2010 at author's alma mater, West University of Timisoara, at University Press, in exclusive Romanian language.

KEYWORDS

- **black-body radiation**
- **dark energy**
- **dark matter**
- **elementary particles**
- **galactic condensation**
- **Hubble law**
- **Planck Universe**

REFERENCES

1. Clarke, C., Carswell, B. (2007). *Astrophysical Fluid Dynamics*, Cambridge University Press.
2. Eisberg, R. M. (1961). *Fundamentals of Modern Physics*, John Wiley and Sons.
3. Feynman, R. P. (1999). *Lectures on Physics*, Perseus Publishing. ISBN 0-7382-0092-1.
4. Feynman, R. P. (1996). *Six Easy Pieces*, Perseus Publishing. ISBN 0-201-40825-2.
5. Gelfand, I. M. (1963). *Calculus of Variations*, Dover. ISBN 0-486-41448-5.
6. Gupta, K. C. (1988). *Classical mechanics of particles and rigid bodies*, Wiley.
7. Jeans, J. H. (1902). The Stability of a Spherical Nebula, *Philosophical Transactions of the Royal Society of London. Series A, Containing Papers of a Mathematical or Physical Character*, 199, 1–53.
8. Landau, L. D., Lifshitz, E. M. (1972). *Mechanics Course of Theoretical Physics, Vol. 1*, Franklin Book Company ISBN 0-08-016739-X.
9. Longair, M. S. (1998). *Galaxy Formation*, Cambridge University Press.
10. Newton, I. (1999). *The Principia Mathematical Principles of Natural Philosophy*, Berkeley: University of California Press. ISBN 0-520-08817-4.Traducere recentă de, I. Bernard Cohen and Anne Whitman, cu ajutorul Julia Budenz.
11. Putz, M. V. Environmental Physics and Universe (in Romanian as "*Fizica Mediului şi Universul*") West University of Timişoara Press: Timişoara, 2010.
12. Thornton, S. T., Marion, J. B. (2003). *Classical Dynamics of Particles and Systems (5th ed.)*, Brooks Cole. ISBN 0-534-40896-6.

13. Dekel, A., Ostriker, J. P. *Formation of Structure in the Universe*, Cambridge University Press (1999). ISBN 0521586321.
14. Eng, A. E. (1985). *A New Approach to Starlight Runs*, Oswego.
15. Freedman, W. L., Madore, B. F., Gibson, B. K., Ferrarese, L., Kelson, D. D., Sakai, S., Mould, J. R., Kennicutt, R. C. Jr., Ford, H. C., Graham, J. A., Huchra, J. P., Hughes, S. M. G., Illingworth, G. D., Macri, L. M., Stetson, P. B. (2001). Final Results from the Hubble Space Telescope Key Project to Measure the Hubble Constant, *The Astrophysical Journal* 553, 47–72. doi: 10.1086/320638.
16. Friedman, A. (1922), Über die Krümmung des Raumes, *Zeitschrift für Physik* 10, 377–386, doi: 10.1007/BF01332580 Traducerea in Engelză a apărut astfel On the Curvature of Space, *General Relativity and Gravitation* 31, (1999). 1991–2000. doi: 10.1023/A:1026751225741
17. Harrison, E. (1992). The redshift-distance and velocity-distance laws, *Astrophysical Journal, Part 1,* 403, 28–31. doi: 10.1086/172179.
18. Hubble, E. P. (1937). *The Observational Approach to Cosmology*, Oxford: Clarendon Press.
19. Keel, W. C. (2007). *The Road to Galaxy Formation* (2nd ed.), Springer, ISBN 3540725342.
20. Kutner, M. (2003). *Astronomy: A Physical Perspective*, Cambridge University Press. ISBN 0521529271.
21. Liddle, A. R. (2003). *An Introduction to Modern Cosmology* (2nd ed.), Chichester: Wiley ISBN 0470848359.
22. Longair, M. S. (2006). *The Cosmic Century*, Cambridge University Press. ISBN 0521474361.
23. Madsen, M. S. (1995). *The Dynamic Cosmos*, CRC Press. ISBN 0412623005.
24. Overbye, D. (1991). *Lonely Hearts of the Cosmos: The Scientific Quest for the Secret of the Universe,* Harper-Collins. ISBN 0-06-015964-2 & ISBN 0-330-29585-3.
25. Padmanabhan, T. (1993). *Structure formation in the universe*, Cambridge University Press, ISBN 0521424860.
26. Putz, M. V. Environmental Physics and Universe (in Romanian as "*Fizica Mediului și Universul*") West University of Timişoara Press: Timişoara, 2010.
27. Sandage, A. R. Current Problems in the Extragalactic Distance Scale, *Astrophysical Journal* 127, (May, 1958) 513–526. doi: 10.1086/146483.
28. Sartori, L. (1996). *Understanding Relativity*, University of California Press. ISBN 0520200292.
29. Weinberg, S. (2008). *Cosmology*, Oxford University Press. ISBN 0198526822.
30. Suyu, S. H., Marshall, P. J., Auger, M. W., Hilbert, S., Blandford, R. D., Koopmans, L. V. E., Fassnacht, C. D., Treu, T. (2010). Dissecting the Gravitational Lens B1608+656. II. Precision Measurements of the Hubble Constant, Spatial Curvature, and the Dark Energy Equation of State, *The Astrophysical Journal*, 711, 201. DOI: 10.1088/0004-637X/711/1/201.
31. Ahluwalia, D. V. (2002). Interface of Gravitational and Quantum Realms, *Mod Phys Lett* A17, 1135. arXiv:gr-qc/0205121v1
32. Annett, J. F. (2004). *Superconductivity, Superfluids and Condensates*, New York: Oxford University Press. ISBN 0198507550.

33. Ashtekar, A. (2007). *Loop Quantum Gravity: Four Recent Advances and a Dozen Frequently Asked Questions*. arXiv:0705.2222.
34. Ashtekar, A. (1987). New Hamiltonian formulation of general relativity, *Phys. Rev.* D36, 1587–1602.
35. Ashtekar, A. (1986). New variables for classical and quantum gravity, *Phys. Rev. Lett.* 57, 2244–2247.
36. Ashtekar, A. (2005). The winding road to quantum gravity, *Current Science* 89, 2064–2074.
37. Ashtekar, A., Lewandowski, J. (2004). Background Independent Quantum Gravity: A Status Report, *Class. Quant. Grav.* 21, R53–R152, arXiv:gr-qc/0404018.
38. Blakemore, J. S. (2002). *Semiconductor Statistics*, Dover. ISBN 978-0486495026.
39. Carlip, S. (2001). Quantum Gravity: A Progress Report, *Rept. Prog. Phys.* 64, 885, arXiv:gr-qc/0108040.
40. Carter, A. H. (2001). *Classical and Statistical Thermodynamics*, Upper Saddle River, NJ: Prentice Hall ISBN 0137792085.
41. Cho, A. (2009). Can Gravity and Quantum Particles Be Reconciled After All?, *Science* 325, 673, doi: 10.1126/science.325_673.
42. Coulomb, C. (1784). Recherches théoriques et expérimentales sur la force de torsion et sur l'élasticité des fils de metal, *Histoire de l'Académie Royale des Sciences* 229–269.
43. Dirac, P. A. M. (1926). On the Theory of Quantum Mechanics, *Proceedings of the Royal Society, Series A* 112, 661–77. doi: 10.1098/rspa.1926.0133.
44. Dirac, P. A. M., (1967). *Principles of Quantum Mechanics* (4th ed.), London: Oxford University Press. ISBN 978-0198520115.
45. Duffin, W. (1980). *Electricity and Magnetism*, 3rd Ed., McGraw-Hill. ISBN 0-07-084111-X.
46. Fermi, E. (1926). Sulla quantizzazione del gas perfetto monoatomico, *Rend. Lincei* 3, 145–149.Lucrarea a fost tradusă ca *On the Quantization of the Monoatomic Ideal Gas* (1999-12-14) http://arxiv.org/PS_cache/cond-mat/pdf/9912/9912229v1.pdf.
47. Feynman, R. P., Leighton, R. B., Sands, M. (1963). *Lectures on Physics, Vol 1*, Addison-Wesley
48. Feynman, R. P., Leighton, R. B., Sands, M. (2006). *The Feynman Lectures on Physics The Definitive Edition Volume II*, Pearson Addison Wesley. ISBN 0-8053-9047-2.
49. Fowler, R. H. On dense matter, *Monthly Notices of the Royal Astronomical Society* 87, (December 1926), 114–22.
50. Fowler, R. H., Nordheim, L. W. (1928). Electron Emission in Intense Electric Fields, *Proceedings of the Royal Society A* 119(781), 173–81. doi: 10.1098/rspa.1928.0091.
51. Griffiths, D. J. (2005). *Introduction to Quantum Mechanics* (2nd ed.). Upper Saddle River, NJ: Pearson, Prentice Hall. ISBN 0131911759.
52. Hawking, S. W., (1987). Quantum cosmology, in Hawking, S. W. & Israel, W. (ed.), *300, Years of Gravitation*, Cambridge University Press. 631–651, ISBN 0-521-37976-8.
53. Hey, A. J., Walters, P. (2003). *The New Quantum Universe*, London: Cambridge University Press ISBN 0521564573.
54. Kiefer, C. (2007). *Quantum Gravity*, Oxford University Press ISBN 019921252X.
55. Kittel, C. (1971). *Introduction to Solid State Physics* (4th ed.), New York: John Wiley & Sons.

56. Kittel, C., Kroemer, H. (1980). *Thermal Physics* (2nd ed.), San Francisco: W. H. Freeman ISBN 978-0716710882.
57. Lämmerzahl, C., ed., (2003). *Quantum Gravity: From Theory to Experimental Search (Lecture Notes in Physics)*, Springer ISBN 354040810X.
58. Leighton, R. B. (1959). *Principles of Modern Physics*, McGraw-Hill ISBN 978-0070371309.
59. Putz, M. V. Environmental Physics and Universe (in Romanian as "*Fizica Mediului și Universul*") West University of Timișoara Press: Timișoara, 2010.
60. Reif, F. (1965). *Fundamentals of Statistical and Thermal Physics*, McGraw–Hill ISBN 978-0070518001.
61. Rigden, J. S. (2005). *Einstein 1905: The Standard of Greatness*. Massachusetts: Harvard University Press ISBN 0674015444.
62. Rovelli, C. (2000). *Notes for a brief history of quantum gravity.* arXiv:gr-qc/0006061.
63. Rovelli, C., (2004). *Quantum Gravity*, Cambridge University Press. ISBN 0521837332.
64. Serway, R. A. (2003). *Physics for Scientists and Engineers*, Philadelphia: Saunders College Publishing. ISBN 0-534-40842-7.
65. Sommerfeld, A. (1927). Zur Elektronentheorie der Metalle, *Naturwissenschaften* 15, 824–32. doi: 10.1007/BF01505083.
66. Townsend, P. K. (1996). *Four Lectures on M-Theory.* arXiv:hep-th/9612121
67. Trifonov, V. (2008). GR-friendly description of quantum systems, *Int. J. Theor. Phys.* 47, 492–510, arXiv:math-ph/0702095
68. Weinberg, S. (1994). *Dreams of a Final Theory*, Vintage Books USA. ISBN 0-679-74408-8.
69. Weinberg, S. (1996), *The Quantum Theory of Fields II: Modern Applications*, Cambridge University Press. ISBN 0-521-55002-5.
70. Zwiebach, B. (2004), *A First Course in String Theory*, Cambridge University Press ISBN 0-521-83143-1.
71. Arrhenius, S. (1908). *Worlds in the Making: The Evolution of the Universe*. New York, Harper & Row.
72. Hoyle, F. *The Intelligent Universe*, Michael Joseph Limited, London (1983). ISBN 0-7181-2298-4.
73. Huang, K. (1967). *Statistical Mechanics*, New York: John Wiley & Sons.
74. Kirchhoff, G. (1896). On the relation between the Radiating and Absorbing Powers of different Bodies for Light and Heat, traducere de, F. Guthrie apărută in *Phil. Mag.* Series 4, 20(130) 1–21. Apariția originală in Poggendorff's *Annalen*, vol. 109, paginile 275, *et seq.*
75. Landau, L. D., Lifshitz, E. M. (1996). *Statistical Physics* (3rd Edition Part 1, ed.), Oxford: Butterworth-Heinemann.
76. Mahan, J. R. (2002). *Radiation heat transfer: a statistical approach* (3rd ed.), Wiley-IEEE ISBN 9780471212706.
77. McKinley, J. M. (1979). Relativistic transformations of light power, *American Journal of Physics* 47, 602. doi: 10.1119/1.11762.
78. Padmanabhan, T. (1993). *Structure formation in the universe*, Cambridge University Press. ISBN 0521424860.
79. Papagiannis, M. D. (1972). *Space physics and space astronomy.* Taylor & Francis ISBN 9780677040004.

80. Planck, M. (1901). On the Law of Distribution of Energy in the Normal Spectrum, *Annalen der Physik* 4, 553.
81. Planck, M. (1914). *The theory of heat radiation*, second edition, traducere de, M. Masius, Blackiston's Son & Co, Philadelphia.
82. Porter, R. (2003). *The Cambridge History of Science*. Cambridge, UK: Cambridge University. Press ISBN 978-0521572439.
83. Putz, M. V. Environmental Physics and Universe (in Romanian as "*Fizica Mediului și Universul*") West University of Timişoara Press: Timişoara, 2010.
84. Robitaille, P. (2003). On the validity of Kirchhoff's law of thermal emission, *IEEE Transactions on Plasma Science* 31, 1263. doi: 10.1109/TPS.2003.820958.
85. Alpher, R. A., Gamow, G. (1948). The Origin of Chemical Elements, *Physical Review* 73, 803. doi: 10.1103/PhysRev.73.803.
86. Alpher, R. A., Herman, R. *Reflections on early work on 'big bang' cosmology*, Physics Today (August 1988) pp. 24–34.
87. Amelin, Y., Krot, A., Hutcheon, I., Ulyanov, A. (2001). Lead isotopic ages of chondrules and calcium-aluminum-rich inclusions, *Science* 297, 5587. 1678–83. doi: 10.1126/science.1073950.
88. Baadsgaard, H., Lerbekmo, J. F. (1988). A radiometric age for the Cretaceous-Tertiary boundary based on K-Ar, Rb-Sr, and U-Pb ages of bentonites from Alberta, Saskatchewan, and Montana, *Canadian Journal of Earth Sciences* 25, 1088–1097.
89. Baadsgaard, H., Lerbekmo, J. F., Wijbrans, J. R. (1993). Multimethod radiometric age for a bentonite near the top of the Baculites reesidei Zone of southwestern Saskatchewan (Campanian-Maastrichtian stage boundary?), *Canadian Journal of Earth Sciences* 30, 769–775.
90. Baker, J., Bizzarro, M., Wittig, N., Connelly, J., Haack, H. (2005). Early planetesimal melting from an age of 4.5662 Gyr for differentiated meteorites, *Nature* 436, 1127–1131. doi: 10.1038/nature03882.
91. Barrow, J. D. (1994). *The Origin of the Universe: To the Edge of Space and Time*. New York: Phoenix. ISBN 0465053548.
92. Bertschinger, E. (2001). Cosmological Perturbation Theory and Structure Formation. arXiv:astro-ph/0101009.
93. Bertschinger, E. (1998). Simulations of Structure Formation in the Universe, *Annual Review of Astronomy and Astrophysics* 36, 599–654. doi: 10.1146/annurev.astro.36.1.599.
94. Boltwood, B. B. (1907). On the ultimate disintegration products of the radio-active elements. Part II. The disintegration products of uranium, *American Journal of Science* 23, 77–88.
95. Bonanno, A., Schlattl, H., Paternò, L. (2002). The age of the Sun and the relativistic corrections in the EOS, *Astronomy and Astrophysics* 390, 1115–1118. doi: 10.1051/0004-6361:20020749. arXiv:astro-ph/0204331.
96. Brookfield, M. E. (2004). *Principles of Stratigraphy*. Blackwell Publishing ISBN 140511164X.
97. Burchfield, J. D. (1998). The age of the Earth and the invention of geological time, *Geological Society, London, Special Publications* 143, 137–143. doi: 10.1144/GSL.SP.1998.143.01.12.

98. Caldwell, R.R, Kamionkowski, M., Weinberg, N. N. (2003). Phantom Energy and Cosmic Doomsday, *Physical Review Letters* 91, 071301. doi: 10.1103/PhysRevLett.91.071301. arXiv:astro-ph/0302506.
99. Christianson, E. (1995). *Edwin Hubble Mariner of the Nebulae*. New York: Farrar, Straus and Giroux ISBN 0374146608.
100. Dalrymple, B. G. (2004). *Ancient Earth, Ancient Skies: The Age of the Earth and Its Cosmic Surroundings*, Stanford University Press. ISBN 978-0804749336.
101. Dalrymple, G. B. (2001). The age of the Earth in the twentieth century: a problem (mostly) solved, *Special Publications, Geological Society of London* 190, 205–221. doi: 10.1144/GSL.SP.2001.190.01.14.
102. Dalrymple, G. B. (1994). *The Age of the Earth*, Stanford University Press. ISBN 0804723311.
103. Davies, P. (1992). *The Mind of God: The Scientific Basis for a Rational World*, New York: Simon & Schuster UK ISBN 0-671-71069-9.
104. Dicke, R. H., Peebles, P. J. E., (1979). The big bang cosmology—enigmas and nostrums, in Hawking, S. W. & Israel, W. (ed). *General Relativity: An Einstein Centenary Survey*. Cambridge University Press. pp. 504–517.
105. Eberth, D. A., Braman, D. (1990). Stratigraphy, sedimentology, and vertebrate paleontology of the Judith River Formation (Campanian) near Muddy Lake, west-central Saskatchewan, *Bulletin of Canadian Petroleum Geology* 38, 387–406.
106. England, P., Molnar, P., Righter, F. (2007). John Perry's neglected critique of Kelvin's age for the Earth: A missed opportunity in geodynamics, *GSA Today* 17(1), 4–9. doi: 10.1130/GSAT01701A.1.
107. England, P. C., Molnar, P., Richter, F. M. (2007). Kelvin, Perry and the Age of the Earth, *American Scientist* 95(4), 342–349. doi: 10.1511/2007.66.3755.
108. Gibson, C. H. (2005). The First Turbulent Combustion. arXiv:astro-ph/0501416.
109. Gibson, C. H. (2001). Turbulence And Mixing In The Early Universe. arXiv:astro-ph/0110012.
110. Goodman, J. (1995). Geocentrism Reexamined, *Physical Review D* 52, 1821. doi: 10.1103/PhysRevD.52.1821.
111. Gradstein, F. M., Agterberg, F. P., Ogg, J. G., Hardenbol, J., van Veen, P., Thierry, J., Huang, Z. (1995). A Triassic, Jurassic and Cretaceous time scale, in Bergren, W. A., Kent, D. V., Aubry, M.-P., Hardenbol, J. (eds.), *Geochronology, Time Scales, and Global Stratigraphic Correlation*. Society of Economic Paleontologists and Mineralogists, Special Publication No. 54, p. 95–126.
112. Guth, A. H. (1998). *The Inflationary Universe: Quest for a New Theory of Cosmic Origins*, Vintage Books ISBN 978-0099959502.
113. Harland, W. B., Cox, A. V., Llewellyn, P. G., Pickton, C. A. G., Smith, A. G., Walters, R. (1982).*A Geologic Time Scale*, Cambridge University Press: Cambridge.
114. Hartle, J. H. (1983). Wave Function of the Universe, *Physical Review D* 28, 2960. doi: 10.1103/PhysRevD.28.2960.
115. Hawking, S. W., Ellis, G. F. R., (1973). *The Large-Scale Structure of Space-Time*, Cambridge (UK): Cambridge University Press. ISBN 0-521-20016-4.
116. Hoyle, F. (1948). A New Model for the Expanding Universe, *Monthly Notices of the Royal Astronomical Society* 108, 372.

117. Ivanchik, A. V. (1999). The Fine-Structure Constant: A New Observational Limit on Its Cosmological Variation and Some Theoretical Consequences, *Astronomy and Astrophysics* 343, 459.
118. Joly, J. (2004). *Radioactivity and Geology: An Account of the Influence of Radioactive Energy on Terrestrial History* (1st ed.), London, UK: Archibald Constable & Co., ltd. (1909). Retipărită de BookSurge Publishing. ISBN 1-4021-3577-7.
119. Khoury1 J., Ovrut, B. A., Seiberg, N., Steinhardt, P. J., Turok, N. (2002). From big crunch to big bang. *Physical Review D* 65(8) 086007. doi: 10.1103/PhysRevD.65.086007.
120. Kolb, E., Turner, M. (1988). *The Early Universe*. Addison–Wesley. ISBN 0-201-11604-9.
121. Komatsu, E. (2009). Five-Year Wilkinson Microwave Anisotropy Probe Observations: Cosmological Interpretation. *Astrophysical Journal Supplement* 180, 330. doi: 10.1088/0067-0049/180/2/330.
122. Kragh, H. (1996). *Cosmology and Controversy*. Princeton (NJ): Princeton University Press. ISBN 0-691-02623-8.
123. Langlois, D. (2002). *Brane Cosmology: An Introduction.* arXiv:hep-th/0209261.
124. Lemaître, G. (1931). The Evolution of the Universe: Discussion, *Nature* 128, 699–701. doi: 10.1038/128704a0.
125. Lemaître, G. (1927). Un univers homogène de masse constante et de rayon croissant rendant compte de la vitesse radiale des nébuleuses extragalactiques, *Annals of the Scientific Society of Brussels* 47A: 41.(in Franceză). Cu traducerea: A Homogeneous Universe of Constant Mass and Growing Radius Accounting for the Radial Velocity of Extragalactic Nebulae, *Monthly Notices of the Royal Astronomical Society* 91, (1931). 483–490.
126. Linde, A. (1986). Eternally Existing Self-Reproducing Chaotic Inflationary Universe *Physics Letters B* 175, 395–400. doi: 10.1016/0370-2693(86)90611-8.
127. Linde, A. (2002). *Inflationary Theory versus Ekpyrotic/Cyclic Scenario* arXiv:hep-th/0205259.
128. Manhesa, G., Allègrea, C. J., Dupréa, B., Hamelin, B. (1980). Lead isotope study of basic-ultrabasic layered complexes: Speculations about the age of the earth and primitive mantle characteristics. *Earth and Planetary Science Letters* 47, 370–382. doi: 10.1016/0012-821X(80)90024-2.
129. Mather, J. C., Boslough, J. (1996). *The very first light: the true inside story of the scientific journey back to the dawn of the Universe*, New York: BasicBooks. ISBN 0-465-01575-1.
130. Milne, E. A. (1935). *Relativity, Gravitation and World Structure*, Oxford (UK): Oxford University Press.
131. Navabi, A. A., Nematollah, R. (2003). Is the Age Problem Resolved? *Journal of Astrophysics and Astronomy* 24, 3. doi: 10.1007/BF03012187.
132. Patterson, C. (1956). Age of meteorites and the earth, *Geochimica et Cosmochimica Acta* 10, (4) 230–237. doi: 10.1016/0016-7037(56)90036-9.
133. Peacock, J. (1999). *Cosmological Physics*. Cambridge University Press. ISBN 0521422701.
134. Peebles, P. J. E., (2003). The Cosmological Constant and Dark Energy, *Reviews of Modern Physics* 75, 559–606. arXiv:astro-ph/0207347.

135. Penrose, R. (1989). Difficulties with Inflationary Cosmology, in Fergus, E. J. (ed), *Proceedings of the 14th Texas Symposium on Relativistic Astrophysics*. New York Academy of Sciences pp. 249–264. doi: 10.1111/j.1749-6632.1989.tb50513.x.
136. Penrose, R. (1979). Singularities and Time-Asymmetry, in Hawking, S. W. & Israel, W. (ed). *General Relativity: an Einstein centenary survey*. Cambridge University Press pp. 581–638.
137. Penzias, A. A., Wilson, R. W. (1965). A Measurement of Excess Antenna Temperature at 4080 Mc/s. *Astrophysical Journal* 142, 419. doi: 10.1086/148307.
138. Powell, J. L. (2001). *Mysteries of Terra Firma: the Age and Evolution of the Earth*, Simon & Schuster. ISBN 0-684-87282-X.
139. Putz, M. V. Environmental Physics and Universe (in Romanian as "*Fizica Mediului şi Universul*") West University of Timişoara Press: Timişoara, 2010.
140. Rutherford, E. *Radioactive Transformations*. London: Charles Scriber's Sons 1906. Retipărită de Juniper Grove 2007. ISBN 978-1-60355-054-3.
141. Sakharov, A. D. (1967). Violation of CP Invariance, C Asymmetry and Baryon Asymmetry of the Universe. *Zhurnal Eksperimentalnoi i Teoreticheskoi Fiziki, Pisma* 5, 32, (in Rusă); Traducerea in *Journal of Experimental and Theoretical Physics Letters* 1967, 5, 24.
142. Singh, S. *Big Bang: The Origins of the Universe*, New York: Fourth Estate (2004). ISBN 0007162200.
143. Stiebing, W. H. *Uncovering the Past*. Oxford University Press US (1994). ISBN 0195089219.
144. Tolman, R. C. *Relativity, Thermodynamics, and Cosmology*. Oxford (UK): Clarendon Press (1934).Republicată in New York (NY): Dover Publications (1987). ISBN 0-486-65383-8.
145. Wilde, S. A., Valley, J. W., Peck, W. H., Graham, C. M. (2001). Evidence from detrital zircons for the existence of continental crust and oceans on the Earth 4.4 Gyr ago, *Nature* 409, 175–178. doi: 10.1038/35051550.
146. Zwicky, F. (1929). On the Red Shift of Spectral Lines through Interstellar Space, *Proceedings of the National Academy of Sciences* 15(10), 773–779. doi: 10.1073/pnas.15.10.773.

CHAPTER 2

INERTON FIELD EFFECTS IN NANOSYSTEMS

VOLODYMYR KRASNOHOLOVETS

CONTENTS

ABSTRACT

The real physical space is derived from a mathematical space constructed as a tessellation lattice of primary balls, or a kind of superparticles. In the tessel-lattice particles are determined as local stable deformations. The motion of a particle generates a cloud of excitations around the particle, which were named *inertons*. An abstract construction of quantum mechanics known as the wave ψ-function is treated as a mapping of the real system "particle + its cloud of inertons." These inertons carry both the particle's inert and gravitational properties. It is shown how inertons manifest themselves experimentally, how they form clusters of particles in diluted cold gases, clusters of electrons, and how they are responsible for the Casimir effect. The inerton field, as well as the electromagnetic field, is a basic field of the universe. The inerton field is able to give a deeper insight into the fundamental nature of things playing an important role in quantum physics, chemical physics, biochemistry, biophysics, and condensed matter in general.

2.1 INTRODUCTION

Nanosystems approaching the fundamental microscopic length scales have demonstrated fundamentally new physical phenomena. New advances have been reached in many basic and enchanced areas of nanophysics, including diluted cold gases, carbon nanotubes, graphene, magnetic nanostructures, composite nanoparticles, transport through coupled quantum dots, spin-dependent electron transport phenomena, optical matrices with doped nanoparticles, molecular electronics and quantum information processing.

Usually new phenomena relating to nanosystems are associated with changes in electric and magnetic behavior of the systems studied in which electrical/magnetic polarizations are locally realized bringing new physical chemical properties, which can be applied to the electronics industry. Research of nanosystems is challenging as it enters the uncharted areas delimited by new nanoelectronic devices and fabrics. Such devices allow us to increase computational density up to 100 times, which extremely amplifies the capabilities of such systems in new areas of application and markets. New dynamic properties revealed in nanosystems enable the possibilities of nano-engines, nano-pumps and nano-propellors with advantages in distinctive sensors, biomedicine and energy savings and in sustainable development in general.

At the same time, in some peculiar situations some nanosystems unveil such new properties that they cannot be directly accounted for by conventional electromagnetism and quantum mechanical laws. To such systems with special properties, in which peculiar quantum fluctuations are realized, belong those that manifest the so-called Casimir effect [1–4], an attractively interacting ensemble of ultracold bosons at negative temperature that is stable against collapse for arbitrary atom numbers [5]; an electron droplet (an aggregation of about 10^{10} electrons in one cluster) [6–12].

In this chapter, the submicroscopic deterministic concept is revealed in detail and its connection to the conventional quantum mechanical formalism is demonstrated. In the next sections unusual effects uncovered in nano- and submicroscopic systems ensue from the submicroscopic standpoint, namely, the effects incorporating quasi-particles named *inertons*, carriers of the inerton field, which, as is shown is a basic physical field of Nature.

2.2 SUBMICROSCOPIC DETERMINISTIC CONCEPT

A sub microscopic approach can be considered as the further development of conventional quantum mechanics, which incorporates the theory of ordinary physical space.

2.2.1 PHYSICAL SPACE

The term **space** is used somewhat differently in different fields of study. In physics **space** is defined via measurement and the standard space interval, called a standard meter or simply meter, is defined as the distance traveled by light in a vacuum per a specific period of time and in this determination the velocity of light is treated as constant. In microscopic physics, or quantum physics, the notion of **space** is associated with an “arena of actions” in which physical processes and phenomena take place. And this arena of actions we feel subjectively as a “receptacle for subjects.” The measurement of **physical space** has long been important.

This “arena of actions” can be completely formalized, because fundamental physical notions (particle, mass, wave ψ-function, etc.) and interactions can be derived from pure mathematical constructions.

It is interesting to read Vernadsky’s work who back in 1920–1930s introduced the notion of *noosphera* (from Greek *nous*—mind and *sphaira*—ball): a sphere of the arena of interaction between people and nature. In particular, he mentioned that Helmholtz probably was the first who noted that geometric space did not embrace all of empirically studied space, which Helmholtz called **physical space**; Helmholtz distinguished physical space from geometric space, as possessing its own properties, such as right-handedness and left-handedness; besides, Poincaré observed that geometry could not have been developed without solids [13]. Further Vernadsky notes: "In discussing the state of space, I will be dealing with the state of empirical or physical space, which has only in part been assimilated by geometry. Grasping it geometrically is a task for the future." Vernadsky introduced such notion as *the state of space*, which in his opinion has to be closely connected with the concept of a physical field, which plays such an important role in contemporary theoretical physics.

Researchers working in the realm of quantum gravity tend to believe that the real space has a cellular structure at the Planck scale. For instance, in the case of loop quantum gravity, basic excitations of the gravitational field are arbitrary and they can describe the quantum spacetime directly at the Planck scale, where the geometry comes by "quanta."

Models of a spin-network in quantum gravity are realized as a one-dimensional graph; spin foams generalize spin networks where instead of a graph one uses a higher-dimensional complex. Such geometry of spacetime corresponds to a kind of a lattice [14]. A "Planck-Lattice," a spacetime cubic lattice with the lattice constant equals to the Planck length $\ell_P \cong 10^{-35}$ m models a ground state of quantum gravity of Wheeler's spacetime foam [15].

Granular space and the problem of large numbers – how many spatial cells may a canonical particle include – have recently been discussed in a simple way [16]. Wilczek says about space the following [17]: "...this is the effervescent Grid... Matter is not what it used to be. It consists of small, more-or-less stable patterns of disturbance in the Grid... Usually the metric field is taken to be fundamental, but in many ways it resembles a condensate, and that view of it may become important... What we ordinarily call matter consists of more-or-less stable patterns of excitation in the Grid, which is more fundamental. At least, that's how things look today."

All this means that **physical space** is a peculiar substrate that is subject to certain laws, which as has been seen below, are purely mathematical. Such a view allows us to completely remove any subjectivity and all the figurants of fundamental physical processes will be 100% defined. So, we can elucidate those something's that form a primordial physical substrate and determinate its mathematical properties.

Modern quantum theories wish to combine all fundamental interactions in a unified theory—*the theory of everything*. However, doing so the theory of everything rests on complete undetermined basic notions, such as mass, particle, charge, lepton, quark, Compton wavelength, de Broglie wavelength, particle-wave, spin, etc. Moreover, the notion of space in which all physical processes occur is also beyond understanding, though the review above illustrates a gradual trend of the researchers to a fine-scale morphology that has to incorporate particles.

How can the situation be clarified? Can we start from *the theorem of something*? Then, having defined basic physics notions, we will be able to pass on to the construction of the theory of everything.

Bearing in mind an idea to build fundamental physics based on mathematical space, we first of all have to answer the question: What is space from the mathematical point of view? Basing on classical mathematics we may start from nothing – a flat space that does not manifest itself. How does real physical space appear? The flatness of the original mathematical space points to the fact that the Poisson brackets and any other forms of noncommutative features should be absent in such a space. In particular, the Heisenberg's uncertainty principle also becomes an alien for the ordinary flat space. This is what tells us the fundamental mathematics…

Mathematical space, as dealing with the notions of *measure*, *distances* and *dimensionality* in a broad topological sense was analyzed and constructed by Michel Bounias [18]. The major results of this work (for the case of totally topologically ordered space) are as follows.

The Jordan and Lebesgue measures involve respective mappings (I) and (M) on spaces which must be provided with operations $\cap, \cup$ and C. In spaces of the R^n type, tessellation by balls is involved, which again demands a distance to be available for the measure of diameters of intervals. Thus, since the intervals can be replaced by topological balls, the evaluation of their diameter still needs an appropriate general definition of a distance. A space E is ordered if any segment owns an infinum and a supremum. Therefore, a distance d between A and B is represented by the relation

$$d(\mathrm{A},\ \mathrm{B}) \subseteq \mathrm{dist}(\inf \mathrm{A},\ \inf \mathrm{B}) \cap \mathrm{dist}(\sup \mathrm{A},\ \sup \mathrm{B}) \tag{1}$$

with the distance evaluated through either classical forms or even the set-distance $\Delta(\mathrm{A}, \mathrm{B})$. Any topological space is metrizable as provided with the set-distance Δ as a natural metrics. All topological spaces are kinds of metric spaces called "delta-metric spaces." Distance $\Delta(\mathrm{A}, \mathrm{B})$ is a kind of an intrinsic case $\left[\Lambda_{(\mathrm{A,B})}(\mathrm{A,B})\right]$ of $\Lambda_{\mathrm{E}}(\mathrm{A,B})$ while the latter is called a "separating distance." The separating distance also stands for a topological metrics. Hence, if a physical space is a topological space, it will always be measurable.

A fundamental segment (A, B) and intervals $L_i = [A_i, A_{(i+1)}]$ allow one to determine similarity coefficients for each interval by $\rho_i = \text{dist}(A_i; A_{(i+1)}) = \text{dist}(A, B)$.

The similarity exponent of Bouligand (e) is such that for a generator with n parts:

$$\sum_{i \in [1, n]} (\rho_i)^{e} = 1 \tag{2}$$

Then, when 'e' is an integer, it reflects a topological dimension

$$e \approx \text{Log}\, n / \text{Log}\, \rho, \tag{3}$$

which means that a fundamental space E can be tessellated with an entire number of identical balls B exhibiting a similarity with E, upon coefficient r.

The measure of the size of tessellating balls as well as that of tessellated space, with reference to the calculation of their dimension is determined through Eqs. (2) and (3). A space may be composed of members, such that not all tessellating balls have identical diameter. Also a ball with two members would have a more complicated diameter. Thus a measure should be used as a probe for the evaluation of the coefficient of size ratio needed for the calculation of a dimension [18].

It is generally assumed [19] that some set does exist. This strong postulate was reduced to the axiom of the existence of the empty set [18]. Supplementing the empty set (Ø) with some operations and rules allowed us to construct a magma, which became an initiating polygon for a spatial lattice.

Recall that in abstract algebra a magma is defined as a set M supplied with a single binary operation interpreted and described usually (but not always) as a form of multiplication "·" and this binary operation is closed by definition. So, a magma is a set S in which the operation "·" forwards any two elements a, b to another element $a{\cdot}b$. To qualify as a magma, the set and operation (M, ·) must satisfy the magma axiom: For all a, b in M, the result of the operation $a{\cdot}b$ is also in M.

So, following ordinary algebraic canons, we [18] in fact could show that providing the empty set (Ø) with operations ($\in$, $\subset$) as the combination rules with the property of complementarity (C) results in the definition of a magma without violating the axiom of foundation if the empty set is seen as a hyperset that is a nonwellfounded set. The magma was defined as $\varnothing^{\varnothing} = \{\varnothing, \mathbf{C}\}$ and it was proved that such construction with the empty hyperset and the axiom of availability is a fractal lattice an these features indeed characterize a fractal object [20].

Writing $\varnothing^{\varnothing}$ denotes that the magma reflects the set of all self-mappings of ($\varnothing$), which emphasizes the forthcoming results. The space constructed with the empty set cells is a Boolean lattice $S(\varnothing)$ and this lattice is provided with a topology of discrete space. The magma of empty hyperset is endowed with self-similar ratios.

Such a lattice of tessellation balls was called a *tessel-lattice* [18, 21]. The magma of empty hyperset is a fractal tessel-lattice.

An abstract tessel-lattice of empty set cells accounts for a primary substrate in a physical space [21, 22]. Space-time is represented by ordered sequences of topologically closed Poincaré sections of this primary space. These mappings are constrained to provide homeomorphic structures serving as frames of reference in order to account for the successive positions of any objects present in the system. Mappings from one to the next section involve morphisms of the general structures, standing for a continuous reference frame, and morphisms of objects present in the various parts of this structure. The combination of these morphisms provides space-time with the features of a nonlinear generalized convolution (and then the process of motion appears as a stack of serial slices, that is, Poincaré sections, which resembles a customary movie). Discrete properties of the lattice allow the prediction of scales at which microscopic to cosmic structures should occur.

The fundamental metrics of space-time is represented by a convolution product where the embedding part D4 is described by the following relation:

$$\mathrm{D4} = \int\left(\int_{dS} (d\mathbf{x}\cdot d\mathbf{y}\cdot d\mathbf{z}) * d\Psi(w)\right) \tag{4}$$

where dS is an element of space-time, and $d\Psi(w)$ a function accounting for the extension of 3D coordinates up to the 4th dimension timeless space through convolution (*) with the volume of space. Thus fractality of space manifests itself through changes in the dimension of geometrical structures. For example, the dimension of a curve exceeds 1D and falls in the interval between 1D and 2D; for a volumetric object the dimension may lay between 3D and 4D.

Thus the real physical space is organized as the tessel-lattice of primary topological balls. The existence of such lattice stands for the universe substrate (or "space"). In a degenerate state the size of a ball, which plays the role of a lattice's cell, is associated with the Planck length ℓ_{P}. Deformations of primary cells by exchange of empty set cells allow a cell to be mapped into an image cell in the next section as far as mapped cells remain homeomorphic.

The tessel-lattice is specified with quanta of distances and quanta of fractality [21, 22]. The sequence of mappings of one into another structure of reference (e.g., elementary cells) represents an oscillation of any cell volume along the arrow of physical time.

A lattice that includes a set with neither members nor parts accounts for both relativistic space and quantic void, since: (i) the concept of distance and the concept of time have been defined on it, and (ii) this space holds for a quantum void since on the one hand it provides a discrete topology with quantum scales, and on the other hand it contains no "solid" object that would stand for a given provision of physical matter.

When a fractal transformation is involved in exchange of deformations between cells, there occurs a change in the dimension of the cell and the homeomorphism is not conserved [21]. Then the fractal kernel (a local deformation of the tessel-lattice) stands for a "particle" and the reduction of its volume is compensated by morphic changes of a finite number of surrounding cells. These morphic changes represent a typical tension of the tessel-lattice around the deformed fractal kernel, that is, particle.

Since we have introduced a particle, we must provide it with physical properties. First of all this is mass: The mass m_A of a particulate ball A is a function of the fractal-related decrease of the volume of the ball:

$$m_A \propto (V^{\text{deg. cell}} / V^{\text{part}}) \cdot (e_{\text{fract}} - 1)_{e_{\text{fract}} > 1} \qquad (5)$$

where $V^{\text{deg. cell}}$ is the typical average volume of a cell in the tessel-lattice in the degenerate state; V^{part} is the volume of the kernel cell of the particle; (e) is the Bouligand exponent, and $(e_{\text{fract}}\ \ 1)$ the gain in dimensionality given by the fractal iteration. Just a volume decrease is not sufficient for providing a ball with mass, since a dimensional increase is a necessary condition (there should be a change in volumetric fractality of the ball) [21, 22].

As follows from Eq. (5), mass appears as a deformation of a cell, that is, at the volumetric fractal contraction of the cell. This is typical for leptons and hence in the case of leptons $V^{\text{deg. cell}} / V^{\text{lepton}} > 1$ (the volume of the particle kernel is less than the volume of the original degenerate cell). In the case of quarks the situation is reciprocal: the quark's kernel cell has a volume bigger than the average volume of a degenerate cell, that is, $V^{\text{quark}} / V^{\text{deg. cell}} > 1$ [22].

Therefore, in the tessel-lattice a lepton is a contracted kernel-cell. Surrounding cells compensate this local deformation by morphic changes (a cell tension) forming a peculiar deformation coat with a radius identified with the particle's Compton wavelength $\lambda_{\text{Com}} = h/(mc)$. Beyond the radius λ_{Com} there is no information about the particle. This hidden radius indeed manifests itself in the experiments on light scattering by particles, though in orthodox quantum mechanics the size of a canonical particle does not play a part in the theory.

In condensed matter physics the availability of a deformation coat is a typical situation. It emerges in a crystal lattice when a foreign particle or isotope defect arises in the solid (for example, small and big polarons); a similar situation occurs in a liquid (a solvate shell forms around an entered ion) and liquid crystal.

2.2.2 MECHANICS OF THE TESSEL-LATTICE

A local stable deformation in the tessel-lattice, that is, a volumetric fractal deformation of a cell of the tessel-lattice, can be treated as a massive particle. The motion of such particle occurs with the interaction with surrounding cells (Fig. 2.1). This situation is fundamentally different from what we have in Newtonian mechanics – because in the latter the particle moves without interacting with space.

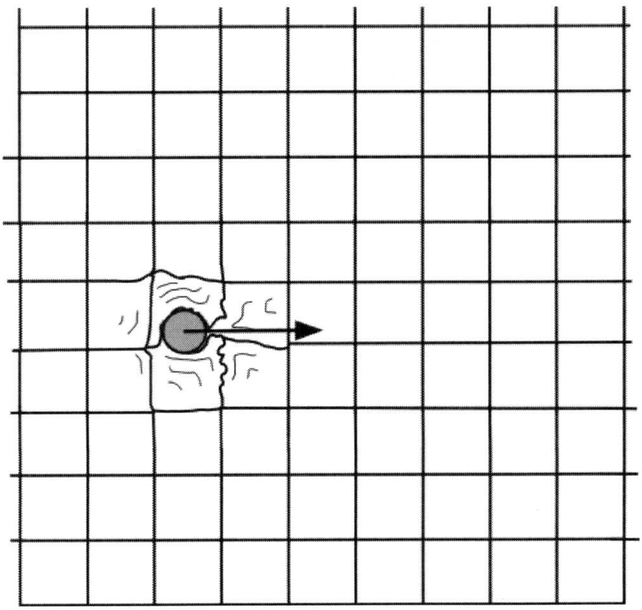

FIGURE 2.1 Motion of the local stable deformation in the tessel-lattice.

Motion of a particle in the tessel-lattice is accompanied by friction, which is obvious. The particle contacting oncoming cells of the tessel-lattice emits excitations, which finally has to stop the particle, as it loses energy. However, if these excitations are reabsorbed by the particle, it will continue to move ahead. No energy is lost in this interaction, that is, all excitations are reabsorbed by the particle and no friction heat is generated. Thus we may assume that a particle in the tessel-lattice moves rectilinearly in such a way that its velocity oscillates between the initial value υ_0 and zero, that is, during odd half a period the particle emits excitations and gradually loses the velocity and during the next even half a period it absorbs the emitted excitations gaining speed, and so on. These spatial excitations were named *inertons* [23, 24] since they reflect the inert properties of matter – a resistance on the side of space to a stimulation of the movement of the object. The principle of motion is shown in Fig. 2.2.

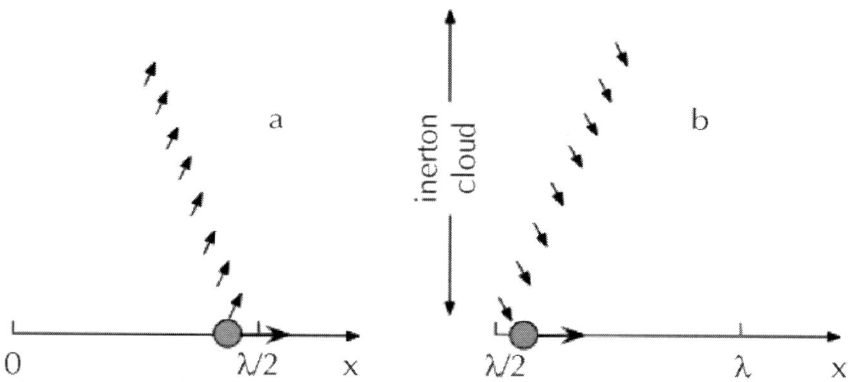

FIGURE 2.2 Motion of the particle in the tessel-lattice, which is accompanied with the emission of inertons. The moving particle periodically emits inertons in the odd part of its spatial period (in the section [0, λ/2], (a); and absorbs them in the second part of the spatial period (in the section [λ/2, λ], (b).

It is interesting that such a pattern of the motion embraces ideas of Poincarè and de Broglie on the principles of motion of canonical particles. Indeed, [25] pointed out that an electron is a singularity in the world ether and it should move surrounded by the ether excitations. De Broglie [26] instinctively put those excitations in order, such that the excitations became a wave guiding the particle (one can see this in Fig. 2.2b). However, in 1927 at the Fifth Solvay International Conference he was persuaded to the probabilistic interpretation of quantum mechanics, which was suggested by Born in 1926 [27, 28]. Nevertheless, later on since 1952 [29]) was firmly searching for the double solution theory, which would be able to reinterpret the wave ψ-function so that it showed its physical interpretation rather than probabilistic discoloration.

The behavior of a particle in the tessel-lattice can be described by the following Lagrangian [23, 24, 30].

$$L = -m_0 c^2 \left\{1 - 1/(m_0 c^2)\left[m_0 \dot{x}^2 + \mu_0 \dot{\chi}^2 - 2\pi / T \sqrt{m_0 \mu_0}\,(x\dot{\chi} - \upsilon_0 \chi)\right]\right\}^{1/2} \quad (6)$$

where m_0 is the particle's mass, x is its position; μ_0 is the mass of the excited cloud of inertons (excitations of the tessel-lattice associated with the field of inertia of the particle), χ is the position of the center mass of the cloud; $1/T$ is the frequency of collisions of the particle with the cloud of inertons; υ_0 is the initial velocity of the particle and c is the speed of light. This Lagrangian is constructed as an inner development of the so-called relativistic Lagrangian of a particle $L = -m_0 c^2 \sqrt{1 - \upsilon_0^2 / c^2}$.

The moving particle is rubbing against the tessel-lattice, which results in the appearance of the particle's cloud of inertons. But this is not a classic friction that stops the particle. Indeed, the Euler-Lagrange equations

$$d / dt\,(\partial L / \partial \dot{q}) - \partial L / \partial q = 0$$

for the particle ($q \equiv x$) and its inerton cloud ($q \equiv \chi$), which are based on the Lagrangian equation (the Eq. (6)), result in:

$$\frac{d^2 x}{dt^2} + \frac{\pi}{T}\frac{\upsilon_0}{c}\frac{dx}{dt} = 0 , \quad (7)$$

$$\frac{d^2 \chi}{dt^2} - \frac{\pi}{T}\frac{c}{\upsilon_0}\left(\frac{d\chi}{dt} - \upsilon_0\right) = 0 . \quad (8)$$

The corresponding solutions to Eqs. (7) and (8) for the particle and the inerton cloud are:

$$\dot{x} = \upsilon_0 \cdot (1 - |\sin(\pi t / T)|) \quad (9)$$

$$x = \upsilon_0 t + \lambda / \pi \cdot \left\{(-1)^{[t/T]} \cos(\pi t / T) - (1 + 2[t / T])\right\} , \quad (10)$$

$$\chi = \Lambda / \pi \cdot |\sin(\pi t / T)|, \tag{11}$$

$$\dot{\chi} = (-1)^{[t/T]}\, c \cos(\pi t / T) \tag{12}$$

$$\lambda = \upsilon_0 T, \qquad \Lambda = cT \tag{13}$$

The Eqs. (9) and (10) show that the particle's velocity periodically oscillates and λ is the amplitude of the particle's oscillations along its path. In particular, λ is the period of oscillation of the particle's velocity that changes between υ_0 and zero. An oscillation of the particle's velocity is a very interesting feature of the tessel-lattice's mechanics (Fig. 2.3). The inerton cloud periodically leaves the particle and then comes back; Λ is the amplitude of oscillations of the cloud and *c* is the velocity of the cloud of inertons.

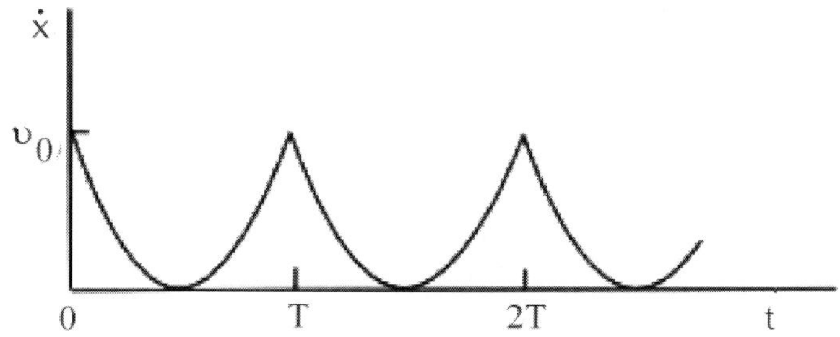

FIGURE 2.3 This is the graphical presentation of the solution (the Eq. (9)) for the behavior of the particle's velocity $\dot{x}$ as a function of time *t*. The time interval *T* of collisions of the particle with its inerton cloud plays the role of the period of the particle's velocity oscillations.

The Lagrangian (6) allows us to introduce an effective Hamiltonian of the particle, which describes its behavior relative to the center of inertia of the particle-inerton cloud system,

$$H_{\text{eff}} = p^2 / (2m) + m\left[2\pi / (2T)\right]^2 X^2 / 2 \tag{14}$$

where $m = m_0 / \sqrt{1 - \upsilon_0^2 / c^2}$. This is the harmonic oscillator Hamiltonian, which means that we can construct the Hamilton-Jacobi equation for a shortened action S_1 of the particle,

$$(\partial S_1 / \partial x)^2 / (2m) + m\left[2\pi / (2T)\right]^2 X^2 / 2 = E \tag{15}$$

where E is the energy of the moving particle. Introduction of the action-angle variables leads to the following increment of the particle action within the cyclic period $2T$,

$$\Delta S_1 = \oint p\,dx = E \cdot 2T \tag{16}$$

The Eq. (16) can be rewritten by using the frequency $\nu = 1/(2T)$. At the same time $1/T$ is the frequency of collisions of the particle with its inertons cloud. Allowance for $E = m\upsilon_0^2/2$ gets,

$$\Delta S_1 = m\upsilon_0 \cdot \upsilon_0 T = p_0 \lambda \tag{17}$$

where $p_0 = m\,\upsilon_0$ is the particle's initial momentum. If we equate the increment of action ΔS_1 per period to Planck's constant h, we obtain instead of Eqs. (16) and (17) the major de Broglie's relationships

$$E = h\nu, \qquad \lambda = h/p, \tag{18}$$

which form the basis of conventional quantum mechanics.

The Eq. (18) allow one to derive the Schrödinger equation [31]

$$-\frac{\hbar^2}{2m}\nabla^2 \psi(r, t) + V(r)\,\psi(r, t) = E\psi(r, t)\,. \tag{19}$$

The submicroscopic concept developed in the real space constructed as the tessel-lattice with the size of a cell equal to the Planck length operates with a particle and the particle's cloud of inertons. Conventional quantum mechanics, which was evolved in an abstract phase space on an atom scale, works with the wave ψ-function. These two approaches can be combined, as the inerton cloud of an entity, which is associated with the entity's field of inertia, is mapped into an abstract phase space of ordinary quantum mechanics in the form of a "mysterious" wave ψ-function (Fig. 2.4).

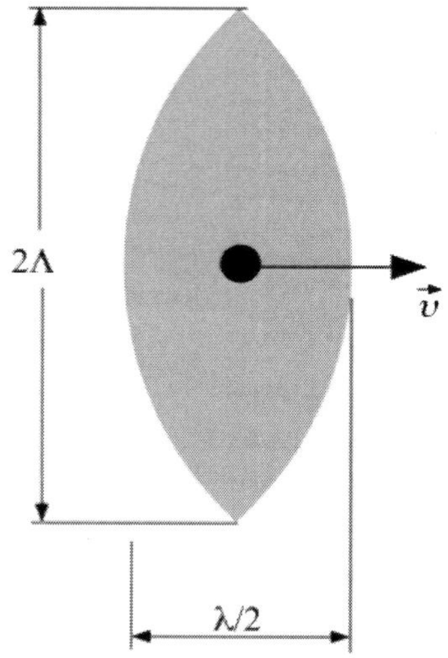

FIGURE 2.4 Moving particle with the velocity $\vec{\upsilon}$, which is surrounded by its cloud of inertons. An approximate size of such complicated system {particle + its inerton cloud} is determined by the cloud's amplitude Λ and the de Broglie wavelength λ. This complicated system determined in the real space is mapped to conventional quantum mechanics constructed in an abstract phase space as the particle's wave ψ-function.

Therefore, in the framework of the submicroscopic concept the cloud's inertons are a substructure of the matter waves. Inertons are field carriers – carriers of the field of inertia of the particle. They transfer mass and fractal properties from the particle to distant points of space. Why inertons emitted by a moving particle comes back to it? This is because the tessel-lattice is a reverberating substrate. (Of course these inertons may interference with other inertons and partly be absorbed by other objects, which correlate the behavior of the particle in question. This allows for entanglement.)

A range of space covered by the particle's inertons (i.e., particle's wave ψ-function) is specified by the amplitude of the inerton cloud,

$$\Lambda = \lambda c / \upsilon_0 , \tag{20}$$

which is spread in transversal directions from the particle's path; the particle's de Broglie wavelength λ is a section of the particle's path, the particle's amplitude; υ_0 is the initial velocity of the particle, and also the component of the velocity of the inerton cloud along the particle's path; c is the speed of light that is the velocity component of migrating inertons in the inerton cloud, which migrate transversal to the particle's path. If the particle is motionless, its inertia and gravitation are restricted in the particle's deformation coat whose radius coincides with the particle's Compton wavelength $\lambda_{\rm Com} = h/(mc)$ where m is the particle's mass.

In a macroscopic object (which is embedded in the tessel-lattice) local oscillations of entities generate inerton clouds that overlap, producing a set of harmonics of inerton waves [32, 33]. Moreover, long-wave inerton harmonics go far beyond the physical size of the object. These oscillating waves bear mass properties to a great distance away from the object. Oscillating waves mean that inertons emitted by the object return to it, which signifies that these inerton waves are standing spherical waves.

Solutions to standing spherical waves are characterized by the inverse dependence on the distance to the wave's front. The standing inerton wave results in a quasi-stationary mass potential field around the object with a mass M_0, which is subjected to spherical symmetry.

$$M \sim M_0 / r \tag{21}$$

This average mass field automatically results in the Newton's gravitational potential [34]

$$U = -GM_0 / r \tag{22}$$

Besides, the notion of a point mass, which is typical for general relativity, cannot be a real point but rather a small macroscopic object whose smallest radius can be estimated at around 1 μm.

The submicroscopic theory has been further developed for the interaction of massive objects. It was exhibited that the gravitational interaction between two objects should be described by a corrected Newton's law [35].

$$U = -G\frac{M_0 M_1}{r} \cdot \left(1 + \frac{\dot{r}_{\rm tan}^2}{c^2}\right) \tag{23}$$

where $\dot{r}_{\rm tan}$ is the tangential velocity of a test mass M_1, which is in line with Poincaré's remark that an expression for the gravitational interaction should include the velocity of a moving object [25]. By using Eq. (23), the submicroscopic approach has successfully been applied to describe four macroscopic phenomena, which are treated as four classical tests for general relativity; they are: the motion of Mercury's perihelion, bending of a light ray by a star, the red shift of spectral lines [35, 32], and the Shapiro time delay effect [36].

Finally in this section, we mention that the submicroscopic approach allowed us to resolve successfully the big cosmological problem related to so-called dark matter [37]. Standing inerton waves oscillate around the massive objects with the speed of light. They are those waves that induce the gravitational attraction between objects. In addition an overlapping of these standing oscillating inerton waves generates an elastic interaction between masses bringing them to a formation of clusters in which masses are characterized by both the Newtonian and elastic interaction. It is this elastic interaction that so far eluded researchers has been associated with the manifestation of dark matter.

The tessel-lattice also enables naturally the introduction of electric and magnetic charges and electromagnetic field [32, 38]

2.3 THE SUBMICROSCOPIC BEHAVIOR OF PARTICLES IN CONDENSED MATTER

The submicroscopic approach allows us to shed new light upon a number of well-known phenomena in condensed matter. This approach makes it possible to investigate how the quantum mechanical field (a substructure of the matter waves, that is, inertons, carriers of the particles' field of inertia) determines the collective behavior of atoms. In submicroscopic mechanics, the momentum p of a particle is decomposed to the mass m and the velocity υ and each of these parameters is characterized by its own behavior in the line of a particle's path. The whole particle path is subdivided by the particle's de Broglie wavelength λ. Along the section λ the particle velocity changes from υ to zero and is then again reinstated to υ, i.e. $\upsilon \to 0 \to \upsilon \to 0$.. Owing to the emission and re-absorption of inertons by the particle, its mass also varies, $m \to 0 \to m \to 0$ …, but of course m does not disappear: the particle's mass state m is periodically passed to a local tension Ξ that is induced on the particulate cell: $\Xi \to 0 \to \Xi \to 0$... (the tension is a displacement of the volumetric fractal deformation from its equilibrium state, periodically changing to a local tension of space). There is nothing extraordinary in such a behavior. An electric charge (e) moves in the same way along its path, which periodically is transformed to the magnetic monopole (g) state: $e \to g \to e \to g$... [32, 38].

Thus, the core of a moving particle is accompanied by a cloud of the particle's inertons ejected from the particle on its collisions with oncoming cells of the space tessel-lattice. In condensed media, atoms/molecules vibrate near their equilibrium positions. The crystal lattice is a good model to reveal major regularities in mass dynamics of condensed matter in general (a solid, liquid or gas) and that is why this model has been used to study the behavior of the mass of atoms in the crystal lattice. In the crystal lattice the behavior of nodes is defined by the Lagrangian equation,

$$L_{\text{vibr.}} = \frac{1}{2}\sum_{\mathbf{n}}\left(m\,\delta\dot{r}_{\mathbf{n}}^2 - \gamma\delta r_{\mathbf{n}}^2\right) \tag{24}$$

where $m_{\mathbf{n}}$ is the mass of the **n***th* node, $\delta r_{\mathbf{n}}$ is the deviation of this node from its equilibrium position and γ is the force constant. It is well known that based on the Lagrangian equation (the Eq. (24)), the Euler-Lagrange equations can be constructed, which completely determine the dynamics of the nodes.

Since massive nodes vibrate near their equilibrium positions, that is, are in motion, they emit and reabsorb clouds of inertons. Therefore inertons periodically remove a part of the mass from vibrating nodes and subsequently bring it back. Such behavior can be described in terms of the Lagrangian equation (the Eq. (25)) (for simplicity of consideration we consider an one-dimensional lattice), which is similar in form to the Eq. (24) but is different in dimension, as its variables are masses [33].

$$L_{\text{mass}} = \sum_{\mathbf{n}}\left\{\frac{1}{2}\dot{m}_{\mathbf{n}}^2 - \frac{\pi}{2T}\left(\dot{m}_{\mathbf{n}}\mu_{\mathbf{n}} + \dot{m}_{\mathbf{n+g}}\mu_{\mathbf{n}}\right) + \frac{1}{2}\dot{\mu}_{\mathbf{n}}^2\right\} \tag{25}$$

where $m_{\mathbf{n}}$ and $\mu_{\mathbf{n}}$ are variations of mass of **n***th* node and its cloud of inertons, respectively, which occur due to the overlapping of inerton clouds of neighboring nodes; g is the lattice vector; T is the period of collision of the mass located in the **n**th node with its inerton cloud. The dot over mass means the derivative in respect to the time *t* treated as a natural parameter.

Instead of variables $m_{\mathbf{n}}$ and $\mu_{\mathbf{n}}$ we may pass on to collective variables $\Phi_{\mathbf{k}}$ and $\phi_{\mathbf{k}}$ by rules

$$m_{\mathbf{n}} = \frac{1}{\sqrt{N}}\sum_{\mathbf{k}}\Phi_{\mathbf{k}}e^{i\,\mathbf{k}\,\mathbf{g}},\ \mu_{\mathbf{n}} = \frac{1}{\sqrt{N}}\sum_{\mathbf{k}}\varphi_{\mathbf{k}}e^{i\,\mathbf{k}\,\mathbf{g}} \tag{26}$$

Substituting Eq. (26) into Eq. (25), we obtain:

$$L_{\text{mass}} = \frac{1}{N}\sum_{\mathbf{k}}\left\{\frac{1}{2}\dot{\Phi}_{\mathbf{k}}\dot{\Phi}_{-\mathbf{k}} - \frac{\pi}{2T}\left(1+\cos\mathbf{kg}\right)\dot{\Phi}_{\mathbf{k}}\,\varphi_{-\mathbf{k}} + \frac{1}{2}\dot{\varphi}_{\mathbf{k}}\,\dot{\varphi}_{-\mathbf{k}}\right\}. \tag{27}$$

The Euler-Lagrange equations for the variables $\Phi_{\mathbf{k}}$ and $\phi_{\mathbf{k}}$ become

$$\ddot{\Phi}_{\mathbf{k}} - \omega(k)\,\dot{\varphi}_{-\mathbf{k}} = 0 \tag{28}$$

$$\ddot{\varphi}_{\mathbf{k}} + \omega(k)\,\dot{\Phi}_{\mathbf{k}} = 0 \tag{29}$$

where we designate $\omega(k) = \frac{\pi}{2T}\left(1 + \cos(\mathbf{kg})\right)$.

Periodical solutions to Eqs. (28) and (29), which satisfy the physical characteristics of the system of varying masses, can be chosen as follows:

$$\Phi_{\mathbf{k}} = \Phi_0 + \Phi_1 \cos\left(\omega(k)t\right), \tag{30}$$

$$\phi_{\mathbf{k}} = -\Phi_1 \cos\left(\omega(\mathbf{k})t\right) \tag{31}$$

where parameters Φ_0 and Φ_1 are proportional to the rest mass of the system's particles and the mass of their inerton clouds, respectively, and inversely proportional to the square root of the total number of particles $N^{-1/2}$. The mentioned arguments point out that the variables Φ_k and ϕ_k represent collective massive excitations in the lattice: Φ_k describes collective mass excitations of the nodes of the lattice; ϕ_k characterizes the mass field of inertons that fill the entire space between the nodes in the lattice, like dust.

It should be emphasized that these mass excitations are completely independent from the phonons of the lattice, because phonons are associated with collective changes of positions of nodes (atoms). Mass excitations described by the variable Φ_k represent the collective mass state of nodes at the moment t and the variable ϕ_k depicts the collective state of the total inerton cloud of the lattice.

The amplitude δm of oscillations of the **n***th* node's mass can crudely be estimated as a ratio of the dispersion of the **n**th node's inerton cloud at the maximal distant object ($r = \Lambda$) and the nearest node ($r = g$).

$$\delta m \approx m_{\mathbf{n}} \frac{g}{\Lambda} \sim 10^{-4} m_{\mathbf{n}} \tag{32}$$

where Λ is the amplitude of the inerton cloud of the **n***th* node. In accordance with Eq. (24), the mentioned amplitude is related to the node's de Broglie wavelength, $\lambda_{\mathbf{n}} \equiv \delta r_{\mathbf{n}}$, the node's velocity υ (sound velocity, as the node participates in acoustic vibrations) and the inerton velocity in the lattice can be equal to the velocity of light c; then

$$\Lambda_{\mathbf{n}} = \delta r_{\mathbf{n}} \frac{c}{\upsilon} \geq \delta r_{\mathbf{n}} \cdot 10^5 \sim 10^4 g. \tag{33}$$

Thus, we can see that in a solid we have an additional physical field, which is the inerton field that so far has not been practically taken into account. In fact, vibrations of atoms result in a series of acoustic waves. But the space between atoms is filled with inertons, which appear owing to the interaction of vibrating atoms with the space organized as the tessel-lattice. Overlapping of inerton clouds and the mobility of atoms, which is the source of these inertons, bring about the formation of inertons waves (with their own harmonics) in the solid as well.

2.4 CLUSTERIZATION OF ATOMS/MOLECULES

[39] studied the behavior of a system of particles with a different character of interaction. The approach makes it possible to describe systems of interacting particles by statistical methods taking into account their nonhomogeneous spatial distribution, that is, cluster formation. For these clusters are evaluated: their size, the number of particles in a cluster, and the temperature of phase transition to the cluster state. The approach developed is very suitable for examination of nanosystems, it allows one to study pair interaction potentials between molecules forming a nanoparticle and also the interaction between nanoparticles. Among these interactions the inerton interaction is also is present, which brings quite new and sometimes unexpected properties to the systems studied.

2.4.1 FORMALISM OF CLUSTER FORMATION

We may start from the construction of the Hamiltonian for a system of interacting particles. The energy for such a system can be written in the general form [40],

$$H = H_0 - \tfrac{1}{2}\sum_{\mathbf{r},\mathbf{r}'} V_{\mathbf{r}\mathbf{r}'}^{\text{att.}}\, c(\mathbf{r})c(\mathbf{r}') + \tfrac{1}{2}\sum_{\mathbf{r},\mathbf{r}'} V_{\mathbf{r}\mathbf{r}'}^{\text{rep.}}\, c(\mathbf{r})c(\mathbf{r}'). \tag{34}$$

Particles occupy knots in Ising's lattice described by the radius vectors $\mathbf{r}$ and $\mathbf{r}$, and the filling number for the ith knot $c_i(\mathbf{r}) = \{0, 1\}$. If the potentials $V_{\mathbf{rr}}^{\text{att.}}, V_{\mathbf{rr}}^{\text{rep.}} > 0$, the second term (comprising $V_{\mathbf{rr}}^{\text{att.}}$) in the right-hand side of the Hamiltonian (34) corresponds to the effective attraction and the third term (comprising $V_{\mathbf{rr}}^{\text{rep.}}$) conforms to the effective repulsion. This allows us to represent the Hamiltonian (34) in the form typical for the model of ordered particles, which is characterized by a certain nonzero order parameter,

$$H(n) = \sum_{s} E_s\, n_s - \tfrac{1}{2}\sum_{s,s'} V_{ss'}^{\text{att.}}\, n_s n_{s'} + \tfrac{1}{2}\sum_{s,s'} V_{ss'}^{\text{rep.}}\, n_s n_{s'}. \tag{35}$$

Here E_s is the additive part of the particle energy (the kinetic energy) in the sth state. So, we have two terms for particle/molecular potential: the attraction and the repulsion components. So, in the Hamiltonian (35) the potential $V_{ss}^{\text{att.}}$ represents the paired energy of attraction and the potential $V_{ss}^{\text{rep.}}$ is the paired energy of repulsion. The potentials take into account the effective paired interaction between particles/molecules located in states s and s'. The filling numbers n_s can run only two values: 1 (the sth knot is occupied) or 0 (the sth knot is not occupied). The signs before positive functions $V_{ss}^{\text{att.}}$ and $V_{ss}^{\text{rep.}}$ in the Hamiltonian (35) directly specify proper signs of attraction (minus) and repulsion (plus).

The statistical sum of the system of interacting particles,

$$Z = \sum_{\{n\}} \exp\left(-H(n) / k_{\text{B}}\Theta\right) \tag{36}$$

can be rewritten via the action $\tilde{S}$, which depends on three functions,

$$Z = \operatorname{Re}\frac{1}{2\pi i}\int D\varphi \int D\psi \oint dz \, \exp\left[\tilde{S}(\varphi,\ \psi,\ z)\right]\cdot \tag{37}$$

The complicated function $\tilde{S}(\phi,\ \psi,\ z)$ was evaluated for extremum [39]. The most stable solution appears when all particles are distributed by clusters, especially if each cluster includes the same number of particles. In this case the action for a cluster of N quantum particles becomes

$$S \approx \tfrac{1}{2}\{a(N) - b(N)\} \cdot N^2 \tag{38}$$

where the functions a and b are defined as follows:

$$a = 3\int_1^{N^{1/3}} V^{\text{rep.}}(gx)x^2\,dx / (k_{\text{B}}\Theta), \tag{39}$$

$$b = 3\int_1^{N^{1/3}} V^{\text{att.}}(gx)x^2\,dx / (k_{\text{B}}\Theta) \tag{40}$$

where g is the lattice constant.

Having known the explicit form of the action (38), one can derive the equation for the number of particles combined in a cluster: $\partial S / \partial N = 0$, which in addition requires holding of the inequality $\partial^2 S / \partial N^2 |_{N=N_{\text{in cluster}}} > 0$.

The formalism described by Eqs. (38)–(40) can be applied to many different physical systems in which particles exhibit interaction. In particular, we described the behavior of electrons on a liquid helium surface; particles interacting by the shielding Coulomb potential, which are found under the influence of an elastic field; and gravitating masses with the Hubble expansion [39].

The formalism Eqs. (38)–(40) was used at studies of several other physical systems, which directly relate to the subject of the present paper.

2.4.2 "FROZEN" MOLECULES

The formalism proposed by Krasnoholovets and Lev [39] allows one to exam a possible cluster formation in a gas, liquid or solid. For the clusterization one needs two pair potentials that consist of a short-range repulsion and a longer-range attraction (or vice versa). In particular, for molecular systems these are potentials of Morse, Lennard-Jones and Buckingham (*see*, for example, Ref. [41]), which respectively look as follows:

$$V(r) = D_0 \cdot \left(e^{-2\alpha(r-r_0)} - 2e^{-\alpha(r-r_0)}\right),$$

$$V(r) = 4\varepsilon \cdot \left[\left(\frac{\sigma}{r}\right)^{12} - \left(\frac{\sigma}{r}\right)^{6}\right], V(r) = Ae^{-\alpha r} - \frac{B}{r^6}. \quad (41)$$

However, following the general scheme presented in Eqs. (37)–(40), we reveal that for these three potentials the equation $\partial S / \partial N = 0$ does not have a solution in which $N >> 1$. In other words, it seems no clusters can be formed in molecular systems.

Notwithstanding this circumstance, the submicroscopic concept allows one to investigate how clusters arise in molecular systems. The problem of cluster formation is solved if we take into consideration vibrations of atoms/molecules near their equilibrium positions. In fact an entity in condensed matter experiences local vibrations that exist even at the absolute zero temperature, which is called the zero point energy.

Therefore, let us try to introduce an additional attracting potential, $\frac{1}{2} m\,\omega^2 \delta r^2$, where the amplitude δr of oscillations plays the role of the de Broglie wavelength of the vibrating entity. This expression can be rewritten as $\frac{1}{2}\gamma \delta r^2$. Such potential is extremely important, because owing to these oscillations the vibrating entity emits and absorbs its inertons, as has been described in the previous sections. Emitted inerton clouds overlap with similar inerton clouds emitted by other entities, which results into the inerton interaction in the physical system. The inerton interaction is additional to the pair interaction of electric nature, which reflects the molecular potentials (41) (and similarly the inerton interaction emerges in systems of magnetically interacting atoms and ions).

Let us examine a model system of N molecules interacting through the Lennard-Jones pair potential taking into account the inerton interaction $\frac{1}{2}\gamma\,\delta r^2$ between molecules. The attraction part of the pair potential is:

$$V_{\text{att}}(gx) = \frac{V_0}{x^6} - \tfrac{1}{2}\gamma(\delta r)^2 x^2 \tag{42}$$

and the repulsion part

$$V_{\text{rep}}(gx) = \frac{V_0}{x^{12}}. \tag{43}$$

Here, on the right hand side of Eq. (42) in the first term the distance r from the node to a distant point is written as $r = gx$ and in the second term the amplitude of oscillations is depicted as $x\delta r$, where g is the lattice constant, δr becomes a parameter and x is the dimensionless variable that describes the distance. On the right hand side of Eq. (42) we also introduce the dimensionless variable x for the distance.

Note in the formalism described above, Eqs. (34)–(40), the sign describing attraction and repulsion is taken out of the values of V_{att} and V_{rep}; that is way the right hand side in the Eq. (42) has the opposite sign to the attraction [37]. In the case of a classical system of interacting particles the action looks as follows [33].

$$S \approx 2[a(N) - b(N)]N + N\ln\xi \tag{44}$$

where the functions a and b are determined in Eqs. (39) and (40), respectively.

For the potentials (the Eqs. (42) and (43)), the action (the Eq. (44)) has the form

$$S = -\tfrac{4}{3}\frac{V_0}{k_{\text{B}}\Theta}N + \tfrac{3}{5}\frac{\gamma\delta r^2}{k_{\text{B}}\Theta}N^{8/3} + N\ln\xi \tag{45}$$

and the equation for the number of molecules in a cluster $\partial S / \partial N = 0$ results in the solution (if we neglect the contribution on the side of the fugacity ξ)

$$N \approx \left(\frac{5}{6} \frac{V_0}{\gamma \, \delta r^2} \right)^{3/5} \tag{46}$$

The dependence of N as a function of δr is shown in Fig. 2.5. Thus the appearance of fractal clusters is governed by local vibrations of entities in condensed media. In particular, we can see the smaller δr, the larger N. However, this is possible only when the amplitude δr of oscillations of particles near their equilibrium positions begins to drop, that is, becomes less than approximately 10^{-11} m (Fig. 2.5).

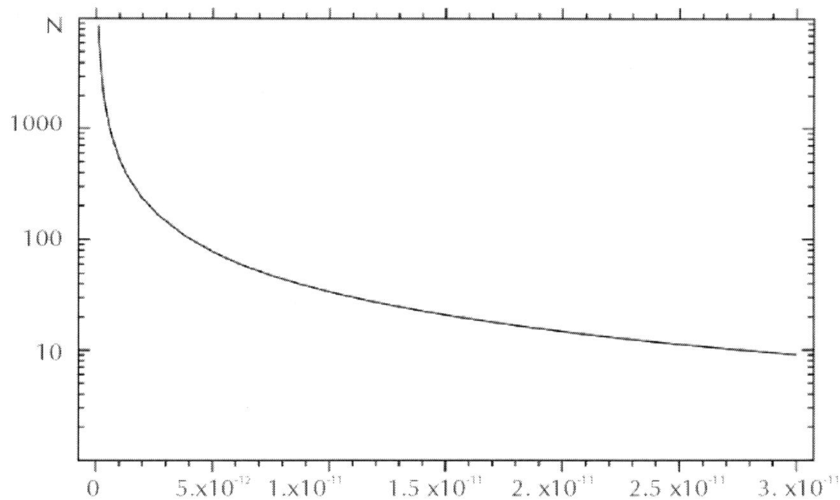

FIGURE 2.5 Numerical solution N versus δr (46). Values of the parameters are $V_0 = 4.25 \times 10^{-20}$ J and the force constant $\gamma = 1$ N/m.

Clusterization of water through the possible influence of an inerton field was also examined [33].

An external inerton signal is able to drop δr even if the temperature does not decrease, which immediately increases the number N of molecules in the cluster (or brings to a cluster in the case when it initially did not exist due to the relatively large value of δr). Indeed, since the mass of entities oscillates together with their oscillating motion, any additional injection of mass in the form of inertons causes absorption of an extra mass leading to the appearance of a mass defect (or surplus defect) Δm in each irradiated entity. Hence, the amplitude of the entity's vibrations decreases:

$$\delta r = h/(mv) \rightarrow \delta r' = h/[(m+\Delta m)\,v], \;\; \delta r > \delta r' \tag{47}$$

The phenomenon of diminution of the amplitude of the entity's vibrations in a condensed matter is similar to decreasing the temperature: the lower the temperature, the lower the amplitude of vibrations. However, the temperature does not change. The phenomenon is stipulated by weighting of atoms, $m \rightarrow m + \Delta m$. The relaxation time, that is, return of the atom's mass to the initial state, $m + \Delta m \rightarrow m$, is very long. The relaxation time lasts for days or even months, which means that classical thermodynamics requires a significant reconsideration, because it must be supplemented by the mass exchange. First attempts to introduce some changes to the thermodynamics potentials caused by the mass exchange have already been taken [42–45].

Typical changes in the viscosity of sorbent samples irradiated by an inerton field are shown in Fig. 2.6. The dynamic viscosity and the shear viscosity of the sample under consideration significantly depend on the exposure time in an inerton field, which is shown in Fig. 2.7 for measurements conducted approximately half an hour after the inerton irradiation. Samples of water irradiated by an inerton field demonstrate an increase in the water density from 1 to about 1.2 kg/L, which was registered by a hydrometer. This indicates a change in both the density and the viscosity of water processed with an inerton field. Liquid substances, gels, hand creams, ointments and similar substances after processing with an inerton field generated by our device (Fig. 2.8) become more fluid demonstrating a kind of a 'superfluidity'. In particular, the rate of penetration of hand creams through the skin increases 20–30%, which was measured by a special facility used in cosmetology.

In Fig. 2.9, one can see our inerton measuring device 'Rudra,' which was designed to measure the intensity and the spectrum of inertons in a range of frequencies from a few Hz to 100 kHz. Two types of the antenna were tested: a ferrite rod with a coil and a piezoceramic sensor. The electric scheme of the device is functioning on the basis of principles described in the present subsection. The antenna's atoms absorbing inertons move from their equilibrium positions trying to arrange short-lived clusters, which results in an induction of local magnetic/electric fields in the antenna. In its turn these local fields impact electrons in the electronic scheme. The generated electric current is further processed, transmitting the received information to an analog-to-digital converter. The Rudra device successfully functions even when its antenna is screened with a metal box, though usual receivers of electromagnetic waves have not been able to catch signals being directed towards this metal box.

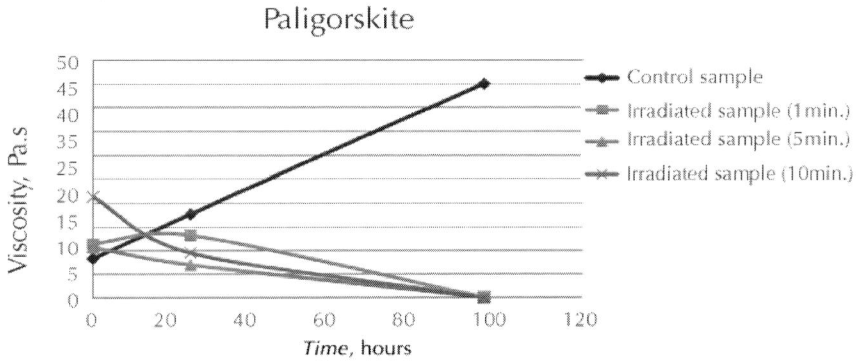

FIGURE 2.6 Behavior of the viscosity of samples of palygorskite (a magnesium aluminum phyllosilicate with formula (Mg, Al)$_2$Si$_4$O$_{10}$(OH)·4(H$_2$O), a type of clay soil) irradiated with an inerton field. The viscosity of the control sample increases under atmospheric conditions with time, though the viscosity of the samples irradiated by the inerton field gradually decreases.

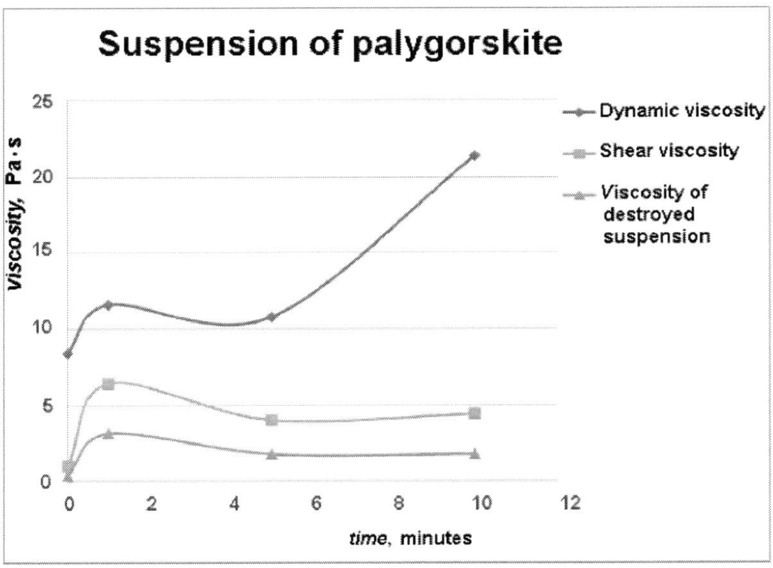

FIGURE 2.7 Viscosity versus time of inerton irradiation of an aqueous suspension of palygorskite.

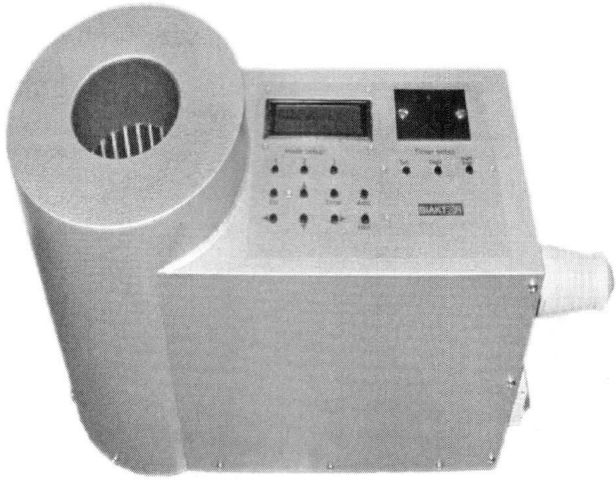

FIGURE 2.8 Biaktor's laboratory batch unit Lx (http://biaktor.com). The device is intended for intensification of chemical physical reactions in liquid and gel substances.

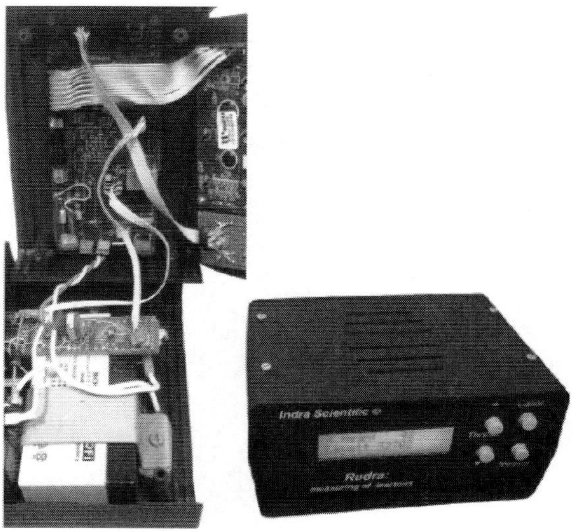

FIGURE 2.9 Device 'Rudra' (interior and exterior) that measures inerton signals.

2.4.3 BOSE-EINSTEIN CONDENSATE

Although the gas cooling mechanism and the Bose-Einstein condensation of dilute gases were studied in detail, the submicroscopic approach allowed one to clarify subtle aspects of these phenomena [33]. Namely, the approach discloses inner interactions and the organization of atoms in the Bose-Einstein condensate.

When atoms are cooled to a temperature when the de Broglie thermal wavelength $\lambda_{\text{th}} = h / \sqrt{3 m k_{\text{B}} \Theta}$ is comparable to the mean distance g between atoms, atomic inerton clouds stop to overlap, but they still touch each other, which synchronizes the atoms. Namely, in such a case the $(\mathbf{n} - \mathbf{g})$th atom emits its inerton cloud that then is fully absorbed by the **n**th atom; the **n**th atom emits its own cloud of inertons, which then is fully absorbed by the $(\mathbf{n} + \mathbf{g})$th atom, etc. In other words, the coherent exchange of inerton clouds by the atoms when an inerton cloud emitted by one atom hops to the neighboring atom and is absorbed by it, we relate with the phenomenon of Bose-Einstein condensation.

Let us investigate whether the occurrence of clusters can be possible in a Bose-Einstein condensate. The action S for the ensemble of N interacting boson particles, which tend to clusterization with N particles in a cluster, has the form [33],

$$S \cong \left\{ \tfrac{1}{2} \left[a(N) - b(N) \right] N^2 - \ln (N+1) + N \ln \xi \right\} \tag{48}$$

where the functions a and b are determined in Eqs. (39) and (40), respectively, ξ is the rugosity, and in the action (the Eq. (48)) next two terms are preserved in contradistinction to the form (38). For simplicity, the repulsion potential can be taken in the form of Eq. (43). The attraction potential should include at least three terms: (i) the dispersion potential of interatomic interaction, which is usually written as $-C_6 / r^6$; (ii) a potential formed by a trap, which can be modeled by a harmonic potential; and (iii) the harmonic potential caused by small spatial oscillations of atoms near their equilibrium positions, that is, inerton elastic interaction. So, the attraction and the repulsion potentials are

$$V^{\text{att.}}(rx) = C_6 / (rx)^6 - \tfrac{1}{2} \gamma_{\text{trap}} \, r^2 x^2 - \tfrac{1}{2} m \,\omega^2 (\delta r)^2 x^2 , \tag{49}$$

$$V^{\text{rep.}}(rx) = V_0 / x^{12} , \tag{50}$$

where m is the mass of an atom, r is the distance between atoms, γ is the effective force constant of the trap, ω is the cyclic frequency of proper oscillations of an atom, δr is the appropriate amplitude and x is the dimensionless distance parameter.

Substituting potentials (the Eqs. (49) and (50)) into functions (the Eqs. (39) and (40)), respectively, we then construct the action (the Eq. (48)), which in the explicit form becomes

$$S = \left\{ \tfrac{1}{6}\frac{V_0}{k_B \Theta} N^2 - \tfrac{3}{6}\frac{C_6}{r^6 k_B \Theta} N^2 + \tfrac{3}{20}\frac{1}{k_B \Theta}\left(\gamma_{\text{trap}}\, r^2 + m\,\omega^2\,\delta r^2\right) N^{11/3} - \ln(N-1) + N \ln \xi \right\} \tag{51}$$

Proper oscillations of atoms, which are characterized by the frequency ω, are produced by the movement of the atoms. In other words, the origin of the frequency ω is produced by collisions of atoms with their inerton clouds [24, 30]: $\omega = 2\pi / 2T$ where T is the period of time between collisions of an atom and its inerton cloud (though in the present case we have to talk about collisions of the inerton cloud of an atom with the neighboring atom); the value of T is related to the amplitude δr of oscillations of an atom and its velocity υ, $1/T = \upsilon / \delta r$ (because the amplitude δr of oscillations of the atom is related to the atom's de Broglie wavelength λ). Since in the case of Bose-Einstein condensation we can put $\mu = 0$ for the chemical potential of atoms, the fugacity $\xi = 1$ and hence the last term in Eq. (51) reduces to zero.

The expression for the number of atoms combined in a cluster comes from the equation $\partial S / \partial N = 0$, or explicitly

$$N \cong \left(\tfrac{20}{33}\, \frac{3C_6 / r^6 - V_0}{\gamma_{\text{trap}}\, r^2 + m\omega^2 \delta r^2} \right)^{3/5}. \tag{52}$$

Here in Eq. (52) in the first approximation the numerator, that is, the difference between the attraction and repulsion energies at an equilibrium distance r of interacting atoms, can be put $3C_6 / r^6 - V_0 = 2.125 \times 10^{-20}$ J. For the case of cesium atoms whose mass $m_{\text{Cs}} = 2.207 \times 10^{-25}$ kg the amplitude δr of oscillations of an atom is associated with its de Broglie wavelength λ: $\delta r = \lambda = h / (m_{\text{Cs}}\, \upsilon)$. The trapping potential $\tfrac{1}{2}\gamma_{\text{trap}}\, r^2$ is the fitting parameter that can be chosen equal to 0, 2.5×10^{-29} and 2.5×10^{-28} J.

The oscillation of atoms is caused only by their thermal motion: $\upsilon \approx \upsilon_{\text{therm}} = \sqrt{3 k_B \Theta / m_{\text{Cs}}} \cong 3\times10^{-3}$ m/s where we put a typical temperature of Bose-Einstein condensate $\Theta = 50$ nK. So, $\delta r = \lambda_{\text{th}} = h / (m_{\text{Cs}}\, \upsilon_{\text{therm}}) \cong 10^{-6}$ m and then the cycle frequency of atom oscillations become $\omega = \pi\, \upsilon / \delta r \cong 9.43 \times 10^3$ s^{-1}. The abovementioned numerical values of the parameters allow the evaluation of the number of atoms that assemble in a Bose-Einstein cluster, as $N \sim 10^5$. The interatomic interaction gathers dilute cold atoms to clusters. But the cluster state is realized when the absolute value of an attraction potential exceeds the thermal energy, $V^{\text{att.}} \geq k_{\text{B}}\Theta$. This inequality holds for the case calculated above: the attraction energy $\frac{1}{2}m_{\text{Cs}}\,\omega^2\delta r^2 + \frac{1}{2}\gamma_{\text{trap}}\,r^2 \approx 1.7\times10^{-30}$ J exceeds the thermal energy $k_{\text{B}}\Theta \approx 7\times10^{-31}$ J.

Thus we have shown that the phenomenon of Bose-Einstein condensation of bosons, which by definition exists in momentum space, represents a stable cluster of these bosons in the real space.

An interesting paper has recently been published in which the authors [5] claim observation of ultracold bosons at negative temperature that is stable against collapse for arbitrary atom numbers. A negative temperature was derived in the framework of the following approach. The starting point was the Bose-Hubbard Hamiltonian

$$\hat{H} = -J\sum_{i,j}\hat{b}_i^{+}\hat{b}_j + \tfrac{1}{2}U\sum_{i}\hat{n}_i(\hat{n}_i - 1) + V\sum_{i}\hat{n}_i\mathbf{r}_i^2\,, \tag{53}$$

which means, all cold atoms were distributed by sites of a lattice. In the Hamiltonian (the Eq. (53)) the first term describes a possible tunneling of atoms from one site to the other, the second term describes the attracting interaction between atoms, and the third term describes a cavity potential.

After some speculations in which Bose-Einstein condensation of atoms is reduced unreasonably to possible condensation of all atoms in one spatial point (though by the definition Bose-Einstein condensate is defined in momentum space), [5] indicate conditions for the existence of negative temperature: stable negative temperature states with bosons can exist only for attractive interactions and an antitrapping potential. This means that in the Hamiltonian (the Eq. (53)) the second term should be negative and the third term positive. Then at negative temperature high-energy states should be more occupied by the atoms than low-energy states.

In pursuit of a negative temperature [5] mixed together momentum space and the ordinary physical space. Besides, by using thermodynamic definition of temperature

$$1/\Theta = \partial S/\partial E \tag{54}$$

where S is the entropy and E is the energy, they say the entropy should decrease with energy if high-energy states are more populated than low-energy ones.

It is obvious that the theoretical considerations [5] are devoid of any physical meaning. In fact, the Hamiltonian (53) is arranged in an abstract space related to a lattice whose sites are allied with the ordinary physical space. At the same time Bose-Einstein condensate is determined in momentum space, that is, all bosons have the same momentum and therefore the condensate cannot be associated with Eq. (53) to the full. By definition the notion of the entropy S is defined only for stationary states, though in the experiments of [5] they dealt with fast transitional processes. The energy E, which is part of the expression (54), also is not defined because in the Hamiltonian (the Eq. (53)) one can see only components of the potential energy, though the authors talk about the population of high-energy states, which is possible only via introduction of the kinetic energy. Why did other researchers who studied upper levels (second, third and so on) occupied by excited atoms not talk about negative temperatures? Moreover, all thermodynamic functions including the entropy S require a revision, which is associated with the presence of a mass field (the same as a temperature field, pressure field, etc.), as has been mentioned in the previous subsection. In addition, objections against the entropy as a fundamental thermodynamic potential, has recently substantiated by Zhang [46].

The above criticism allows us to reinterpret the interesting experimental data presented by Braun et al. [5] in the following way. Momentum distributions in the atomic cloud were measured for the cases of the rest atomic cloud and that excited by the laser beam. The measurement meant a series of captured images at intervals of a few milliseconds. Braun et al. distributed these images arranged in the final states of the atomic cloud between positive and negative temperatures. Images of the rest atomic cloud were related to positive temperatures and the images of excited atomic cloud were associated with negative temperatures.

If we look at the system of cold ^{39}K atoms studied by Braun et al. from the submicroscopic viewpoint and the mechanism of clustering described above, we will see that the Bose-Einstein condensate (or in other words the cluster state of cold atoms) is preserved as long as the kinetic energy of excited atoms remains less than the potential energy U in Eq. (52). Switching antitrapping potential, which decreases the trapping potential V in Eq. (52), creates excitations in the motion of atoms. Clearly, the temperature Θ should also be less than $|U$.

What is occurring in the cluster when some atoms become excited? First of all the anti-trapping potential partly diminishes the trapping potential $\frac{1}{2}\gamma_{\text{trap}}r^2$ in Eqs. (49) and (51). Besides, since the anti-trapping potential is distributed in 3D space, we have to anticipate similar momentum distributions of atoms. Namely, the initially spherical cluster with the solution (the Eq. (51)) for the number of atoms must change the shape followed by the symmetry of anti-trapping potential. In fact, in the general case the variable N is subdivided into N_x, N_y and N_z. The corresponding changes will appear in expressions for the functions $a(N)$, $b(N)$ and also the action S (the Eq. (51)). Then the final solution for the number of atoms in a cluster will represent three different expressions along the axes X, Y and Z. Herewith $N_x + N_y + N_z = N$. So a local disturbance in the lattice creates the situation that [5] interpreted as negative energy but if we look at the whole system, there is no reason to assume negative temperatures below 0 K.

2.4.4 A CLUSTER OF ELECTRONS

A series of reports informed about observation of electron droplets [6–11]. Those droplets were generated through electric discharges. Interesting studies of charge droplets punching various materials were carried out by Shoulders and Shoulders [47].

Kukhtarev and Kukhtareva [12] carried out an experiment in which the formation of electron droplets was observed after illumination of a $LiNbO_3$ crystal by a focused laser beam (CW green laser, $\lambda_{\text{laser}} = 532$ nm, $P = 100$ mW). In the course of the experiments, we observed a usual recording of holographic gratings. An interference pattern was formed between the pump laser beam and the scattered waves producing moving space-charge waves inside the crystal. Due to the electro-optic effect these space-charge waves modulated the refractive index, which resulted in the formation of holographic gratings. Thus, the pump laser beam diffracted on the holographic gratings visualizing the space-charge waves.

In addition we observed a very unusual phenomenon. Owing to the illumination of the crystal, a photon bunch induced electron emission in the impact area of the crystal boundary. The emitted electrons manifested themselves via an emergence of specular reflection spots (Fig. 2.10). An analysis showed that those enigmatic droplets had the size of about 100 μm and each droplet included around 10^{10} electrons. Those bright droplets slowly moved in the space along the crystal surface.

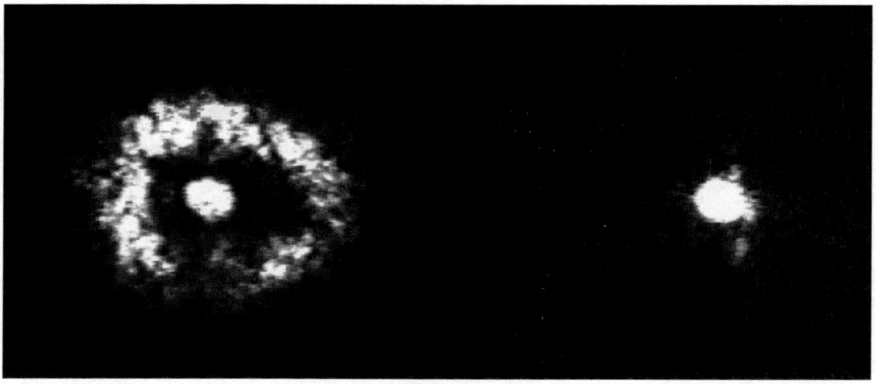

FIGURE 2.10 Left image: Enhanced back-reflected scattering from the surface of the ferroelectric crystal $LiNbO_3$ covered with a thin metal film as seen on the screen. Right image: The specular reflected beam with the droplet.

Free electrons have never been observed at a velocity smaller than 10^5 m s^{-1}. At the same time in the experiment studied a droplet of electrons had a velocity only about 0.5 cm s^{-1}. But how naked electrons could gather together in such a small space? A 10^{10} electrons in a droplet with a radius of 50 μm results in an electron density of about 10^{24} m^{-3}, which means that the mean distance between electrons $\bar{r} \approx 10^{-8}$ m. Hence the repulsion energy between nearest electrons in the droplet becomes

$$E_{\text{repulsion}} = \frac{e^2}{4\pi\varepsilon_0 \bar{r}} \approx 2.3\times10^{-20} \text{ J} \tag{55}$$

where e is the elementary electric charge and ε_0 is the dielectric constant. It seems the repulsion energy (the Eq. (54)) should lead to the scattering of electrons with the same kinetic energy, $E_{\text{repulsion}} = E_{\text{kinetic}} = mv^2/2$, which gives the electron velocity $\upsilon \approx 2\times10^5$ m×s^{-1}. However, electrons were not released from the droplets. So, they are strongly kept by a mysterious force!

To unravel the mystery, we must assume that the echo pulse transferred not only acoustic waves by also a flow of inertons knocked out from the vibrating crystal lattice of $LiNbO_3$. In the vicinity of the surface of the crystal photoelectrons absorbing inertons significantly change their behavior. Namely, the inerton field ties the electrons together. This means that in the droplet electrons are characterized by two kinds of interactions: the Coulomb repulsion (55) and an elastic interaction caused by the overlapping of electrons' inerton clouds (or wave ψ-functions in terms of the conventional quantum mechanical formalism). The absorption of inertons strengthens the elastic interaction between electrons, which is able to compensate the electrons' Coulomb repulsion.

Let us introduce a dimensionless distance x. Then the repulsive paired potential for electrons can be rewritten as

$$V^{\text{rep.}} = \frac{1}{4\pi\varepsilon_0}\frac{e^2}{\overline{r}\,x} \tag{56}$$

The elastic interaction of electrons through the inerton field may be presented in the form of a typical harmonic potential

$$V^{\text{att.}} = \tfrac{1}{2} m\omega^2 \cdot (\delta\overline{r}\ x)^2 \tag{57}$$

where m is the mass of an electron in the droplet, ω is the cyclic frequency of its oscillations and $\delta\overline{r}$ is the amplitude of the electron displacement from its equilibrium state (note that this amplitude $\delta\overline{r}$ is directly connected with the de Broglie wavelength of the electron, $2\,\delta\overline{r} = \lambda$).

Then calculating the functions a (39) and b (40) and substituting them into the action (Eq. (38)) we obtain

$$S \approx \left(\frac{3e^2}{8\pi\varepsilon_0 \overline{r}\, k_{\text{B}}\Theta} N^{2/3} - \frac{3\,m\,\omega^2\,\delta\overline{r}^{\,2}\,N^{5/3}}{10\,k_{\text{B}}\Theta} \right) N^2 \tag{58}$$

The minimum of action (Eq. (58)) is reached at the solution of the equation $\partial S/\partial N = 0$ (if the inequality $\partial^2 S/\partial N^2 > 0$ holds). The corresponding solution is

$$N \approx \frac{20}{11}\,\frac{e^2/(4\pi\varepsilon_0\,\overline{r})}{\frac{1}{2}m\omega^2\,\delta\overline{r}^{\,2}} \tag{59}$$

Therefore the quantity of electrons involved in a droplet is determined by the ratio of repulsive and attractive paired potentials.

The value of the frequency $2\pi\omega$ can be estimated as 1 MHz, as evidenced by radiosignals recorded from droplets. Then substituting numerical values of all the parameters into Eq. (59), we will find that the quantity $N \approx 10^{10}$ of electrons in a droplet is reached at a non-realistically large size of the amplitude $\delta\overline{r} \sim 10^{-5}$ m of oscillations of electrons near their equilibrium positions. However, the conflict can be easily overcome, if we introduce an effective mass m^* for the electron and put a reasonable size $\delta\overline{r} \approx 10^{-10}$ to 10^{-9} m for the amplitude of the electron oscillations. The quantity $N \approx 10^{10}$ is satisfied when $m^* \approx 2\times(10^{-22}$ to $10^{-24})$ kg, which exceeds the rest mass of electrons millions of times.

Hence the variation in the mass of electrons allows us to resolve the problem of the electron droplet stability.

2.5 CASIMIR EFFECT

Investigators studying the Casimir effect are confident that Casimir forces can drive the operation of nanomachinery and therefore such research is very important [48]. The majority of researchers associate these forces with the fractal energy $E_0 = \frac{1}{2}\hbar\omega$, or zero-point energy, which is treated as the inner energy of vacuum fluctuations of the electromagnetic field [49]. As a result all mechanical effects observable in mesoscopic physics are connected with effects occurring in the physical vacuum itself. In quantum field theory, the London – van der Waals forces, Casimir-Polder forces and the Casimir and Lifshitz forces are physical forces arising from a quantized field.[1] considered two conducting square plates with the size $\mathcal{L}\times\mathcal{L}$ separated by a distance a. One plate is movable and in the first situation the distance a is small and in the second situation it is large.

Casimir [1] considered two

$$\delta E = (\tfrac{1}{2}\sum_i \hbar\omega_i)_I - (\tfrac{1}{2}\sum_i \hbar\omega_i)_{iI} \tag{60}$$

between two summations that extend over all possible resonance frequencies in the vacuum cavity confined by these two plates. The geometric size of the cavity $0 \le x \le \mathcal{L}, 0 \le y \le \mathcal{L}$ and $0 \le z \le a$ determines its possible vibrating modes: $k_x = n_x\,\pi/\mathcal{L}$, $k_y = n_y\,\pi/\mathcal{L}$ and $k_z = n_z\,\pi/a$. To every k_x, k_y, k_z correspond two standing waves. In an explicit form these standing electromagnetic waves spontaneously excited in vacuum have the form

$$\mathbf{A} = \sum_k c\sqrt{\frac{\pi\hbar}{\omega\mathcal{L}^2}}\,\mathbf{e}(\mathbf{k})\times\left\{A_k e^{-(\omega t-\mathbf{kr})} + A_k^+ e^{(\omega t-\mathbf{kr})}\right\} \tag{61}$$

For large $\mathcal{L}$ wave numbers k_x and k_y can be regarded as continuous variables. Then Casimir presented the difference δE (60) in an integral form, which allowed him to obtain the final result – an attractive energy between the two plates:

$$\delta E / \mathcal{L}^2 = -\hbar c \frac{\pi^2}{24 \times 30} \cdot \frac{1}{a^3} \tag{62}$$

The calculated result of Eq. (62) was verified experimentally by different researchers. Moreover, the Casimir effect occurs also in the case of dielectrics [50], which points to its universality. Nowadays both theorists and experimentalists continue to intensively study the Casimir effect. Theoreticians have been developing complicated generalized mathematical approaches based on fluctuations of electromagnetic field, virtual photons and other things that would appear from the zero-point energy and the physical vacuum in general [2, 3, 48–50].

At the same time Refs. [51] and [52] presented a very different viewpoint describing the same phenomenon, i.e. the attraction of two plates. The starting point is the consideration of a scalar field of mass m that satisfies the wave equation $(\nabla^2 + k^2)\phi(x) = 0$. The Casimir energy is written as an integral over the difference between the density of states $\delta\rho(k)$ in a domain of conducting planes and the vacuum,

$$\delta E = \tfrac{1}{2}\hbar \int_0^\infty dk\, \omega(k) \frac{2k}{\pi} \operatorname{Im} \int_D d^3x\, \tilde{G}(x,\ x,\ k + i\varepsilon) \tag{63}$$

where $\omega = \sqrt{c^2k^2 + m^2c^4 / \hbar^2}$ and $\tilde{G} = G - G_0$ is the difference between the Green's function in the background of conducting plates and the Green's function in vacuum. To resolve the equation (63), [51] chose the Green's function typical for classical geometric optics when G is defined by the sum over optical paths, which includes a combination of abstract factors used in classical ray optics. Then the mass m is approaching zero and they finally acquire Casimir's result (62).

Thus, the Casimir effect can be considered without reference to zero point energies. It can be originated from relativistic, quantum forces between fluctuations of charges and currents in borderline material plates [52].

We can further develop the Jaffe's view; namely, materialize his fictitious scalar massive field $\phi(x)$ and combining it with the mathematical method suggested by Casimir [1].

In the previous sections, we introduced submicroscopic mechanics starting from the constitution of the real space in the form of the tessel-lattice. It has been shown that the motion of an object in the tessel-lattice leads to the generation of inertons around it. In particular, neutral atoms and ions oscillating at their equilibrium positions in a solid generate inertons. Acoustic vibrations of entities in condensed matter produce long inerton waves that spread out of the matter as standing inerton waves. These are standing inerton waves of a body that form its Newtonian gravitational potential [32, 34, 37].

A linear spectrum of acoustic phonons breaks at the Debye frequency. At higher frequencies the dispersion relation is no longer linear (Fig. 2.11.). Round the edge of the Brillouin zone the spectrum of large wave numbers k becomes increasingly closer to discrete, because the distance between the wave numbers grows towards the edge of the Brillouin zone [53].

Therefore, since the spectrum of phonons at the edge of the Brillouin zone is specific, we may anticipate that the edge phonons will contribute to the inerton spectrum distinctively as well. Especially it can be justified in the case of near-surface atoms of the material body. Near-surface atoms may have a more obvious discrete spectrum at the edge of the Brillouin zone than the bulk atoms and the difference of these spectra may give an additional distribution of inertons out of the body's surface.

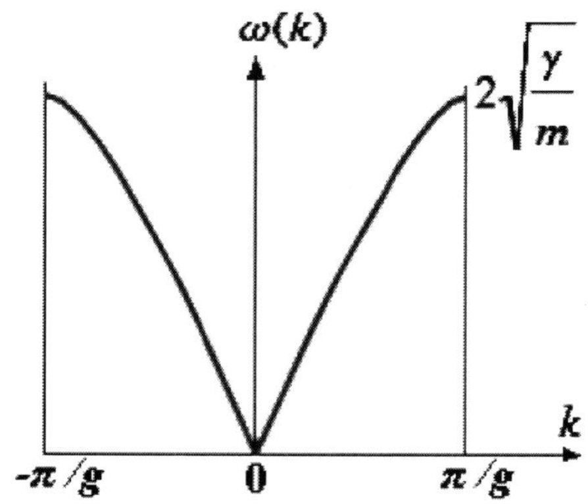

FIGURE 2.11 Spectrum of acoustic phonons that is destroyed near the end of the Brillouin zone.

Indeed, if the amplitude of a vibrating atom is δr, which is the atom's de Broglie wavelength λ, than the amplitude of its inerton cloud is $\Lambda = \lambda c / \upsilon$ (20), where the phase speed is $\upsilon = (2g/\pi) \times \sqrt{\gamma / m}$ (for 1D chain). For example, in the case of gold, its force constant γ may be estimated as varying from 4 to 9 N×m^{-1} [54]. Let us put $\gamma = 4$ N×m^{-1}. The lattice constant of gold is $g = 4.08\times10^{-10}$ m; the mass of the atom of gold is $m_{\text{Au}} = 3.27\times10^{-25}$ kg. Knowing these parameters, we valuate the phase velocity as $\upsilon \approx 9.08\times10^{2}$ m×s^{-1}. Then the amplitude of vibrations of the pair of atoms at the edge of the Brillouin zone, that is, the atom's de Broglie wavelength, becomes $\delta r \equiv \lambda = h/(m_{\text{Au}}\upsilon) = 2.23\times10^{-12}$ m. The appropriate amplitude of the atom's inerton cloud is $\Lambda = \lambda\, c/\upsilon = 0.74$ μm. This value of Λ should be slightly lengthened owing to contributions on the side of other discrete modes and the thermal smearing. Nevertheless, the estimated value of Λ accounts for an order of the radius of the boundary effects that can specifically manifest themselves in the vicinity of the body.

It is intuitively clear these boundary inerton effects are logically linked to the Casimir effect. To proof this, we need to calculate the energy of attraction caused by short wave inertons.

Let us imagine a rectangular cuboid whose base $\mathcal{L}\times\mathcal{L}$ is located on the surface of a material body and it sticks outward having the height a. Let another cuboid with the same size be disposed inside the material body. What is the difference in the vibrating inerton energy for these two cuboids? Then we follow the computational scheme proposed by Casimir [1]:

$$\delta E = \hbar\omega\frac{1}{(\pi/\mathcal{L})^2}\frac{1}{(\pi/a)} \times\left\{\sum_{n=0}^{\infty}\int_0^{\infty}\int_0^{\infty}\sqrt{k_x^2+k_y^2+\frac{\pi^2 n^2}{a^2}}\,dk_x dk_y - \int_0^{\infty}\int_0^{\infty}\int_0^{\infty}\sqrt{k_x^2+k_y^2+\frac{\pi^2 n^2}{a^2}}\,dk_x dk_y dn\right\}. \tag{64}$$

Now turn to polar coordinates ($\sqrt{k_x^2+k_y^2} = \kappa$)

$$\delta E = \hbar 2\pi v a\frac{\mathcal{L}^2}{\pi^2}\frac{1}{\pi}\left\{\sum_{n=0}^{\infty}\frac{\pi}{2}\int_0^{\infty}\sqrt{\kappa^2+\frac{\pi^2 n^2}{a^2}}\,\kappa d\kappa - \frac{\pi}{2}\int_0^{\infty}\int_0^{\infty}\sqrt{\kappa^2+\frac{\pi^2 n^2}{a^2}}\,\kappa d\kappa dn\right\}$$

$$= \hbar c\pi\frac{\mathcal{L}^2}{\pi^2}\left\{\sum_{n=0}^{\infty}\int_0^{\infty}\sqrt{\frac{a^2\kappa^2}{\pi^2}+n^2}\,\frac{\pi}{a}\kappa d\kappa - \int_0^{\infty}\int_0^{\infty}\sqrt{\frac{a^2\kappa^2}{\pi^2}+n^2}\,\frac{\pi}{a}\kappa d\kappa dn\right\} \tag{65}$$

Let us introduce a new variable, $u = \kappa^2/(\pi/a)^2$, which changes Eq. (65) to the form

$$\delta E = \hbar c \pi \frac{\mathcal{L}^2}{\pi^2} \cdot \frac{1}{2}\left(\frac{\pi}{a}\right)^3 \left\{ \sum_{n=0}^{\infty} \int_0^{\infty} \sqrt{u+n^2}\, du - \int_0^{\infty}\int_0^{\infty} \sqrt{u+n^2}\, du\, dn \right\}, \tag{66}$$

where under the integrands we have only dimensionless variables. Now we need to get rid of divergences in the integrals. Following Ref. [1], we introduce a cutoff function $f(k/k_{\mathrm{m}})$, which equals to 1 when $k << k_{\mathrm{m}}$, and 0 when $k >> k_{\mathrm{m}}$. Besides, imputing a new variable, $w = \sqrt{u+n^2}$, we get instead of Eq. (66)

$$\begin{aligned} \delta E &= \frac{\hbar c \pi^2 \mathcal{L}^2}{2a^3} \cdot \left\{ \sum_{n=0}^{\infty} \int_n^{\infty} 2w^2 f\left(\frac{\pi w/a}{k_{\mathrm{m}}}\right) dw - \int_0^{\infty}\int_n^{\infty} 2w^2 f\left(\frac{\pi w/a}{k_{\mathrm{m}}}\right) dw\, dn \right\} \\ &= \frac{\hbar c \pi^2 \mathcal{L}^2}{a^3} \left\{ \sum_{n=0}^{\infty} F(n) - \int_0^{\infty} F(n)\, dn \right\}, \end{aligned} \tag{67}$$

where

$$F(n) = \int_0^{\infty} w^2 f[\pi w/(a k_{\mathrm{m}})]\, dw \tag{68}$$

If $F(n)$ and all its derivatives tend to 0 as $n \to \infty$, the curly braces in Eq. (67) can be presented as follows [55].

$$\sum_{n=0}^{\infty} F(n) - \int_0^{\infty} F(n)\, dn = \frac{1}{2}F(0) - \frac{1}{12}F'(0) + \frac{1}{24 \cdot 30}F'''(0) + \ldots \tag{69}$$

Since $F'(n) = -n^2 f[\pi n/(a k_{\mathrm{m}})]$, $F''(n) = -2n f[\pi n/(a k_{\mathrm{m}})]$, $F'''(n) = -2 f[\pi n/(a k_{\mathrm{m}})]$ and the function f is the step function (zero and unity), it is reasonable to put here its mean value, $f = \langle f \rangle = 1/2$. Then we get: $F(0) = 0$, $F(0) = 0$, $F(0) = -1$. Substituting these values into the right-hand side of Eq. (69) we obtain the result: $-1/(24 \cdot 30)$. Substituting this value into the expression for δE (the Eq. (67)), we immediately arrive at the Casimir outcome (the Eq. (62)).

Thus, the Casimir attraction energy is caused by the inerton deformation of space in the vicinity of the plate's surface. Inertons carry mass and hence they induce a peculiar distribution of the gravitational potential around a body. Shortwave inertons generated by vibrating atoms at the edge of the

Brillouin zone are responsible for the gravitational potential energy (the Eq. (62)), which produces a gravitational force $\mathcal{F} = \hbar c \pi^2 / (240 a^4)$ per square unit of the area. Longwave inertons generated by atoms in the acoustic spectrum induce the conventional Newtonian gravitational potential. Hence, at a short distance (about 1 μm or so) the gravitational potential is proportional to $1/r^3$ and at macroscopic distances it is proportional to $1/r$.

2.6 CONCLUDING REMARKS

In this chapter, we have shown that the real physical space is constructed as the tessel-lattice of primary topological balls, superparticles of the nature. In the tessel-lattice particles are determined as local stable deformations. The motion of a particle generates a cloud of Poincaré like excitations around the particle, which were named *inertons*. It has been argued that a particle together with its cloud of inertons emerges in conventional quantum mechanics as an abstract construction called the wave ψ-function. These particle's inertons carry both particle's inert and gravitational properties.

The inerton field, as well as the electromagnetic field, are basic fields of the nature. The inerton field is able to give a deeper insight into fundamental nature of things playing an important role in quantum physics, chemical physics, biochemistry, biophysics, and condensed matter. The inerton field assembles neutral and charged particles, changes physical chemical properties of substance, such as viscosity, density, microhardness, conductivity, thermodynamic potentials, etc.

There are a large variety of nanosystems and in some of them inerton effects are quite significant. We have to mention also the recent finding of a new bosonic quasi-particle named 'bondon,' which aggregates distant quantum particles [56, 57]. What is the reason for coupling of distant quantum particles? We may suggest that it is through the particles' clouds of inertons. Because the wave ψ-function of such an aggregation allows for a separate study of the ψ-function's core and its tail [58].

Fuchs et al. [59] presented a measurement of the excited state spin levels of a single Nitrogen-vacancy center in diamond. It was found that strain plays a critical role in the spin dynamics significantly influencing transverse anisotropy and hyperfine splitting. Therefore, the spin levels can be adjusted through strain engineering. Not only magnetic and electric field, but also an inerton field can induce strain. Hence the inerton field also is able to control high-speed physical reactions, the more so that it is able to deeper penetrate materials in comparison with the electromagnetic field.

In nanophotonics, optical properties of thin films are determined by diffractive elements, which represent a quantum confinement needed to tune the properties of electrons and photons together [60]. An inerton field is able to affect the absorption cross section of quantum dots, which can be used in nanophotonics. In fact the author of Ref. [61] has shown that at a submicroscopic level inertons are responsible for the phenomenon of diffraction.

A few research teams could generate a hydrogen atom of the nuclear size. This phenomenon was called differently by them: 'hydrex' (for hydrogen) and 'deutrex' (for deuteron) [62–65] 'hydrino' [66–71], and 'pseudoneutron' [72–74]. To understand the principles of operation of such an exotic system, the researchers suggest new ideas, which seem very different in each special case. Nevertheless, from the author's point of view, these nano- and submicroscopic systems can be unified and accounted for in the framework of the deterministic submicroscopic concept of physics, which includes inerton field as a basic field of nature. Besides, though a number of experiments were conducted in the area of low energy nuclear reaction, so far still no reasonable theory was reported [75]. A theory of could fusion can be developed on the basis of the present submicroscopic concept, which will open a gateway for optimization of crucial experiments in this important area of research.

KEYWORDS

- **Casimir effect**
- **electromagnetic field**
- **frustrated matter**
- **quasi-particles**
- **space excitation**
- **space particles' gravitation**
- **inerton field effects**

REFERENCES

1. Casimir, H. B. G. (1948). On the attraction between two perfectly conducting plates, *Proc. K. Ned. Akad. Wet.* 51, 793–795.
2. Genet, C., Intravaia, F., Lambrecht, A., Reynaud, S. (2004). Electromagnetic vacuum fluctuations, Casimir and Van der Waals forces, *Ann. Fond. L. de Broglie* 29, 311–328; arXiv:quant-ph/0302072.
3. Denardo, B. C., Puda, J. J; Larraza, A. S. (2009). A water wave analog of the Casimir effect, *Americ. J. Phys.* 77(12), 1095–1101.

4. Lambrecht, A., Reynaud, S. (2011). Casimir Effect: Theory and Experiments, arXiv:1112.1301 [quant-ph].
5. Braun, S., Ronzheimer, J. P., Schreiber, M., Hodgman, S. S., Rom, N., Bloch, I., Schneider, U. (2013). Negative absolute temperature for motional degrees of freedom, Science 339, no. 6115, 52–55; arXiv:1211.0545 [cond-mat.quant-gas].
6. Beckmann, P. (1990). Electron clusters, *Galilean Electrodynamics*, Sept./Oct. 1(5), 55–58.
7. Shoulders, K. R. Energy conversion using high charge density, U.S. patent 5,018,180, May, 1991.
8. Shoulders, K., Shoulders, S. (1996). Observations of the role of charge clusters in nuclear cluster reactions, *J. New Energy* 1(3), Fall.
9. Ziolkowski, R. W., Tippett, M. K. (1991). Collective effect in an electron plasma system catalyzed by a localized electromagnetic wave, *Phys. Rev. A* 43(6), pp. 3066–3072.
10. Mesyats, G. A. (1996). Ecton processes at the cathode in a vacuum discharge, Proc. XVIIth Int. Symposium on Discharges and Electrical Insulation in Vacuum, Berkeley, CA, July 21–26; pp. 721–731.
11. Lisitsyn, I., Akiyama, H., Mesyats, G. A. (1998). Role of electron clusters - ectons - in the breakdown of solids dielectrics, *Phys. Plasma* 5(12), 4484–4487.
12. Krasnoholovets, V., Kukhtarev, N., Kukhtareva, T. (2006). Heavy electrons: Electron droplets generated by photogalvanic and pyroelectric effects, *International Journal of Modern Physics B* 20(16), 2323–2337; arXiv:0911.2361 [quant-ph].
13. Vernadsky, V. On the states of physical space, *21th Century Science and Technology*, Winter 2007–2008 Issue, pp. 10–22, http://www.21stcenturysciencetech.com/Articles%202008/States_of_Space.pdf
14. Perez, A. (2008). Spin foam models for quantum gravity, *Class. Quant. Grav.* 20, R43.
15. Preparata, G. (1995). Quantum gravity, the Planck lattice and the standard model, arXiv:hep-th/9503102.
16. Konushko, V. I. (2011). Granular space and the problem of large numbers, *J. Mod. Phys.* 2, 289–300.
17. Wilczek, F. (2009). What is space? *Physics @MIT* 30, http://web.mit.edu/physics/people/faculty/docs/wilczek_space06.pdf
18. Bounias, M., Krasnoholovets, V. (2003). Scanning the structure of ill-known spaces: Part 1. Founding principles about mathematical constitution of space, *Kybernetes: The Int. J. Systems and Cybernetics* 32, no. 7/8, 945–75, Eds.: L. Feng, B. P. Gibson and Yi Lin; also arXiv:physics/0211096.
19. Schwartz, L. (1991). *Analyse I: théorie des ensembles et topologie* (Hermann, Paris), p. 24.
20. James, G., James, R. C. (1992). *Mathematics Dictionary* (Van Nostrand Reinhold, New York), pp. 267–68.
21. Bounias, M., Krasnoholovets, V. (2003). Scanning the structure of ill-known spaces: Part 2. Principles of construction of physical space, *Kybernetes: The Int. J. Systems and Cybernetics* 32, no. 7/8, 976–1004, Eds.: L. Feng, B. P. Gibson and Yi Lin; also arXiv:physics/0212004.
22. Bounias, M., Krasnoholovets, V. (2003). Scanning the structure of ill-known spaces: Part 3. Distribution of topological structures at elementary and cosmic scales,

Kybernetes: The Int. J. Systems and Cybernetics 32, no. 7/8, 1005–20, Eds.: L. Feng, B. P. Gibson and Yi Lin; also arXiv:physics/0301049.

23. Krasnoholovets, V., Ivanovsky, D. (1993). Motion of a particle and the vacuum, *Phys. Essays* 6(4), 554–563; arXiv:quant-ph/9910023.
24. Krasnoholovets, V. (1997). Motion of a relativistic particle and the vacuum, *Phys. Essays* 10(3), 407–416; arXiv:quant-ph/9903077.
25. Poincaré, H. Sur la dynamique de l'électron, *Comptes Rendus* (1905). 140, pp. 1504–1560, *Rendiconti del Circolo matematico di Palermo* (1906). 21, 129–176 (1906); also: Oeuvres, t. IX, pp. 494–550 (and also in Russian translation: *Selected Transactions*, Ed.: N. N. Bogolubov (Nauka, Moscow, 1974), 3, 429–486).
26. De Broglie, L. Recherches sur la théorie des quanta, *Ann. De Phys.*, 10e série, t. III (Janvie-Février 1925); translation by A.F. Kracklauer: *On the theory of quanta*, Lulu.com; Morrisville, NC; 2007 (ISBN: 978–1-84753–358–6).
27. De Broglie, L. (1987). Interpretation of quantum mechanics by the double solution theory, *Ann. de la Fond. L. de Broglie* 12, 399–421.
28. Born, M. Das Adiabatenprinzip in der Quantenmechanik, *Zeitschrift für Physik* (1926). 40, 167–192.
29. Born, M. The statistical interpretation of quantum mechanics. *Nobel Lecture*, December 11, 1954.
30. Krasnoholovets, V. (2002). Submicroscopic deterministic quantum mechanics, *Int. J. Computing Anticipatory Systems* 11, 164–179; arXiv:quant-ph/0109012.
31. De Broglie, L. *Les incertitudes d'Heisenberg et l'interprétation probabiliste de la méchanique Ondulatoire* (Gauthier-Villars, Bordas, Paris, 1982); Chapter 2, sect. 4.
32. Krasnoholovets, V. (2010). Inerton fields: Very new ideas on fundamental physics, *American Inst. Phys. Conf. Proc.* - December 22, vol. 1316, 244–268. *Search for fundamental theory: The VII International Symposium Honoring French Mathematical Physicist Jean-Pierre Vigier* (12–14 July 2010, Imperial College, London), Eds.: R. L. Amoroso, P. Rowlands and S. Jeffers; doi:10.1063/1.3536437.
33. Krasnoholovets, V. (2010). Variation in mass of entities in condensed media, *Applied Physics Research* 2(1), 46–59; http://www.ccsenet.org/journal/index.php/apr/article/view/4287.
34. Krasnoholovets, V. (2008). Reasons for the gravitational mass and the problem of quantum gravity. In *Ether, spacetime and cosmology. Vol. 1. Modern ether concepts and geometry*; Duffy, M., Levy, J., Krasnoholovets, V., Eds., PD Publications: Liverpool, Vol. 1; pp. 419–450; arXiv: 1104.5270 [physics.gen-ph].
35. Krasnoholovets, V. (2009). On microscopic interpretation of phenomena predicted by the formalism of general relativity, in Ether space-time and cosmology, Vol. 2: New insights into a key physical medium. Eds.: M. C. Duffy and J. Levy (Publisher: C. Roy Keys Inc. Apeiron, 2009); also: *Apeiron* 16(3), 418–438, http://redshift.vif.com/JournalFiles/V16NO3PDF/V16N3KRA.pdf.
36. Krasnoholovets, V. (2013). On the gravitational time delay effect and the curvature of space, *Int. J. Computing Anticipatory Systems. Proceedings of the Tenth International Conference CASYS'11 on Computing Anticipatory Systems, Liège, Belgium, August 8–13, 2011, D. M. Dubois (Ed.),* Publ. by CHAOS, 2012. ISSN 1373–5411.
37. Krasnoholovets, V. (2011). Dark matter as seen from the physical point of view, *Astrophys. Space Science* 335(2), 619–627.

38. Krasnoholovets, V. (2009). Inerton fields: A new approach in fundamental physics, *Proc. Bath Royal Literary and Scientific Institution,* Vol. 11, Sep 2006-Aug 2007, Bath Royal Literary and Scientific Institution, Bath, pp. 266–272 (ISSN 1465–8496).
39. Krasnoholovets, V., Lev, B. (2003). Systems of particles with interaction and the cluster formation in condensed matter, *Cond. Matt. Phys.* 6, 67–83; arXiv:cond-mat/0210131.
40. Khachaturian, A. G. (1974). *The theory of phase transitions and the structure of solid solutions,* Nauka, Moscow, p. 100 (in Moscowian).
41. Melker, A. I. (2009). Potentials of interatomic interaction in molecular dynamics, *Rev. Adv. Mater. Sci.* 20, 1–13.
42. Krasnoholovets, V., Tomchuk, P. M., Lukyanets, S. P. (2003). Proton transfer and coherent phenomena in molecular structures with hydrogen bonds, in *Advances in Chemical Physics,* Vol. 125, Eds.: Prigogine, I. and Rice, S. A. John Wiley & Sons, Inc.
43. Krasnoholovets, T., Tane, J.-L. (2006). An extended interpretation of the thermodynamic theory, including an additional energy associated with a decrease in mass, *Int. J. Simulation and Process Modeling* 2(1/2), 67–79; arXiv:physics/0605094.
44. Tane, J.-L. (2009). Unless connected to relativity, the first and second laws of thermodynamics are incompatible, arXiv:0910.0781 [physics.gen-ph].
45. Tane, J.-L. (2012). A simple illustration of the need for relativity in thermodynamics, *The General Science J.* http://gsjournal.net/Science-Journals/Research%20Papers-Relativity%20Theory/Download/4312 .
46. Zhang, S. (2012). Entropy: A concept that is not a physical quantity, *Phys. Essays* 25(2), 172–176.
47. Shoulders, K., Shoulders, S. (1999). Charge clusters in action, http://www.svn.net/krscfs/Charge%20Clusters%20In%20Action.pdf
48. Milton, K. A. (2011). Resource Letter VWCPF-1: Van der Waals and Casimir-Polder forces arXiv: 1101.2238 [cond-mat.other].
49. Milonni, P. W. (1994). *The quantum vacuum. An introduction to quantum dynamics.* Academic Press.
50. Bordag, M., Mohideen, U., Mostepanenko, V. M. (2001). New developments in the Casimir effect, *Physics Reports* 353, 1–205.
51. Jaffe, R. L., Scardicchi, A. (2004). The Casimir Effect and Geometric Optics, *Phys. Rev. Lett.* 92: 070402, arXiv:quant-ph/0310104.
52. Jaffe, R. L. (2005). The Casimir effect and the quantum vacuum, *Phys. Rev. D* 72(2), 021301, arXiv:hep-th/0503158.
53. Kosevich, A. M. (2005).*The crystal lattice: Phonons, solitons, dislocations, superlattices*. Willey-VCH Verlag GmbH & Co. KGaA, Weinheim, p. 20.
54. Valkering, A. M. C., Mares, A. I., Untiedt, C., Babaei Gavan, K., Oosterkamp, T. H., van Ruitenbeek, J. M. (2005). A force sensor for atomic point contacts, *Rev. Scientific Instruments* 76, 103903 (5 pages).
55. Kac, V., Cheung, P. (2001). *Quantum Calculus*, Springer; p. 93 and 87.
56. Putz, M. V. (2010). The bondons: The quantum particles of the chemical bond, *Int. J. Mol. Sci.* 11, 4227–4256.
57. Putz, M. V., Ottorino, O. (2012). Bondonic characterization of extended nanosystems: Application to graphene's nanoribbons, *Chem. Phys. Lett.* 548, 95–100.

58. Putz, A.-M; Putz, V. M. (2012). Spectral inverse quantum (spectral-IQ) method for modeling mesoporous systems: application on silica films by FTIR, *Int. J. Mol. Sci.* 13, 15925–15941.
59. Fuchs, G. D., Dobrovitski, V. V., Hanson, R., Batra, A., Weis, C. D., Schenkel, T., Awschalom, D. D. (2008). Excited-State Spectroscopy Using Single Spin Manipulation in Diamond, *Phys. Rev. Lett.* 101, 117601 [4 pages].
60. Flory, F., Escoubas, L., Berginc, G. (2011). Optical properties of nanostructured materials: a review, *J. Nanophoton.* 5(1), 052502. http://nanophotonics.spiedigitallibrary.org/article.aspx?articleid=1226099
61. Krasnoholovets, V. (2010). Sub microscopic description of the diffraction phenomenon, *Nonlin. Optics, Quant. Optics* 41(4), 273–286.
62. Dufour, J. (1993). Cold fusion by sparking in hydrogen isotopes, *Fusion Technol.* 24, 205–228.
63. Dufour, J., Foos, J., Millot, J. P., Dufour, X. (1997). Interaction of palladium/hydrogen and palladium/deuterium to measure the excess energy per atom for each isotope, *Fusion Technol.* 31, 198–209.
64. Dufour, J., Murat, D., Dufour, X., Foos, J. (2000). Hydrogen triggered exothermal reaction in uranium metal, *Phys. Lett. A* 270(5), 254–264.
65. Dufour, J., Murat, D., Dufour, X., Fos, J. (2004). Exothermic reaction induced by high density current in metals – Possible nuclear origin, *Ann. Fond. L. de Broglie* 29(3), 1081–1093.
66. Mills, R. L. (2002). The grand unified theory of classical quantum mechanics *Int. J. Hydrogen Energy* 27, 565–590.
67. Mills, R., Dhandapani, B., Greenig, N., He, J. (2000). Synthesis and characterization of potassium iodo hydride, *Int. J. Hydrogen Energy* 25(12), 1185–1203.
68. Mills, R., Dhandapani, B., Nansteel, M., He, J., Voigt, A. (2001). Identification of compounds containing novel hydride ions by nuclear magnetic resonance spectroscopy, *Int. J. Hydrogen Energy* 26(9), 965–979.
69. Mills, R., Nansteel, M., Ray, P. (2002). Argon-hydrogen-strontium discharge light source, *IEEE Transact. Plasma Sci.* 30(2), 639–653.
70. Mills, R. L., Ray, P., Dhandapani, B., Mayo, R. M., He, J. (2002). Comparison of excessive Balmer alpha line broadening of glow discharge and microwave hydrogen plasmas with certain catalysts, *J. Appl. Phys.* 92, 7008–7022.
71. Mills, R., Ray, P., Dhandapani, B., Good, W., Jansson, P., Nansteel, M., He, J., Voigt, A. (2004). Spectroscopic and NMR identification of novel hydride ions in fractional quantum energy states formed by an exothermic reaction of atomic hydrogen with certain catalysts, *Eur. Phys. J. App. Phys.* 28, 83–104.
72. Borghi, C., Giori, C., Dall'Ollio, A. A. (1993). Experimental evidence of emission of neutrons from cold hydrogen plasma, *American Institute of Physics (Phys. At. Nucl.)* 56(7).
73. Conte, E., Pieralice, M. (1999). An experiment indicates the nuclear fusion of the proton and electron into a neutron, *Infinite Energy* 4(23), 67.
74. Santilli, R. M. (2006). Confirmation of Don Borghi's experiment on the synthesis of neutrons from protons and electrons, Preprint IBR-EP-39 of 12–25–06; arXiv.org: physics/0608229.
75. Storms, E. (2007). *The science of low energy nuclear reaction. A comprehensive compilation of evidence and explanations about could fusion*, World Scientific.

CHAPTER 3

NON-CLASSICAL PROPERTIES OF CLASSICAL NANOSTRUCTURES

JAN C. A. BOEYENS

CONTENTS

ABSTRACT

Structures of the major types of nanoparticle, including nanogold, graphene, fullerene, boron nitride, semiconductors, and their derivatives are described. The methods of quantum chemistry, widely used to rationalize the electronic structure, physical properties and molecular structure of nanomaterials are critically reviewed and found inadequate. The neglect of nonlinearity and the four-dimensional nature of nonclassical systems are highlighted.

A fresh look at nanomaterials as intermediate between the molecular and macroscopic states of matter recognizes significant aspects of property-variation with particle size, traditionally ignored in their blanket characterization as "quantum effects," without further elucidation. Of special interest is the variation in melting point and surface plasmonics. Based on these observations an alternative mechanism for the growth of nanoparticles is proposed.

Fibonacci patterns that emerge in the surface structure of Ag/silica core shell particles suggest reexamination of other nanoparticles for similar features. This resulted in the identification of a holistic interaction pattern for the C_{60} fullerene, in line with the projective topology of atomic structure, elemental periodicity and space-time curvature. The alternative suggested approach towards the understanding and modeling of the nonclassical properties of nanomaterials is *via* number theory and universal self-similarity.

3.1 INTRODUCTION

Although nanomaterials have been around, and have been recognized, for centuries, the nano-revolution in chemistry and material science only took off during the last two decades. It has much in common with the development of synthetic organic chemistry in response to newly discovered natural products during the early part of the twentieth century. It was the medicinal properties of natural products that stimulated the chemical response and to a very large extent it is the medicinal applications of nanosized composites that again drive the explosive research activities in nanoscience.

The nanoworld consists of all materials and structures intermediate between molecules and materials in bulk. Scientific interest in nanomaterials arose from the unusual properties of nanosized objects such as their intensified electrical, magnetic, optical and mechanical characteristics, popularly identified as quantum effects, very often with the connotation of being too mysterious for analysis. The implied nonclassical nature of nanosized entities is generally, but noncritically, accepted by most without trying to stipulate the quantum-mechanical rules at play. The real danger exists that an artificial distinction between macroscopic and microscopic systems, based on more than optical resolution, could result in sensational interpretations of mundane effects.

Theoretical models for nanosystems assumed to be quantum-based could be, not only inappropriate, but also misleading. Standard wave mechanics, applied to atomic and molecular systems, is based on linear differential equations, which allow the superposition of elementary solutions. All quantum-chemical simulations are based on this assumption and therefore apply only approximately to physically real systems. Moving into the nonlinear regime of nanoparticles, the assumed linearity becomes increasingly inappropriate. Considerations, commonly based on concepts such as quantum confinement, therefore need careful reassessment as a universal guiding principle in the analysis of nanosystems.

Important nanostructures such as diatomaceous earth and various colloids had been in common use before the discovery of the fullerenes, which stimulated worldwide research interest in the new field of nanoscience. Many concepts, familiar to colloid chemists for ages, acquired new meaning in the new nanoscience, routinely claimed to be of a nonclassical nature, different from the adequate models of colloid chemistry. The mechanisms of biological growth, which have been studied by botanists for centuries, are being rediscovered as nanoscience, but the principles remain the same and all new nanomaterials have their counterparts in Nature. The formation of composite structures by self-assembly on a substrate is equivalent to the interaction between aggregates and biological membranes and mesophases.

The appearance of nanomaterials in the manufacture of electronic devices provided a major impetus for the development of nanoscience. Many nanoparticles are binary semiconductors formed between chemical elements from periodic groups $(4 - n)$ and $(4 + n)$, $(n = 0–3)$ such as GaAs, CdS, AgBr, etc. Carbon compounds, epitomized by graphene, are of central importance in this scheme. The revolutionary electric and optical properties of such nanosystems are theoretically simulated by elementary models of traditional wave mechanics of atoms and the band theories of macroscopic semiconductors. Such models may not be entirely appropriate and it may be of benefit to reconsider the details and applicability of the basic theory.

The field of nanoscience is too wide to cover in a single review. Even the single aspect of nanomaterial structures is too wide. This topic will be reviewed very briefly, especially to put the more detailed discussion of nanoparticles into proper perspective.

3.2 NANOSTRUCTURES

The large number and variety of nanomaterials all derive from a small number of elementary stable forms, the most versatile of these being nanogold, graphitic carbon entities and silica. Using standard techniques of surface chemistry, sol-gel processes, passivation, metal galvanic exchange, polymer coating, uncontrolled monolayer self-assembly, molecular recognition, hydrodynamic jetting, metathesis polymerization, colloidal lithography, intercalation, exfoliation, chemical vapor deposition, pulsed laser deposition, ionbeam techniques and combinations of these, an amazing variety of new composite nanomaterials and structures are being produced around the world, with new techniques being reported all the time. Although the synthesis of target products is often carefully planned, the majority of experiments rely on the characterization of unexpected products for their success. The emphasis is still on the serendipitous discovery of novel nanostructures by trial and error, with almost assured technological application as immediate reward. The incentive for synthesis by design is minimal. Many of these techniques have been reviewed elsewhere [1–7] and will not be further discussed here.

3.2.1 NANOGOLD DERIVATIVES

Colloidal gold represents best what became known as nanoparticles. Surviving artifacts, such as the Lycurgus cup in the British museum [1, 8], attest to the use of colloidal gold to make ruby glass and for coloring ceramics as early as 500 BCE in the Middle East and China. In the form of "soluble" gold it was equally famous as a cure against various diseases and for the diagnosis of syphilis.

The modern synthesis of gold clusters, in which 12 nearest neighbors surround each atom, was largely pioneered by Schmid [9, 10]. A cluster consists of layers with $10n^2 + 2$ atoms in the nth layer. The two-layer cluster of 55 (13+42) atoms could be characterized in the compound $Au_{55}(PPh_3)_{12}Cl_6$. Larger clusters of 147 → 2057 atoms ($n = 3 \rightarrow 8$) have been isolated [11–14].

The stability of nanogold clusters depends on a protective organic ligand shell, typically a long-chain thiol, such as $C_{12}H_{25}SH$, to form a monolayer-protected cluster. Figure 3.1 shows the monolayer when a spherical gold nanoparticle, protected by an organic thiol is cut diametrically.

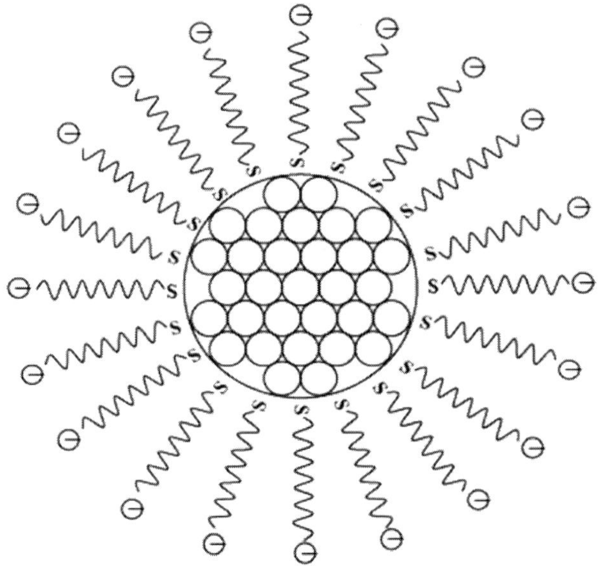

FIGURE 3.1 Schematic drawing of a monolayer-protected cluster.

Three-dimensional structure of the particle is recovered by revolution of the monolayer about any diameter (*see* Section 3.2.5.1).

Adsorption of layer after layer of oppositely charged polyelectrolytes on the thiol shell, followed by dissolution of the gold core shell, results in the production of hollow polymer capsules [15].

Bimetallic nanoparticles are obtained by reacting protected gold nanoparticles in aqueous solution with solutions such as H_2PdCl_4, producing core shell nanoparticles which can be further modified, by galvanic exchange, into nanocages [2] and multiwalled nanoshells.

Composite materials that result from chemical modification of protective shells are obtained in almost endless variety for use as specialized sensors and carriers [1].

3.2.2 GRAPHENE AND DERIVATIVES

Carbon-based nanomaterials such as the fullerenes, nanofibers, single-, double and multiwalled nanotubes, nested fullerenes, icosahedral nanocages, spherical graphitic onions, nanoribbons and modifications of these forms, are all derived from a graphitic monolayer, known as graphene.

Historically, interest in carbon nanoparticles started with the synthesis of C_{60} fullerene [16] and the subsequent syntheses of several variant forms.

The earlier synthesis of graphene by the reduction of graphitic oxide [17] attracted little interest until it became possible to mechanically peel off a monolayer from graphite surface, for study [18]. Although planar graphene in the free state had been presumed to be unstable with respect to the formation of curved structures such as soot, fullerenes and nanotubes, it was shown that exfoliated films could be stably supported on oxidized silicon substrates.

It is fairly obvious that rolling a flat graphene sheet into a cylinder would produce the equivalent of a nanotube if the hexagonal rings on the edges could be joined coherently.

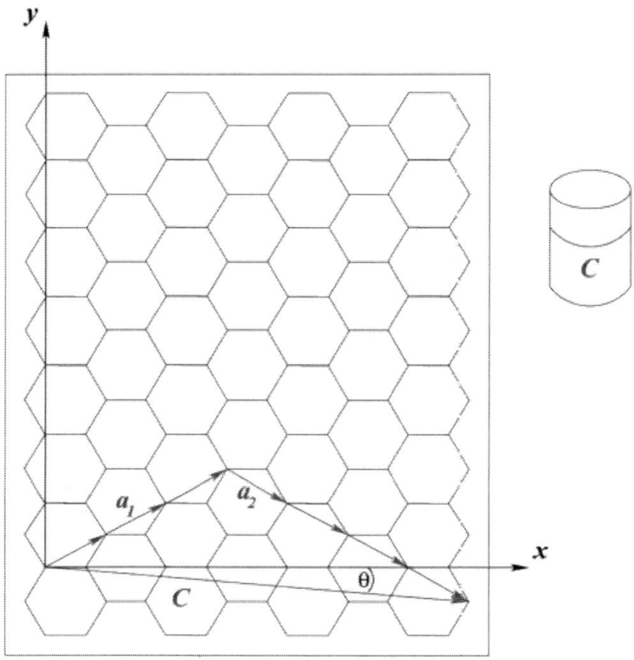

FIGURE 3.2 Construction of a nanotube from graphene.

This is precisely the model, which has been used to analyze the structure of the different types of nanotube. With reference to Fig. 3.2, three modes of producing nanotubes from a graphene sheet suggest themselves.

It appears equally logical to produce a cylinder with its axis either along x or y. As shown schematically in Fig. 3.3 (a and b) the resulting tubes are fundamentally different with respect to the orientation of the carbon hexagons. The same two types of nanotube are formed around cylindrical axes inclined at 30° from the y and x axes, respectively.

a

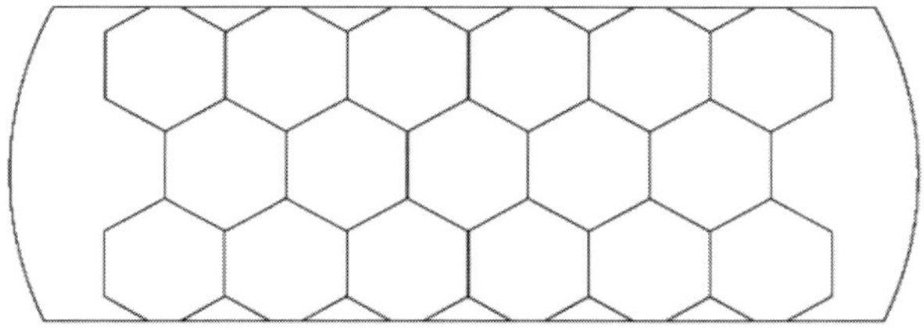

b

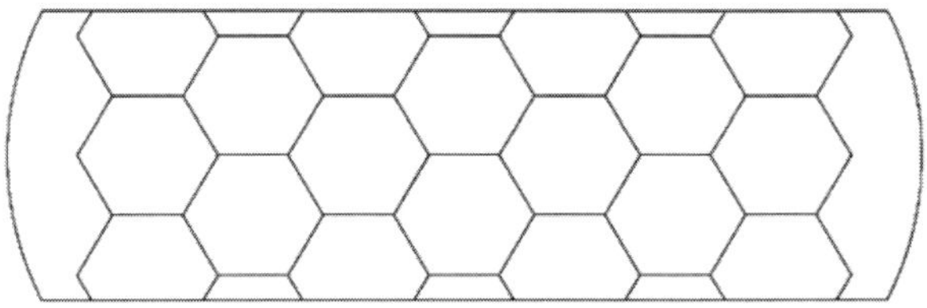

c

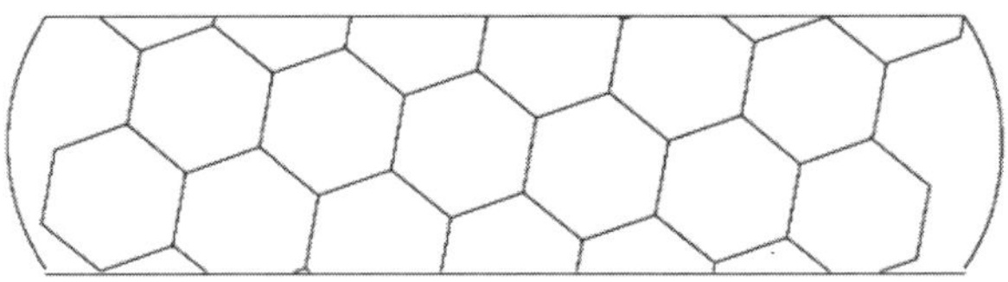

FIGURE 3.3 Three types of nanotube produced by cylindrical deformation of graphene: (a) armchair configuration, (n, n), $\theta = 30°$; (b) zig-zag arrangement, $(n, 0)$, $\theta = 0$; (c) helical form, (n, m), $0 < \theta < 30°$.

A third type of nanotube may thus be produced with cylinder axis at an angle of helicity $\theta < 30°$. The various ways to roll graphene into tubes are therefore mathematically defined by the angle θ and a vector of helicity C.

In terms of the unit vectors,

$$a_1 = a\left(\frac{\sqrt{3}}{2}x + \frac{1}{2}y\right),\ a_2 = a\left(\frac{\sqrt{3}}{2}x - \frac{1}{2}y\right),$$

and

$$a = \sqrt{3}\cdot(C-C),\ C = na_1 + ma_2$$

where *n* and *m* are integers, all possible configurations are readily specified.

The set of numbers *(n, m)* is sufficient to define the structure type of any nanotube, and its diameter,

$$d = a\sqrt{m^2 + mn + n^2}\ /\ \pi = |C|\ /\ \pi$$

The modulus of the helicity vector defines the circumference of the nanotube. The angle of helicity is defined as

$$\theta = \tan^{-1}\left(\frac{\sqrt{3}n}{2m+n}\right).$$

The geometrical and other properties of a large variety of carbon nanotubes have been the subject of several recent reviews [19, 20].

Although multiwalled carbon nanotubes, consisting of concentric cylindrical nanotubes, are perhaps not uncommon, the spontaneous formation of needle-like helical microtubules that resemble concentric MWNT appears more likely. The structure of such needles was studied with electron microscopy and diffraction by Iijima [21]. The needles are best understood as the product of continuous rolling of a graphene sheet around a scroll axis as shown schematically in Fig. 3.4 with the crystallographic orientation of the needle axis defined.

The pitch of the helix depends on the orientation of the sheet with respect to the needle axis. If hexagons A and B coincide with those labeled C and D, the needle axis will be along [010] and hence there will be no spiral rows of hexagons.

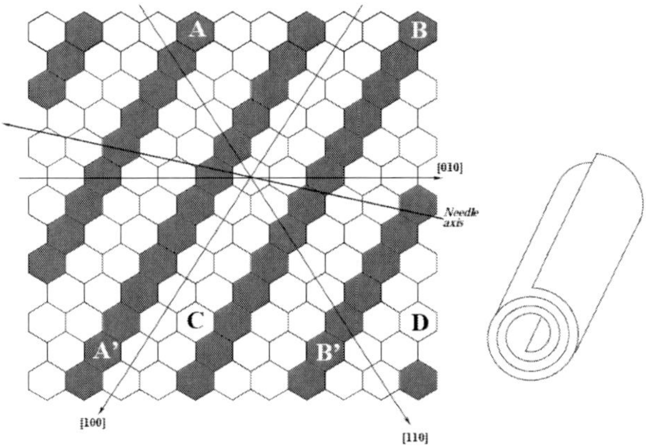

FIGURE 3.4 Schematic drawing of nested helical nanotubes, on the right. The unrolled graphene sheet is shown on the left. The needle axis is indicated by the heavy line. Superposition of the hexagons A and B over A' and B', respectively, generates the tube and the row of red hexagons defines a helix that spirals out through the nested tubules.

A variety of forms, described as graphite whiskers, cones and polyhedral crystals, can formally be considered as graphene sheets rolled into helical cones, but, which more likely occur as the products of a growth mechanism that resembles a screw dislocation [22], as shown in Fig. 3.5.

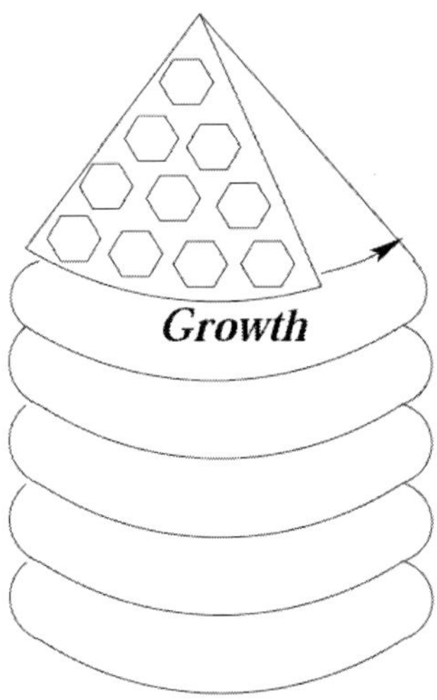

FIGURE 3.5 Schematic illustration of a growth mechanism for columnar graphitic pencils and whiskers [22].

By way of demonstrating their relationship to graphene it was shown how to snip nanotubes by various chemical and electronic means [23] into graphene and graphitic nanoribbons, which are the likely starting materials for the process demonstrated in Fig. 3.5.

3.2.3 BORON NITRIDE

Borazine, $B_3N_3H_6$, is isoelectronic and isostructural with benzene, but because of the difference in polarity of the B and N atoms, its chemistry is completely different from that of benzene. Not surprisingly, boron nitride, BN, has a graphite-like structure, but once more, because of the polarity of a B–N unit the mutual orientation of successive layers is different for graphite and boron nitride. Because of the feeble interaction between them the neighboring layers in graphite appear in a staggered arrangement, shown in Fig. 3.6. whereas in BN they are eclipsed, with boron atoms over nitrogen, and vice versa. However, in the rhombohedral crystalline form of BN a different mode of overlap, illustrated in Fig. 3.7 is observed.

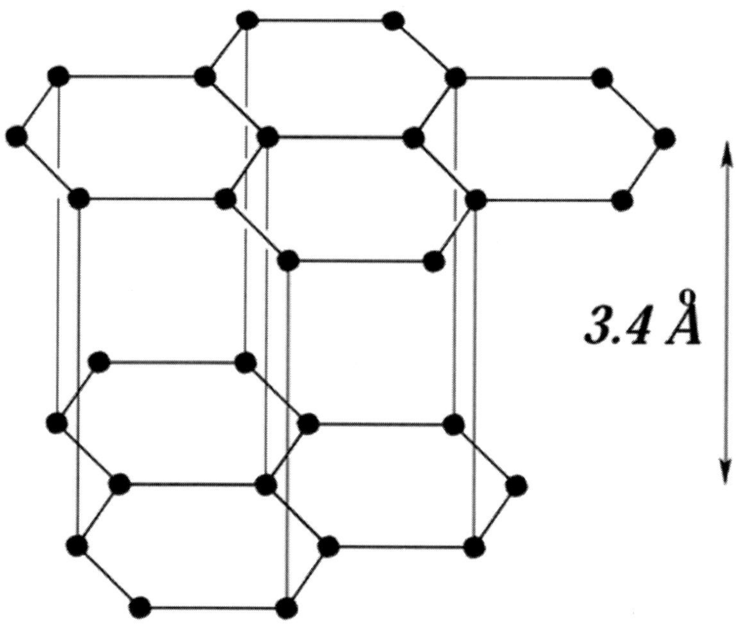

FIGURE 3.6 The mode of overlap of graphene sheets in crystalline graphite.

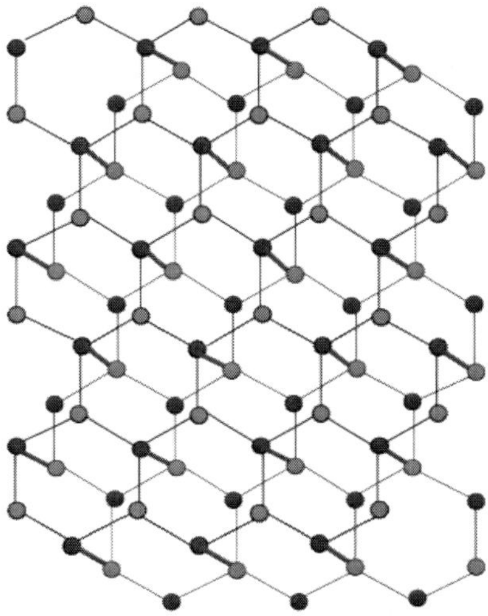

FIGURE 3.7 A double sheet of BN with the layers superimposed in rhombohedral style. Boron atoms (blue) in the top layer are directly above the nitrogen atoms of the bottom layer.

Like graphene, BN layers are also readily transformed into nanotubes. Because of the polar B–N interactions between neighboring layers, single-walled BN nanotubes are rare. Boundary conditions after one rotation around a cylinder axis limit the possible modes of helicity [24]. The most common are the zig-zag and armchair modes, illustrated schematically in Fig. 3.8 of which the former is the more common.

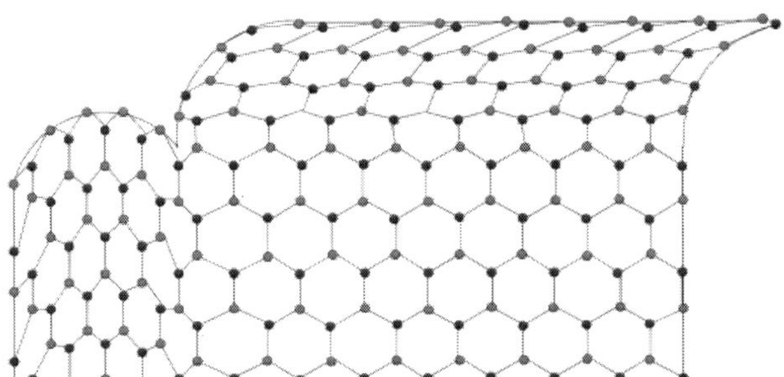

FIGURE 3.8 A BN sheet, partially rolled up in zig-zag mode on the left and armchair mode on top.

The geometry of multiwalled nanotubes are also conditioned by B–N interactions between neighboring layers and the same variety of graphene MWNT is not observed in the case of BN.

It is noted that boron nitride is of the same family as III–V semiconductors with increasing importance as nanomaterials.

3.2.4 FULLERENES

The reversible transformation of a graphene sheet into nano-tubes, nano-rods, nano-ribbons and nano-helices is readily understood by noting that little distortion is required to roll it into a cylinder or a cone[1]. A cylindrical or a conical surface is said to be of the same topology as a flat surface.

The differential that characterizes a nonEuclidean surface is known as curvature. In two dimensions it describes how a smooth curve deviates from linearity. The curvature, which varies from point to point, is specified in terms of the osculating circle of radius R and centered on the perpendicular to the tangent at p, and which follows the curve in the vicinity of p, as shown in Fig. 3.9. On an infinitesimal scale each point on a curve has a unique osculating circle.

FIGURE 3.9 The osculating circle that defines the curvature of a surface.

A straight line has zero curvature and is considered to be its own osculating circle of infinite radius. Curvature in general is therefore defined in terms of the radius of the osculating circle as $K = 1/R$. The larger the curvature, the smaller is the osculating circle, and therefore, the faster the curve is turning.

[1]Notice how a flat sheet of paper can be rolled into a cylinder or a cone without getting crumpled.

To distinguish between positive and negative curvature a unit vector is defined normal to the curve. A curve of positive curvature turns toward the unit vector; if it turns away the curvature has a negative sign.

In the case of a two-dimensional surface, embedded in R^3 the curvature at a point is described by two numbers, called the principal curvatures. As an example, consider a regular cylinder. The two principal curvatures at p are clearly $\kappa_1 = 0$ and $\kappa_2 = 1/R$. More complicated expressions are obtained by different choice of principal directions. We give without proof Gauss's *Theorema Egregium* which states that irrespective of direction, the product $K = \kappa_1 \times \kappa_2$ on a curved two-dimensional surface is a constant. This means that the Gaussian curvature is an intrinsic property of the surface itself.

Another simple example is a sphere of radius R. The principal curvatures at any point $K = \pm 1/R$ and the Gaussian curvature, $\kappa_1 \times \kappa_2 = 1/R^2$, which is a positive constant. The only other quadric of constant curvature is the projective plane. The curvature of any dome-shaped surface, although not necessarily constant, is always positive. On the other hand, the curvature of a saddle-shaped surface is negative, because the principal curvatures are of opposite sign. Examples of surfaces with negative Gaussian curvature are the torus, the catenoid and the helicoid.

For surfaces with more than one nonzero radius of curvature, the situation is more complicated and graphene sheets would have to be modified more seriously to assume this type of shape.

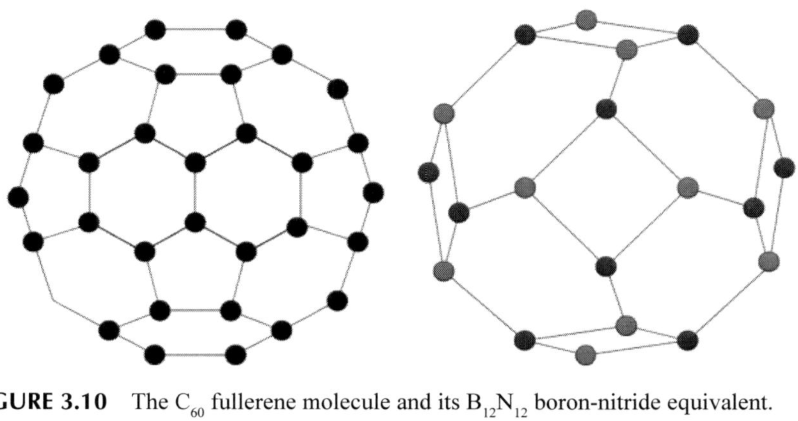

FIGURE 3.10 The C_{60} fullerene molecule and its $B_{12}N_{12}$ boron-nitride equivalent.

The spherical fullerene (C_{60}) is the most familiar nanoparticle of this type. As seen from Fig. 3.10 the modification of graphene required to produce the fullerene consists of the conversion of some six- into five-membered rings. Figure 3.11 shows, on a larger scale, the formation of a cap on a nanotube by the insertion of pentagons.

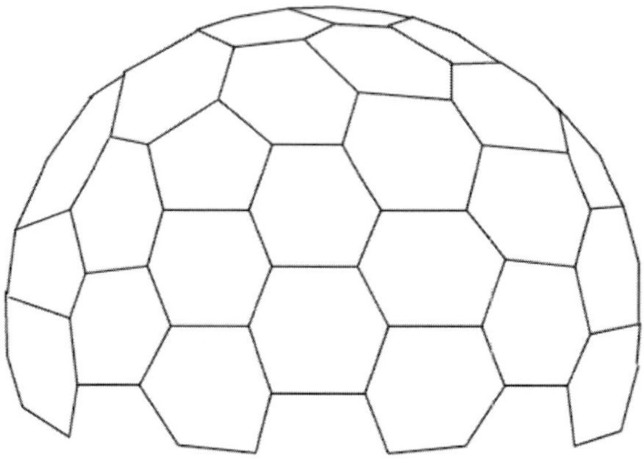

FIGURE 3.11 A graphene cap on a nanotube created by the insertion of pentagonal rings.

C_{70} is the smallest nanotube; compared to C_{60} it contains an extra belt of hexagons normal to the five-fold axis of the hemi-C_{60} caps. The structure [25] is shown in Fig. 3.12. Addition of more belts leads to longer tubes of the metallic armchair category.

A comprehensive review [26] of graphitic distortions also details the effect of heptagons and octagons on the curvature of graphene sheets in an analysis of onions, helical coils and other distorted forms. The principles involved in the formation of giant octahedral fullerenes, spherical graphitic onions and nested fullerenes [5] are the same as for smaller counterparts.

3.2.5 NANOSIZED SEMICONDUCTORS (NANODOTS)

The development of nanoscience received a major thrust from the realization that the properties of semiconductor materials depended critically on size factors, with a gradual transition from single molecules, through macro molecular clusters to materials in bulk.

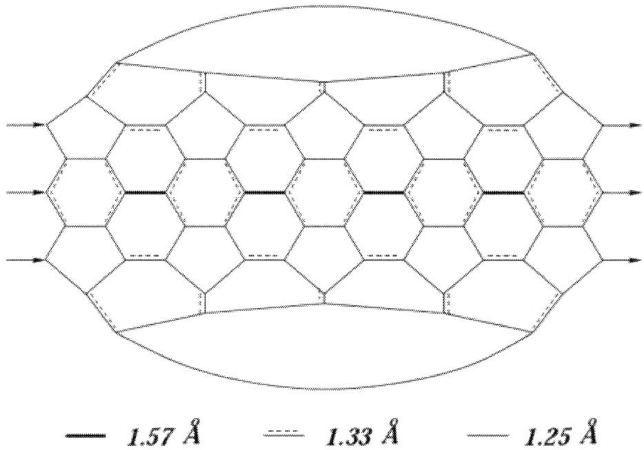

FIGURE 3.12 The C_{70} molecule unrolled. The interatomic distances of 1.57, 1.33 and 1.25 Å are of bond order 1, 1.25, and 1.5, respectively

In particular, the variation of band gap with crystallite size was categorized as a quantum effect, leading to concepts such as *quantum confinement* and *quantum dots*, now in general use.

One of the most thoroughly analyzed examples is the semiconductor CdSe [27] with an optical band gap at 690 nm in the bulk, and continuous optical absorption at lower wavelengths. However, CdSe crystallites with a diameter of 35–40 Å, containing some 1500 atoms, exhibit a series of discrete excited states, with the lowest at 530 nm. With increasing size, these states shift towards the red and merge to form the optical absorption of the bulk crystal. The crystallites have the same crystallographic structure as the bulk semiconductor. The band gap can be tuned from deep red (1.7 eV) to green (2.4 eV) by reducing the cluster diameter from 200 to 20 Å [28].

The "quantum" behavior of the nanosized crystallites is typically modeled by simple free-electron methods, such as the Hückel description of linear conjugated systems. In terms of a linear combination of atomic orbitals (LCAO) applied to a chain of atoms these methods lead to the orbital energies of the form [29]:

$$E_k = \alpha + 2\beta\cos(ka)$$

For an infinite chain, k becomes continuous in the range 0 to π/a, the first Brillouin zone.

The explanation of a "quantum size effect" [27] in terms of this model is not convincing, for reasons to be outlined in a later section. In the interim, concepts such as "quantum dot" and "quantum confinement" will be used with caution[2], without trying to invalidate the exciting experimental results being reported in a flood of publications, almost impossible to keep track of. The immediate emphasis will be on the structures of nanoparticles with semiconductor cores.

It is important to realize that isolated nanosized clusters, also known as 'naked clusters,' are formed in molecular or cluster beams, but can never be isolated as stable materials. Interest in such nanoparticles is based on the dramatic change in the electronic and other physical properties on the reduction in size from the bulk to the nano size. In practice most of the clusters and colloids of this type are protected by a shell of ligand molecules, or they are embedded in cages or matrices like polymers or other solids.

As pointed out before [30] ligand molecules and cages chemically interact with the surface atoms of the cluster and hence have a remarkable influence on their electronic character. This means that the properties of 'naked' clusters must be considerably different from those of ligated or otherwise fenced crystallites.

In reviewing the influence of the surface on the optical and electrical properties of semiconductors [31] the need to embed semiconducting clusters in a passivating medium is pointed out. Despite the seeming perfection of a pure cluster in the gas phase, from the point of view of semiconductor physics, it is in many ways a highly defective system.

Passivation is the chemical process by which surface atoms are bonded to another material of a much larger band gap, eliminating all of the energy levels inside of the gap. The resulting jump in the chemical potential for electrons at the interface confines the electron inside the cluster, likened to the elementary particle-in-the-box. The passivation of a CdS core covered by a layer of HgS and a final monolayer of CdS is discussed [31] as a typical example. Because HgS has a much smaller band gap than CdS, electronic excitations remain confined in the well.

Because nanoconfined semiconducting clusters (quantum dots) exhibit discrete optical spectra and electrical characteristics, they are also referred to as artificial atoms. This analogy is particularly useful when such are connected into artificial molecules, typically through an organic molecule. All inorganic semiconductors may in principle be used to construct such nano molecules.

---

[2]The terms, nanodot and nano confinement, will be used preferentially in this work.

The only requirements are the ability to control the size and achieve surface passivation [31].

The number of surface atoms and hence the surface free energy of nano crystals is a large fraction of that for the whole system. The effect of this is the possible modification of its crystal structure compared to the macroscopic crystal. One example is the stabilization of a rock-salt structure for CdS, which is the high-pressure form in the bulk.

The idea of artificial atoms extends into the formation of super lattices [32]. Colloidal CdSe nanocrystals, passivated with organic surfactants, were precipitated from solution in the form of three-dimensional arrays.

3.2.5.1 SURFACE TOPOLOGY

Nanostructures of ZnO are one of the most splendid families of nanomaterials [33], based on the hexagonal wurtzite structure which consists of alternating planes of tetrahedrally coordinated O^{2-} and Zn^{2+} ions, perpendicular to the polar c-axis, defining positively charged (0001) layers of zinc ions parallel to negatively charged oxide layers. The formation of a wide range of nanostructures, such as nanosprings, nanobows and nanohelices, is ascribed to this polar arrangement, all derived from a nanobelt as shown in Fig. 3.13.

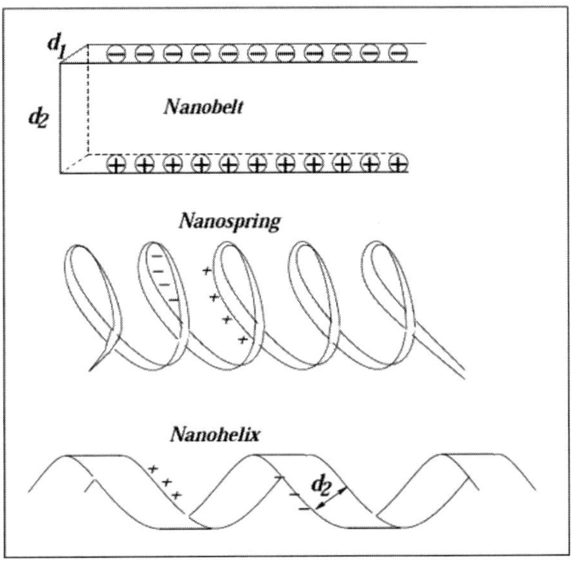

FIGURE 3.13 Nanobelts and wires curl up spontaneously to form rings, springs and helices.

The polar direction is labeled d_2. As $d_2 \rightarrow d_1$ the belt is better described as a nanocolumn, or nanowire. Interaction between the polar surfaces may cause the belt or column to curl up and form either a nanohelix or a nanospring. The formation of nanocylinders and nanorings is also known to happen.

ZnO nanowire can grow on various substrates in the form of flat surfaces or cylindrical rods. The alignment of nanowires remains in register to either generate a sheet with all wires in vertical alignment, or a cover of radially aligned nanowires, which in cross-section resembles the thiol-covered nanosphere shown in Fig. 3.1. Such an arrangement of nanowires, all in register, cannot be extended to completely cover a spherical substrate. In order to find the three-dimensional arrangement of nanowires, or linear molecular chains, that may cover a spherical particle, the section, shown in Fig. 3.1 needs to be rotated around any diameter through 180°. Only the wires, initially aligned with the rotation axis, are not affected by the rotation and maintain their orientation. These special positions are known as topological fixed points and they differ from all other points in the surface.

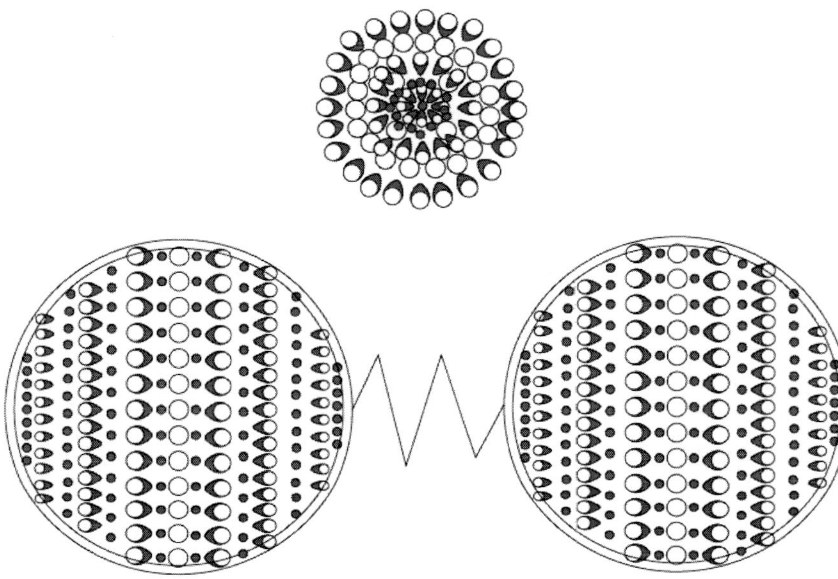

FIGURE 3.14 Fixed point on a core/shell nanoparticle (top) and a pair of rippled particles connected by an organic molecule.

The orientation of surface wires around a fixed point is shown in Fig. 3.14 at the top. Such an arrangement around fixed points is expected to occur on all surfaces with positive Gaussian curvature. It is well known that hair on a disc, a sphere, or a human head cannot be brushed to lie in the same direction, but develops a crown. Precisely this type of rippled arrangement has been observed [34] on metallic cores coated with self-assembled monolayers. The polar fixed points show increased chemical reactivity and could be used to link nanoparticles into strings through long-chain chemicals. A binary unit of CdSe in nanocrystals of 4 nm diameter, linked in this way has been observed [31].

3.3 THE QUANTUM-CHEMICAL MODEL

The reasonable expectation that the theory of quantum physics, formulated a century ago, could provide a fundamental basis for chemistry has developed, by the empirical incorporation of well-established notions about molecular structure, into a computational scheme known as *Quantum Chemistry*. The axioms of quantum physics are assumed to hold without change and all molecular and atomic systems to be correctly described in terms of this formalism.

As one of the seminal concepts it was assumed that the classical and quantum theories merge into one at some classical limit that depends on Planck's constant, h, in the same way that classical and relativistic mechanics merge in the limit of c, the speed of light. By this hypothesis it could be inferred that quantum theory applies to submicroscopic entities, as opposed to the classical theory of macroscopic objects.

The way in which quantum theory was shown to depend on non-commuting variables and complex algebra branded it as counterintuitive and at variance with the classical concepts of reality, causality and predictability. One of the strangest quantum notions was the property of superposition. For any operation with two possible outcomes a linear superposition of these presents an equally valid outcome.

The most popular example of this concept goes under the label of *Schrödinger's cat*. A cat, sealed into an enclosure with a source of poisonous gas to be released at an unspecified time, has the same chance of being dead or alive, as long as the box remains closed. The quantum description of the cat, according to the superposition principle, is said to be half-dead and half-alive. Such is taught as serious science, with the result that several generations of scientists have been conditioned to identify any unexpected observation as a *quantum effect*.

In order to rationalize the structure of nanomaterials the methods of quantum chemistry inevitably came into play. These methods were considered as particularly appropriate applied to graphene, nanodiamond and their derivatives. However, more recent assessment of concepts such as orbital hybridization as essentially classical in nature, militates against the use of quantum chemistry to simulate nanostructures. More worrisome is the realization that the classical notion of an isolated molecule with a robust molecular structure and shape has no counterpart in quantum theory. In the jargon of quantum mechanics, each observable is associated with a quantum operator. No such operator that relates to molecular structure could however, be identified.

The use of traditional wave mechanics in the analysis of nanoparticles is equally misleading. Based on the solution of a linear differential equation it provides a fair approximation of electron behavior, but has no validity in the study of nonlinear nanosystems.

Finally, whereas the all-important physical properties of technological important nanomaterials depend exclusively on the nonclassical spin variable, traditional wave mechanics in three dimensions cannot provide the full answer.

A brief analysis of these aspects must necessarily precede discussion of the nonclassical properties of classical nanostructures.

3.3.1 ORBITAL HYBRIDIZATION

The structural model of quantum mechanics is based on Schrödinger's solution of the three-dimensional matter-wave equation

$$-i\frac{\partial\Psi}{\partial t}=\frac{\hbar^2}{2m}\nabla^2\Psi$$

that reduces to Schrödinger's amplitude equation

$$\nabla^2\psi+\frac{2m}{\hbar^2}(E-V)\psi=0\,, \tag{1}$$

on substituting $\Psi=\psi e^{-iE/\hbar}$.

The amplitude equation, also known as the time-independent wave equation, has the peculiar property of specifying time-dependent variables, such as momentum, angular momentum and kinetic energy of a particle of mass m in a potential field defined by the classical potential energy V. This feature is explained by the assumption that all attributes of the quantum system are carried by the wave function ψ. Specific information is extracted by applying a suitable operation on ψ. This way linear and angular momenta are produced by the operators $p_x \to -i\hbar(\partial/\partial x)$ and $L_z = -i\hbar(\partial/\partial\phi)$, respectively.

In order to formulate a molecular-structure model wave mechanically, the first step is to solve Eq. (1). Solutions that describe the motion of the electron associated with a stationary proton in an isolated spherically symmetrical hydrogen atom were obtained by Schrödinger. This central field problem is most conveniently formulated in terms of spherical polar coordinates. As a second-order linear differential equation mathematical solutions, obtained by the separation of the variables, were known at the time. Separation in terms of spherical polar coordinates depends on the assumption that the wave function is equivalent to the product function,

$$\psi(r,\theta,\phi) = R(r)\cdot\Theta(\theta)\cdot\Phi(\phi)$$

of orthogonal functions R, Θ and Φ. The three resulting equations are solved by the Frobenius power-series procedure, which yields three sets of Eigen function solutions characterized by integers, known as quantum numbers. In the final analysis the quantum numbers n, l and m_l are interpreted to specify the eigenvalues of total energy, E, angular momentum, L and angular momentum in the polar direction, L_z, for the hydrogen electron.

In summary:

$$E_n = -\frac{2me^4}{n^2\hbar^2}(esu), \quad n = 1,2,3...$$

$$L^2 = l(l+1)\hbar^2, \quad l = 0,...(n-1)$$

$$L_z = m_l\hbar, \quad m_l = -l,...,l$$

Each value of m_l is interpreted to define an "orbital," associated with a pair of electrons of opposite spin, which is empirically assumed.

3.3.1.1 THE SOMMERFELD MODEL

Solution of Schrödinger's equation for the hydrogen electron provided a firm basis for the atomic theories of Bohr and Sommerfeld. As a computational tool it simplified the interpretation of spectroscopy, but it complicated the accepted model of molecular structure.

The Sommerfeld atomic model assigned extranuclear electrons to elliptic orbits with well-defined spatial orientation. For the carbon atom the four valence electrons were assigned to four degenerate orbits directed towards the vertices of a tetrahedron, at alternate corners of a cube that surrounds the nucleus. This arrangement is in excellent agreement with the tetrahedral carbon atom postulated by van't Hoff and the cubic arrangement of electron pairs at the valence level, according to Lewis.

3.3.1.2 THE SCHRÖDINGER SOLUTION

The results of the Schrödinger analysis were in conflict with the harmonious state of affairs as the four denegerate valence orbits were replaced by two orthogonal sublevels corresponding to the nondegerate s and three-fold degenerate spectroscopic p states.

3.3.1.3 THE PAULING–COULSON SCHEME

The scheme developed by Pauling and Coulson to overcome the discrepancy, although fatally flawed in several respects, was enthusiastically embraced by the chemical community and turned into the notorious concepts of orbital hybridization and resonance, still in use today.

The ultimate objective was to demonstrate that the four functions:

$$\psi(2s) = Y_{0,0} = \frac{1}{\sqrt{4\pi}} \qquad (n = 1, l = 0, m_l = 0)$$

$$\psi(2p_z) = Y_{1,0} = \sqrt{\frac{3}{4\pi}}\cos\theta = \sqrt{\frac{3}{4\pi}}\cdot\frac{z}{r} \qquad (n = 1, l = 1, m_l = 0)$$

$$\psi(2p_{\pm 1}) = Y_{1,\pm 1} = \mp\sqrt{\frac{3}{8\pi}}\cos\theta e^{-i\phi} = \mp\sqrt{\frac{3}{8\pi}}\cdot\frac{x \pm iy}{r} \qquad (m_l = \pm 1)$$

could by some appropriate transformation, be reformulated as a degerate set, equivalent to the Sommerfeld ellipses. Notice how the quantum number $m_1 = 0$ generates a real function, but a complex pair for $m_1 \neq 0$.

The hybridization strategy relies on the linear combination of complex pairs, that is:

$$Y_{1,1} + Y_{1,-1} = \frac{1}{\sqrt{2}} \cdot \sqrt{\frac{3}{8\pi}} \left(\frac{x+iy}{r} + \frac{x-iy}{r} \right) = \sqrt{\frac{3}{4\pi}} \cdot \frac{x}{r}$$

$$Y_{1,1} - Y_{1.-1} = \sqrt{\frac{3}{4\pi}} \frac{iy}{r} \qquad (2)$$

to produce one real and one imaginary function. This operation is seen to result in a 90° rotation of the coordinate axes to yield:

$$\psi(2p_x) = \sqrt{\frac{3}{4\pi}} \cdot \frac{x}{r}$$

$$\psi(2p_{\pm 1}) = \frac{1}{\sqrt{2}} \cdot \sqrt{\frac{3}{4\pi}} \left(\frac{z \mp iy}{r} \right)$$

as before, with polar axis z rotated to x.

However, the hybridization authors preferred a different interpretation. Using an imaginary coefficient in Eq. (2) they arrive at a set of real functions,

$$2p_x = \sqrt{\frac{3}{4\pi}} \cdot \frac{x}{r}$$

$$2p_y = \sqrt{\frac{3}{4\pi}} \cdot \frac{y}{r} \qquad (3)$$

$$2p_z = \sqrt{\frac{3}{4\pi}} \cdot \frac{z}{r}$$

which constitutes the basis of the hybridization scheme. A detailed critical analysis of the scheme has been published [35], but its most devastating defect relates to the exclusion principle that demands a unique set of the three quantum numbers and the spin quantum number, $m_s \pm 1/2$, for each electron in an atom. Since the three real functions (the Eq. (3)) each requires $m_l = 0$, the exclusion condition cannot be satisfied for the postulated set of three p-electrons, required in the hybridization procedure. The hallmark of quantum theory is the appearance of discrete quantum numbers. Without that the theory reverts to the classical.

It is interesting to note that the sp^3 hybridization scheme, considered as a classical construct, reduces to the geometrical synthesis of the tetrahedral carbon structure proposed by van't Hoff [37].

3.3.2 NANOMODELS

Further details of orbital hybridization are considered too well known and too depressing to warrant further discussion. In the present context a number of attempts to rationalize the electronic structure [38, 27], conformation [39, 40] and topology [23] of nanomaterials in terms of hybrid orbitals are noted for their misleading content. It is generally agreed that [38]:

Carbon nanotubes … consist of either one … or multiple … shells of carbon in sp^2 bonding configuration.

This is clearly intended as a precise quantum description of carbon nanotubes, although it strictly applies only to a rigid planar system.

A more precise formulation is attempted elsewhere [23]:

The angle $\theta\sigma\pi$ between the [π orbital axis vector] and a σ direction (i.e., a bond) indicates the degree of "pyramidalization" and the hybridization. For $\theta\sigma\pi = 90°$ (planar system), the σ orbitals are in a sp^2 hybridization and the π orbital is a pure p_z orbital. For a folded graphene sheet, $\theta\sigma\pi$ has an intermediate value which decreases as the inverse of the radius of curvature of the folding, and reaches 90° at the limit $R \rightarrow \infty$.

This is quantum chemistry gone berserk, but perhaps not as wild as the ingeneous affirmation of the quantum-mechanical nature of nanomaterials [40] that follows:

From a theoretical point of view, the Dirac equation – which replaces the Schrödinger equation for electrons in graphene – has to be modified when defects are in the lattice. … The overlap of p_z-orbitals determines the electronic properties but is altered in the vicinity of structural defects. … defects lead to a local rehybridization of sigma and pi-orbitals, which … changes the electronic structure.

The different bonding models proposed for nanocarbon compounds are all based on the assumption that the in-plane interactions in graphite are of sp^2 type. According to the theory, right or wrong, all bond lengths must be the same, d(C–C)=1.42 Å, with each atom trigonally surrounded by three others in the plane. According to hybridization theory such a graphite sheet, like extended aromatic systems, is stabilized by a delocalized grid of π-electrons that ensures sterically rigid planarity as a function of the combined barriers of rotation for each bond. The interplanar spacing of 3.35 Å is ascribed to van der Waals forces [41].

As already demonstrated the theory has no quantum-mechanical basis and its predictions simply paraphrase the stereochemical principles proposed by Kekulé and van't Hoff. Not surprisingly, the hybridization model for graphite breaks down completely when applied to graphene and its derivatives. The rigid graphene sheet predicted by the theory has never been observed. In fact, the theory already fails when applied to polycyclic aromatic compounds.

The structure of the smallest nanotube, C_{70}, as derived crystallographically [25] is projected in Fig. 3.12. The first-neighbor interatomic distances observed here are of three types, *d*(C–C)=1.57(3), 1.50(3) and 1.33(3) Å, representing proposed bond orders of 1, 1.25 and 1.5, respectively. Note how the bond orders add up to the classical tetravalency of four at each atom. It is also of interest to note that the pentagons have strict five-fold symmetry, whereas the six-membered rings have no chemically meaningful symmetry.

This result is in satisfying accord with the definition of bond order as a fundamental characteristic of wave-like covalent interactions [42]. Hybridization and resonance are powerless in this situation. Not that molecular orbital methods, also based on linear combination of real atomic orbitals, can perform any better. The only useful results, pertaining to the folding and distortion of graphene sheets, obtained until now, are based on empirical analysis of curved surfaces. Even the nearly free-electron studies of semiconductor band structures of nanomaterials are of limited value, especially those that start from hybridization levels.

3.3.2.1 ELECTRONIC PROPERTIES

The quantum-chemical models for nanosized semiconductor crystals are also based on hybridization and resonance, according to the general prescription of Coulson [43] for general intermetallic compounds.

All interactions are modeled by overlapping sp^3 orbitals, augmented by a host of empirical quantities, including effective nuclear charge, Madelung constant, net atomic charge, effective charge per bond, bond length for first neighbors and distance between second neighbors – all of these variationally and empirically adjusted.

An ambitious analysis of semiconductor clusters by this procedure [27] is concluded with remarks, such as:

- The calculated discrete electronic spectra may need modification.
- Quantitative calculations will require improvement.

- The atomic nature and structure of intrinsic surface states, and the general question of possible surface electronic bands and reconstruction, remain largely unexplored.
- Experiments on CdSe crystallites clearly demonstrate the reality of the electronic quantum size effect.

It is fair to infer that the methods of quantum chemistry have not contributed in a meaningful way to an improved understanding of size effects in nano-sized semiconductors, apart from the assumption that it is a quantum effect.

More recent attempts (e.g., [44]) to obtain more meaningful results at a "higher level of theory" resorted to DFT calculations. As no experimental data are available for the hypothetical clusters selected for study, the quality of the calculated results is hard to assess. However, what is beyond dispute is that density-functional theory is based on the same real basis sets as orbital hybridization and other variations of quantum-chemical computations. Although the classical nature of the formalism does not necessarily invalidate the results, there is no justification to designate the calculated trends as quantum effects.

3.3.3 MOLECULAR STRUCTURE

The representation of molecules by line drawings or mechanical models, familiar to all, conveys the impression of essentially robust, albeit flexible, molecular structures and shapes. These notions are so intimately associated with, and well established by experimental studies in molecular spectroscopy and X-ray crystallography that theoretical support is often considered self evident. However, beyond the assumptions there is no understanding of chemical bonds or molecular conformation as quantum-mechanical concepts.

One of the declared aims of quantum chemistry is the prediction of molecular structure and shape from first principles. The common strategy is based on the Born–Oppenheimer prescription for the separation of nuclear and electronic motion. Because of the large difference in mass, atomic nuclei are assumed to remain essentially stationary on the scale of electronic motion. The philosophy is that an initial trial structure in the form of an assumed static nuclear framework may be used for quantum-mechanical optimization of the associated electron distribution, subsequently to be used also for optimization of the trial structure, until self-consistency is achieved in an iterative process.

The well-known failure of this pursuit is due to two fundamental problems – the initially assumed framework is a classical construct, which cannot be optimized quantum-mechanically, and a vectorial molecular structure cannot be predicted by the minimization of energy, which is a scalar quantity [45].

However, structural molecular formulae are of unquestionable importance to practicing chemists, but these are often seriously at variance with known chemical properties. In the present context the most obvious example of such discrepancy occurs in aromatic systems, supposedly based on the resonance formulation of the benzene structure,

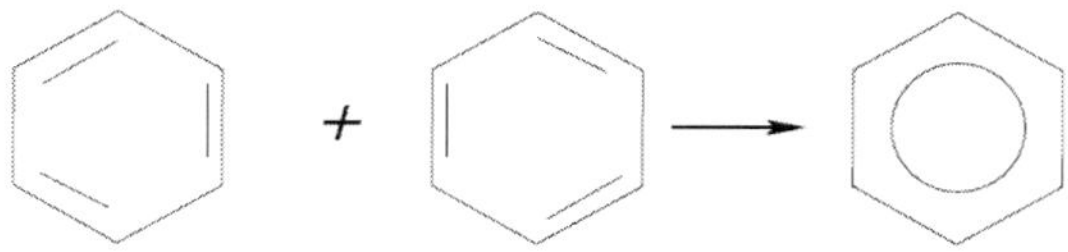

said to be a nonclassical modification of the classical Kekulé structure. To explain why resonance does not operate for the molecule (I),

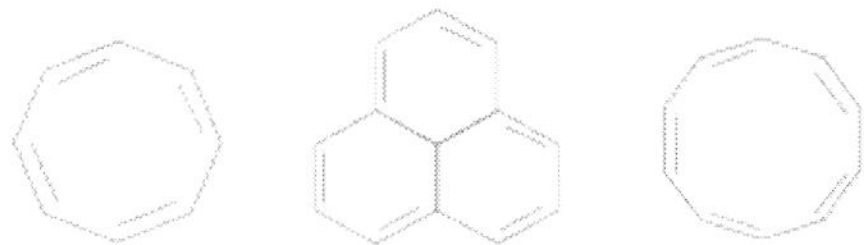

I II III

which lacks the planarity associated with aromaticity, a quantum-chemical molecular-orbital analysis is invoked to postulate the secondary principle that restricts aromaticity to systems with $4n + 2$ π-electrons. Although the rule holds for naphthalene, anthracene and other cata-condensed systems, it fails for peri-condensed systems [37], such as (II) and the $4n + 2$ compound (III).

The common assertion that the structure of graphene and its nanoderivatives is adequately explained as extended aromatic, or sp^2 systems with local variation must therefore be seriously questioned. Not only in view of the failure of orbital hybridization, but also because of the inadequacy of free electron simulation of aromaticity.

As concluded in a recent review [46], free molecules are structureless in three dimensions and only acquire structure in condensed phases. Nanoparticles are of this type, albeit at the lower limit, and their three-dimensional structure is to be understood in a classical sense. However, like molecules, their true nature may depend on nonclassical attributes that only become apparent in four-dimensional analysis.

3.3.4 NONLINEAR EFFECTS

The most efficient mathematical modeling of physical phenomena is in terms of differential equations, which can be either linear or nonlinear. An equation is said to be linear if each term is either linear in all the dependent variables and their various derivatives, or does not contain any of them. Any linear equation is of degree one. A term such as $y\frac{dy}{dx}$ is of degree two in y, together with its derivative, and therefore nonlinear. The way in which the dependent variable enters the equation is immaterial.

The mathematically essential difference between linear and nonlinear equations exists therein that any two solutions of a linear equation can be added together to form a new solution. In contrast such superposition fails for nonlinear equations. For this reason there is no general analytic approach for solving nonlinear equations and applied mathematicians tend to describe physical systems as far as possible with linear differential equations. However, on dealing with essentially nonlinear behavior this approach is an oversimplification that may obscure the actual characteristics of the system.

The behavior of nanomaterials is essentially nonlinear. The difference between linear and nonlinear physical systems is epitomized by fluids in laminar and turbulent flow. As the rate of laminar flow, characterized by a linear velocity gradient, exceeds a critical value, nonlinear effects set in, eddies develop and the flow becomes turbulent [48]. Likewise, on reducing the size of a perfect crystal until the relative number of surface particles exceeds a critical limit, nonlinear effects emerge and signal the appearance of a host of anomalous physical properties.

Any effort to describe the behavior of colloids and nanoparticles by linear equations is an exercise in futility. Even the confinement effect, resulting from the restriction on electronic motion, is nonlinear, which means that the linear nearly free electron models used for bulk-size semiconductors must be poor approximations at best. Although the nonlinear properties of nanomaterials are clearly nonclassical, they can only be properly described by a nonlinear wave equation. In particular, the simulation of their band structure by linear superposition of sp^3 orbitals [27] is no more informative than the general statement [49]:

In semiconductor particles of nanometer size, a gradual transition from solid-state to molecular structure occurs as the particle size decreases. Consequently, a splitting of the energy bands into discrete, quantized levels occurs.

It goes without saying that computational quantum chemistry, exclusively based on linear superpositions in all of its forms, holds no promise of contributing any insight into the nature of nonlinear nanostructures.

The rapid progress with computer experiments in the analysis of nonlinear lattices [50] points at an alternative approach that could be of considerable benefit in the development of nanotechnology.

3.3.5 THE SPIN FUNCTION

A bright future is anticipated for nanoscience and technology in medicinal applications and the production of electronic devices at the molecular level. Both of these aspects are closely related to electron distributions and interactions in nanomaterials. As shown already the traditional quantum-chemical approximation is, for several reasons, inappropriate to address this issue. As an alternative approach without ad hoc simplications is clearly required, a critical reassessment of the entire process is indicated. One of the most important concerns to be addressed is the absence of a spin variable in the traditional formulation of wave mechanics, noting that spin is the single most significant nonclassical concept in quantum theory.

The underlying problem has been identified [51] as arising from separation of space and time variables in the derivation of Schrödinger's equation. The theory of relativity demands that the space and time variables be treated on an equivalent footing, as embodied in d'Alembert's equation

$$\nabla^2\Psi = \left(\frac{\partial^2}{\partial x_0^2} + \frac{\partial^2}{\partial x_1^2} + \frac{\partial^2}{\partial x_2^2} + \frac{\partial^2}{\partial x_3^2}\right)\Psi = 0 \tag{4}$$

where $x_0 = ict$, with $x(1, 3)$ given by the traditional Cartesian x, y, z. This equation is commonly formulated in three dimensions, that is,

$$\nabla^2\Psi = \frac{1}{c^2}\frac{\partial^2\Psi}{\partial t^2}$$

Solution of this equation by separation of the variables destroys the Lorentzian structure of (4) and fails to identify four-dimensional solutions of the type $\Psi(x_0, x_1, x_2, x_3)$. It is precisely these solutions that define the spin function.

An assumed remedy of this defect is by addition of the spin variable to Schrödinger's equation. The problem with this is that a four-dimensional function, known as a quaternion, cannot be represented by the superposition of one complex and two real functions. Instead of a four-dimensional undulation one ends up with a two-dimensional rotation and two one-dimensional vibrations, or worse, with four vibrations as in the quantum-chemistry approximation.

3.4 THE NANOSTATE

The nanostate is generally defined as intermediate between the molecular regime and the macrocrystal, judging by the gradual variation in electrical and optical properties with increasing cluster size. Equally dramatic, but less often emphasized, is change in physical properties, such as melting temperature and the applied pressure, needed to induce phase transitions. In the case of CdS an increase in melting point from 400°C to 1600°C is observed and a general lowering of pressure by a factor 2 is routinely observed [52, 53]. This increase in melting temperature may be considered as a transition from liquid to solid state [54].

The efficiency of inert metals like Pt as catalysts in nanoform may relate to their effective liquid structure at room temperature. The pyrophorous property of several metal powders, including Al, Mg, Zr, Ti, Ta, Pb and Th, is another manifestation of the same effect.

As further evidence of the liquid-like nature of nanoparticles it is noted [54] that metals in this form become more malleable whereas ceramics are known to show dramatically reduced melting temperatures, superplastic behavior and increased mechanical strength.

Another intriguing aspect, also reviewed [54], but unfortunately considered beyond the scope of the present treatise, is the modified structure and properties of liquid water in nanoconfinement. It may be tempting to consider this response as a quantum effect, but it more logically relates to the limited size and the conformation of optimal molecular clusters. In this sense the nanobehavior reflects no more than modified chemical reactivity and capacity to form hydrogen bonds. Admittedly, all of these special interactions that emerge as a result of the increased percentage of surface species are of quantum-mechanical nature, but the same is true of the normal interactions between macroscopic crystals and their environment.

Following the process one step further it is noted [32] that:

Uniform particles, whether atomic, molecular, or colloidal, organize to form ordered solids when attractive and repulsive forces are properly balanced.

The thermodynamic and kinetic factors that control the formation of superclusters are no different from what occurs in the crystallization of atomic or molecular materials. The special term, self-organization, frequently used to describe the process is unnecessarily misleading. There is no mysterious driving force at play.

However, not all colloidal suspensions are strictly monodispersive and during assembly into superstructures, preferential incorporation of larger clusters, at the expense of smaller particles, occurs in a process known as Ostwald ripening, which is standard procedure in the aging of precipitates [55].

Nanoparticle building blocks are considered as artificial atoms and the superclusters are also known as "artificial solids" [56], but natural opals, examples of such ordered super lattices, are hardly artificial.

3.4.1 PLASMONICS

The optical properties of metallic nanoparticles remain one of the most actively studied aspects of nanotechnology. It is generally referred to as plasmonics, a description that derives from the similarity between metallic films and plasmas. A plasma is a medium with equal concentrations of positive and negative charges, of which at least one charge type is mobile. In a solid the negative charges of the conduction electrons are balanced by an equal number of positively charged ion cores.

A plasma oscillation results from the displacement of electrons as shown in Fig. 3.15 [57]. A Plasmon is a quantum of plasma oscillation, which may be excited by passing an electron through a thin metallic film or by reflecting an electron or a photon from a film.

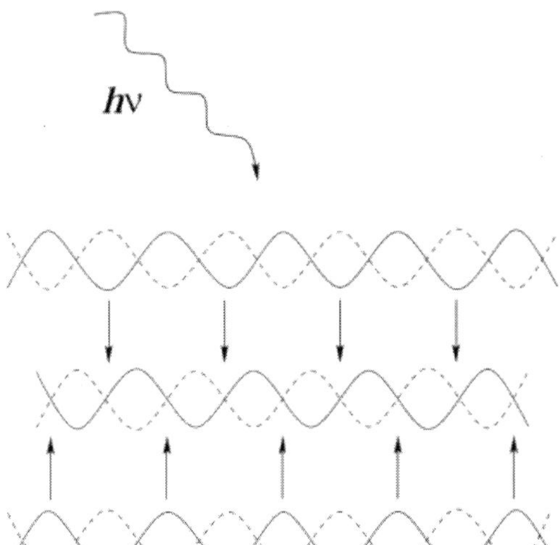

FIGURE 3.15 Creation of surface Plasmon's as electronic charge couples with the fluctuating field of electromagnetic radiation. The arrows indicate electronic displacement.

Plasmon excitations have been observed in surfaces, interfaces and spherical metallic colloids. The deep red color of gold nanoparticle sols in water and glasses reflects a surface Plasmon band; a broad absorption band in the visible region around 520 nm [1]. The phenomenon is interpreted in Mie's theory, first formulated in 1908 [58]. The resonances of electric and magnetic oscillations in the surface were described by solving Maxwell's equations for spherical particles with the appropriate boundary conditions. The Plasmon band may therefore be ascribed to the dipole oscillations of free electrons in the conduction band at energy levels immediately above the Fermi level [59].

An intriguing characteristic of the surface Plasmon band is its disappearance from particles with core diameter less than 2 nm, as well as from gold in bulk. This behavior becomes hard to rationalize as a simple quantum effect. Phenomenologically there is no mystery. With increased clustering beyond the molecular level the 6*s* conduction electrons accumulate in the surface. On reaching a critical size Mie-type resonance occurs and persists until the particle has grown to a size where normal metallic conduction sets in. This happens when the particle size has increased to match the de Broglie wavelength of the valence electrons, which now propagate as matter waves through the periodic lattice [1].

3.4.2 GROWTH OF NANOSTRUCTURES

The endless variety in which matter occurs in Nature consists almost exclusively of variable composites. Pure forms, exemplified by perfect crystals, although extremely rare, are at the basis of practically all structure models. It is important to note that crystallographic analysis reveals the structure of highly condensed phases rather than that of free molecules, presumed to constitute the building blocks of nanoparticles. The crystal structures of gold metal and graphite must therefore be considered as poor models for nanogold and graphene respectively. In fact, the observed variation of melting temperature with cluster size [54] favors a liquid-like structure for nanometals and the unfettered curving of unsupported graphene sheets demonstrates spontaneous deviation from the graphite structure.

The carbon network in graphite follows a strict triangular pattern exemplified by hexagons, some of which may be converted into nonhexagonal rings in graphene-derived structures, for instance by 90° rotation of a C–C bond [40],

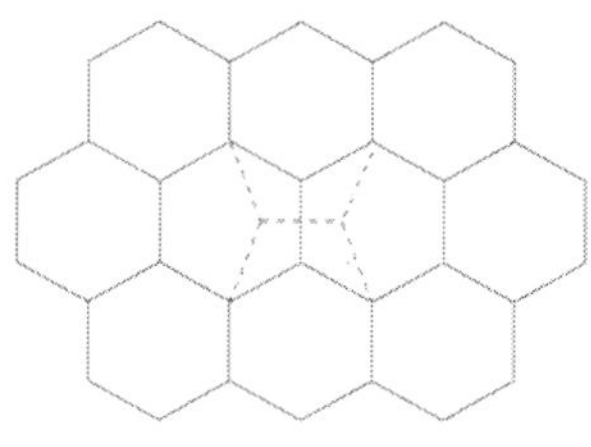

in the formation of spherical particles (Fig. 3.10), or the capping of nanotubes (Fig. 3.11).

3.4.3 FIBONACCI PATTERNS

The stress pattern in the surface of Ag/SiO_x core/shell microstructures [60] appears in the same style as arrays of triangles and pentagons. In addition, the spherules line up along Fibonacci spirals as discussed in Fig. 3.16.

This observation brings the formation of nanoparticles into line with the growth of biological structures and the conjectured mechanism that controls the shape of polymers and macromolecules in terms of space-time curvature [46].

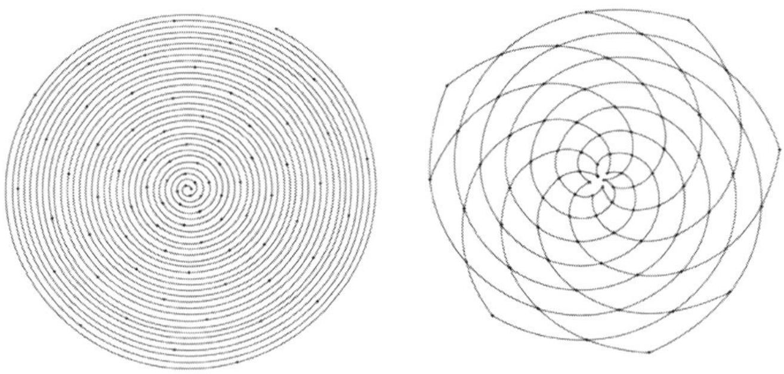

FIGURE 3.16 Points on an Archimedean spiral ($r = a\theta$) with a convergence angle of 1 radian. The second frame shows the same points connected by 8 (F6) and 5 (F5) secondary Fibonacci spirals, curving clocks and anticlockwise, respectively. Similar Fibonacci patterns (F5, F6) and (F7, F8) occur in the surface of Ag/SiO_x core/shell spherules [60].

FIGURE 3.17 Fibonacci phyllotaxis on a growing cactus plant with approximate spherical shape.

Computerized nonlinear optimization of the arrangement of outward pointing, freely rotating magnetic dipoles in a linear stack was shown [47] to simulate the phyllotaxis observed in a growing cactus, and proposed as a model for curved nanostructures.

To explore the likelihood that all nanostructures are shaped along the same lines, the atomic positions in the spherical fullerene are shown schematically in Fig. 3.18 to be consistent with the Fibonacci phyllotaxis of spherical growing cacti, shown in Fig. 3.17.

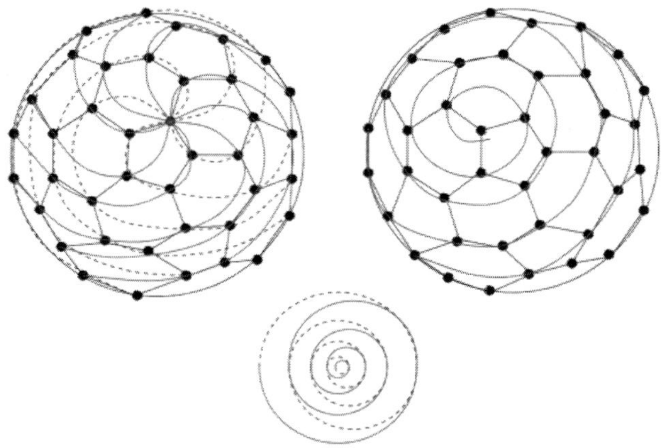

FIGURE 3.18 Schematic drawing to show the atomic positions of fullerene connected by 8.5 Fibonacci spirals (left) and a single spiral (right). In three dimensions it spirals from pole to pole as shown in the lower frame.

The exciting conclusion is that nanostructures are the natural intermediates in the growth of macroscopic bodies from the atomic and molecular building blocks that assemble in a fashion dictated by space-time curvature, which is quantified by the golden parameter $\tau = \frac{1}{2}(\sqrt{5}-1) = 0.61803...$. The most prominent golden growth spiral in Nature is inscribed within a golden rectangle with sides of τ and 1. All of these constructs share the five-fold symmetry imposed by τ [62]. The growth points on a golden spiral are described equally well by secondary spirals with a Fibonacci number (F_n) of arms [63], such as the F_5 and F_6 arms as shown in Fig. 3.18.

From this point of view the C_{60} fullerene molecule consists of 12 interconnected pentagons in a three-dimensional arrangement with six fivefold alternating $\bar{5}$ symmetry axes that intersect at a center of symmetry, as shown in Fig. 3.19.

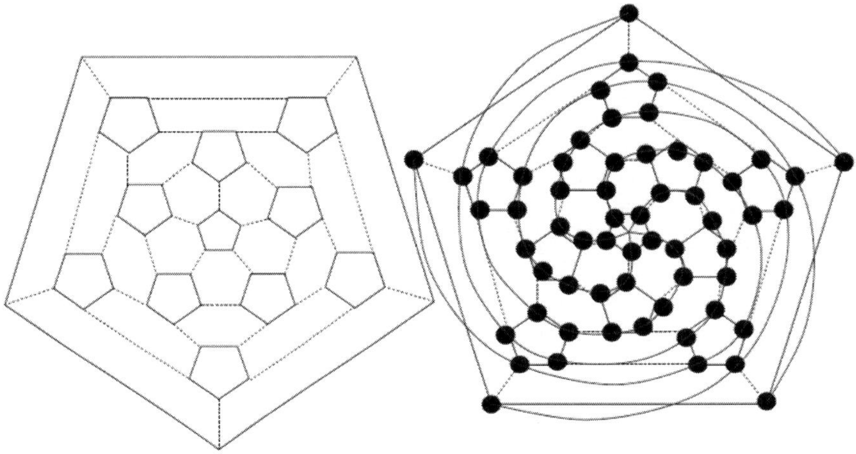

FIGURE 3.19 Polar projection of the C_{60} fullerene molecule along one of the six equivalent $\overline{5}$ axes, with a set of closed F_5 Fibonacci spirals shown on the right.

Distortion of the polygons that lie in the spherical surface is inevitable when drawn in a flat surface. In Fig. 3.20, where five-fold symmetry is strictly preserved, the F_5 Fibonacci spirals are less obvious.

A stereoscopic view of the molecule is shown in Fig. 3.21.

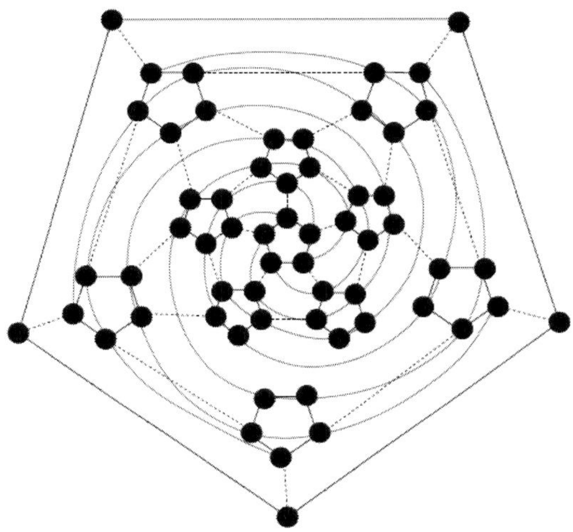

FIGURE 3.20 A third alternative way of identifying F_5 spirals that connect all atoms in C_{60}. This projection preserves the symmetry of all pentagons.

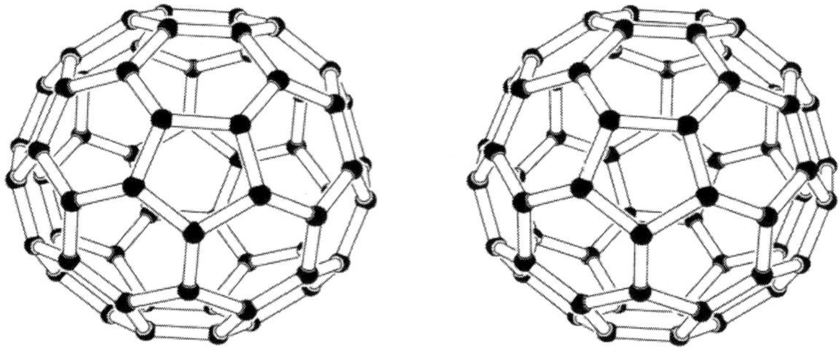

FIGURE 3.21 Stereoscopic view of the C_{60} fullerene structure as found crystallographically [61].

3.4.4 NON-CLASSICAL STRUCTURE

The 60 carbon atoms that make up the molecule are completely equivalent in all respects and together they define a holistic unit. Even the appearance of 12 equivalent pentagons is of minor importance compared to the arrangement of all atoms on a closed spiral as shown at the bottom of Fig. 3.18. Starting with any atom as pole, a spiral connects the 30 atoms in its hemisphere and continues to spiral in towards the opposite pole.

By considering the two poles as a single point the double spiral becomes topologically equivalent to a closed single spiral in elliptic, or projective, space, embedded in four dimensions. This is the same structure that was proposed for the periodic table of the elements [63] and for universal space-time [64].

The Fibonacci spirals that catch the eye in Fig. 3.19 provide a clue to the understanding of the way in which a fullerene molecule grows from smoky vapors. It resembles the botanical process known as Fibonacci phyllotaxis [65] in which growing florets with increasing size maintain a close-packed arrangement on a curved surface. In the case of fullerene the initial embryonic growth point is a C_5 unit. On addition of more pentagons a growth surface expands and in order to preserve the five-fold symmetry for all interconnected units the planar surface must curve into a spherical shape.

Instead of the pressure from a centralized growth point as in botanical growth, the pressure that drives the fullerene to curve develops at the periphery, until it closes on itself. The two processes are controlled by the same geometrical principle.

The proposed mechanism is fully consistent with the two bond orders, representing interatomic distances of 1.38(2) Å and 1.45(1) Å measured by atomic force microscopy [66], without the need to invoke alternating aromatic bond orders.

Evidence [67] that fullerene occurs in carbonaceous chondrites points at its possible existence in interstellar space, presumably as a four-dimensional molecule with an involuted hyperspherical structure. The well-known three-dimensional shape only emerges in condensed phases. By the same argument it may be argued that embryos of $\overline{1}$, $\overline{2}$ and $\overline{3}$ symmetry, all Fibonacci numbers, would lead to the growth of amorphous C, nanodiamond [68] and graphene derivatives.

There is no other known molecule that reflects the three-dimensional projection of space-time structure as well as C_{60}. The simple reason is that five-fold symmetry is incommensurate with three-dimensional translational symmetry. Since traditional structural chemistry is based almost exclusively on three-dimensional crystallography the analysis of incommensurate structures and "quasi-crystals" that depends on four-dimensional translational symmetry has largely been neglected until fairly recently.

An interesting variety of structures, including those of viruses and quasicrystals such as icosahedral $Al_{70}Pd_{2}Mn_{9}$, of this type, are reviewed in two recently published chapters [36, 69, 70]. A central feature of this work is the prominence of the golden ratio in the metrics that characterize these structures. The theoretical implications are discussed in the same volume [35] and elsewhere [46]. The implied self-similarity, ranging from atoms, molecules, through nanostructures, to astronomical systems, strongly suggests an intimate relationship of nanomaterials with botanical growth structures and biological composite materials, as widely recognized empirically [71, 72]. It signals the need for a serious new approach to nanoscience.

The lame description of nanomaterials in terms of traditional valence-bond and molecular-orbital theories leads nowhere. The nature of the C_{60} fullerene provides a pertinent example. The appearance of hexagons is clearly an incidental feature, arising from the linkage of pentagons, rather than sp^2 hybridization. Only when forced into regular three-dimensional solids by environmental factors, is the preferred $\overline{5}$ symmetry modified into structures with either $\overline{4}$ (diamond) or $\overline{6}$ (graphite) symmetries. Compared to fullerene these are classical structures. Non-classical nature resides in the four-dimensional and nonlinear aspects of C_{60} and its derivatives, associated with composite materials. A convenient approach towards the understanding and modeling of the nonclassical properties of such nanomaterials is via number theory and self-similarity – the recommended new approach to nanoscience and technology.

ACKNOWLEDGMENT

The valuable assistance of Prof. D.C. Levendis, at many levels, is gratefully acknowledged.

KEYWORDS

- **atoms and molecules**
- **fullerene**
- **golden ratio**
- **graphene**
- **hybridization**
- **nanomodels**
- **quantum models**
- **quantum principles**

REFERENCES

1. Daniel, M.-C., Astruc, D. (2004). Gold Nanoparticles: *Assembly, Supramolecular Chemistry, Quantum-Size-Related Properties, and Applications toward Biology, Catalysis and Nanotechnology*, Chem Rev. 104, 293–346.
2. Skrabalak, S. E., Chen, J., Sun, Y., Lu, X., Cobley, C. M., Xia, Y. (2008). *Gold Nanocages: Synthesis, Properties and Applications*, Acc. Chem. Res. 41, 1587–1595.
3. Kwon. S. G., Hyeon, T. (2008). *Colloidal Chemical Synthesis and Formation Kinetics of Uniformly Sized Nanocrystals of Metals, Oxides and Chalcogenides,* Acc. Chem. Res. 41, 1696–1709.
4. Hughes, B. K., Luther, J. M., Beard, M. C., (2012). The Subtle Chemistry of Colloidal, Quantum-Confined Semiconductor Nanostructures, ACS Nano, 6, 4573–4579.
5. Terrones, M., Terrones, G., Terrones, H., (2002). Structure, Chirality, and Formation of Giant Icosahedral Fullerenes and Spherical Graphitic Onions, Struct. Chem. 13, 373–383.
6. Sumper, M., Kröger, N., (2004). Silica formation in diatoms: the function of long-chain polyamines and silaffins, J. Mater. Chem. 14, 2059–2065.
7. Palmer, L. C., Stupp, S. I., (2008). Molecular Self-Assembly into One- Dimensional Nanostructures, Acc. Chem Res. 41, 1674–1684.
8. Hallas, N. J., (2008). Nanoscience under Glass: The Versatile Chemistry of Silica Nanostructures, ACS Nano 2, 179–183.
9. Schmid, G., Pfeil, R., Boese, R., Bandermann, F., Meyer, S., G. H. M., Calis, J. W.A van der Velden, (1981). [Au55{P(C6H5)3}12Cl6] – A Gold Cluster of Unusual Size, Chem. Ber. 114, 3634–3642.
10. Schmid, G., (1992). Large Clusters and Colloids, Chem. Rev. 92, 1709–1727.

11. Schmid, G., Bäumle, M., Geerkens, M., Heim, I., I., Osemann, C., Sawitowski, T., (1999). Current and Future Applications of Nanoclusters, Chem. Soc. Rev. 28, 179–185.
12. Schmid, G., Corain, B. (2003), Nanoparticulated Gold: Syntheses, Structures, Electronics, and Reactivities, Eur. J. Inorg. Chem. 3081–3098.
13. Zhang, H., Schmid, G., Hartmann, U., (2003). Reduced Metallic Properties of Ligand-Stabilized Small Metal Clusters, Nano Lett. 3, 305–307.
14. Schmid, G., Harms, M., Malm, J. O., Bovin, J. O., J. van Ruitenbeck, Zandbergen, H. W., Fu, W. T., (1993). Ligand-stabilized Giant Palladium Clusters: Promising Candidates in Heterogeneous Catalysis, J. Am. Chem. Soc. 115, 2046–2048.
15. Gittins, D. I., Caruso, F., (2001). Tailoring the Polyelectrolyte Coating of Metal Nanoparticles, J. Phys. Chem. B, 105, 6846–6852.
16. Kroto, H. W., Heath, J. R., S. C. O'Brien, Curl, R. F., Smalley, R. E., (1985). Buckminsterfullerene, Nature 318, 162–163.
17. V. H. P., Boehm, Clauss, A., Fischer, G. O., Hofmann, U. Z., (1962). Dünnste Kohlenstoff-Folien, Naturforsch. 17b, 150–153.
18. Novoselov, K. S., Geim, A. K., Morozov, S. V., Jiang, D., Zhang, Y., Dobonos, S. V., Grigorieva, I. V., Firsov, A. A., (2004). Electric Field Effect in Atomically Thin Carbon Films, Science 306, 666–669.
19. Krüger, A., Neue Kohlenstoffmaterialien, Teubner Verlag, Wiesbaden, 2007.
20. Arezoomand, M., Taeri, B., (2009). The full symmetry and irreducible representations of nanotori, Acta Cryst. A65 249–252.
21. Iijima, S., (1991). Helical microtubules of graphitic carbon, Nature, 354, 56–58.
22. Disnovski, S., Gogotsi, Y., Graphite Whiskers, Cores and Polyhedra, in, Y. Gogotsi (ed.) Carbon Nanomaterials, CRC Press, Boca Raton, 2006.
23. Terrones, M., (2010). Sharpening the Chemical Scissors to Unzip Carbon Nanotubes: Crystalline Graphene Nanoribbons, ACS Nano 4, 1775–1781.
24. Goldberg, D., Bando, Y., Huang, Y., Terao, T., Mitome, M., Tang, C., Zhi, C., (2010). Boron Nitride Nanotubes and Nanosheets, ACS Nano 4, 2979–2993.
25. Boeyens, J. C. A., Ramm, M., Zobel, D., Luger, P. (1995). Static disorder and packing in two orthorhombic crystal structures of fullerene inclusion compounds, S. Afr. J. Chem. 50, 28–33.
26. Terrones, M., Terrones, H., (1996). The role of defects in graphitic structures, Fullerene Sci. Tech. 4, 517–533.
27. Bawendi, M. G., Steigerwald, M. L., Brus, L. E., (1990). The quantum mechanics of larger semiconductor clusters ("Quantum dots"), Ann. Rev. Phys.Chem. 41, 477–496.
28. Murray, C. B., Norris, D. J., Bawendi, M. G., (1993). Synthesis and characterization of nearly monodisperse CdE (E=sulfur, selenium, telurium) semiconductor nanocrystallites, J. Am. Chem. Soc. 115, 8706–8715.
29. Cox, P. A., The Electronic Structure and Chemistry of Solids, Univ. Press, Oxford, 1987.
30. Schmid, G., Clusters and Colloids, VCH, Weinheim, 1994.
31. Alivisatos, A. P., (1996). Semiconductor Clusters, Nanocrystals and Quantum Dots, Science 271, 933–937.

32. Murray, C. B., Kagan, C. R., Bawendi, M. G., (1995). Self-organization of CdSe nanocrystallites into three-dimensional quantum dot super lattices, Science 270, 1335–1338.
33. Wang, Z. L., (2008). Splendid one-dimensional nanostructures of zinc oxide: A new nanomaterial family for nanotechnology, ACS Nano 2, 1987–1992.
34. G. A. de Vries, Brunnbauer, M., Hu, Y., Jackson, A. M., Long, B., Neltner, B. T., Uzun, O., Wunsch, B. J., Stellacci, F., (2007). Divalent metal nanoparticles, Science 315, 358–361.
35. Boeyens, J. C. A., The holistic molecule. 36.
36. J. C. A., Boeyens and, J. F. Ogilvie (eds) Models, Mysteries and Magic of Molecules, Springer.com, 2008.
37. Boeyens, J. C. A., Chemistry from First Principles, Springer.com, 2008.
38. Javey, A., (2008). The 2008 Kavli prize in Nanoscience: Carbon nanotubes, ACS Nano 2, 1329–1335.
39. Stone, A. J., Wales, D. J., (1986). Theoretical studies of icosahedral C60 and some related species, Chem. Phys. Lett. 128, 501–503.
40. Banhart, F., Kotakoski, J., Krasheninnikov, A. V., (2011). Structural defects in graphene, ACS Nano 5, 26–41.
41. Coulson, C. A., Valence, 2nd ed., Univ. Press, Oxford, 1961.
42. Boeyens, J. C. A., (2013). Covalent Interaction, Struct. Bond. 148, 93–135.
43. Coulson, C. A., Rédei, L. B., Stocker, D., (1962). The electronic properties of tetrahedral intermetallic compounds, I. Charge distribution, Proc. Roy. Soc. A270, 357–372.
44. Inerbaev, T. M., Masunov, A. E., Khondaker, S. I., Dobrinescu, A., A-Plamadă, V., Kawazoe, Y., (2009). Quantum chemistry of quantum dots: Effects of ligands and oxidation, J. Chem. Phys. 131, 044106–1, 5.
45. Boeyens, J. C. A., (2005). Quantum theory of molecular conformation, C. R. Chimie 8, 1527–1534.
46. Boeyens, J. C. A., (2010). A Molecular-Structure Hypothesis, Int. J. Mol. Sci. 11, 4267–4284.
47. Nisoli, C., Gabor, N. M., Lammert, P. E., Maynard, J. D., Crespi, V. H., (2009). Static and dynamic phyllotaxis in a magnetic cactus, Phys. Rev. Lett. 102, 186103: 1–4.
48. Stephenson, R. J., Mechanics and Properties of Matter, 2nd ed., Wiley, NY, 1960.
49. Weller, H., (1993). Colloidal semiconductor Q-particles: Chemistry in the transition region between solid state and molecules, Ang. Chem. Int. Ed. Eng. 32, 41–53.
50. Toda, M., Nonlinear Waves and Solitons, Kluwer, Dordrecht, 1989.
51. Boeyens, J. C. A., (2013). Calculation of Atomic Structure, Struct. Bond. 148, 71–91.
52. Yin, Y., Alivisatos, P., (2005). Colloidal nanocrystal synthesis and the organic–inorganic interface, Nature 437, 664–670.
53. Goldstein, A. N., Echer, M. C., Alivisatos, A. P., (1992). Melting in semiconductor nanocrystals, Science 256, 1425–1427.
54. Usković, V., (2008). Nanomaterials and nanotechnologies: Approaching the crest of the big wave, Curr. Nanosci. 4, 119–129.
55. Mullin, J. W., Crystallization, 2nd ed., Butterworths, London, 1972.
56. Rogach, A. L., Talapin, D. M., Shevchenko, E. V., Kornowski, A., Haase, M., Weller, H. (2002). Organization of matter on different size scales: monodispersive monocrystals and their superstructures, Adv. Funct. Mater. 12, 653–664.

57. Kittel, C., Introduction to Solid State Physics, 5th ed., Wiley, NY, 1976.
58. Mie, G., (1908). Beiträge zur Optik Trüber Medien, Speziell Kolloidaler Metallösungen, Ann. Phys. 25, 377–445.
59. Alvarez, M. M., Khoury, J. T., Schaaf, F. G., Shafigullin, M. N., Vezmar, I., Whetten, R. L., (1997). Optical absorption spectra of nanocrystal gold molecules, J. Phys. Chem B101 3706–3712.
60. Li, C., Zhang, X., Cao, Z., (2005). Triangular and Fibonacci Number Patterns Driven by Stress on Core/Shell Microstructures, Science 309, 909–911.
61. Ramm, M., Luger, P., Zobel, D., Duczek, W., J. C. A., (1996). Boeyens, Static disorder in hexagonal crystal structures of C60 at 100, K and 20, K, Cryst. Res. Technol. 31, 43–53.
62. Boeyens, J. C. A., Comba, P., Chemistry by Number Theory, Struct. Bond. 148, (2013). 1–24.
63. Boeyens, J. C. A., Levendis, D. C., Number Theory and the Periodicity of Matter, Springer.com, 2008.
64. Boeyens, J. C. A., Chemical Cosmology, Springer.com, 2010.
65. Dixon, R., (2000). Fibonacci phyllotaxis: mathematically speaking, in, Hargittai, I., T. C. Laurent (eds), Symmetry Portland Press, London, 2002.
66. Gross, L., Mohr, F., Moll, N., Schuler, B., Criado, A., Guitián, E., Peña, D., (2012). A Gourdon and, Meyer, G., Bond-order discrimination by atomic force microscopy, Science 337, 1326–1329.
67. Rehder, D., Chemistry in Space, Wiley-VCH, Weinheim, 2010.
68. Ho, D., (2009). Beyond the sparkle. The impact of nanodiamonds as biolabeling and therapeutic agents, ACS Nano, 3, 3825–3829.
69. Janner, A., Mysterious Crystallography: From snow flake to virus, 36, 233–254.
70. Papadopolos, Z., Gröning, O., Widmer, R., Clusters in F-phase icosahedral quasicrystals, 36, 255–282.
71. Lin, J., Cates, E., Bianconi, P. A., (1994). A synthetic analog of the biomineralization process, J. Am. Chem. Soc. 116, 4738–4745.
72. Munch, E., Launey, M. E., Alsem, D. H., Saiz, E., Tomsia, A. P., Ritchie, R. O., (2008). Tough bio-inspired hybrid materials, Science 322, 516–520.

CHAPTER 4

EXOTIC MULTI-SHELL NANOSTRUCTURES

MIRCEA V. DIUDEA

CONTENTS

ABSTRACT

Multi-shell nanostructures populate the Nanoworld in a wide variety of structures, spongy or filled ones. As constructive units, small cages can be used to design complex structures, of rotational or translational symmetry. In this chapter, the attention was focused on the design of multishell cages based on the Platonic solids and their transforms, obtained by using simple map operations. It is shown that the primary hyperstructures can self-arrange in even more complex arrays, expanded linearly (with 1-periodicity) or spherically, the majority of which belonging to quasicrystals. To prove the consistency of such molecular constructions, the calculation of genus (for spongy structures) and figure sum (for filled ones) was performed. Topological characterization of some 3-periodic networks is given in terms of Omega polynomial.

4.1 INTRODUCTION

Generalization of a polygon (2D) or a polyhedron (3D) to *n*-dimensions is a polytope. By using the Schlăfli symbols, convex regular polyhedra, also known as the Platonic solids: tetrahedron T, cube C, dodecahedron Do, octahedron Oct and icosahedron Ico, can be written as [3,3]; [4,3]; [5,3]; [3,4] and [3,5], respectively. The space filling by cubes can be written as [4,3,4], with four cubes meeting at every edge. In general, the [p, q, r] symbol means: *r* cages (of [*p*, *q*] type, p being the size of polygon while q is the vertex degree) meet at any edge of the net. For example, the polytope [3,3,5], that can be considered a 4D-analog of icosahedron, consists of 600 tetrahedral cells. Its dual is the regular polytope [5,3,3], the analog of dodecahedron, whose 120 cells are dodecahedra. These 4D polytopes, also called polychora, have been first described by Swiss mathematician Schlăfli [1].

To investigate the *n*-dimensional polytopes, a generalization of Euler's formula, also due to Schlăfli, is used

$$\sum_{i=0}^{n-1}(-1)^i f_i = 1-(-1)^n \tag{1}$$

In Eq. (1), f_i are elements of the *f*-vector ($f_0, f_1, \ldots, f_{n-1}$) of a convex polytope. In case $n = 4$, f_i are identified to vertices, edges, rings and cages, respectively, and Eq. (1) is written as

$$Sum(f_i) = f_0 - f_1 + f_2 - f_3 = 0 \tag{2}$$

In case $n = 3$, Eq. (2) is the well-known Euler (1758) formula [2]

$$v - e + f = 2 \tag{3}$$

Either related to the number of holes (in spongy structures, also called multi-tori) by the genus or counting the constituting figures (i.e., polytopes) in space filling structures, the above relations are useful in description of the spatiality of rather complex structures. In spongy structures, built up by tube junctions, the following theorem holds [3] demonstration coming out from construction.

Theorem 1: *The genus of a structure, built up from u tube junction units, of genus g_u, is calculated as: $g=u(g_u-1)+1$, irrespective of the unit tessellation.*

TOPO GROUP CLUJ has developed several software programs, dedicated to polyhedral tessellation (based on map operations) and embedment in surfaces of various genera, either as finite or infinite structures: TORUS, CageVersatile_CVNET, JSCHEM, OMEGA counter and NANO-Studio [4–8].

Structure design can be done either by using map operations [9–16] or by inflation–deflation operation, matching rules, the grid method, strip projection, cut projection or generalized dual method [17–20].

Structures are formally built up by joining small units/cages, eventually called fullerenes, of tetrahedral, octahedral or icosahedral symmetry. Recall the five Platonic solids, as dual pairs and the symbols used herein: Tetrahedron T (self-dual); Cube C and Octahedron Oct; Dodecahedron Do and Icosahedron Ico (Fig. 4.1).

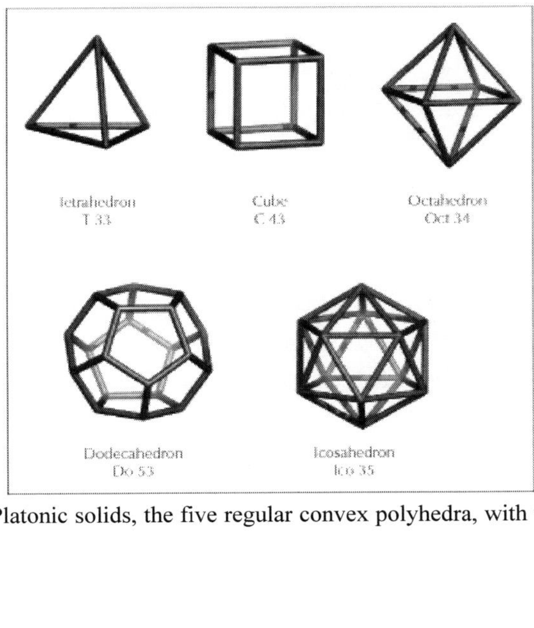

FIGURE 4.1 Platonic solids, the five regular convex polyhedra, with their symbols and vertex type.

Majority of the herein described multishell structures belong to quasicrystals that are highly ordered materials, with atomic clusters repeated in a complex, nonperiodic pattern. Even quasicrystals are aperiodic structures, they still possess long-range positional and orientational order. Among the rotational symmetries, 2-, 3-, 4- and 6-fold axes are allowed in crystals, while 5-, 7- and all higher rotations are disallowed but can appear in quasicrystals. Quasicrystals occur in various systems, such as metal alloys [21–23], binary systems (Fe_2O_3, Fe_3O_4, PbS, etc.) [24–26], liquid crystal dendrimers [27], and copolymers [28]. Very recently, the naturally occurring quasicrystalline mineral icosahedrite, $Al_{63}Cu_{24}Fe_{13}$, has been identified in a sample from the Khatyrka River in Chukhotka, Russia [29, 30].

There are also multishell structures that show translational symmetry, as in classical crystals. The 3-periodic networks are characterized in crystallographic terms but also by Omega polynomial [31]; to define this polynomial, some additional notions have to be introduced.

Let $G(V, E)$ be a connected graph, with the vertex set $V(G)$ and edge set $E(G)$. Two edges $e = uv$ and $f = xy$ of G are called *codistant e co f* if they obey the following relation [32].

$$d(v,x) = d(v,y) + 1 = d(u,x) + 1 = d(u,y) \tag{4}$$

which is reflexive, that is, *e co e* holds for any edge *e* of *G*, and symmetric, if *e co f* then *f co e*. In general, relation *co* is not transitive; an example showing this fact is the complete bipartite graph $K_{2,n}$. If "*co*" is also transitive, thus an equivalence relation, then *G* is called a *cograph* and the set of edges $C(e) := \{f \in E(G); f \ co \ e\}$ is called an *orthogonal cut oc* of *G*, *E*(*G*) being the union of disjoint orthogonal cuts: $E(G) = C_1 \cup C_2 \cup ... \cup C_k,\ C_i \cap C_j = \varnothing, i \neq j$.

Klavžar [33] has shown that relation *co* is a theta [34, 35] relation. Two edges *e* and *f* of a plane graph *G* are in relation *opposite, e op f,* if they are opposite edges of an inner face of *G*. Note that the relation *co* is defined in the whole graph while *op* is defined only in faces. Using the relation *op* we can partition the edge set of *G* into *opposite* edge *strips, ops*. An *ops* is a quasi-orthogonal cut *qoc*, since *ops* is not transitive.

Let *G* be a connected graph and $S_1, S_2, \ldots S_k$ be the *ops* strips of *G*. Then the *ops* strips form a partition of *E*(*G*).

Denote by *m*(*G*, s) the number of *ops* of length s. The Omega polynomial [31, 36–39] is defined as:

$$\Omega(G,x) = \sum_s m(G,s) \cdot x^s \tag{5}$$

Its first derivative (in x=1) equals the number of edges in the graph:

$$\Omega'(G,1)=\sum_{s} m(G,s)\cdot s=e=|E(G)| \tag{6}$$

On Omega polynomial, the Cluj-Ilmenau index [32], $CI=CI(G)$, was defined

$$CI(G)=\left\{[\Omega'(G,1)]^2-[\Omega'(G,1)+\Omega''(G,1)]\right\} \tag{7}$$

There are graphs with a single *ops*, of length $s=e=|E(G)|$, which is precisely a cycle. At the opposite side, there are graphs with s=1, namely graphs with either odd rings or with no rings, that is, tree graphs. For such graphs, minimal and maximal value, respectively, of CI is calculated [40].

$$\Omega(x)=1\cdot x^e;\, CI_{\min}=e^2-(e+e(e-1))=0 \tag{8}$$

$$\Omega(x)=e\cdot x^1;\, CI_{\max}=e^2-(e+0)=e(e-1) \tag{9}$$

The length of *ops* is taken as maximum. It depends on the size of the maximum fold face/ring F_{max}/R_{max} considered, so that any result on Omega polynomial will have this specification.

The chapter is organized as follows. After a short introduction, structures consisting of C_{20} building blocks are presented in the Section 4.2. The Section 4.3 provides examples of multishell cages derived from C_{24}. Section 4.4 describes in detail the use of tetrahedron and its truncated derivative in building of a variety of structures, expanded linearly or spherically (as multishell cages) and belonging to quasicrystals. Section 4.5 presents results on Omega polynomial count in four 3-periodic networks here investigated. Conclusions and references close the chapter.

4.2 STRUCTURES INCLUDING THE FULLERENE C_{20}

Fullerene C_{20}, or Do (by its associate Platonic solid), can be used as a constructive unit in the design of either spongy or filled structures, of formula $M_1(nM_2)$ and belonging to the three main point group symmetries. The design of such structures can be rationalized in terms of map operations (a map M is a discretized surface) as follows:

The *envelope* is obtained by S_2(M) map operation [41] while the hollow *core* can be drawn by P4TRS(M). Next, the two cages: S_2(M) (outside) and P_4TRS(M) (inside) are interconnected (by joining the closest pair atoms) to give SP4TRS@S2(M). The procedure is a general way to build a double-shell (spongy) structure by Do-units and it works on any tri-connected map. For the three cubic Platonic solids, the objects are shown in Fig. 4.2.

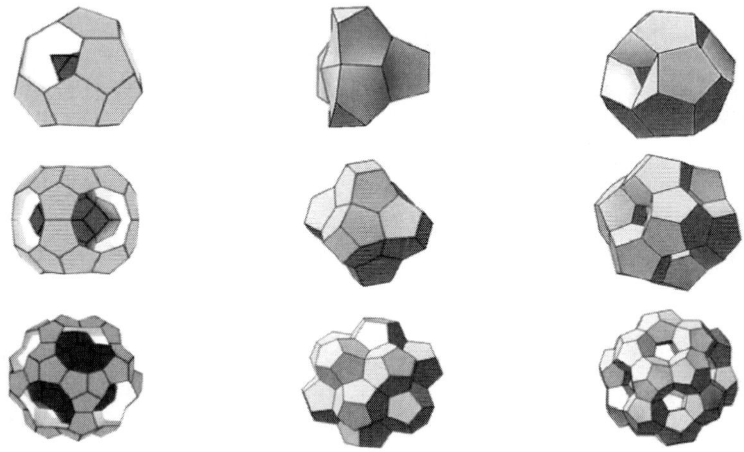

FIGURE 4.2 Top row: The envelope TS_2_28 (left) and the core TP_4TRS_44 (middle) of which joining leads to the spongy TP4TRS@TS2_50 = T(Do_4)_50 (right); Middle row: The envelope CS_2_56 (left), the core CP_4TRS_44 (middle) and the joining spongy CP4TRS@CS2_100 = C(Do_8)_100 (right); Bottom row: The envelope DoS_2_140 (left), the core DoP_4TRS_110 (middle) and the joining spongy DoP4TRS@DoS_2_250 = Do(Do_{20})_250 (right).

The structure Do(Do_{20})_250 is a spongy hyper-dodecahedron, 6-nodal 3,4-c $\{(5^6)_{60}[(5^5)_{30}(5^5)_{60}][(5^3)_{20}(5^3)_{20}(5^3)_{60}]\}$, made from 20 dodecahedral cells, a face-regular $5R_5$ map of genus g=11, its core being the 110-keplerate. The packing fraction $\phi = 20/33 \approx 0.6060$ is calculated with respect to the filled structure Do(Do_{20})_270 (i.e., Do(Do)(12Do)(Do_{20})_270, see Fig. 4.3 and Table 4.2). Comparing with the spheres maximum fraction ($\varphi = \pi/\sqrt{18} \approx 0.7405$ [42], it is clearly a spongy, nonconvex structure. Its five-fold symmetry is evident, even the pentagonal faces show some distortion (and strain) to the regular pentagon. For general regular polytopes the reader in invited to consult Refs. [43–46].

Note, the subscript numbers in round brackets give the number of units (from the core to envelope) consisting a multishell structure while the last number counts the heavy atoms in the whole structure.

The filled structures corresponding to the spongy ones shown in Fig. 4.2 are illustrated in Fig. 4.3. They are filled with the corresponding Platonic objects, so that only the icosahedral structure $Do(Do_{33})\_270$ has both the building blocks and core the same Do (Fig. 4.3, right).

Filled structures have been described by several authors [47–51] and they are connected to the sphere-packing problem [52–57].

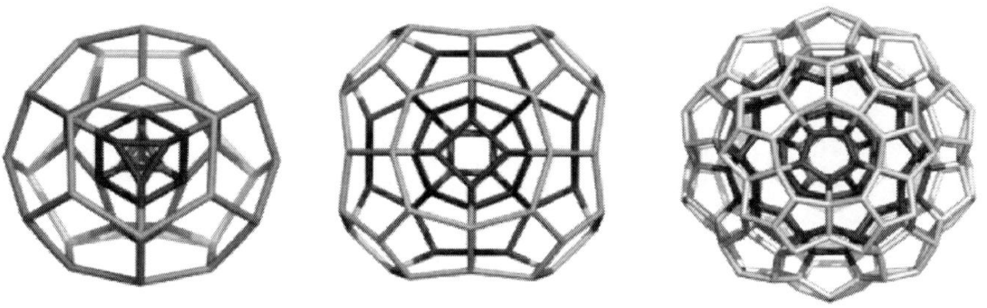

FIGURE 4.3 Filled structures: $T@T(Do_4)\_54$ (left), $C@C(Do_8)\_108$ (middle), and $Do@Do(Do_{20})\_270$ (right).

Genus calculation for the spongy structures of Fig. 4.2 is given in Table 4.1. For the corresponding filled structures, the figure calculation is shown in Table 4.2. As hydrocarbons, the spongy structures are more stable than the filled ones [58].

TABLE 4.1 Genus Calculation in Spongy Structures of Fig. 4.2.

Spongy structure	v	e	f_5	$f_w(s)$	f	χ	g	g_u	u	$g=f_{op\text{-}}1$
TP_4TRS	22	36	12	4(3)	12	-2	2			
CP_4TRS	44	72	24	6(4)	24	-4	3			
DoP_4TRS	110	180	72	12(5)	60	-10	6			
$T(Do_4)\_50$	50	90	42	4(3)	36	-4	3	1.5	4	3
$C(Do_8)\_100$	100	180	84	6(4)	72	-8	5	1.5	8	5
$Do(Do_{20})\_250$	250	450	222	12(5)	180	-20	11	1.5	20	11

TABLE 4.2 Figure Calculation in Radial Space Filling Structures of Fig. 4.3 cf

$$Sum(f_i) = f_0 - f_1 + f_2 - f_3 = 0\, ;\, f_3 = c = u + 2$$

Space filling structure	v	e	u	r_3	r_4	r_5	c	$Sum(f_i)$
T@T(Do_4)_54	54	100	(4+4)	8	0	48	10	0
C@C(Do_8)_108	108	200	(8+6)	0	12	96	16	0
Do@Do(Do_{20})_270	270	500	(20+12)	0	0	264	34	0

The structures in Fig. 4.2 the middle row, can form 3-periodic nets that join to each other in the same manner as the finite units, to form a CP4TRS@CS2_Id network (in short C(Do)Id, Fig. 4.4) or, by considering the optimized unit in Fig. 4.4 right, an optimized C(Do)Opt-type lattice (Fig. 4.5) can be designed. These represent new crystal-type networks.

C(Do)Id: group *Im*-3 *m*; Point symbol for net: $\{4.5^8.8\}3\{5^3\}2\{5^5.8\}3$; 3,4,5-c net with stoichiometry (3-c)2(4-c)3(5-c)3; 3-nodal net.

C(Do)Opt: group *Pm*-3 *m*; Point symbol for net: $\{4.5^{10}.6.7^2.8\}3\{4.5^5\}6\{5^3\}4\{5^5.8\}6$; 3,3,4,4,4,6-c net with stoichiometry (3-c)4(4-c)12(6-c)3; 6-nodal net.

FIGURE 4.4 Connecting of 3-Periodic networks to form an infinite C(Do)Id-type lattice: CS_2_333_1008 (left); CP_4TRS_333_1080 (middle) and CP4TRS@CS2_Id_333_2052 (right).

FIGURE 4.5 Net (left) and co-net (middle) units of the optimized C(Do)Opt-type lattice: CP4TRS@CS2Opt_100 (left); CP4TRS@CS2Opt_co_124 (middle) and CP4TRS@CS2Opt_333_2268 (right).

The case of arrays C_{20}&C_{28} (Fig. 4.6) is related to the *mtn* network, called ZSM-39, or clathrate II, or *fcc*_C_{34} or diamond D_5 [59–62]. The hollow core of Do($(C28)_{20}$)_380 is identical to that of Do(Do_{20})_250, supporting the idea of the space filling by Do($(C28)_{20}$)_380 can evolve either linearly or radially, as shown in Fig. 4.10 bottom. Note that Do($(C28)_{20}$)_*k* represents a new 1-periodic 21-nodal 3,4-c net, belonging to the group *Pmma*; the point symbol for this net is $\{5^2.6\}12\{5^3\}12\ \{5^4.6.8\}19\ \{5^5.6\}11$.

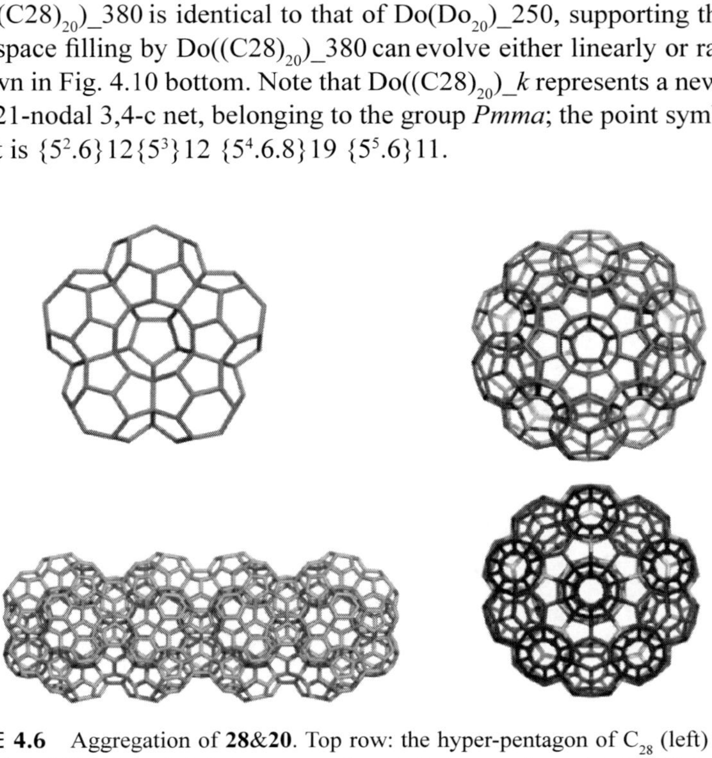

FIGURE 4.6 Aggregation of **28&20**. Top row: the hyper-pentagon of C_{28} (left) and the Do($(C28)_{20}$)_380 unit (right); Bottom row: a 1-periodic (spongy) array Do($(C28)_{20}$)_4_1190 (left) and an icosahedral (filled) array C20@((C20)12)@((C28)20;(C20)12)_460 (right).

4.3 STRUCTURES INCLUDING THE FULLERENE C_{24}

Truncated octahedron TO/C_{24} shares hexagons with itself to form the C_{60}@ $((C_{10})12;(C_{24})_{20})$_300 unit (Fig. 4.7, top, right), of which windows are the pentagonal prism P_5 (C_{10}) while the core is the fullerene C_{60}. Examples of linear and radial evolution of C_{60}@$((C_{10})_{12};(C_{24})_{20})$_300 are given in Fig. 4.7, bottom. C_{60}@$((C_{10})_{12};(C_{24})_{20})$_k represents a new 12-nodal 3,4-c 1-periodic net of the *Cmcm* group; its net symbol is $\{4.6^2\}6\{4^2.5.6^3\}11\{4^2.6^4\}4$. In such arrays, the component cages C_{60} do not touch to each other. The radial space filling in this case is presented in Table 4.3.

TABLE 4.3 Radial Space Filling by Cages C60&C24&C10. Structures: C60@((C10)12;(C24)20)_300; C60@((C10)12;(C24)20)@((C10)30;(C24)20;(C60)12); $Sum(f_i) = f_0 - f_1 + f_2 - f_3 = 0$

Shell	v	e	r_4	r_5	r_6	r	24	10	C_{24}	C_{10}	c+1	$Sum(f_i)$
0	60	90	0	12	20	32	0	0	0	1	2	0
1	300	540	120	24	130	274	20	12	32	1	34	0
2	900	1620	270	156	390	816	40	42	82	13	96	0

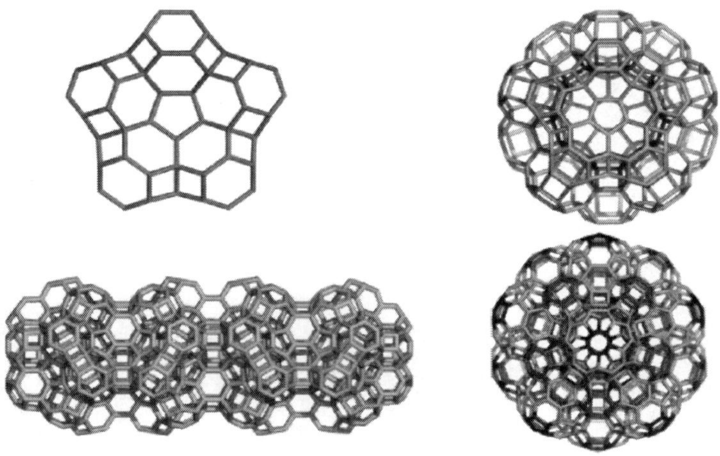

FIGURE 4.7 Aggregation of C60&C24&C10. Top row: the hyper-pentagon of C_{24}=TO (left) and the C_{60}@$((C_{10})_{12};(C_{24})_{20})$_300 unit (right); Top row: a 1-periodic, linear array C_{60}@$((C_{10})_{12};(C_{24})_{20})$_4_930 (left) and a spherical, icosahedral array C_{60}@$((C_{10})_{12};(C_{24})_{20})$@$((C_{10})_{30};(C_{24})_{20};(C_{60})_{12})$_900 (right).

4.4 STRUCTURES INCLUDING THE TETRAHEDRON

Space filling by the tetrahedron (C_4=T) and truncated tetrahedron (C_{12}=TT=Friauf polyhedron) has been discussed in the literature [48, 63–65] Tetrahedrally close packed structures are encountered in Ref. [47] phases (of intermetallic compounds) and in zeolites as well (with SiO_4 or $SiAlO_4$ tetrahedral). A cell of the 600-cells icosahedron-like 4D [3,3,5] polytope consists of five tetrahedral joined around a line.

A structure ID@T20_50 (Fig. 4.8, left) can be easily designed by stellateing the icosidodecahedron (i.e., the medial of icosahedron and dodecahedron – see also [16]. It can be truncated (Fig. 4.8, middle) to form a new unit C60@TT20_150 (Fig. 4.8, right). The 1-periodic structures derived from these units are illustrated in Fig. 4.9. They represent linear aggregation of C_{30}=ID and C_{60} fullerenes, respectively.

The structure ID@T20_*k* is a new 4-nodal, 3,6,6,6-c net of the *P*2/*m* group; its point symbol is: {3^3}2{3^6.5^2.6^6.7}4{3^6.5^3.6^6}. The structure C60@TT20_*k* is a new 8-nodal 3,4-c net of the *Pmma* group; it has the symbol net $\{3.5.6^4\}11\{3.6^2\}6\{3.6^4.10\}4$.

In the idea of 4D icosahedron, Fig. 4.10 illustrates arrays of the Friauf polyhedron (i.e., truncated tetrahedron TT), within the icosahedral symmetry. The structure Ico@TT20_84 has the hollow core an icosahedron but no windows/open faces to visit the core; its envelope is a truncated icosahedron (Fig. 4.10, left - see also the Samson's 104 cluster). An experimental realization of this structure was brought by Breza et al. [66] with a beautiful icosahedral diamond. A rod-like structure Ico@TT20_4_240 was designed [67] in this respect (Fig. 4.10, right; see also Ref. [50].

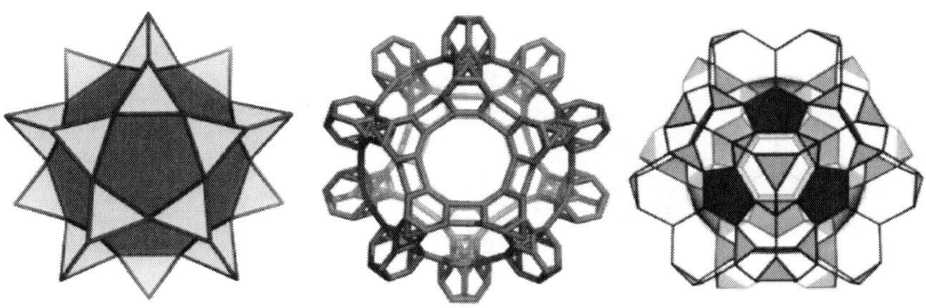

FIGURE 4.8 Unit ID@T20_50 (5-fold sym. left) is truncated (5-fold sym. middle) to form C60@TT20_150 (3-fold sym. right) by identifying trigonal faces.

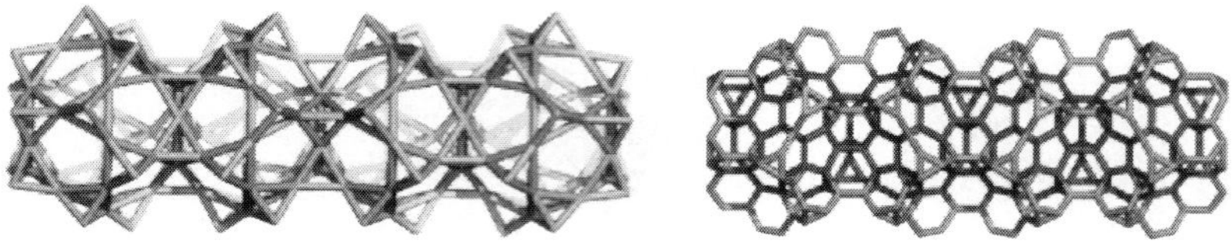

FIGURE 4.9 Rod-like structures derived from C30&C4 and C60&C12: ID@T20_4_155 (left) and C60@TT20_4_465 (right), respectively.

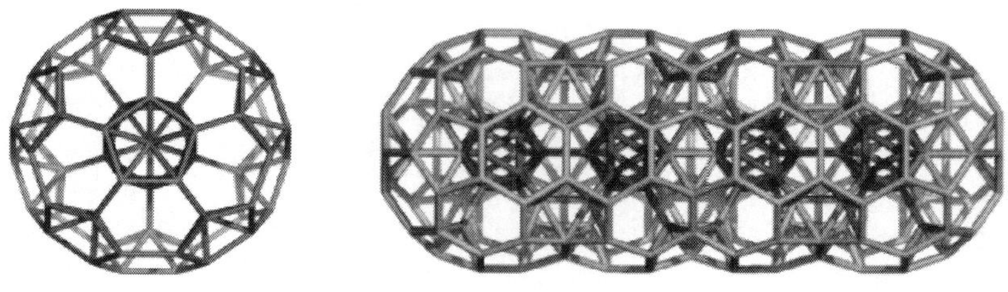

FIGURE 4.10 Icosahedral diamond unit Ico@TT20_84 (left) and a corresponding rod structure Ico@TT20_4_240 (right).

Note, Ico@TT20_*k* is a new 1-periodic net, of the group *P21/m*. The point symbol for this 5-nodal 4,5,6-c net is: $\{3^2.4.6^3\}10$ $\{3^3.4^2.5.6^4\}5\{3^5.4^5.6^5\}11$.

Radial aggregation of C30&C4 system is illustrated in Fig. 4.11. The unit ID@T20_50 forms two hyper-faces F_5 and F_6, respectively (Fig. 4.11, top). These substructures arranges as the atoms and faces, respectively, in the Buckminster fullerene C_{60} to form the hyper-structure C60((ID@T20)50)_1650 (Fig. 4.11, bottom), of which substructure content is given in Table 4.4. By faces, there are the following classes of atoms: (3^3)=300 (the external ones; atom degree 3); $(3^6.5^2)$=720; $(3^6.5^3)$=510; $(3^6.5^2.6)$=120; all these with atom degree 6. By the layer matrix of faces (and centric index: 1×{30} (of type $3^6.5^3$, counting the 15 C_2 axes); 13×{60} and 7×{120} classes are distinguished.

FIGURE 4.11 Unit ID@T20_50 (two-fold symmetry, top, left) and two hyper-faces F_5-175 (top, middle)and F_6-210 (top, right) aggregate to the icosahedral C_{60}-like structure C60((ID@T20)50)_1650 (bottom).

TABLE 4.4 Figure Counting for the Radial Aggregation of ID@T20_50 (structure 1) to the Icosahedral Hyper-structure C60((ID@T20)50)_1650 (structure 4) Intermediated by the Hyper-faces F_5-175 (Structure 2) and F_6-210 (Structure 3)

$$Sum(f_i) = f_0 - f_1 + f_2 - f_3 = 0$$

Structure	v	e	f_3	f_5	f_6	T	Str. 1	R	C_all	Sum(f_i)
1	50	120	80	12	0	20				
2	175	450	300	56	0	75				
3	210	540	360	66	1	90				
4	1650	4500	3000	642	20	750	60	3662	812	0

Medial transform of the multishell dodecahedron-containing structures (Fig. 4.12, left and middle) contain multitetrahedral substructures (Fig. 4.12, right).

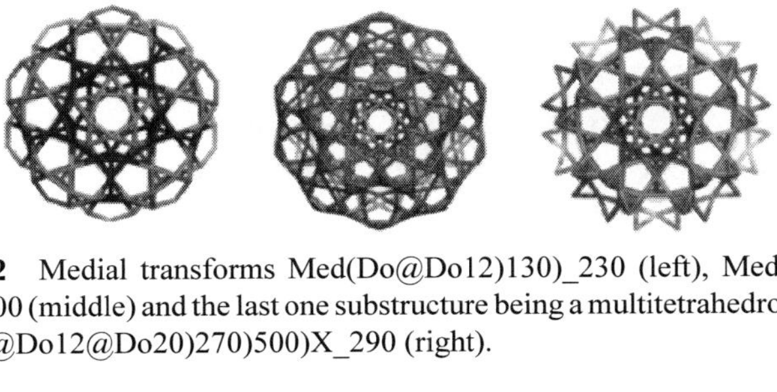

FIGURE 4.12 Medial transforms Med(Do@Do12)130)_230 (left), Med(Do@Do12@Do20)270)_500 (middle) and the last one substructure being a multitetrahedron substructure MT(Med(Do@Do12@Do20)270)500)X_290 (right).

Radial aggregation of C60&C12 system (i.e., C_{60} and truncated tetrahedron TT) can be easily designed by applying the map operations, for example, *Le*, on the shell-structures (see Fig. 4.13).

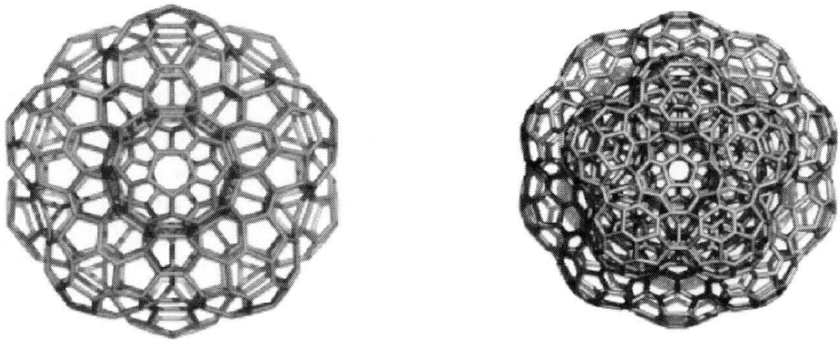

FIGURE 4.13 Radial aggregation by C60&C12, within leapfrog *Le* map operation: Le((Do@Do12)130)_570=C60@(C12)20@((C12)20;(C60)12_570 (left) and Le((Do@Do12@Do20)270)_1320 = C60@(C12)20@((C12)20;(C60)12)@((C12)30;(C12)60)@(C60)20_1320.

Energies of (hydrogenated) carbon structures designed by (C60&C24&C10) and (C60&C12) (Figs. 4.6 to 4.13) have been computed by the DFTB+ method [68]; the most stabilization by aggregation (reference molecule C_{60}) among all the structures herein discussed was shown by the (C60&C24&C10)-structures, where the C_{60}-units behave almost as the isolated icosahedral fullerene C_{60} [58].

Let truncate a double-shell tetrahedron DT (Fig. 4.14, left), the external one, and we have a tetrahedron in a truncated tetrahedron, for example, TDT_16 (Fig. 4.14, right). This could be used to build various structures, as shown in Fig. 4.15 to 4.17. Following the idea used in building Ico@TT20_84, one can design a truncated pentagonal bipyramid containing inside tetrahedral (Fig. 4.15, left), which empty corresponding structure is shown in Fig. 4.15, right.

The structures in Fig. 4.15 can self-assembly by identifying a hexagonal face, once above, once under, to form circular structures of 5-fold symmetry (Fig. 4.16). These circular structures can form cylindrical/tubular arrays, as shown in Fig. 4.17.

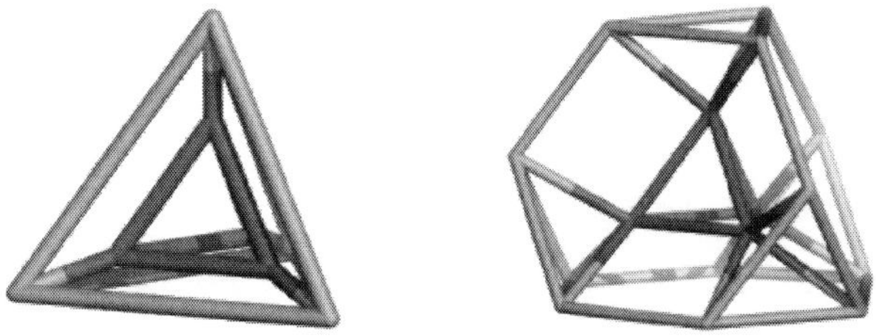

FIGURE 4.14 A double-shell tetrahedron DT_8 (left) and its truncated derivative TDT_16 (right).

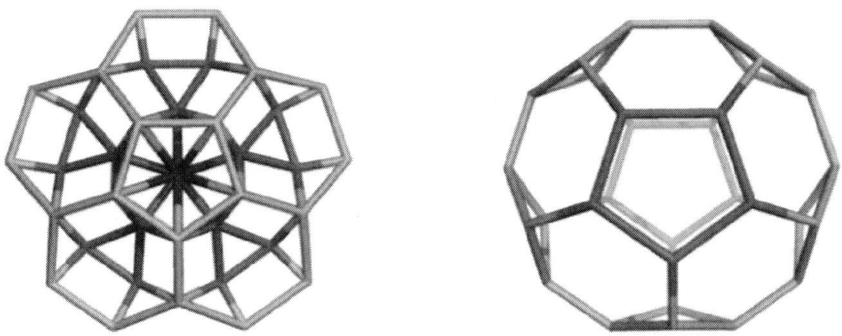

FIGURE 4.15 A truncated pentagonal bipyramid, filled by tetrahedral TT_5_52 (left) and its empty structure TT_5_30 (right).

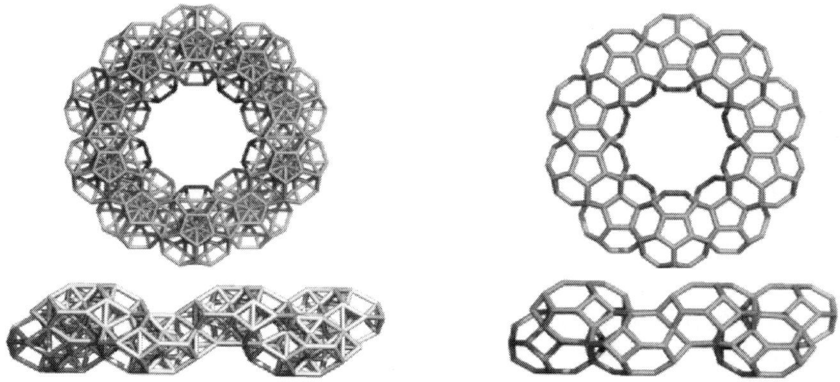

FIGURE 4.16 Circular alternating structures constructed by using the units in Fig. 4.15. TT_5F10_460 (left) and TT_5X10_240 (right).

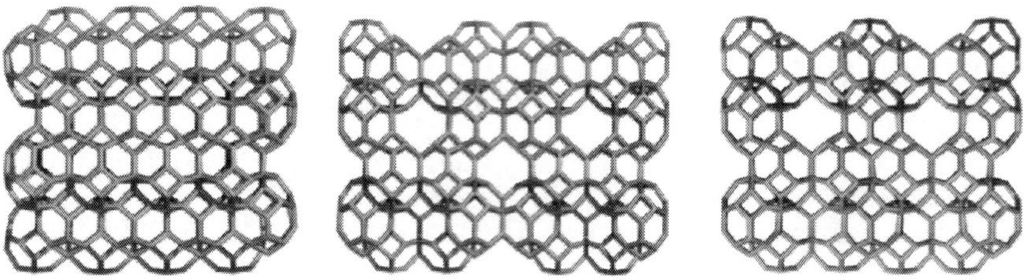

FIGURE 4.17 Tubular arrays of circular structures in Fig. 4.16 TT_5XA-5,1,4_750 (left); TT_5XB-5,1,4_885 (middle) and TT_5XC-5,1,4_755 (right).

Stability of these structures, patched by the coranulenne flower, both in hydrogenated and nonhydrogenated forms, approaches (by our calculations) that of C_{60}, known being the stability of the corannulene molecule. They could be challenges for the future syntheses.

4.5 OMEGA POLYNOMIAL IN 3-PERIODIC NETWORKS

Results on Omega polynomial calculation in the two 3-periodic networks illustrated in Figs. 4.4 and 4.5 (C(Do)Id and C(Do)Opt, respectively) and in some additional ones, designed as follows. Take the unit of the C(Do)Opt-co-net (Fig. 4.5, middle) and change the pentagon triple (lying in each of the eight corners of the cube) by a hexagon, next delete all the edges, except the squares on the cube faces, already connected to the new hexagons; it results in the cage named CCOpt_72 (Fig. 4.18, left). This unit forms a 3-periodic net (Fig. 4.18, middle), known as *ssd-e* (topos&RCSR.ttd), of the group *Pm*-3 *m*. The point symbol for net is: $\{4^2.6^4\}2\{4^3.6^3\}$, a 2-nodal 4-c net, with stoichiometry (4-c)3. This net is an aggregation of the truncated octahedron.

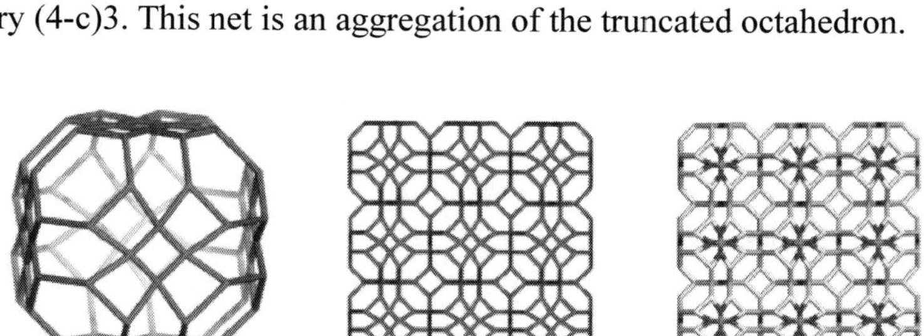

FIGURE 4.18 Unit CCOpt_72 (left) forms a 3-periodic net, known as *ssd-e* (in our symbols CCOpt_333_1296, middle) by identifying the cross-decorated faces; it can be modified by deleting the edges of the face central square and joining the 2-connected atoms to a new central atom (C(HPM)Opt_333_ 1404, right).

The next network is C(HPM)Opt, where H and P stands for hexagon and pentagon, respectively while M is a metal atom closing the pentagons (Fig. 4.18, right). It is derived from the CCOpt/*ssd-e* net by deleting the edges of the face central square while its atoms are being connected to an additional atom, as in plane-square metal complexes in inorganic chemistry. This is a new 3-nodal 3,4-c net, of the group *Pm*-3 *m*, with the point symbol $(4.5.6^4)_8(5^2.6)_4(5^4.8^2)$ and the repeating unit C(HPM)Opt_78.

The unit CCOpt_72 (Fig. 4.18, left) can self-arrange by identifying the hexagonal faces, in "syn" and "anti" modes, respectively, thus resulting a 3-periodic C(6_{anti})Opt net (Fig. 4.19, left) and a hyper-dodecahedron Do(6 sin) Opt_1260 (Fig. 4.19, middle), of which hyper-pentagon Cy5(6 syn)Opt_330 is also illustrated in Fig. 4.19, right). This is a clear example of a quasicrystal (in the five-fold symmetry view) and its 3-periodic approximant. Note, C(6_{anti}) Opt is a new 9-nodal, 3,4,5-c net of *Cmmm* group, point symbol for net: $(4.6^2)_4(4^2.6^3.8)_3(4^3.6^3)_5(4^3.6^6.8)(4^4.6^2)$.

A Poincaré dual of CCOpt_72, namely CCDuOpt_42 (Fig. 4.20, left), by identifying the cross-face and deleting the cross, form a 3-periodic net, known as *rhr* (or sqc5544 – topos&RCSR.ttd) and symbolized here CCDuX (Fig. 4.20, middle), a uninodal net of the group *Im*-3 *m*, and having the point symbol $(4^2.6^2.8^2)$; its repeating unit is detailed in (Fig. 4.20, right).

In case of hexagonal face identification, the intermediate "6_{syn}"-dimer can form a hyper-pentagon (similar to that in Fig. 4.19, right) and next the corresponding hyper-dodecahedron Do(DuX(6_{syn}))_540 (Fig. 4.21, left and middle). The five-fold symmetry is clearly seen in the core (Fig. 4.21, right) of this hyper-dodecahedron. By the layer matrix of faces (and centric index [11]), nine classes of atoms are recognized: 1×{120}; 6×{60} and 2×{30}.

This is another example of translational-rotational pair structures derived from one and the same repeating unit (in this case CCDuX_36, Fig. 4.20, right).

For the 3-periodic structures, Omega polynomial was computed by our software Nano-Studio [7]; analytical formulas, derived by a numerical analysis, are given in Tables 4.5 to 4.16. Since the length of *ops* depends on the maximum size of ring searched, R_{max} will be always indicated. The calculations are made on cuboids of dimensions (*k*, *k*, *k*), where *k* is the number of repeating units along the edge of such a cubic domain and the formulas are function of this net parameter. Examples are also given.

FIGURE 4.19 The 3-periodic net C(6_{anti})Opt_1656 (left), can be seen as the "approximant" of the hyper-dodecahedron Do(6_{syn})Opt_1260 (middle); the corresponding hyper-pentagon Cy_5(6_{syn})Opt_330 is also shown (right).

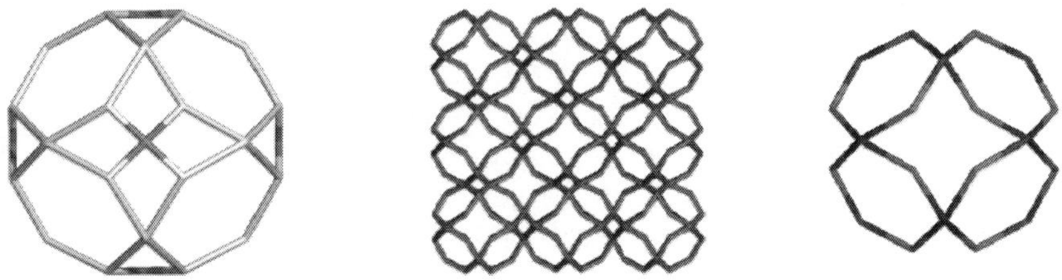

FIGURE 4.20 A dual of CCOpt_72 (the unit CCDuOpt_42, left) and the 3-periodic *rhr*-net (i.e., CCDuX_333_756, middle) derived by deleting the cross in the identified faces of the above unit; the repeating unit of *rhr*-net (i.e., CCDuX_36) is detailed (right).

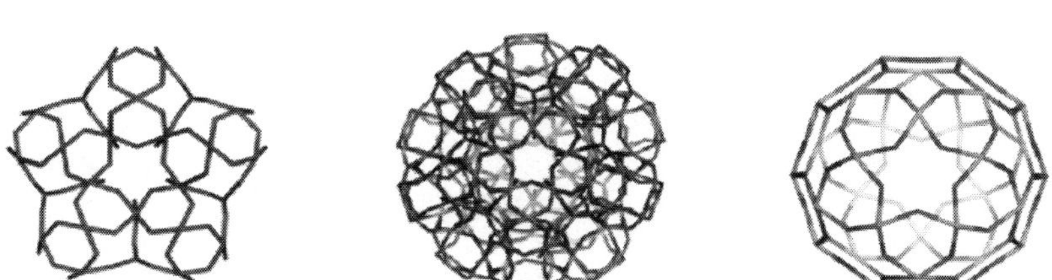

FIGURE 4.21 A hyper-pentagon Cy5DuX(6_{sin})_150 (left) in the hyper-dodecahedron, Do(DuX(6_{syn}))_540 (middle) with its detailed Do(DuX(6_{syn}))Core_90 (right).

TABLE 4.5 Formulas for Omega Polynomial and Related Topological Data in C(Do) Opt Net

Structure	Formulas
C(Do)Opt R_{max}=5	$\Omega(C(Do)Opt) = 12k(10k^2+4k-1)\cdot x^1 + 2\cdot f_4 \cdot x^2$ $= 12k(10k^2+4k-1)\cdot x^1 + 6k(3k^2-2k+1)\cdot x^2$ $e(C(Do)Opt) = 12k^2(13k+2)$ $CI(C(Do)Opt) = 12k(2028k^5+624k^4+48k^3-16k^2-1)$ $v(C(Do)Opt) = 4k^2(19k+6)$ $f_4(C(Do)Opt) = 6k^3+3k(k-1)^2 = 3k(3k^2-2k+1)$ $f_5(C(Do)Opt) = 4k^2(18k+3)$

TABLE 4.6 Omega Polynomial and Related Data for the C(Do)Opt Net.

C(Do)Opt net; R_{max}=5						
k	**v**	**e**	**f_4**	**f_5**	**Omega**	**CI**
1	100	180	6	84	$156x+12x^2$	32,196
2	704	1344	54	624	$1128x+108x^2$	1,804,776
3	2268	4428	198	2052	$3636x+396x^2$	19,601,964
4	5248	10,368	492	4800	$8400x+984x^2$	107,483,088
5	10,100	20,100	990	9300	$16,140x+1980x^2$	403,985,940
6	17,280	34,560	1746	15,984	$27,576x+3492x^2$	1,194,352,056

TABLE 4.7 Formulas for Omega Polynomial and Related Topological Data in (C(Do)Id Net

Structure	Formulas
(C(Do)Id R_{max}=5	$\Omega(C(Do)Id) = 12k(9k^2+5k-1)\cdot x^1 + 2\cdot f_4 \cdot x^2$ $= 12k(9k^2+5k-1)\cdot x^1 + 6k(2k^2-k+1)\cdot x^2$ $e(C(Do)Id) = 12k^2(11k+4)$ $CI(C(Do)Id) = 12k(1452k^5+1056k^4+192k^3-13k^2-1)$ $v(C(Do)Id) = 4k^2(16k+9)$ $f_4(C(Do)Id) = 3k(2k^2-k+1)$ $f_5(C(Do)Id) = 4k^2(18k+3)$

TABLE 4.8 Omega Polynomial and Related Data for the C(Do)Id Net

C(Do)Id net; R_{max}=5						
k	**v**	**e**	**f_4**	**f_5**	**Omega**	**CI**
1	100	180	6	84	$156x+12x^2$	32,196
2	656	1248	42	624	$1080x+84x^2$	1,556,088
3	2052	3996	144	2052	$3420x+288x^2$	15,963,444
4	4672	9216	348	4800	$7824x+696x^2$	84,924,048
5	8900	17,700	690	9300	$14,940x+1380x^2$	313,269,540
6	15,120	30,240	1206	15,984	$25,416x+2412x^2$	914,422,536

TABLE 4.9 Formulas for Omega Polynomial and Related Topological Data in CCOpt Net

Structure	Formulas
CCOpt R_{max}=6	$\Omega(CCOpt)=12\sum_{i=1}^{k-1}x^{i(8k+4)}+6x^{k(8k+4)}+4x^{6k^2(k+1)}$ $e(CCOpt)=24k^2(3k+2)$ $CI(CCOpt)=16k(315k^5+398k^4+119k^3-12k^2-8k-2)$ $v(CCOpt)=36k^2(k+1)$ $f_4(CCOpt)=3k(7k^2+k+2)$ $f_6(CCOpt)=2k(7k^2+3)$

TABLE 4.10 Omega Polynomial and Related Data in CCOpt Net

CCOpt; R_{max}=6						
k	***v***	***e***	**f_4**	**f_6**	**Omega**	**CI**
1	72	120	30	20	$6x^{12}+4x^{12}$	12,960
2	432	768	192	124	$12x^{20}+6x^{40}+4x^{72}$	554,688
3	1296	2376	612	396	$12x^{28}+12x^{56}+6x^{84}+4x^{216}$	5,369,376
4	2880	5376	1416	920	$12x^{36}+12x^{72}+12x^{108}+6x^{144}+4x^{480}$	27,637,632

5	5400	10,200	2730	1780	$12x^{44}+12x^{88}+12x^{132}+12x^{176}+6x^{220}+4x^{900}$	99,812,640
6	9072	17,280	4680	3060	$12x^{52}+12x^{104}+12x^{156}+12x^{208}+12x^{260}+6x^{312}+4x^{1512}$	287,085,120

TABLE 4.11 Formulas for Omega Polynomial and Related Topological Data in C(HPM) Opt Net

Structure	Formulas
C(HPM)Opt	
$R_{max}=6$	$\Omega(C(HPM)Opt) = 12k^2(k+1)x + 24kx^2 + 24k(k-1)x^3 + 6k(k-1)^2x^4 + 12(2k-1)x^{3k+1} + 6(k-1)^2x^{2(3k+1)}$
	$e(C(HPM)Opt) = 72k^3 + 48k^2$
	$CI(C(HPM)Opt) = 5184k^6 + 6912k^5 + 2088k^4 - 36k^3 - 24k^2 - 24k - 12$
	$v(C(HPM)Opt) = 39k^2(k+1)$
	$f_4(C(HPM)Opt) = 6k(k-1)^2$
	$f_5(C(HPM)Opt) = 12k^2(k+1)$
	$f_6(C(HPM)Opt) = 2k(7k^2+3)$

TABLE 4.12 Omega Polynomial and Related Data for the C(HPM)Opt Net

C(HPM)Opt R_{max}=6							
k	**v**	**e**	**f_4**	**f_5**	**f_6**	**Omega**	**CI**
2	468	768	12	144	124	$144x+48x^2+48x^3+12x^4+36x^7+6x^{14}$	585,924
3	1404	2376	72	432	396	$432x+72x^2+144x^3+72x^4+60x^{10}+24x^{20}$	5,626,608
4	3120	5376	216	960	920	$960x+96x^2+288x^3+216x^4+84x^{13}+54x^{26}$	28,843,284
5	5850	10,200	480	1800	1780	$1800x+120x^2+480x^3+480x^4+108x^{16}+96x^{32}$	103,899,768
6	9828	17,280	900	3024	3060	$3024x+144^2+720x^3+900x^4+132x^{19}+150x^{38}$	298,309,668

TABLE 4.13 Formulas for Omega Polynomial and Related Topological Data in $C(6_{anti})$ Opt Net

Structure	Formulas
$C(6_{anti})$Opt $R_{max}=6$	$\Omega(C(6_{anti})Opt) = 2k^3x^{12} + 5k^2x^{10k+2} + k^2x^{24k} + kx^{4k(2k+1)}$
	$e(C(6_{anti})Opt) = 2k^2(53k+7)$
	$CI(C(6_{anti})Opt) = 4k^2(2809k^4 + 726k^3 - 236k^2 - 126k - 5)$
	$v(C(6_{anti})Opt) = 8k^2(16+7(k-2))$
	$f_4(C(6_{anti})Opt) = 2k^2(31+16(k-2))$
	$f_6(C(6_{anti})Opt) = 4k^2(9+4(k-2))$

TABLE 4.14 Omega Polynomial and Related Data in $C(6_{anti})$Opt Net

					$C(6_{anti})$Opt; $R_{max}=6$	
k	***v***	***e***	f_4	f_6	**Omega**	**CI**
2	512	904	248	144	$16x^{12}+20x^{22}+4x^{48}+2x^{40}$	792,816
3	1656	2988	846	468	$54x^{12}+45x^{32}+9x^{72}+3x^{84}$	8,806,464
4	3840	7008	2016	1088	$128x^{12}+80x^{42}+16x^{96}+4x^{144}$	48,722,112
5	7400	13,600	3950	2100	$250x^{12}+125x^{52}+25x^{120}+5x^{220}$	183,984,000
6	12,672	23,400	6840	3600	$432x^{12}+180x^{62}+36x^{144}+6x^{312}$	545,475,312

TABLE 4.15 Formulas for Omega Polynomial and Related Topological Data in CCDuX/*rhr* Net

Structure	Formulas
CCDuX/*rhr* $R_{max}=6$	$\Omega(CCDuX)=\sum_{i=1}^{k}(24k\cdot x^{4(i-1)+2})$ $e(CCDuX)=48k^3$ (all edges belong to hexagons that are edge-disjoint) $CI(CCDuX)=32k^2(72k^4-4k^2+1)$ $v(CCDuX)=12k^2(3+2(k-1))$ $f_4(CCDuX)=12k^2(k-1)$ $f_6(CCDuX)=8k^3$

TABLE 4.16 Omega Polynomial and Related Data in CCDuX/*rhr* Net

					CCDuX/rhr; R_{max}=6	
k	*v*	*e*	f_4	f_6	**Omega**	CI
1	36	48	—	8	$24x^2$	2208
2	240	384	48	64	$48x^2+48x^6$	145,536
3	756	1296	216	216	$72x^2+72x^6+72x^{10}$	1,669,536
4	1728	3072	576	512	$96x^2+96x^6+96x^{10}+96x^{14}$	9,404,928
5	3300	6000	1200	1000	$120x^2+120x^6+120x^{10}+120x^{14}+120x^{18}$	35,920,800
6	5616	10,368	2160	1728	$144x^2+144x^6+144x^{10}+144x^{14}+144x^{18}+144x^{22}$	107,330,688

4.6 CONCLUSION

Multi-shell nanostructures were designed by using small cages as constructive units. The discussed complex structures show either rotational symmetry, as in quasicrystals, or translational symmetry, as in classical crystals. In this chapter, the attention was focused on the design of multishell cages based on the Platonic solids and their transforms using simple map operations. It was shown that the primary hyperstructures, either spongy or filled ones, can self-arrange in even more complex arrays, expanded linearly or spherically (as multishell cages). The consistency of such molecular constructions is proved by the calculation of genus (for spongy structures) and figure sum (for filled ones).

Genus/figure calculation and topological symmetry were the main goals of this study while the operations on maps, implemented in our CVNET software, were used in the structure design. The study on topology (equivalence classes, counted by the layer matrix of faces around each atom) and structure assembling were performed by the Nano Studio software package. Omega polynomial was calculated, as additional information, for the 3-periodic lattices, by using the Nano Studio software.

ACKNOWLEDGMENTS

Thanks are addressed to Professor Davide Proserpio, University of Milano for the crystallographic data while for the design of some structures, to Dr. Csaba L. Nagy and Ing. Adrian Militaru.

KEYWORDS

- **nanostructures**
- **Omega polynomials**
- **periodic networks**
- **Platonic related solids**
- **spongy structures**
- **topological data**

REFERENCES

1. Schläfli, L., (1901). Theorie der vielfachen Kontinuität Zürcher und Furrer, Zürich, (Reprinted In: Ludwig Schläfli, 1814–1895, Gesammelte Mathematische Abhandlungen, Band 1, 167–387, Verlag Birkhäuser, Basel, 1950).
2. Diudea, M. V., Schefler, B. (2012). Nanotube junctions and the genus of multitori. Phys Chem Chem Phys 14, 8111–8115.
3. Diudea, M. V., Parv, B., Ursu, O. (2003). TORUS. Babes-Bolyai Univ, Cluj.
4. Stefu, M., Diudea, M. V. (2005). CageVersatile_CVNET, Babes–Bolyai Univ, Cluj.
5. Nagy, C. L., Diudea, M. V. (2005). JSChem, Babes–Bolyai Univ, Cluj.

6. Cigher, S., Diudea, M. V. (2006). Omega Counter. Babes-Bolyai Univ, Cluj.
7. Nagy, C. L., Diudea, M. V. (2009). NANO–Studio, Babes–Bolyai Univ, Cluj.
8. Diudea, M. V. (2003). Capra-a leapfrog related operation on maps. Studia Univ "Babes-Bolyai," 48, 3–16.
9. Diudea, M. V. (2004). Covering forms in nanostructures. Forma (Tokyo) 19, 131–163.
10. Diudea, M. V. (2005b), Nanoporous carbon allotropes by septupling map operations. J Chem Inf Model 45, 1002–1009.
11. Diudea, M. V. (2010a) Diamond D_5, a novel allotrope of carbon, Studia Univ Babes-Bolyai, Chemia, 55, 11–17.
12. Diudea, M. V., Ştefu, M., John, P. E., Graovac, A. (2006). Generalized operations on maps. Croat Chem Acta 79, 355–362.
13. Eberhard, V. (1891). Zur Morphologie der polyeder. B. G. Teubner, Leipzig.
14. Fowler, P. W. (1986). How unusual is C_{60}? Magic numbers for carbon clusters. Chem Phys Lett 131, 444–450.
15. Goldberg, M. (1937). A class of multisymmetric polyhedral. Tôhoku Math, J., 43, 104–108.
16. Pisanski, T., Randić, M. (2000). Bridges between geometry and graph theory, Geometry at Work. MAA Notes 53, 174–194.
17. de Bruijn, N. G. (1981). Algebraic theory of Penrose's nonperiodic tilings of the plane, I, II. Proc Nederl Acad Wetensch Proc Ser, A., 84, 39–66.
18. Kramer, P. (1982). Non–periodic central space filling with icosahedral symmetry using copies of seven elementary cells. Acta Cryst, A., 38, 257–264.
19. Bak, P. (1986). Icosahedral crystals, Where are the atoms? Phys Rev Lett 56, 861–864.
20. Socolar JES, Steinhardt, P. J., Levine, D. (1986). Quasicrystals with arbitrary orientational symmetry. Phys Rev, B., 32, 5547–5551.
21. Shechtman, D., Blech, I., Gratias, D., Cahn, J. W. (1984). Metallic phase with long–range orientational order and no translational symmetry. Phys Rev Lett 53, 1951–1953.
22. Tsai, A. P., Inoue, A., Masumoto, T. (1987). A stable quasicrystal in Al–Cu–Fe system, Jpn, J., Appl Phys 26, L1505–L1507.
23. Tsai, A. P., Guo, J. Q., Abe, E., Takakura, H., Sato, T. J. (2000). A stable binary quasicrystal. Nature 408, 537–538.
24. Shevchenko, E. V., Talapin, D. V., Kotov, N. A., O'Brien, S., Murray, C. B. (2006a) Structural diversity in binary nanoparticle superlattices. Nature 439, 55–59.
25. Shevchenko, E. V., Talapin, D. V., Murray, C. B., O'Brien, S. (2006b) Structural characterization of self–assembled multifunctional binary nanoparticle superlattices. J Am Chem Soc 128, 3620–3637.
26. Talapin, D. V., Shevchenko, E. V., Bodnarchuk, M. I., Ye, X., Chen, J., Murray, C. B. (2009). Quasicrystalline order in self–assembled binary nanoparticle superlattices. Nature 461, 964–967.
27. Zeng, X, Ungar, G, Liu, Y, Percec, V, Dulcey, A E, Hobbs, J K (2004). Supramolecular dendritic liquid quasicrystals. Nature 428, 157–160.
28. Hayashida, K., Dotera, T., Takano, T., Matsushita, Y. (2007). Polymeric quasicrystal, mesoscopic quasicrystalline tiling in ABS star polymers. Phys Rev Lett 98, 195502.
29. Bindi, L., Steinhardt, P. J., Yao, N., Lu, P. J. (2009). Natural Quasicrystals. Science 324, 1306–1309.

30. Bindi, L., Steinhardt, P. J., Yao, N., PJ Lu (2011). Icosahedrite, $Al_{63}Cu_{24}Fe_{13}$, the first natural quasicrystal. Amer Mineralog, 96, 928–931.
31. Diudea, M., V (2006). Omega polynomial. Carpath, J., Math 22, 43–47.
32. John, P., Vizitiu, A., Cigher, S., Diudea, M. (2007). CI Index in Tubular Nanostructures. MATCH Commun Math Comput Chem 57, 479–484.
33. Klavžar, S. (2008). MATCH Some comments on co graphs and, C. I., inde4. Commun Math Comput Chem 59, 217–222.
34. Djoković, D. (1973). Distance preserving subgraphs of hypercubes. Combin, J., Theory Ser, B., 14, 263–267.
35. Winkler, P. (1984). Isometric embedding in products of complete graphs. Discrete Appl Math 8, 209–212.
36. Diudea, M. V., Cigher, S., John, P. (2008). Omega and Related Counting Polynomials. MATCH Commun Math Comput Chem 60, 237–250.
37. Diudea, M. V., Cigher, S., Vizitiu, A., Florescu, M., John, P. (2009). Omega polynomial and its use in nanostructures description. J Math Chem 45, 316–329.
38. Diudea, M. V., Klavžar, S. (2010). Omega polynomial revisited. Acta Chim Sloven 57, 565–570.
39. Diudea, M. V., Ilić, A., Medeleanu, M. (2013). Omega polynomial in hyperdiamonds. In: Diudea, M. V., Nagy, C. L. (Eds), Carbon Materials, Chemistry and Physics. Vol. 6, Diamond and Related Nanostructures. Springer, Dordrecht, Chapter 9, pp. 169–190.
40. Diudea, M. V. (2010a) Nanomolecules and Nanostructures – Polynomials and Indices, MCM, No 10, Univ Kragujevac, Serbia.
41. Diudea, M. V. (2005a), Nanoporous carbon allotropes by septupling map operations. J Chem Inf Model 45, 1002–1009.
42. Hales, T. C. (2005). A proof of the Kepler conjecture. Ann Math 162, 1065–1185.
43. Grünbaum, B. (1967). Convex polytopes. Wiley, New York.
44. Coxeter HSM (1973). Regular polytopes. 3rd edn. Dover Publications, New York.
45. Wells, A. F. (1977). Three–dimensional nets and polyhedral. Wiley, New York.
46. Ziegler, G. M. (1995). Lectures on polytopes. Springer–Verlag, New York.
47. Frank, F. C., Kasper, J. S. (1958). Complex alloy structures regarded as sphere packings. Definitions and basic principles. Acta Cryst 11, 184–190.
48. Haji-Akbari, A., Engel, M., Keys, A. S., Zheng, X., Petschek, R. G., Palffy–Muhoray, P., Glotzer ShC (2010). Disordered, quasicrystalline and crystalline phases of densely packed tetrahedral. arXiv, 10125138.
49. Dandoloff, R., Döhler, G., Bilz, H. (1980), Bond charge model of amorphous tetrahedrally coordinated solids. J Non–Cryst Solids 35–36, 537–542.
50. Shevchenko VYa (2011). Search in chemistry, biology and physics of the nanostate. Lema, St Petersburg.
51. Zeger, L., Kaxiras, E. (1993). New model for icosahedral carbon clusters and the structure of collapsed fullerite. Phys Rev Lett 70, 2920–2923.
52. Mackay, A. L. (1962). A dense non–crystallographic packing of equal spheres. ActaCrystallogr 15, 916–918.
53. Coxeter HSM (1958). Close-packing and so forth. Illinois, J., Math 2, 746–758.
54. Coxeter HSM (1961). Close packing of equal spheres. In: Coxeter HSM (ed) Introduction to Geometry. 2nd edn. Wiley, New York, pp 405–411.

55. Goldberg, M. (1971). On the densest packing of equal spheres in a cube. Math Mag 44, 199–208.
56. Hales, T. C. (1992). The sphere packing problem. J Comput Appl Math 44, 41–76.
57. Hales, T. C. (2006). Historical overview of the Kepler conjecture. Discrete Comput Geom 36, 5–20.
58. Bende, A., Diudea, M. V. (2013). Energetics of multishell cages. In: Diamond and related nanostructures. Springer, Dordrecht, Chapter 6, pp. 105–117.
59. Benedek, G., Colombo, L. (1996). Hollow diamonds from fullerenes. Mater Sci Forum, 232, 247–274.
60. Blasé, X., Benedek, G., Bernasconi, M. (2010). Structural, mechanical, and superconducting properties of clathrates. In: Colombo, L., Fasolino, A. (eds) Computer-based modeling of novel carbon systems and their properties. Carbon Materials, Chemistry and Physics 3, Springer, Dordrecht.
61. Diudea, M. V. (2010b) Diamond D_5, a novel allotrope of carbon, Studia Univ Babes-Bolyai, Chemia, 55, 11–17.
62. Diudea, M. V., Nagy, C. L. (2012). All pentagonal ring structures related to the C_{20} fullerene, Diamond D_5. Diam Relat Mater 23, 105–108.
63. Pearson, W. B. (1972). The crystal chemistry and physics of metals and alloys. Wiley, New York.
64. Conway, J., H, Torquato, S. (2006). Packing, tiling and covering with tetrahedral. Proc Natl Acad Sci 103, 10612–10617.
65. Kallus, Y., Elser, V., Gravel, S. (2009). A dense periodic packing of tetrahedral with a small repeating unit. arXiv, 09105226.
66. Breza, J., Kadlečikova, M., Vojs, M., Michalka, M., Vesely, M., Danis, T. (2004). Diamond icosahedron on a TiN-coated steel substrate. Microelectron, J., 35, 709–712.
67. Diudea, M. V. (2013). Quasicrystals, between spongy and full space filling. In: Diudea, M. V., Nagy, C. L. (Eds), Carbon Materials, Chemistry and Physics. Vol. 6, Diamond and Related Nanostructures. Springer, Dordrecht, Chapter 19, pp. 333–383.
68. Aradi, B., Hourahine, B., Frauenheim, T. (2007). $DFTB^+$, a sparse matrix-based implementation of the DFTB method. J Phys Chem, A., 111, 5678–5684.

CHAPTER 5

THE SELF-ASSEMBLY OF PORPHYRIN DERIVATIVES INTO 2D AND 3D ARCHITECTURES

MIHAELA BIRDEANU and EUGENIA FAGADAR-COSMA

CONTENTS

ABSTRACT

Nowadays the Atomic Force Microscopy (AFM) might be considered the most applied SPM technique, used mainly for the topographic analysis of the surfaces. Even if the AFM technique is recently invented (only in 1986) is a very high-resolution probe-scanning microscope, at least 1000 times better than the optical microscope resolution. With the help of AFM, areas between 1 nm^2 and 100 μm^2 can be scanned. The AFM can operate in various modes depending on the specimen application and/or characteristics. Usually, the AFM modes, which will produce the imagistic surface, can be split into static (also known as "contact") and other types of variations modes ("non-contact"). The thin films of porphyrin derivatives (bare or incorporated in diverse hybrid nanomaterials) increased in attention due to their amazing applications in medicine (photosensitizers in PDT), as catalysts, chemical sensors, optical actuators and in optoelectronic devices. Due to existing potential for balancing hydrophobic/hydrophilic properties of porphyrins by different peripheral substitution with functional groups, the porphyrin derivatives are among the best candidates for self-assembling by weak Van der Waals forces or hydrophobic effects, hydrogen bond and π–π stacking interactions that finally generates a large variety of geometries. These are often topologically closed systems that might properly be analyzed by AFM technique. Based on AFM images, this chapter emphasizes the influence of the specific structure of the porphyrin macrocycle, its concentration and the nature of the solvent on the self-aggregation and organization of porphyrins. In this respect, several symmetrical and unsymmetrical substituted A_3B *meso* and β-porphyrins were studied varying the used solvents (CH_3Cl, THF and EtOH). AFM put into evidence that the J aggregates (edge-to-edge stacked) and H-bonding interactions are both responsible for generating of different architectures, that are sometimes coexisting together, such as: triangles type sheets, concave-convex surfaces, columns, pyramids, roofs and even rings. Taking into consideration that modern research is deeply concerned about the air-solvent interface of new multifunctional micro and nano-materials, AFM is one of the most important techniques to reveal deep inside the topography of surfaces, especially porphyrins.

5.1 INTRODUCTION IN ATOMIC FORCE MICROSCOPY

5.1.1 BRIEF HISTORY AND MAIN CHARACTERISTICS

Nowadays, the most often used methods for investigating surfaces are the Scanning Electron Microscopy (SEM) and the Scanning Probe Microscopy (SPM) [1–8].

The design and construction of Scanning Electron Microscopy (SEM) was preceded by the STM (Scanning Tunneling Microscopy) invention in 1981 at Zurich Research Laboratory by Binning and Rohrer, the winners of Physics Nobel Prize in 1986 for their contribution to science development.

Scanning probe microscopy of materials (SPM) is a revolutionary technique that allows scientists to obtain 3D images of surfaces at atomic scale, from 100μm to 10pm [9].

The SPM is a measurement family of techniques, which imply the scanning of a surface with a very sharp scanning tip and the monitoring of the scanning tip – surface interaction in order to create a high-resolution image of the investigated material. Many other SPM techniques were developed in order to provide information related to friction force, adherence, elasticity, hardness, electric field, magnetic field, electrical carrier concentration, temperature distribution, electrical resistance and conductivity, etc., due to the fact that the access to the physical properties of the investigated surfaces is fast.

The Atomic Force Microscopy (AFM) became the most applied SPM technique, used mainly for the topographic analysis of the surfaces [9–11]. The AFM technique was invented in 1986 and made available commercially in 1989 by the developers of Digital Instruments. [12].

The AFM is a very high resolution probe scanning microscope, a resolution demonstrated onto nanometer fractions, that is, at least 1000 times better than the optical microscope resolution [13]. Figure 5.1 presents the AFM positioning between the most often used surface imagistic techniques.

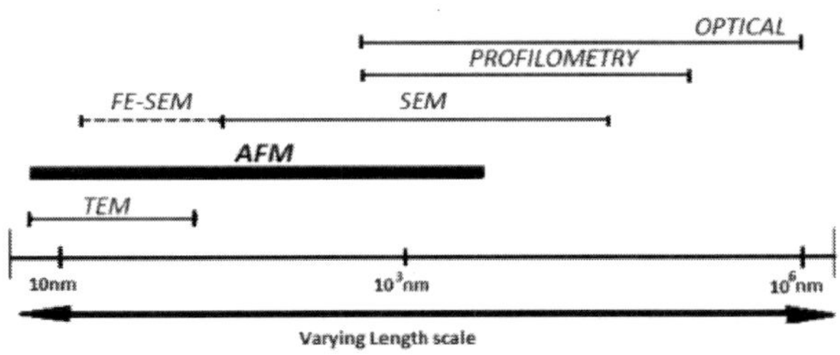

FIGURE 5.1 Scanning domain and the most used imagistic techniques.

With the help of AFM, areas between 1 nm^2 and 100μm^2 can be scanned. Generally, the AFM is equipped with two types of scanners: one that can scan up to 5μm and the second one that scans up to 100μm. The scanning speed can be adjusted through the specialized software, between 0.1 Hz and 100 Hz.

When selecting the scanning speed, one should know that an optimum response time exist for each system. The number of included points contained in the resulting image after the scanning can be also selected through the specialized software (e.g., 128 × 128, 256 × 256, 512 × 512, or 1024 × 1024).

Beside the response time, one should also know that there are other factors that can influence the measurement and the consequent results: dust, grease layer thickness, the water that can accumulate, under certain circumstances, onto the studied surface, etc.

Another important factor that can have an important influence onto the measurement and the obtained results is the noise.

The use of the "microscope" term in AFM is rather improper since it implies observation, while for AFM the information is gathered through "sensing" the surface with a mechanical probe.

A set of piezo-electrical elements facilitate the slow but correct and precise movement, having as result an uniform and detailed scanning of the surface [1–5].

5.1.2 BASIC PRINCIPLES

The AFM contains a console microscale, which includes a tip (peak) whose end is used to scan the surface. The console is usually made of silica or silica nitrate and has a peak's curvature radius in the nanometric domain.

When the tip is in contact with the surface, various forces appear between the tip and the sample, forces which in turn will produce a deviation of the console, according with the Hooke's law.

Depending of the actual situation, the forces measured by the AFM are mechanical, Van der Waals, capillary, chemical, electrostatic, magnetical, etc. Usually, the console deviation is measured by using a laser beam, which reflects onto the upper part of the console into a photodiode system.

Other methods used for evaluating the deviation are based onto optical interferometry, capacitative or piezoresistive phenomena.

These latter ones are realized of piezoresistive elements which are acting as tenso-metrical measurement system. A Wheatstone bridge is used for measuring the console deviation but this method is not as sensitive as the use of laser beam deviation or the interferometrical deviation [1–6].

If the tip scans at a constant height, the risk of touching the surface exists and thus to degrade / affect the surface, but to minimize this risk, a feedback mechanism is used to constantly adjust the tip – specimen distance and a constant interaction between the two.

Usually, the tip is located into a piezoelectric tube which restricts its movements to the z axis only with a constant force, while on x and y axis, respectively; the scanning of the specimen will be realized.

At the same time, a tripod configuration can be used, including three piezoelectrical crystals which will provide the necessary movements onto the x, y and z axis, respectively. This method does avoid some of the distortions which appear when using the piezoelectric tube only.

The resulting map $s = f(x, y)$ represents the topography of the investigated specimen. Figure 5.2 presents the block diagram of the AFM device.

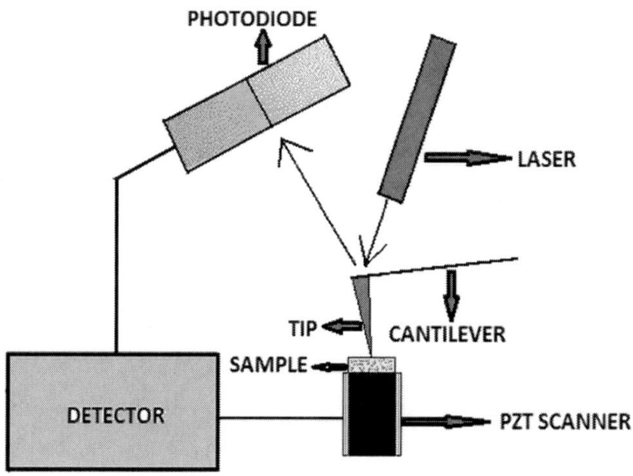

FIGURE 5.2 AFM block diagram.

The main components of an AFM are:

- a cantilever which includes a very sharp tip (peak).
- a scanner which controls the x-y-z axis positioning.
- a semiconductor laser.
- a photodetector.
- a reaction control circuit.

In terms of design, **the peak** is sharp, usually having a diameter of 5 μm and 10 nm in height, and is located at the free end of a **cantilever**, 100–500 μm long. Interaction forces between the tip and sample surface causes deflection or deformation of cantilever and a detector is used to measure cantilever deflection as the peak scans the sample. The signal received by the **(photo) detector** is transmitted to a computer that will generate a map of surface topography [1].

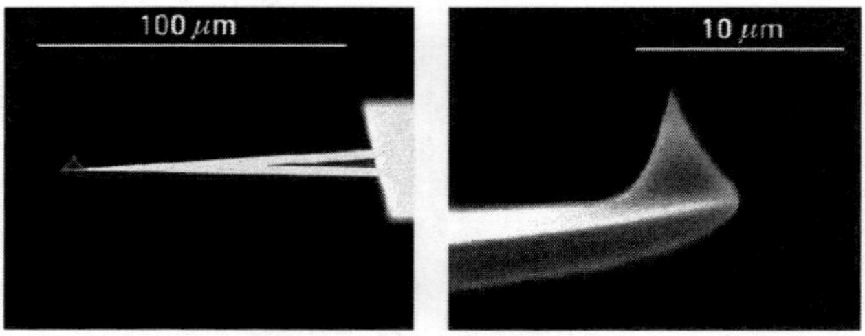

FIGURE 5.3 Images of scanning probe (cantilever and tip) made with SEM (Scanning Electron Microscopy) [8].

The **scanner** device provides precise control of the peak position and is made of piezoelectric ceramic, whose geometry changes when a voltage is applied, and applied voltage is proportional to the resulting mechanical movement.

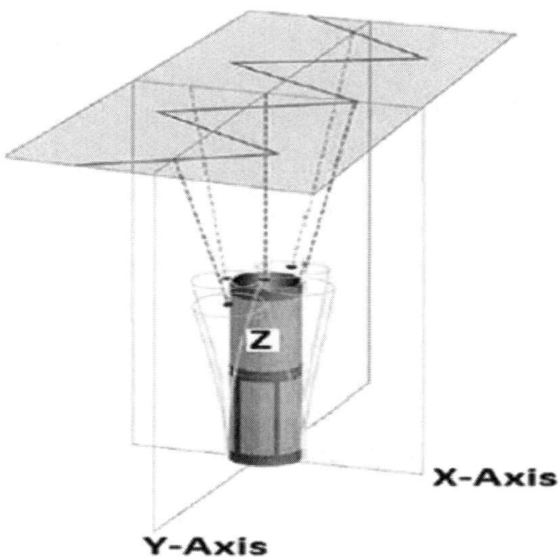

FIGURE 5.4 The piezoelectric scanner [2].

Most of the atomic force microscopes are using optical techniques to detect the cantilever position. A ray of light emanating from a laser diode is projected on the cantilever surface and its reflection onto a position sensitive photodetector. As the cantilever sags, the laser radiation detector position is changing. The signal amplification is given by the ratio of the size of the detector distance from the cantilever and cantilever length. As a result, the system can detect movements in the open end of the cantilever as small as 1 Å, from the position where the tip is located.

5.1.3 WORKING MODES OF THE AFM

The AFM can operate in various modes depending on the specimen application and/or characteristics. Usually, the AFM modes which will produce the imagistic surface can be split into static (also known as "contact") and other types of variations modes ("noncontact").

5.1.3.1 CONTACT MODE AFM

Is the most common operation mode of the AFM and as its name suggests, the peak and the specimen remain in close contact from the start of the scanning until the end of the scanning. Through "contact" one understands a "rejection" regime of the intermolecular force curvature, as presented in Fig. 5.5.

The rejection region of the curvature is located above the x axis. One of the main disadvantages of remaining in close contact with the specimen is the fact there are a large number of lateral forces which will exert their couple onto the cantilever, while it is moved onto the specimen's surface.

The main characteristics of the contact mode are [13]:

- it provides the 3d information visualization about a specimen's surface in a nondestructive way
  - 1.5 nm lateral resolution
  - 0.05 nm vertical resolution
- Powerful rejection forces act between the peak and the specimen
- Requires a rather simple preparation procedure of the specimen
- Dielectrics and conductors can be analyzed by using this mode
- It provides information about the physical properties of the specimen
  - Elasticity, adhesion, hardness, friction, etc.

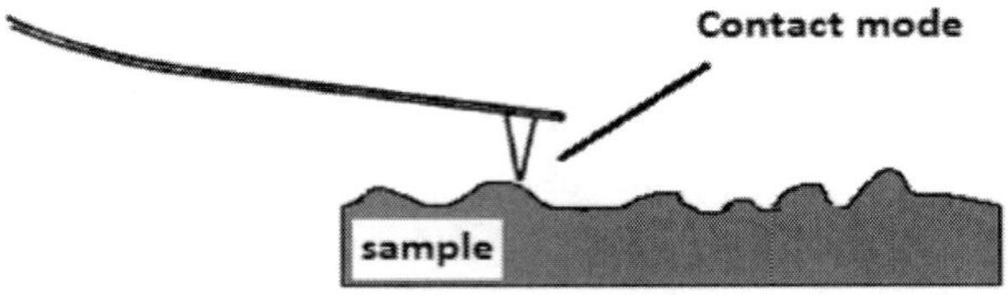

FIGURE 5.5 Contact mode.

5.1.3.2 TAPPING MODE AFM

Is the second most often used mode, following the contact mode. When operating in air or in other media, the cantilever oscillates at its own specific resonance frequency (hundreds of KHz) and it is positioned above the surface is such a way that it will get into contact with it only a small fraction of time of the oscillation period.

This mode is also a contact mode, as the one described previously, but with considerably reduced lateral forces appearing during the peak's movement over the surface due to the reduced time of direct contact.

This mode is a better solution against the contact mode when one investigates specimens who cannot be fixed firmly or specimens with a reduced hardness for visualizing the topographical map of the specimen.

By using constant mode force, the reaction is adjusted in such a way that the cantilever oscillation remains constant.

An image can be obtained through this amplitude signal, due to the fact that small variation of oscillation's amplitude will take place due to the fact that the electronic circuits will not react instantaneously according to the changes appearing on the specimen's surface.

Recently, a special attention was given to the phases imagistic – a procedure that works by measuring the phase difference between the piezo device which controls the cantilever and the detected oscillations.

One considers that this obtained image contrast is derived from properties such as rigidity and viscoelasticity.

The main characteristics of the tapping mode [13]:

- the phase difference can measure properties related to composition, adhesion, friction, elasticity.
- the structure of the polymers mixture can be identified.

- it is less damaging for the specimens for low hardness specimens compared with the contact mode.

Figure 5.6 shows Tapping mode scanning.

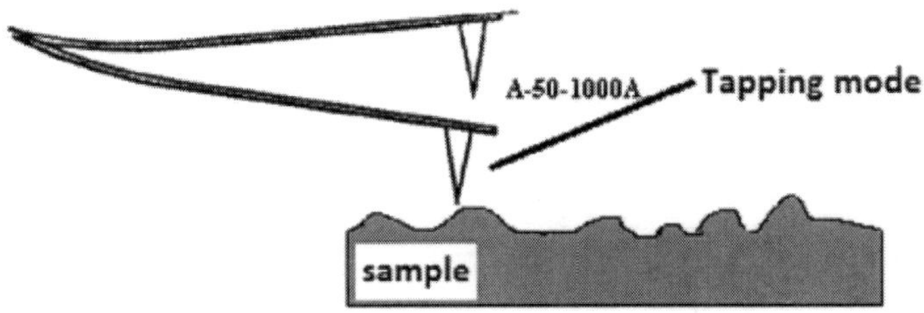

FIGURE 5.6 Tapping mode.

5.1.3.3 NON-CONTACT MODE AFM

This mode represents another choice of scanning method for investigating a specimen with the AFM.

The cantilever scans above the specimen's surface, at a distance at which the rejection regime is not established any more, which makes difficult the operation in usual ambient conditions. In the ambient conditions, a thin layer of water will be produced on the specimen's surface, which will continuously try to form a capillary bridge between the peak and the specimen causing the peak to be "tempted" to jump into the contact mode (Fig. 5.7.) [13].

Even with liquid or in vacuum, the fact that the peak may jump from time to time in the contact mode with the specimen presents an important probability, and thus the obtained image may be finally an image obtained through a tapping mode derivative method. The successful application of this method depends greatly onto the specimen's consistency and the medium in which the investigation takes place.

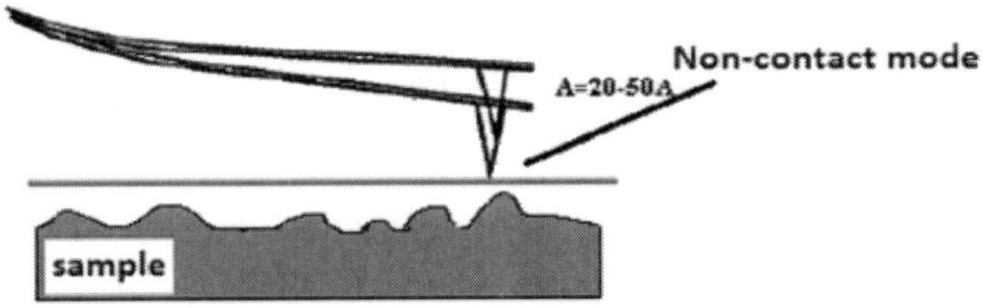

FIGURE 5.7 Non-contact mode.

5.1.4 IMAGE FORMATION METHODS

The main operation modes are: static force and dynamic force mode, respectively.

5.1.4.1 STATIC FORCE MODE

In the *static force operation mode* the static deviation of the peak is used as a feedback signal. For clearing out the signal of noises and signal displacements more rigid consoles are used. Anyway, in the nearby area of the specimen's surface, strong enough attraction forces exist which make the tip of the console to get in contact with it.

This is the reason why the static AFM mode is used only in contact with surfaces that produce repulsion interactions. That is also the reason why, sometimes, the static mode is called also the contact mode. With this mode, the interaction between the peak and the surface is constant during the scanning, by means of maintaining a constant deviation.

5.1.4.2 DYNAMIC FORCE MODE

In the *dynamic force operation mode*, the console oscillates near the resonance value. The variation of the amplitude, the phase and the resonant frequency is influenced by the peak – specimen interaction, the modification of the oscillation (taking into account the reference oscillation) is actually providing information about the specimen's characteristics.

The schematics for operating into dynamic mode include frequency modulation and most often amplitude modulations. The frequency can be measured precisely and its modulation permits the use of very rigid consoles.

The rigid consoles offer better stability very close to the specimen's surface and, as a result, this AFM technique was used first to offer information at a true atomic resolution, in ultra-high vacuum conditions.

In terms of amplitude modulation and the amplitude or phase oscillations – these are used to provide the feedback signal in order to obtain the image. The phase oscillation modifications are used to determine the various materials types present on the surface.

The modifications of the amplitude can be determined even if there is no contact or in an intermediary contact regime. Keeping the peak very close to the surface without touching it represents a major issue, which was solved by using the dynamic contact mode. In this mode, the console oscillates as it would come in direct contact with the surface of the specimen at every cycle, but the recovery force will keep it from actually getting in contact with the specimen's surface.

The amplitude modulation was used in a noncontact regime to obtain atomic resolution by using very rigid consoles and very small amplitudes in ultra-high vacuum [1–5].

5.1.4.3 PHASE CONTRAST MODE

In the *phase contrast mode*, the console oscillates up and down near the resonant frequency due to a small piezoelectric element mounted inside the AFM's peak stand. The amplitude of this oscillation is bigger than 10 nm, reaching values between 100 and 200 nm.

Due to the peak – specimen's surface interaction, the Van der Waals forces or the dipole-dipole interactions, or the electrostatic forces make the oscillation amplitude to decrease as the peak is closer to the surface.

In order to control the height between the console and the specimen a piezoelectric element is used and controlled through an electronic system. This mode is an improvement of the conventional contact AFM mode where the console will scan the surface at a constant force, with the risk of damaging the specimen. The *phase contrast mode* can be used also for investigating polymers submerged in liquids [1–5].

5.1.4.4 FORCE MODULATION MODE

Another important application of AFM beside imagistic is the measurement of the force-distance curvatures. In this case, the peak of the AFM is moved closer to the surface and than withdrawn, while the statically deviation of the console is monitored, according to the piezoelectric element displacement.

The possible problems by using this technique do not depend onto the direct measurement of the peak – specimen separation and the console rigidity which have the tendency to "pinch" the specimen's surface.

This pinching phenomena can be reduced by doing the analysis of the specimens in liquid submersions or by using more rigid consoles, or, as a final solution, by using a more sensible sensor for the deviation. By applying a small agitation for the peak, one can measure also the rigidity (the force gradient) [1–5].

5.1.5 AFM IMAGE GENERATION PRINCIPLE

In order to generate an AFM image, the scanner moves the peak near the sample surface for it to bring high-sample interaction.

Once in this regime, all art – cantilever produces a signal that is proportional to the size of the interaction because the signal is usable to establish a benchmark called *set point*.

The scanner moves the peak over the sample in a precise, well known as a defined *raster* consisting in making zigzag movements shaped area covering a square (Fig. 5.8). As changes in topography senses tip sample distance tip-sample changes, occurring as a corresponding variation of the detector signal.

The data necessary to create an AFM image is calculated by comparing the value recorded by the detector signal baseline.

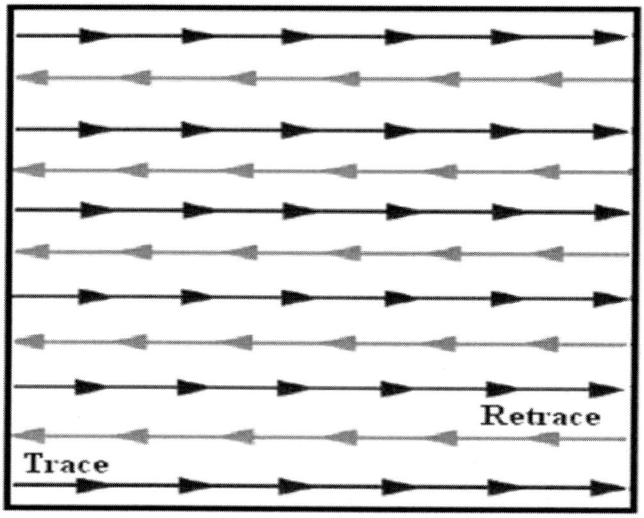

FIGURE 5.8 Representation of scanned directions. Quick scan direction, slow scans direction.

AFM image software downloads the data and presents it in a useful way for interpretation of the data. The number of points per line scan and the number of scan lines covering the image area will determine the image resolution in fast scan direction, respectively slow.

5.1.6 ANALYSIS AND DETERMINATIONS DONE WITH AFM

The AFM can provide useful analysis and data in various domains, like:

- materials and nanomaterials
- biology
- cosmetics
- nanolithography
- semiconductor industry

The main determinations that can be done with the AFM are:

- high resolution study of (nano)materials' surfaces
- roughness analysis both on a line and a certain area of the material's surface
- microstructural analysis of the semiconductors, polymers, ceramics and metals
- analysis of the surface defects
- determination of the number of the particles on the scanned surface
- determination of the volume of a surface defect
- graphical representation of the measured data, most of them in 3D

The surface roughness is calculated using the equations [14], for the average roughness is:

$$S_a = \frac{1}{MN} \sum_{k=0}^{M-1} \sum_{l=0}^{N-1} \left| z\left(x_k, y_l\right) \right| \tag{1}$$

and the mean square root roughness:

$$S_{rms} = \sqrt{\frac{1}{MN} \sum_{k=0}^{M-1} \sum_{l=0}^{N-1} \left(z\left(x_k, y_l\right) \right)^2} \tag{2}$$

where N and M is the number of crystal axes x and y, respectively; z is the average height of crystallites; x_k and y_l are maximum and minimum deviations from the average crystallite.

According to the current standards, the roughness is considered to be a third degree geometric deviation (when it has a periodical or pseudoperiodical character: striations, thread patterns), or a fourth degree (when it has a nonperiodical character: material pulling, patterns, voids, pores, etc.).

The roughness is caused by the oscillating movements of the peak, the friction between the peak and the surface, the high frequency vibrations of the processing machines, etc.

5.1.7 ADVANTAGES AND DISADVANTAGES

The AFM does present a number of advantages while compared with the Scanning Electron Microscopy (SEM). Unlike the SEM, which gives a bi-dimensional projection or a bi-dimensional image of a specimen, the AFM offers a tridimensional profile of the investigated surface. In addition, the specimens analyzed with AFM does not necessarily require a special preparation (e.g., metals and carbon layers), which in turn would modify or deteriorate irreversibly the specimen's surface.

While the electron microscope needs high vacuum for a good functioning, the AFM can function in air or even in liquid medium.

This characteristic makes possible the studying of the biological macromolecules and even the study of living organisms. In principle, the AFM can offer resolutions higher than the ones provided by the SEM, that is, the real atomic resolutions in ultra-high vacuum (UHV), comparably with the Scanning Tunnel Microscope (STM) or the Transmission Electron Microscope (TEM) [1–5].

A disadvantage of the AFM compared with the SEM is the dimension of the image. The SEM can offer images in the order of millimeter by millimeter with a depth of millimeter order while the AFM can give an image with a maximum height in the micrometer order and can scan maximum a surface of approximately 150 x 150 micrometers [5–7].

Another inconvenient of the AFM method is that the incorrect choice of the peak can lead to a low resolution of the obtained image.

A traditional AFM can not scan the images as fast as a SEM and it needs more minutes for a typical scan, while a SEM is capable of an almost real-time scanning (although with a lower quality in terms of precision and/or resolution and quality) after the vacuum conditions were established.

Beside the relatively slow scanning speed, the AFM produces also a thermal deviation of the image, and thus the AFM method is less appropriate for the exact measurement of the distances between various points on the obtained image.

The AFM images can be also affected by the hysteresis curve of the piezoelectric materials and due to the three scanning axes, one needs superior filtering software compared with other methods.

This kind of filtering software can level the topographic characteristics, although currently the AFM uses real-time correction software, or closed-loop feedback scanning which practically eliminate these problems.

Due to its specific nature of the AFM, normally, the abrupt of very high surfaces can not be investigated but only with special consoles which can be laterally modulate but also on up and down direction (with a dynamic contact and without contact) which in turn will imply a higher investment in respect to the acquisition.

The specimens presented in the following subchapter were obtained with AFM equipment available at National Institute for Research and Development in Electrochemistry and Condensed Matter, Timisoara: Nanosurf ® EasyScan type 2 Advanced Research (Fig. 5.9).

FIGURE 5.9 AFM equipment "Nanosurf® EasyScan type 2 Advanced Research."

5.2 AFM STUDY

5.2.1 INTRODUCTION

Due to existing potential for balancing hydrophobic/hydrophilic properties of porphyrins by different peripheral substitution with functional groups, the porphyrin derivatives are among the best candidates for self-assembling by weak Van der Waals forces or hydrophobic effects, hydrogen bond and π–π stacking interactions [15, 16].

A large variety of geometries, such as: pyramids, rings, lame have been generated by self-assembling and self-organization of porphyrins. These are often topologically closed systems [17].

Reported data stated [17] that the J aggregates (edge-to-edge stacked) might layer on different substrates, such as glass, quartz or silica to give porphyrin uniform films. H-bonding interactions are usually responsible for generating crystalline systems.

The main factors that influence the aggregation of porphyrins are the specific structure of the macrocycle, the concentration [18] and the nature of the solvent, as determined by UV-visible spectroscopy [19].

5.2.2 SELF EXPERIMENTS

The samples were formed by THF (or $CHCl_3$, or EtOH) evaporation from a solution containing the porphyrin, on silica plates. The orientation and aggregation of porphyrins at the air–THF/$CHCl_3$/ EtOH interfaces was studied.

Atomic force microscopy (AFM) investigations were carried out with Nanosurf® EasyScan 2 Advanced Research AFM.

Atomic force microscope measurements (quantitative on all three dimensions) were made with sample preparation onto a silica plate from THF or $CHCl_3$, or EtOH solutions. A stiff (450 μm × 50 μm × 2 μm) piezoelectric ceramic cantilever (spring constant of 0.2 Nm^{-1}), with an integral tip oscillated near its resonance frequency of about 13 kHz was used in the measurements.

Areas of maximum 9 μm × 9 μm were scanned, with lateral and vertical resolution of 20 nm and respectively 2 nm. AFM images are displayed by color mapping: in a gray scale dark tones represent the low and light ones high features, respectively.

All AFM measurements were done in contact mode or tapping mode, in environmental conditions (temperature: 23±2°C; relative humidity: 50–70%).

5.2.3 THE MORPHOLOGY AND TOPOGRAPHY OF THE PORPHYRIN SURFACES DEPOSITED BY LANGMUIR BLODGETT AND MAPLE TECHNIQUES

Several methods were performed in order to immobilize the porphyrins thin films [20]. Solvent casting or its variant spin-coating are the most facile and the more elaborated are: electropolymerization, and Langmuir–Blodgett or Langmuir-Schäfer techniques. The Langmuir–Blodgett (LB) method is used for the deposition monolayer after monolayer, consisting in organized structures with targeted thickness and designed architecture. Besides, the LB technique affords extended conjugation when using electroactive systems.

Nowadays, the Langmuir–Blodgett (LB) films of porphyrin derivatives increased in attention due to their amazing applications in catalysts, chemical sensors, and optical actuators and in optoelectronic materials [21].

Porphyrin derivatives are usually behaving as an electron donor when involved in donor–acceptor systems.

Porphyrin derivatives are intensively used for the detection of the toxic gases like HCl, NO_2, Cl_2, NH_3, due to their faster response and higher sensitivity. One of the major issues for using porphyrins for the detection of the toxic gases is to fabricate the thin film and control the aggregation and orientation of the porphyrin derivatives in order to achieve highly ordered film structures.

Strong π–π interaction between porphyrin molecules could produce important aggregation, which would not be beneficial to gas diffusion.

From AFM images corroborated with the knowledge that the length of the porphyrin ring together with the substituted phenyl group is about 1.5 nm, the published data support the result according to which the porphyrin derivatives are almost vertical to the water surface.

The hydrophilic groups have significant influence on the orientation and organization of the porphyrin rings to form supramolecular systems formed at the air/water interface [22]

Water soluble porphyrins with the strongest hydrophilic groups might exist as ions. These porphyrins can form mixed monolayers with other surfactants. The Soret band is in these cases bathochromically shifted. The side to-side or the head-to-tail configurations are both possible.

Regarding porphyrins that possess $-NO_2$, –NHCO–R–C(O)OR and –C(O) OH, substituents on the *meso-* or β- position, the red-shift of the Soret bands would be observed only for the spectra of the monolayer of *meso*-porphyrins [23, 24].

In case of β -substituted porphyrins the Soret band may be blue-shift or red-shift function of peripheral substituents nature. This kind of porphyrin rings can twist at the air/water interface.

Most of porphyrins without hydrophilic groups cannot form stable monolayer by themselves at the air/water interface.

MAPLE technique is suitable for the immobilization of functionalized porphyrin, namely: 5-(4-carboxyphenyl)-10,15,20-tris(4-phenoxyphenyl)-porphyrin (Fig. 5.10) as thin film on screen-printed electrodes (SPE) as a free enzyme glucose amperometric sensors with large limits of detection.

AFM investigations have shown that thin films have a uniform and continuous morphology dependent on incident laser fluence.

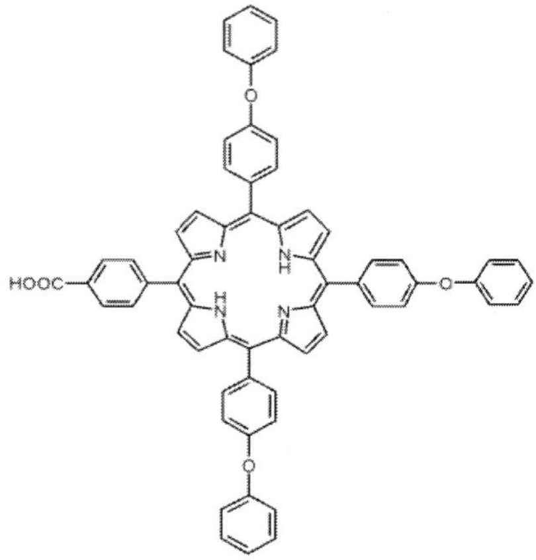

FIGURE 5.10 The chemical structure of 5-(4-carboxyphenyl)-5,10,15-tris(4-phenoxyphenyl)-porphyrin.

Porphyrin thin film surface is formed by droplets generating platelet-like stacking, at small fluencies (200 and 300 mJ/cm^2), these nanostructures being perpendicular onto the substrate surface. At higher fluence of 400 mJ/cm^2, the porphyrin aggregates becomes randomly distributed but at values of fluence of 500 mJ/cm^2, the platelet-like stacking suffer a rearrangement in oriented rows on the substrate surface [25]. The thickness has reached 200 nm for 200 mJ/cm^2 fluence, 400 nm for 300 mJ/cm^2 and 400 mJ/cm^2 laser fluencies, while for 500 mJ/cm^2 fluence, the thickness was 800 nm.

5.2.4 THE MORPHOLOGY AND TOPOGRAPHY OF THE PORPHYRIN SURFACES DEPOSITED BY ELECTROPOLIMERIZATION

Adsorption and electropolymerization of protoporphyrin IX, Fe(III)-protoporphyrin IX and Zn(II)-protoporphyrin IX on highly oriented pyrolytic graphite was reported with in situ atomic force microscopy. AFM images showed that the porphyrins adsorb spontaneously onto pyrolytic graphite and self-assemble into ordered monolayer with the sole molecules lying planar on the surface. By increasing the potentials to the oxidation ranges, some islands are rising on the monolayer. These islands finally unified and formed a continuous film on the surface of graphite electrode [26]. The porphyrin-coated electrodes have sensitivity for detecting dopamine.

5.2.5 THE MORPHOLOGY AND TOPOGRAPHY OF THE PORPHYRIN SURFACES DEPOSITED BY CASTING THE INFLUENCE OF SOLVENT

Investigation of AFM-images of deposited TPP layers shows that their shape change as the deposited porphyrin concentration is higher. The roughness of the surface is increasing and this generates an enlargement of the total area of the deposited TPP. Besides, the average height of 'islands' remain constant with the growth of the TPP concentration; Islands smaller than 50 nm are dominating at any concentration [27].

The formation of single layer of zinc metalloporphyrin deposited on mica substrate and the transformation upon thermal annealing in vacuum are reported [28].

By doing the deposition from a solution, two major advantages over vacuum sublimation can occur: fragmentation of the molecule is not possible and the presence of the solvent molecules provide an extra dimension in the growth of the molecular monolayer. According to the AFM images, the growth of the zinc porphyrin needles follows a layer-by-layer mode.

Our previous work [29] with 5,10,15,20-tetrapyridyl-21H,23H-porphine (Fig. 5.11), entrapped within a silica matrix derived from a two steps acid/base sol-gel process gave the architectures discussed below (Fig. 5.12).

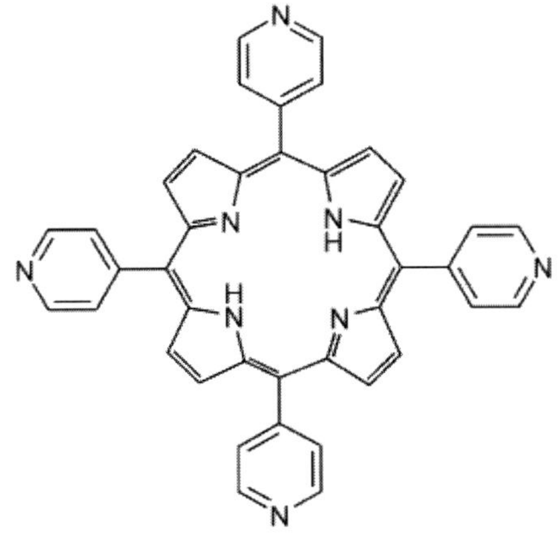

FIGURE 5.11 Molecular structure of the 5,10,15,20-tetrapyridyl-21H,23H-porphine.

FIGURE 5.12 3D AFM image of 5,10,15,20-tetrapyridylporphyrin-silica hybrid deposited on mica from THF.

From 3D AFM image measurements of hybrid 5,10,15,20-tetrapyridyl-21H,23H-porphine -silica material it can be seen that the aggregates have a height ranging around 20–25 nm above the mica substrate and these structures are varying between 300–500 nm in width and up to 2 μm in length.

In agreement with the literature data [28] specifying that small substituents on the 4-position of tetraaryl porphyrins favor π stacking, whereas those on 2 or 3 position prevent significant π stacking, and also in connection with our conclusions over UV-vis study, these aggregates are supposed to be cofacial assemblies.

AFM imaging of another nanomaterial surfaces put into evidence that the aggregation of Zn (II) 5,10,15,20-tetrapyridylporphyrin (Fig. 5.13) is due to π–π cofacial stacking (Fig. 5.14 a and b).

The intensity of the emission spectra of both nanomaterials is significantly increased in comparison with the silica matrix control sample, slightly higher when using HCl, proving that such materials, containing an organic photosensitizer, can be used in design of photocurrent generating hybrid devices in order to optimize light adsorption, surface binding, and charge transfer, with promising efficiencies in building of solar cells.

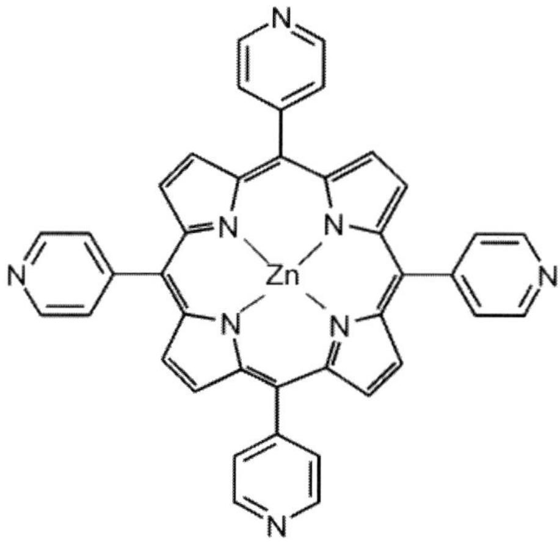

FIGURE 5.13 Molecular structure of the Zn (II) 5,10,15,20-tetrapyridyl-21H,23H-porphine.

The orientation of the molecules is influenced by a stronger π–π interaction between the macrocycles more than that between the molecule and the substrate.

Zinc porphyrin needles randomly dispersed are surrounded by regions of bare mica 1 micrometer across. Nevertheless there is also evidence of alignment of neighboring zinc porphyrin needles, parallel to one another.

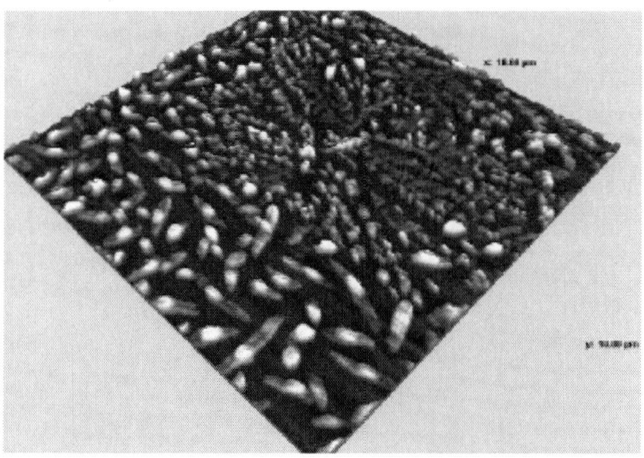

FIGURE 5.14 *(Continued)*

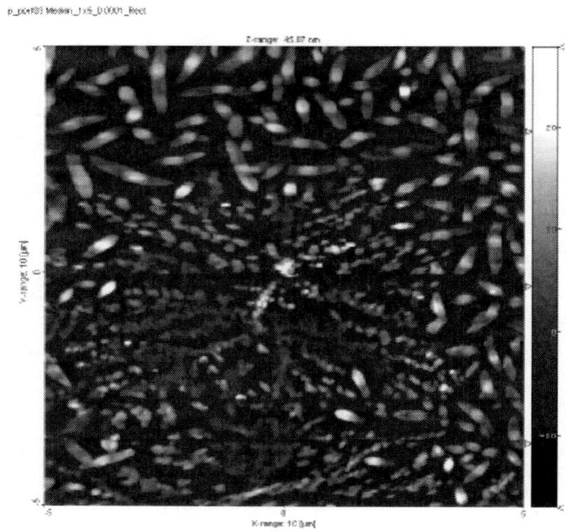

FIGURE 5.14 2D (b) and 3D (a) AFM images of Zn (II) 5,10,15,20-tetrapyridylporphyrin deposited on mica from THF.

The sensitization of semiconductor materials using dyes started already in the late 1960s [30]. The photo conversion process of dye-sensitized solar cells is analogous to natural photosynthesis, being also named artificial photosynthesis. The maximum reported yield of 12.3% was reported with a CoIII/II(bpy)3 complex based electrolyte that was mixed with nanoparticles of π-acceptor zinc porphyrin dye [31] to increase the light absorption.

Corroborating AFM with UV-vis spectra (Fig. 5.15) it is suggested that the red shifts of Soret and Q bands absorptions are produced by the fact that a face-to-face stacking arrangement is adopted in the direction of compression of the monolayer [32]. It is also suggested that the face-to-face packing is stabilized by the intermolecular interactions including hydrogen-bonding interactions between the conjugated porphyrin rings.

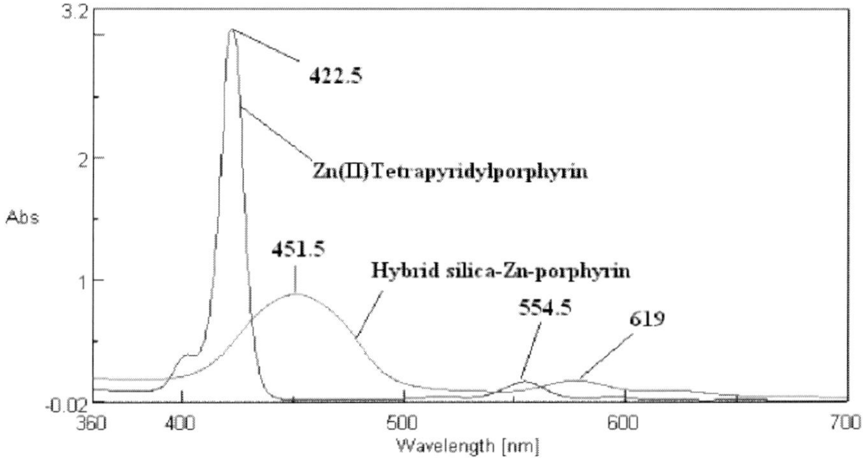

FIGURE 5.15 Red shift of Soret and Q bands due to face to face stacking.

Additionally, the preference of porphyrin to self- assembly and self-organize [33] leads to the decrease of their absorption intensity with time, as shown in the Fig. 5.15.

Self-assembly morphology of three tetrakis(hydroxyphenyl) porphyrins on the substrates of silicon wafer and mica were investigated [34]. Tetrakis(*p*-hydroxyphenyl) porphyrin was reported to form disk-like morphology on both silicon wafer and mica substrates but the porphyrin substituted in the *meta* position of the phenyl groups with hydroxyphenyl performed quite different. It organizes in ring structures on the silicon wafer, while the morphology of the surface of the aggregates on the mica substrate was platelet like randomly dispersed.

In the structure of tetrakis(*p*-hydroxyphenyl) porphyrin (Fig. 5.16) the C-O bond is in the same plane with that of porphyrin planar ring. Instead, the C-O bond in the structures of tetrakis(*m*-hydroxyphenyl) porphyrin and tetrakis (*o*-hydroxyphenyl) porphyrin is located above or under the planar ring of the porphyrin macrocycle.

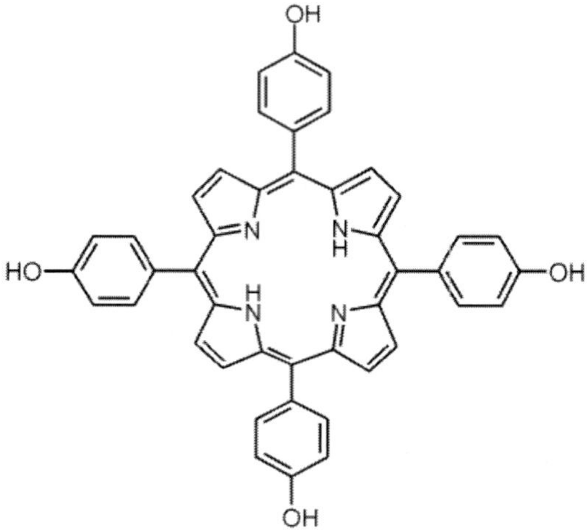

FIGURE 5.16 Structure of tetrakis (p-hydroxyphenyl) porphyrin.

As a consequence the π–π interaction between the molecules is significant in case of the tetrakis(*p*-hydroxyphenyl) porphyrin, which can form a face-to-face staking, but this type of interaction should be weaker for the other two isomers.

Tetrakis(*p*-hydroxyphenyl) porphyrins will generate disk-like structures due to the face to face staking.

In case of tetrakis(*m*-hydroxyphenyl) porphyrin and tetrakis(*o*-hydroxyphenyl) porphyrin the molecules can interact by the hydrogen bond between the hydroxyls and might form a flexible side-by-side structure that is appropriate to reorganize during the drying process and to form rings.

The major explanation for the difference in the morphology of the aggregates should be the position of the neighbor hydroxyls and their possibility to interact by hydroxyl bonds.

Very interesting situation takes place in case of casting of a water soluble metalloporphyrin, 5,10,15,20-tetrakis(N-methyl-4-pyridyl)porphyrin-Zn(II) tetrachloride from different solvents, possessing different polarities on mica substrates.

The solvents used were chloroform and THF, and as can be seen from Figs. 5.17a and 5.17b, chloroform is favoring ring shape structures of various diameter sizes, in the range 250–1800 nm.

Instead, the casting from THF is producing only straw or pyramidal structures with tailored sizes of 375 nm, uniformly dispersed and oriented.

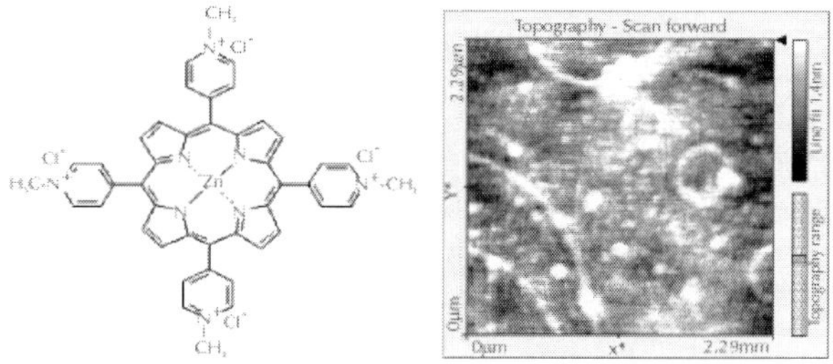

a)

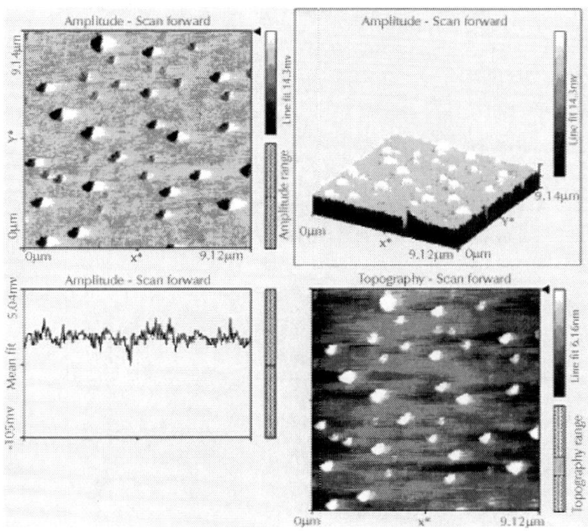

b)

FIGURE 5.17 2D AFM images of 5,10,15,20-tetrakis(N-methyl-4-pyridyl)porphyrin-Zn(II) tetrachloride deposited from Chloroform (a) and THF (b).

Another porphyrin which was studied to see its aggregation from different solvents is 5,10,15,20-tetrakis-(4-allyloxyphenyl)-porphyrin (Fig. 5.18). Regarding 5,10,15,20-tetrakis-(4-allyloxyphenyl)-porphyrin behavior we can state that it behaves in a similar way with 5,10,15,20-tetrakis(N-methyl-4-pyridyl)porphyrin-Zn(II) tetrachloride.

Using as solvent THF (Fig. 5.19), some triangular shapes of 347 nm in width are formed, all uniform oriented and similar in sizes.

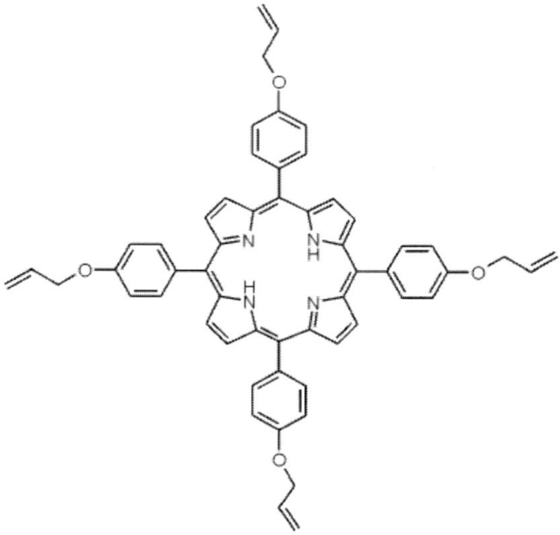

FIGURE 5.18 Molecular structure of 5,10,15,20-tetrakis-(4-allyloxyphenyl)-porphyrin (TAPP).

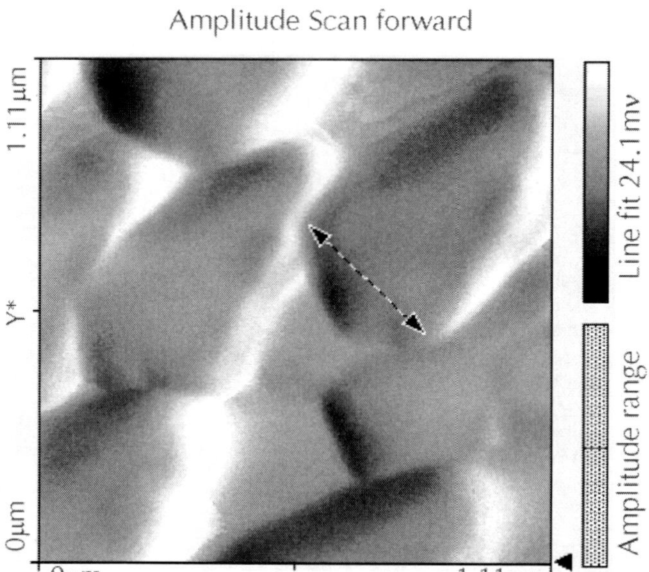

FIGURE 5.19 2D AFM image of 5,10,15,20-tetrakis-(4-allyloxyphenyl)-porphyrin from THF.

On the other hand, when the deposition of porphyrin was done from chloroform, the same ring shaped architectures as in case of 5,10,15,20-tetrakis(N-methyl-4-pyridyl)porphyrin-Zn(II) tetrachloride have been formed with more narrow distribution of the diameters, in the range of 2000–2500 nm. Columnar architectures are also visible.

In case of 5,10,15,20-tetrakis-(4-allyloxyphenyl)-porphyrin the initial sandwich type compounds are the building blocks for rings formed by self-assembly of J-aggregates of porphyrins (Fig. 5.20).

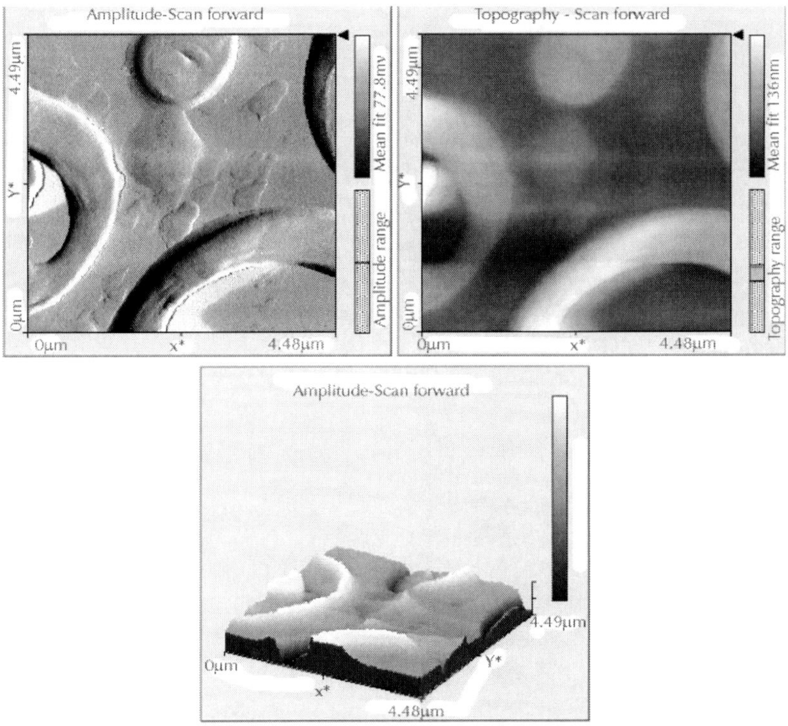

FIGURE 5.20 2D AFM image of 5,10,15,20-tetrakis-(4-allyloxyphenyl)-porphyrin from CHCl3.

AFM measurements revealed interesting features in case of 5-(4-pyridyl)-10,15,20-tris-(4-phenoxyphenyl)-porphyrin (Fig. 5.21).

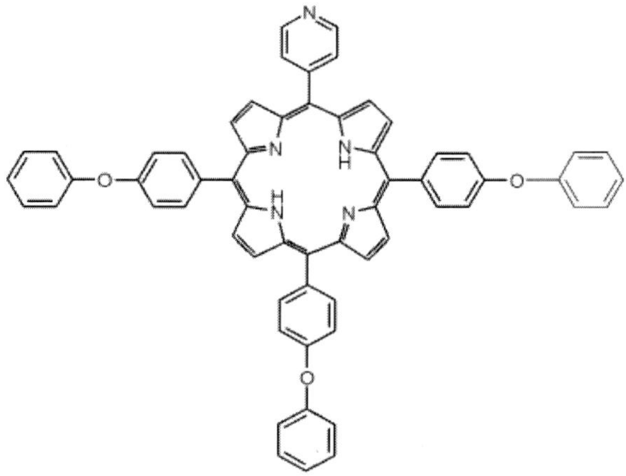

FIGURE 5.21 Structure of 5-(4-pyridyl)-10,15,20-tris-(4-phenoxyphenyl)-porphyrin.

The particles of spherical shape that are called *kvatarons*, originated from Abkhazian word kvatar, meaning ball, are formed [35]. The spherical shape of the particles is produced by the unique properties of liquid water structure. Spherical particles with diameters of 150–400 nm were produced and the whole morphology is looking like different aria concave surfaces (Fig. 5.22).

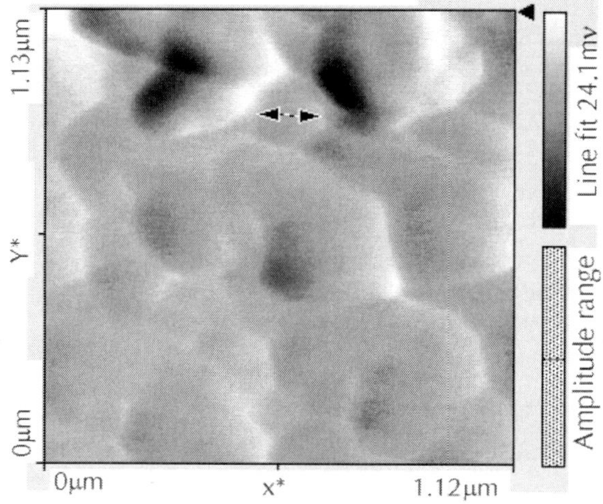

FIGURE 5.22 2D AFM image of 5-(4-pyridyl)-10,15,20-tris-(4-phenoxyphenyl)-porphyrin from CHCl3.

Another solvent, EtOH, gave roof like morphology of 5-(4-pyridyl)-10,15,20-tris-(4-phenoxyphenyl)-porphyrin, as in Fig. 5.23 due to π–π stacking interactions.

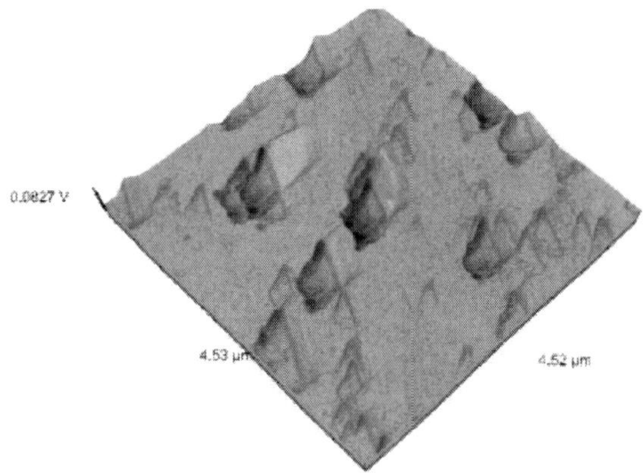

FIGURE 5.23 3D AFM image of 5-(4-pyridyl)-10,15,20-tris-(4-phenoxyphenyl)-porphyrin from EtOH.

A novel A_3B mixed substituted porphyrin with both hydrophilic and hydrophobic groups, namely: 5-(3-hydroxy-phenyl)-10,15,20-tris (3,4-dimethoxy-phenyl)-porphyrin (Fig. 5.24) generates crystalline pyramidal structure (Fig. 5.25), probably favored by interaction of hydroxyl bonds.

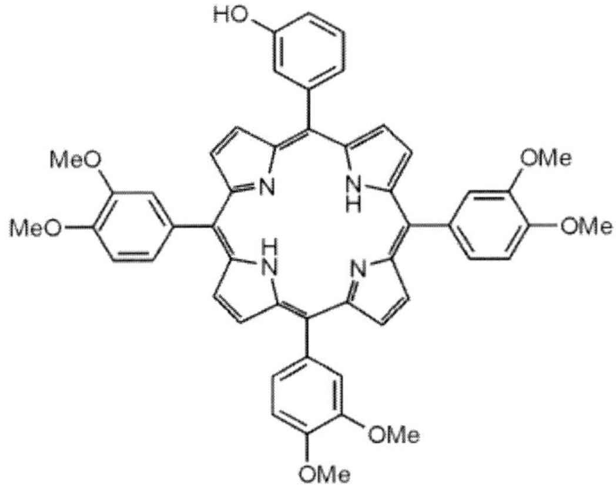

FIGURE 5.24 Structure of 5-(3-hydroxy-phenyl)-10,15,20-tris (3,4-dimethoxy-phenyl)-porphyrin.

FIGURE 5.25 3D AFM image of crystalline pyramidal structures of 5-(3-hydroxy-phenyl)-10,15,20-tris (3,4-dimethoxy-phenyl)-porphyrin.

Aggregation of 2,3,7,8,12,13,17,18-octaethyl-porphyrin (Fig. 5.26) deposited on silica pure plates at the air–CHCl3 interface produces densely packed lamellar triangles with similar dimensions, around 400 nm. The main characteristic of these multilayers is that they are uniformly oriented.

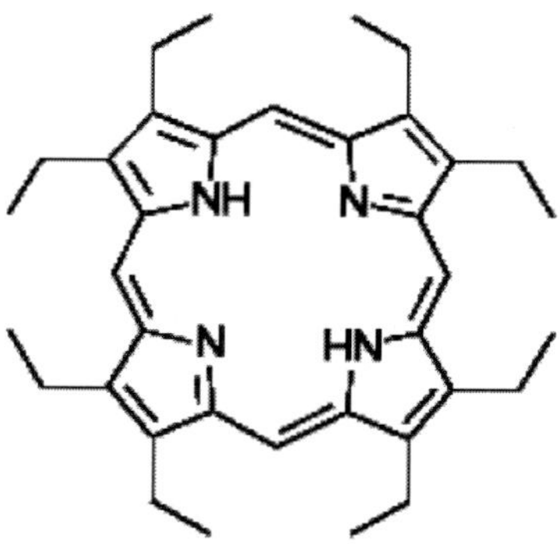

FIGURE 5.26 Structure of 2,3,7,8,12,13,17,18-octaethyl-porphyrin.

Aggregation of the above mentioned β-porphyrin at the air-$CHCl_3$ interface generates rings, from the initial triangular architectures, as in case of air-THF interface, previously discussed. $CHCl_3$ solvent favors the spiral organization of the triangles, into pyramids, so that rings (diameter 4062 nm) are finally formed (Fig. 5.27) by distortion of the local order of the porphyrin layer.

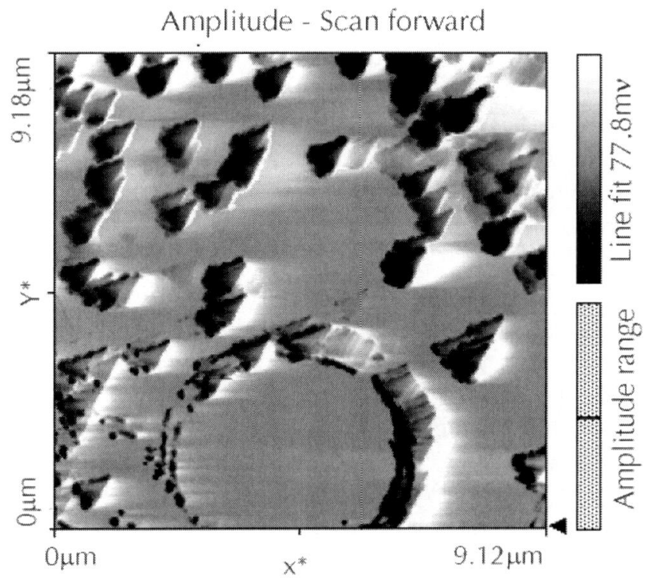

FIGURE 5.27 2D AFM image of 2,3,7,8,12,13,17,18-octaethyl-porphyrin, different aggregates, lamellar triangles and rings, coexisting together.

In this case lamellar triangles, pyramids and rings formed by sandwich type and side-by-side type aggregation are coexisting.

From AFM measurements (more images for each sample) the roughness parameters: average surface roughness (Sa, nm) was found to vary between 18 and 95 nm; maximum peak height (Sp, nm) in the range of 110 nm up to 270 nm; maximum valley depth (Sv, nm) in the limits of −50 nm to −22 nm were determined. In completion, the diameter of the particles of the porphyrin was also estimated and besides, the number of islands and the mean surface of islands. The mean surface area of the peaks is 16.8 μm^2, and the volume is around 0.198 μm^3.

5.3 CONCLUSION

Due to the importance of the Atomic Force Microscopy (AFM) in topographic analysis of the surfaces (areas between 1 nm^2 and 100 μm^2 can be scanned) this technique becomes more and more important in analyzing self – assembled porphyrin structures. Usually, the AFM modes which will produce the imagistic surface can be split into static (also known as "contact") and other types of variations modes ("noncontact"). Porphyrins are offering self assembles that are often topologically closed systems and can be properly analyzed by AFM. Based on AFM images, this chapter emphasizes the influence of the specific structure of the porphyrin macrocycle, its concentration and the nature of the solvent on the self-aggregation and organization of porphyrins in connection with an up-to-date literature description. In this respect, several symmetrical and unsymmetrical substituted A3B *meso* and β-porphyrins were studied varying the used solvents (CH_3Cl, THF and EtOH). AFM put into evidence that the J aggregates (edge-to-edge stacked) and H-bonding interactions are both responsible for generating of different architectures, that are sometimes coexisting together, such as: triangles type sheets, concave-convex surfaces, columnar stacks, pyramids, roofs and even rings. The nature of the solvent can influence the type of architectures as follows: CH_3Cl favors the spiral organization of the triangles, into pyramids and even rings, sometimes coexisting together; EtOH helps with roof like structures formation by J-aggregation of pyramids; THF favors the assembling into triangles and sandwich type organization into multilayers. Taking into consideration that modern research is deeply concerned about the air-solvent interface of new multifunctional micro and nano-materials, AFM is one of the most important techniques to reveal deep inside the topography of porphyrins surfaces.

KEYWORDS

- **chemical nano-synthesis**
- **H-interaction**
- **nanochemsitry**
- **porphyrins**
- **self-asenbling compounds**
- **topographic nano-analysis**

REFERENCES

1. Chen. C. J., Introduction to Scanning Tunneling Microscopy, Oxford University Press. New York. 1993.
2. Bonnell, D., Scanning Probe Microscopy and Spectroscopy: Theory, Techniques, and Applications. USA: Wiley-VCH, 2000.
3. Boley, B. A., Weiner, J. H., Theory of Thermal Stress, John Wiley and Sons, New York, 1960.

4. Timoshenko, S. P., Goodier, J. N., Theory of Elasticity, Auckland; Singapore: Mc-Graw-Hill, 1982.
5. Hansma, P. K., Elings, V. B., Marti, O., Bracker, C. E., (1988). Scanning Tunneling Microscopy and Atomic Force Microscopy: Application to Biology and Technology, *Science*, 242, 209–216.
6. Yao, N., Wang, Z. L., Handbook of Microscopy for Nanotechnology, Kluwer (now Springer): Boston, MA, 2005.
7. Meyer, E., Hug, H. J., Bennewitz, R., Scanning Probe Microscopy. Springer, Berlin, 2004.
8. Binning, G., Quate, C. F., Gerber Ch., (1986). Atomic force microscope. *Phys. Rev. Lett.* 56(9), 930–933.
9. Wisendanger, R., *Scanning probe microscopy and spectroscopy: methods and applications*, Cambridge England.; New York: Cambridge University Press, 1994.
10. Binnig, G., Rohrer, H., Gerber, C., Weibel, E., (1983). (111) facets as the origin of reconstructed Au(110) surfaces, *Surf. Sci.*, 131(1), L379-L384..
11. Howland, R., Benatar, L., A practical guide to SPM. Scanning Probe Microscopy, ThermoMicroscopes USA, 2000.
12. Hutter, J. L., Bechhoefer, J., (1994). Measurement and Manipulation of Van der Waals Forces in Atomic Force Microscopy, *J. Vac. Sci. Technol., B*, 12, 2251–2253.
13. Binnig, G., Rohrer, H., Gerber, C., Weibel, E., (1983). 7, × 7, Reconstruction on Si(111) Resolved in Real Space, *Phys. Rev. Lett.*, 50(2), 120–123.
14. Kapaklis, V., Poulopoulos, P., Karoutsos, V., Manouras Th., Politis, C., (2006). Growth of thin Ag films produced by radio frequency magnetron sputtering, *Thin Solid Films*, 510–138–142.
15. Natale, D., Monti, C., Paolesse, D. R. (2010). Chemical sensitivity of porphyrin assemblies. *Mater. Today*, *13*(7–8), 46–52.
16. Ghosh, A., Mahato, P., Choudhury, S., Das, A. (2011). Comparative study of porphyrin derivatives in monolayers at the air–water interface and in Langmuir–Blodgett films. *Thin Solid Films*, 519, 8066–8073.
17. Drain, C. M., Varotto, A., Radivojevic, I. (2009). Self-Organized Porphyrinic Materials. *Chem Rev.*, *109*(5), 1630–1658.
18. Guo, L. (2008). Side-chain-controlled H- and J-aggregation of amphiphilic porphyrins in CTAB micelles. *J. Colloid. Interf. Sci.*, *322*(1), 281–286.
19. Fagadar-Cosma, E., Fagadar-Cosma, G., Vasile, M., Enache, C., (2012). Synthesis, Spectroscopic and Self-Assembling Characterization of Novel Photoactive Mixed Aryl-Substituted Porphyrin, *Curr. Org. Chem.*, 16, (24), 931–941.
20. Giancane, G., Valli, L., (2012). State of art in porphyrin Langmuir–Blodgett films as chemical sensors, *Adv. Colloid Interface Sci.*, 171–172, 17–35.
21. Cao, Z., Chen, Q., Wang, C., Ren, Y., Liu, H., Hu, Y., (2011). Interfacial behaviors and aggregate structure of atropisomers of "picket-fence" porphyrin at the air/water interface, *Colloids Surf., A*, 377, (1–3), 130–137.
22. Liu, H. G., Feng, X. S., Zhang, L. J., Ji, G. L., Qian, D. J., Lee, Y. I., Yang, K. Z., (2003). Influences of hydrophilic and hydrophobic substituents on the organization of supramolecular assemblies of porphyrin derivatives formed at the air/water interface, Mater. Sci. Eng. C., 23, 585–592.

23. Cruz, F. D., Armand, F., Albouy, P.-A., Nierlich, M., Ruaudel-Tiexier, A., (1999). *Langmuir* 15, 3653–3660.
24. Facci, P., Fontana, M. P., Dalcanale, E., Costa, M., Sacchelli, T., (2000). *Langmuir* 16, 7726–7730.
25. Iordache, S., Cristescu, R., Popescu, A. C., Popescu, C. E., Dorcioman, G., Mihailescu, I. N., Ciucu, A. A., Balan, A., Stamatin, I., Fagadar-Cosma, E., Chrisey, D. B., (2013). Functionalized porphyrin conjugate thin films deposited by matrix assisted pulsed laser evaporation, *Appl. Surf. Sci.,* Online. http://dx.doi.org/10.1016/j.apsusc.2013.01.080. (accessed 18, January 2013).
26. Duong B, Arechabaleta R, Tao, N. J. (1998). In situ AFM/STM characterization of porphyrin electrode films for electrochemical detection of neurotransmitters, *J. Electroanal. Chem.*, 447, (1–2), 63–69.
27. Solovieva, A., Vstovsky, G., Kotova, S., Glagolev, N., Zav'yalov, B. S., Belyaev, V., Erina, N., Timashev, P., (2005). The effect of porphyrin supramolecular structure on singlet oxygen photogeneration, *Micron*, 36, (6), 508–518.
28. George, H., Palmer, R. E., Guo, Q., Bampos, N., Sanders, J. K. M., (2006). Needles and clusters of zinc porphyrin molecules on mica, *Surf. Sci.*, 600, (16), 3274–3279.
29. Fagadar-Cosma, E., Enache, C., Dascalu, D., Fagadar-Cosma, G., Gavrila, R., (2008). FT-IR, Fluorescence and Electronic Spectra for Monitoring the Aggregation Process of Tetra-Pyridylporphyrine Entrapped in Silica Matrices. *Optoel. Adv. Mat.-Rapid Communications*, 2(7), 437–441.
30. Grätzel, C., Zakeeruddin, S. M., (2013). Recent trends in mesoscopic solar cells based on molecular and nanopigment light harvesters, *Mater. Today*, 16, 11–18.
31. Yella, A., Lee, H. W., Tsao, H. N., Yi, C., Chandiran, A. K., Nazeeruddin, M. K., Diau, E. W. G., Yeh, C. Y., Zakeeruddin, S. M., Grätzel, M., (2011). Porphyrin-Sensitized Solar Cells with Cobalt (II/III)–Based Redox Electrolyte Exceed 12, Percent Efficiency, *Science*, 334, 629–634.
32. Qian, X., Taia, Z., Sun, X., Xiao, S., Wu, M., Lu, Z., Wei, Y., (1996). Molecular packing in LB films of a new porphyrin investigated by atomic force microscopy, *Thin Solid Films*, 284–285, 432–435.
33. Udal'tsov, A. V., Tosaka, M., Kaupp, G., (2003). Microscopy of large-scale porphyrin aggregates formed from protonated TPP dimers in water–organic solutions, *J. Mol. Struct.*, 660, (1–3), 12, 15–23.
34. Zhang, H., Ma, Y., Lu, Z., Guthe, Z.-Z., (2005). Self-assembly films of tetrakis(hydroxyphenyl) porphyrins, *Colloids Surf., A*, 257–258, 291–294.
35. Askhabov, A. M., Yushkin, N. P. (1999). The kvataron mechanism responsible for the genesis of noncrystalline forms of nanostructures, *Doklady Earth Sci.,* (1999). 368, (7), 940–942. *Translated from Doklady Akademii Nauk*, 368(1), 84–86.

CHAPTER 6

RECENT TRENDS IN NANO-OPTOMECHANICAL SYSTEMS

ARANYA B. BHATTACHERJEE, NEHA AGGARWAL, and SONAM MAHAJAN

CONTENTS

ABSTRACT

Nano scale optomechanical systems are emerging as new generation quantum devices, which can support coherent energy exchange between multiple modes. In this chapter, we review three optomechanical systems: (1) Two optical modes coupled to one mechanical mode. This system belongs to the next generation of quantum communications and quantum information processing units. This hybrid optomechanical system is useful as the photonic modes can exchange energy over long distances and phononic modes can store energy for a longer duration. (2) A Bose Einstein condensate coupled to magnetic mechanical oscillator. This system is shown to act like a phonon laser. (3) A Bose Einstein condensate coupled to a optical cavity mode and a mechanical oscillator. This novel system is shown to be sensitive to weak external force by cooling the mechanical oscillator to its quantum ground state.

6.1 INTRODUCTION

Quantum optomechanics is the modern branch of optical engineering, which aims to achieve complete control of the quantum mechanical interaction between electromagnetic radiation and nano-mechanical resonators. Coherent control, quantum measurement and state preparation are the three essential features of the technology related to quantum optomechanics [1]. Pioneering theoretical work of Braginsky [2] and Caves [3] have led to a rapid development in this field. A large variety of quantum systems, particularly nanomechanical cantilever [4–9], vibrating microtoroids [10, 11], ultracold atoms [12–17] and membranes [18], have been investigated recently in various experiments and theoretically. The coherent control over the properties of the optomechanics in the quantum regime has been made possible due to these achievements.

New quantum mechanical behavior of macroscopic quantum systems has been observed due to electromagnetic coupling between nano-mechanical resonators and optical cavities [19]. At the single photon level, such novel quantum systems have been executed leading to significant experimental progress [14, 15, 20–23]. A single optical mode strongly coupled to a mechanical mode has been studied in various theoretical works. Quantum effects are studied in these hybrid systems if the interaction between the mechanical mode and light mode is much larger than the cavity decay rate and the frequency of the mechanical mode. By introducing a Kerr medium inside the optical cavity, quantum nonlinearities arise into the system [24].

The theoretical concepts of phonon laser and experiments to control and study about single phonons [25] are the strong activities going on in the field of quantum optics with phonons instead of photons. Several physical systems based on paramagnetic ions in a lattice [26], paraelectric crystals [27], isolated trapped ions [28], quantum wells [29], semiconductors [30–33], nanomechanic systems [34], nanomagnets [35], and ultracold matter [36] have proposed the phonon analog of the optical laser. Recently a compound microcavity system [37], harmonically bound magnesium ions [38], semiconductor super lattices [39], cryogenic Al_2O_3:Cr^{3+} [40–42], Al_2O_3: V^{4+} [43] are the experiments which have demonstrated phonon laser action.

Quantum information processing is also one of the potential application of the optomechanical systems [44]. A quantum phase gate can be engineered for photonic or phononic qubits using all optical switch which is formed on the basis of Kerr type nonlinearity. Quantum memory can also be served by mechanical mode [45]. A quantum interface between atomic, solid-state and optical qubits arises due to the coupling between the mechanical and optical degrees of freedom [46]. To architect a quantum communication and quantum information processing unit, a hybrid design can be made consisting of optical modes and mechanical modes that can be fruitfully used to store and transfer information coherently. Acoustic modes (phonons) in quantum information processing units can store information for an extended period whereas optical modes can transfer information over long distances.

Systems involving trapped atoms, ions, nanospheres [47–51] and mechanically compliant optical cavity structures [52] have all been considered in the case of phonon-photon state transfer. The motional state of an atom, ion or macroscopic mirror is being mapped by the state of an incoming optical field through an exact timing of control pulses by turning the coupling between the mechanical and light motion on and off. An optomechanical crystal (OMC) structures can be simultaneously implemented using phononic and photonic waveguides simultaneously [53].

Bose-Einstein Condensate [12–15, 54–61] and atomic ensembles [62–67] have made the research in the field of cavity optomechanics more attractive. Generally, a cavity opto-mechanical system is composed of an optical cavity with one movable end mirror. To achieve the quantum ground state of a nanomechanical resonator, such a hybrid system is used by exerting the pressure on the movable mirror by cavity light field. The studies of atom-oscillator entanglement, quantum state transfer, and quantum control of mechanical force sensors is possible under the coherent and strong coupling regime as shown by the research in the field of cavity optomechanics of atoms. The coherent control over nanomechanical oscillators has been done in recent experiments. Rational quantum control over all degrees of freedom can be accomplished using magnetically coupled ultracold atoms to mechanical oscillators [68].

Our work is strongly motivated by recent work on radiation pressure effects in nano-scale mechanical systems [69, 70]. We have shown that the dynamics of linearized Hamiltonian of a mechanical mode coupled to two optical modes is similar to that of Dicke model [71]. The analytical expressions for the normal modes of the system have been calculated. It has also been shown that the system can undergo a Dicke-Hepp-Lieb type super radiant phase transition for different coupling constants. We have also depicted the possibility of selective energy transfer between any two modes. The squeezing variances of the three modes near the quantum critical point have also been investigated. For this system, we have also demonstrated the splitting of normal mode into three modes in the displacement spectrum of the mechanical resonator. Further, we have proposed a phonon laser operating like a two-level optical laser by using a Bose Einstein Condensate of ^{87}Rb atoms coupled to a magnetic cantilever [16]. We also show that the same system can undergo a Dicke-Hepp-Lieb type phase transition [71–74] into a phonon super radiance regime for strong coupling. We also see the importance of atomic two-body interaction in this system dynamics. Moreover, we have discussed how the hybrid optomechanical system, involving a BEC confined in an optical cavity with movable end mirror, can be cooled using different cooling techniques. We have also shown the effect of condensate two body interaction in cooling the oscillating mirror effectively.

6.2 BASIC PHENOMENON

Here, we discuss some basic phenomenon used in our nano-optomechanical systems.

6.2.1 RADIATION PRESSURE

Radiation pressure, that is, force due to the momentum of light has been known as long as from seventeenth century when scientists predicted that the inclination of comet tails is because of solar radiation pressure [75, 76]. Later in twentieth century, Experimentalists Hull [77] and Lebedev [78] studied the strength of radiation pressure. The application of radiation pressure in the context of gravitational wave detection was first studied by Braginsky and co-workers.

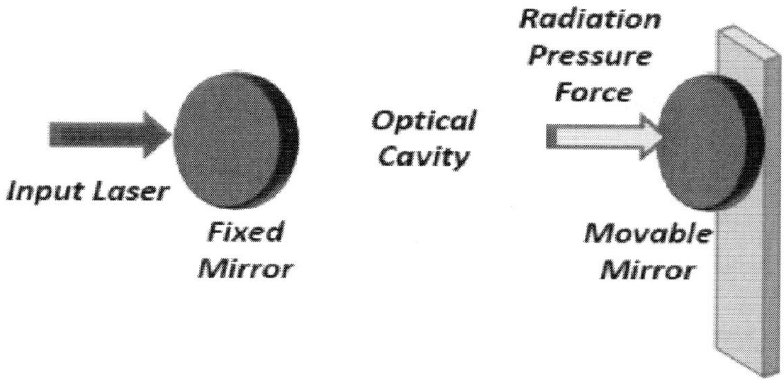

FIGURE 6.1 Schematic representation of radiation pressure cooling.

They predicted that the radiation pressure inside a cavity with finite decay time gives rise to dynamic back action which is the underlying phenomena for cooling of mechanical oscillators and parametric instabilities as shown in Fig. 6.1. Braginsky et al. also studied the sensitivity limits due to quantum nature of light inside a cavity [79, 80]. Later in 1980, other groups mainly of Caves [3] and Meystre [81] analyzed the radiation pressure noise in interferometers. After this a great surge of interest creped in to find out the potential application of radiation pressure. Many theoretical models were also predicted including the possible application of radiation pressure such as quantum non demolition measurement of photon numbers [82, 83], generation of squeezed light [84, 85], feedback cooling of mechanical oscillator [86], optical and mechanical mode entanglement [87–89], quantum state transfer from light to mechanical oscillator [52], etc. Also very recently some very interesting phenomena have been realized experimentally such as strongly coupled optomechanics [90], cavity cooling of mechanical oscillator [5, 6, 11, 91], parametric amplification of mechanical motion [92], measurement of motion of mechanical oscillator [93–95], which can pave the way towards the measurement of weak forces. There have been a number of review articles discussing in greater details the possible application of optical cavities and radiation pressure [96–100]. The radiation pressure interaction has also been used to study the dynamics of the mechanical oscillator such as cooling of mechanical oscillator in resolved side band regime [101]. The cooling in the resolved sideband is possible because due to radiation pressure. The optical mode (photons) and the mechanical mode (phonons) get entangled which results in energy exchange between the two modes. As a result if a photon is created in the side band then this results in the annihilation of phonon of the mechanical mode and vice-versa.

6.2.2 OPTICAL LATTICE

Optical lattice has rapidly become a vital tool for atomic and condensed matter physicists. Optical lattice provides a way to realize the simplified models in condensed matter theory. A number of research articles and review papers can be found which discuss in greater details the applications of optical lattice to condensed matter system [102, 103]. Optical lattice is the perfect example of Richard Feynman's notion of 'Quantum Simulation,' that is, using one quantum system to study the another one because from the results predicted from optical lattice model theorists can test which of their models is more realistic. Optical lattice can be formed from two counter propagating laser beams as shown in Fig. 6.2. This result in a standing wave interference pattern where the lattice spacing is just half the wavelength of the laser light used. In such a configuration, atoms can be trapped in the intensity maxima or intensity minima depending upon whether the laser light used is blue detuned or red detuned respectively. The light induces an electric dipole moment on the atom, which in-turn modifies their energy. By using additional lasers, two- and three-dimensional optical lattices can also be constructed.

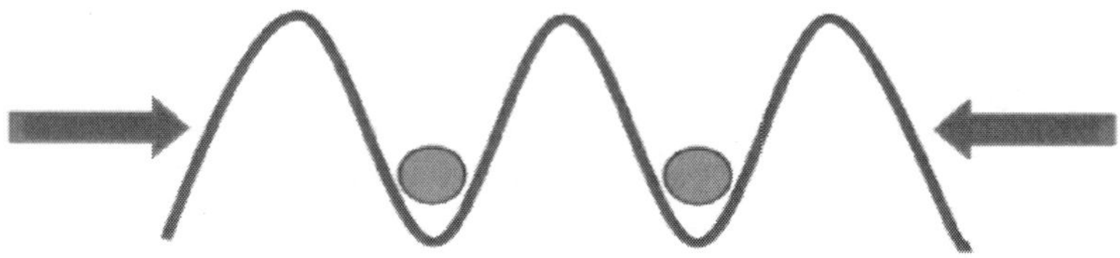

FIGURE 6.2 Schematic representation of a one-dimensional optical lattice with standing waves.

Neutral atoms interact with light in two ways: dissipative way via the scattering force and conservative way via the dipole force. The conservative interaction is because of the interaction of the light field with the induced dipole moment of the atom. This causes a shift in the potential energy called a.c.-Stark shift. On the other hand, the dissipation comes due to the absorption of photons followed by spontaneous emission. Because of conservation of momentum, the net effect is a dissipative force on the atoms caused by the momentum transfer to the atom by the absorbed and spontaneously emitted photons. Laser cooling techniques make use of this light force. Optical dipole traps are based on the interaction between an induced dipole moment in an atom and an external electric field, which can be provided by the oscillating electric light field of the laser. This induces an oscillating dipole moment in the atom while at the same time interacts with this dipole moment in order to create a trapping potential [102, 103]. Therefore, for large detuning spontaneous emission processes can be neglected and the energy shift can be used to create a conservative trapping potential. This is how the optical lattice works in cooling and trapping of neutral atoms.

6.2.3 BOSE EINSTEIN CONDENSATES

A Bose Einstein Condensate (BEC) is a state of dilute gas of a weakly interacting Bosons confined in an external potential and cooled to temperature very near to absolute zero (0 Kelvin). Under such conditions, a large fraction of the Bosons occupy the lowest quantum state of the external potential, at which quantum effects become apparent at macroscopic level.

The Bose Statistics were first predicted by Satyendra Nath Bose in 1924, where he described the phenomena of Black Body Radiation [104]. Later in 1925, Einstein extended the work of Bose to noninteracting particles [105]. The resulting statistics that came out of the work of Bose and Einstein is known as Bose – Einstein Statistics. A very interesting behavior came out of Bose Einstein statistics that, at very low but finite temperature, all the atoms behave identically and a large number of atoms occupy the lowest quantum energy states [105, 106]. The particle density at the center of the cloud of BEC is of the order of 10^{13}–10^{15} cm^{-3} whereas the density of solids and liquids is of the order of 10^{22} cm^{-3}. To observe quantum mechanical phenomena at such a low density, temperature should be of the order of 10^{-3} K. As we know that de-Broglie wavelength depends on temperature therefore in other words we can say that Bose-Einstein Condensation takes place when the de-Broglie wavelength, λ_{dB}, of each atom is large enough, such that the wavelength of individual atoms overlap and atom particles cannot be distinguished from one another. Later in 1930, Fritz Landon and Laszata Tisza [107, 108] assumed that the BEC is the basic phenomena responsible for the super fluidity in Liquid ^{4}He. This was the first time, which paved the way to the idea that BEC can display quantum phenomena at macroscopic level.

Though BEC was predicted way back in 1924–25 but for many decades, scientists were only able to make BEC from ^{4}He. ^{4}He Bose condenses from a liquid rather than from a gas because in ^{4}He atoms interact strongly with each other. But due to technological limitations, it took researchers and scientists nearly 70 years to create first BEC in laboratory in a gaseous system after the invention of laser cooling [109–111] and evaporative cooling techniques. In 1995, Eric Cornell and his team at Colorado, JILA, achieved for the first time BEC in dilute rubidium vapors [112] that lead him along with C. E. Wieman and W. Ketterle, the noble prize of physics in 2001. After that in a series of experiments, BEC was also achieved in Sodium [113] and lithium vapors [114, 115]. Nowadays Bose Einstein Condensate is produced and studied in number of laboratories across the world.

To achieve BEC experimentally, atoms are first trapped and cooled to a temperature as low as up to micro Kelvin, using the technique of laser cooling and trapping [109–111]. To further cool the atoms, lasers are switched off and a magnetic trap is switched on where the atoms are trapped because of the Zeeman interaction of the electron spins of the atoms with inhomogeneous magnetic field. In this configuration, atoms having their electronic spin parallel to the applied magnetic field are attracted to the minimum of the magnetic field and atoms with electronic spin anti parallel to the field are repelled. Here evaporative cooling of the atoms is also achieved by using a radio frequency magnetic field. Radio Frequency field flips the electronic spin of more energetic atoms from parallel to anti parallel, thereby repelling the most energetic atoms from the BEC cloud. The thermal cloud rethermalizes and the temperature falls to nano Kelvin regime. After this a critical phase space density is reached in the atomic vapors and BEC is said to be achieved.

Ever since the discovery of BEC, research in this field has flourished very rapidly both theoretically and experimentally. Discovery of BEC has initiated some new topics such as atom optics, atom interferometry, quantum computation atom cavity electrodynamics (Cavity QED), precision measurement, low dimensional physics etc. Realization of BEC is also very important from the fact that relatively weak two-particle interaction in dilute alkali atoms allows them to be used as theoretical and experimental tool to study coherent matter wave properties. Although all the atoms in BEC occupy the same wave function, BEC does not quite behave like a single atom wave function. Since a BEC is a Superfluid, atoms in a BEC move with zero viscosity, suppressing collisions. However, collisions do occur within a BEC. At high densities phase collisions may destroy the BEC due to three-body recombination process [112] but three body recombination process can be suppressed at low atomic densities [116].

6.3 OPTOMECHANICAL SYSTEMS

In this section, we study various optomechanical systems involving mechanical mode, optical mode and Bose-Einstein Condensates with their applications in nano-systems.

6.3.1 QUANTUM DYNAMICS OF TWO-OPTICAL MODES AND A SINGLE MECHANICAL MODE

We are considering an optomechanical setup composed of an optical cavity with two optical modes (represented by the operators a_1 and a_2) and one mechanical mode (represented by b) as shown in Fig. 6.3.

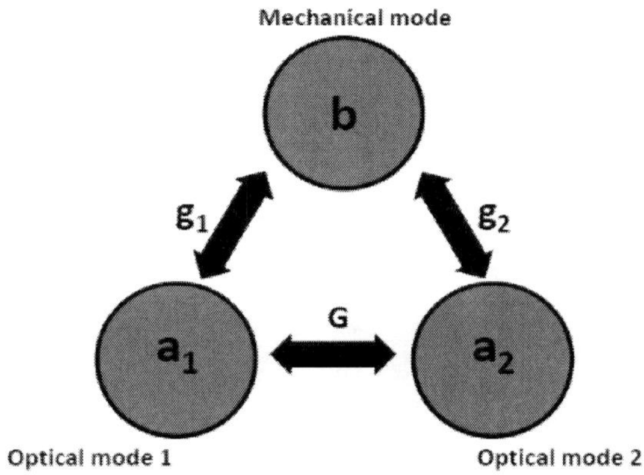

FIGURE 6.3 Schematic view of the optomechanical setup comprising of single mechanical mode (denoted by b) and two optical modes (denoted by a_1 and a_2) with different coupling rates.

The Hamiltonian, describing this optomechanical system, is given by Refs. [117, 118]:

$$H_1 = H_0 + H_{int}, \tag{1}$$

where

$$H_0 = \hbar\omega_1 a_1^\dagger a_1 + \hbar\omega_2 a_2^\dagger a_2 + \hbar\omega_m b^\dagger b, \tag{2}$$

$$H_{\text{int}} = -[\hbar g_1 a_1^\dagger a_1 + \hbar g_2 a_2^\dagger a_2 - \hbar G(a_1 a_2^\dagger + a_2 a_1^\dagger)](b + b^\dagger) \tag{3}$$

H_0 denotes the bare energies of the mechanical mode having frequency ω_m and that of the two optical modes having frequencies ω_1 and ω_2. H_{int} denotes the interaction between all the three modes with coupling rates g_1 and g_2. G is the coupling between the two optical modes via the mechanical oscillator. The space dependence of the cavity mode frequencies results in this interaction Hamiltonian. This type of Hamiltonian is usually found in a setup having membrane in the center of the cavity [119].

Using the Hamiltonian of Eq. (1), the Quantum Langevin Equations for the operators' a_1, a_2 and b are given as:

$$\dot{a}_1 = -i\omega_1 a_1 + i g_1 a_1 (b + b^\dagger) - iG a_2 (b + b^\dagger) - \frac{\gamma_{c1}}{2} a_1 + \sqrt{\gamma_{c1}}\, a_{in1}, \tag{4}$$

$$\dot{a}_2 = -i\omega_2 a_2 + i g_2 a_2 (b + b^\dagger) - iG a_1 (b + b^\dagger) - \frac{\gamma_{c2}}{2} a_2 + \sqrt{\gamma_{c2}}\, a_{in2}, \tag{5}$$

$$\dot{b} = -i\omega_m b + i g_1 a_1^\dagger a_1 + i g_2 a_2^\dagger a_2 - iG(a_1 a_2^\dagger + a_2 a_1^\dagger) - \gamma_m b + \sqrt{\gamma_m}\, \xi, \tag{6}$$

Where ξ represents the noise operator arising as the mechanical mode is also affected by a random Brownian force. a_{in1} and a_{in2} are the input noise operators for the two optical modes.

6.3.1.1 DYNAMICS OF FLUCTUATIONS: NORMAL-MODE SPLITTING

Now in order to study the fluctuation dynamics of the system, we rewrite the operator's a_1, a_2 and b around their steady state values as $a_1 \rightarrow \alpha_1 + a_1$, $a_2 \rightarrow \alpha_2 + a_2$ and $b \rightarrow \beta + b$. Hereα_1, α_2 and β are the different steady state parameters for the two optical modes and the mechanical mode, respectively. Their values can be evaluated by factorizing the nonlinear algebraic Eqs. (4)–(6) and putting their time derivatives to zero, given as:

$$\alpha_1 = \frac{iG\alpha_2(\beta + \beta^\dagger)}{\left[i g_1(\beta + \beta^\dagger) - \left(i\omega_1 + \frac{\gamma_{c1}}{2} \right) \right]}, \tag{7}$$

$$\alpha_2 = \frac{iG\alpha_1(\beta + \beta^\dagger)}{\left[ig_2(\beta + \beta^\dagger) - \left(i\omega_2 + \frac{\gamma_{c2}}{2} \right) \right]}, \tag{8}$$

$$\beta = \frac{ig_1\alpha_1^\dagger\alpha_1 + ig_2\alpha_2^\dagger\alpha_2 - iG(\alpha_1\alpha_2^\dagger + \alpha_2\alpha_1^\dagger)}{(i\omega_m + \gamma_m)} \tag{9}$$

It gives a new effective Hamiltonian for the system after retaining the bilinear terms only.

$$H_1 = \hbar\Omega_1 a_1^\dagger a_1 + \hbar\Omega_2 a_2^\dagger a_2 + \hbar\omega_m b^\dagger b - [\hbar G_1(a_1^\dagger + a_1) + \hbar G_2(a_2^\dagger + a_2)](b + b^\dagger) + \hbar\lambda(a_1 a_2^\dagger + a_2 a_1^\dagger) \tag{10}$$

where $\Omega_1 = \omega_1 - 2\beta g_1$, $\Omega_2 = \omega_2 - 2\beta g_2$, $G_1 = g_1\alpha_1 - G\alpha_2$, $G_2 = g_2\alpha_2 - G\alpha_1$ and $\lambda = 2G\beta$.

The following conditions: $\omega_1\alpha_1 + 2\beta G\alpha_2 - 2\beta g_1\alpha_1 = 0$, $\omega_2\alpha_2 + 2\beta G\alpha_1 - 2\beta g_2\alpha_2 = 0$ and $\omega_m\beta + 2G\alpha_1\alpha_2 - g_1\alpha_1^2 - g_2\alpha_2^2 = 0$ can eliminate the linear terms in the fluctuation operators. We can also evaluate the steady state values of the fluctuations using these conditions.

The linearized equations of motion obtained by using the Hamitonian equation (10) are given as follows:

$$\dot{a}_1(t) = \left[i\Delta_1 - \frac{\gamma_{c1}}{2} \right] a_1(t) + iG_1\left(b(t) + b^\dagger(t)\right) - i\lambda a_2(t) + \sqrt{\gamma_{c1}}a_{in1}(t), \tag{11}$$

$$\dot{a}_2(t) = \left[i\Delta_2 - \frac{\gamma_{c2}}{2} \right] a_2(t) + iG_2\left(b(t) + b^\dagger(t)\right) - i\lambda a_1(t) + \sqrt{\gamma_{c2}}a_{in2}(t), \tag{12}$$

$$\dot{b}(t) = -(i\omega_m + \gamma_m)b(t) + iG_1\left(a_1(t) + a_1^\dagger(t)\right) + iG_2\left(a_2(t) + a_2^\dagger(t)\right) + \sqrt{\gamma_m}\xi(t) \tag{13}$$

where Δ_1=$-\Omega_1$, Δ_2=$-\Omega_2$. Now, using the amplitude and phase quadratures for the system with $X_1(t) = [a_1(t) + a_1^\dagger(t)]$, $Y_2(t) = i[a_2^\dagger(t) - a_2(t)]$, $Y_2(t) = i[a_2^\dagger(t) - a_2(t)]$, $Y_2(t) = i[a_2^\dagger(t) - a_2(t)]$, $Q(t) = [b(t) + b^\dagger(t)]$, $X_{in1}(t) = [a_{in1}(t) + a_{in1}^\dagger(t)]$, $X_{in1}(t) = [a_{in1}(t) + a_{in1}^\dagger(t)]$, $X_{in2}(t) = [a_{in2}(t) + a_{in2}^\dagger(t)]$,

$X_{in2}(t)=[a_{in2}(t)+a_{in2}^{\dagger}(t)]$ and $Y_{in2}(t)=i[a_{in2}^{\dagger}(t)-a_{in2}(t)]$, the above linearized equations can be rewritten as follows:

$$\dot{X}_1(t)=-\Delta_1 Y_1(t)+\lambda Y_2(t)-\frac{\gamma_{c1}}{2}X_1(t)+\sqrt{\gamma_{c1}}X_{in1}(t) \tag{14}$$

$$\dot{Y}_1(t)=\Delta_1 X_1(t)+G_1 Q(t)-\lambda X_2(t)+\sqrt{\gamma_{c1}}Y_{in1}(t)-\frac{\gamma_{c1}}{2}Y_1(t), \tag{15}$$

$$\dot{X}_2(t)=-\Delta_2 Y_2(t)+\lambda Y_1(t)-\frac{\gamma_{c2}}{2}X_2(t)+\sqrt{\gamma_{c2}}X_{in2}(t), \tag{16}$$

$$\dot{Y}_2(t)=\Delta_2 X_2(t)+G_2 Q(t)-\lambda X_1(t)-\frac{\gamma_{c1}}{2}Y_2(t)+\sqrt{\gamma_{c2}}Y_{in2}(t), \tag{17}$$

$$\dot{Q}(t)=\omega_m P(t) \tag{18}$$

$$\dot{P}(t)=-\omega_m Q(t)-\gamma_m P(t)+G_1 X_1(t)+G_2 X_2(t)+W(t) \tag{19}$$

where $W(t)=i\sqrt{\gamma_m}(\xi^{\dagger}(t)-\xi(t))$ is the Hermitian Brownian noise operator, which satisfies the following correlation [120]:

$$<W(t)W(t')>=\frac{\gamma_m}{\omega_m}\int\frac{d\omega}{2\pi}e^{-i\omega(t-t')}\omega\left[1+\coth\left(\frac{\hbar\omega}{2k_B T}\right)\right]$$

However, in the Fourier space, it satisfies the following correlation [121]:

$$<W(\omega)W(\omega')>=2\pi\omega\frac{\gamma_m}{\omega_m}\left[1+\coth\left(\frac{\hbar\omega}{2k_B T}\right)\right]\delta(\omega+\omega'), \tag{20}$$

where T is the temperature of the reservoir and k_B is the Boltzmann constant.

By solving Eq. (14)–(19) in the frequency domain, the position spectrum for the mechanical mode of the system can be obtained as:

$$S_Q(\omega)=\frac{1}{4\pi}\int d\omega' e^{-i(\omega+\omega')t}\left\langle Q(\omega)Q(\omega')+Q(\omega')Q(\omega)\right\rangle, \tag{21}$$

where $Q(\omega)=\dfrac{A_1(\omega)+A_2(\omega)+A_3(\omega)+A_4(\omega)+A_5(\omega)}{B(\omega)}$, (22) with the help of the correlations

$$< X_{in1}(\omega)Y_{in1}(\omega') >=< X_{in2}(\omega)Y_{in2}(\omega') >= 2i\pi\delta(\omega+\omega'),$$

$$< X_{in1}(\omega)X_{in1}(\omega') >=< X_{in2}(\omega)X_{in2}(\omega') >=< Y_{in1}(\omega)Y_{in1}(\omega') >=< Y_{in2}(\omega)Y_{in2}(\omega') >= 2\pi\delta(\omega+\omega'),$$

$$< Y_{in1}(\omega)X_{in1}(\omega') >=< Y_{in2}(\omega)X_{in2}(\omega') >= -2i\pi\delta(\omega+\omega').$$

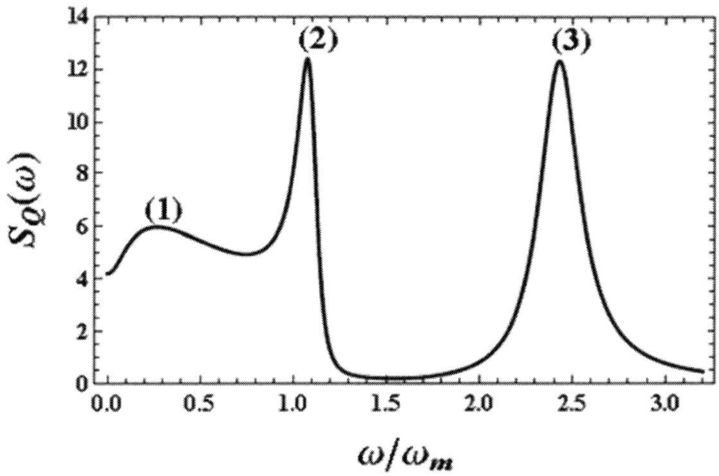

FIGURE 6.4 Plot of position spectrum $S_Q(\omega)$ as a function of dimensionless frequency (ω/ω_m) for $\Upsilon_m=10^{-4}\omega_m$, $\Delta_1=-1.3\omega_m$, $\Delta_2=-1.5\omega_m$, $G=1.5\omega_m$, $G_1=1.5\omega_m$, $G_2=6\omega_m$, $\Upsilon_{c1}=0.2\omega_m$, $\Upsilon_{c2}=0.6\omega_m$, $\beta=0.06$ and $k_BT/\hbar\omega_m=10^5$.

The values of $A_1(\omega)$, $A_2(\omega)$, $A_3(\omega)$, $A_4(\omega)$, $A_5(\omega)$ and $B(\omega)$ are given as:

$$A_1(\omega) = W(\omega)\omega_m\left(\Delta_2^2 - \omega^2 + \frac{\gamma_{c2}^2}{4} + i\omega\gamma_{c2}\right)C_1(\omega)C_2(\omega), \tag{23}$$

$$A_2(\omega) = X_{in1}(\omega)\sqrt{\gamma_{c1}}\left(\Delta_2^2 - \omega^2 + \frac{\gamma_{c2}^2}{4} + i\omega\gamma_{c2}\right)C_1(\omega)C_3(\omega) \tag{24}$$

$$A_3(\omega) = Y_{in1}(\omega)\sqrt{\gamma_{c1}}\left(\Delta_2^2 - \omega^2 + \frac{\gamma_{c2}^2}{4} + i\omega\gamma_{c2}\right)\begin{bmatrix} \lambda\omega_m G_2\left\{i\omega + \frac{\gamma_{c2}}{2}\right\}C_2(\omega) \\ + \left\{\lambda^2\Delta_2 - \Delta_1\left(\Delta_2^2 - \omega^2 + \frac{\gamma_{c2}^2}{4} + i\omega\gamma_{c2}\right)\right\}C_3(\omega) \end{bmatrix}, \tag{25}$$

$$A_4(\omega) = X_{in2}(\omega)\sqrt{\gamma_{c2}}\left(\Delta_2^2 - \omega^2 + \frac{\gamma_{c2}^2}{4} + i\omega\gamma_{c2}\right)\begin{bmatrix} \omega_m G_2 C_2(\omega)\left\{i\omega + \frac{\gamma_{c1}}{2}\right\}\left\{i\omega + \frac{\gamma_{c2}}{2}\right\} \\ +\lambda\omega_m G_1 C_1(\omega)\begin{Bmatrix} \Delta_1\left(i\omega + \frac{\gamma_{c2}}{2}\right) \\ +\Delta_2\left(i\omega + \frac{\gamma_{c1}}{2}\right) \end{Bmatrix} \\ +\lambda^2\omega_m G_2\left\{\Delta_1\left(i\omega - \frac{\gamma_{c2}}{2}\right) + \Delta_2\left(i\omega + \frac{\gamma_{c1}}{2}\right)\right\}^2 \end{bmatrix}, \quad (26)$$

$$A_5(\omega) = Y_{in2}(\omega)\sqrt{\gamma_{c2}}\begin{bmatrix} -\omega_m G_2 \Delta_2 C_2(\omega)\left(\Delta_2^2 - \omega^2 + \frac{\gamma_{c2}^2}{4} + i\omega\gamma_{c2}\right)\left(i\omega + \frac{\gamma_{c1}}{2}\right) \\ +\lambda C_3(\omega)\begin{Bmatrix} C_1(\omega)\left(i\omega + \frac{\gamma_{c2}}{2}\right) + \lambda^2\Delta_2^2 \\ -\Delta_1\Delta_2\left(\Delta_2^2 - \omega^2 + \frac{\gamma_{c2}^2}{4} + i\omega\gamma_{c2}\right) \end{Bmatrix} \end{bmatrix}, \quad (27)$$

$$B(\omega) = C_4(\omega) - C_5(\omega), \quad (28)$$

where $C_1(\omega) = \left(\Delta_2^2 - \omega^2 + \frac{\gamma_{c2}^2}{4} + i\omega\gamma_{c2}\right)\left(i\omega + \frac{\gamma_{c1}}{2}\right) + \lambda^2\left(i\omega + \frac{\gamma_{c2}}{2}\right)$ (29)

$$C_2(\omega) = \left(\Delta_2^2 - \omega^2 + \frac{\gamma_{c2}^2}{4} + i\omega\gamma_{c2}\right)\left(\Delta_1^2 - \omega^2 + \frac{\gamma_{c1}^2}{4} + i\omega\gamma_{c1}\right) + \lambda^4 + 2\lambda^2\left[\left(i\omega + \frac{\gamma_{c2}}{2}\right)\left(i\omega + \frac{\gamma_{c1}}{2}\right) - \Delta_1\Delta_2\right] \quad (30)$$

$$C_3(\omega) = \omega_m G_1\left[\left(i\omega + \frac{\gamma_{c1}}{2}\right)\left(\Delta_2^2 - \omega^2 + \frac{\gamma_{c2}^2}{4} + i\omega\gamma_{c2}\right) + \lambda^2\left(i\omega + \frac{\gamma_{c2}}{2}\right)\right] + \lambda\omega_m G_2\Delta_1\left(i\omega + \frac{\gamma_{c2}}{2}\right) + \lambda\omega_m G_2\Delta_2\left(i\omega + \frac{\gamma_{c1}}{2}\right) \quad (31)$$

$$C_4(\omega) = C_2(\omega)\left(\Delta_2^2 - \omega^2 + \frac{\gamma_{c2}^2}{4} + i\omega\gamma_{c2}\right)\begin{bmatrix} (\omega_m^2 - \omega^2 + i\omega\gamma_m)C_1(\omega) + \omega_m G^2\Delta_2\left(i\omega + \frac{\gamma_{c1}}{2}\right) \\ -\lambda\omega_m G_1 G_2\left(i\omega + \frac{\gamma_{c2}}{2}\right) \end{bmatrix} \quad (32)$$

$$C_5(\omega)=C_3(\omega)\begin{bmatrix} -\left\{\lambda G_2\Delta_2+G_1\left(\Delta_2^2-\omega^2+\frac{\gamma_{c2}^2}{4}+i\omega\gamma_{c2}\right)\right\} \\ \left\{\Delta_1\left(\Delta_2^2-\omega^2+\frac{\gamma_{c2}^2}{4}+i\omega\gamma_{c2}\right)-\lambda^2\Delta_2\right\} \\ +\lambda G_2 C_1(\omega)\left(i\omega+\frac{\gamma_{c2}}{2}\right) \end{bmatrix} \tag{33}$$

We have shown the plot of displacement spectrum $S_Q(\omega)$ as a function of dimensionless frequency (ω/ω_m) in Fig. 6.4. The coupling between the two optical mode fluctuations and the mechanical mode fluctuations around the steady state results in the splitting of normal modes into three modes (Normal mode splitting (NMS)). The NMS is observed in the fluctuation spectra of mirror displacement not in the steady state spectra. It involves driving three parametrically coupled nondegenerate modes out of equilibrium. Moreover, Fig. 6.4. further shows the coherent energy exchange between the three modes. This energy exchange should take place on a time scale faster than the decoherence of each mode in order to observe the NMS. The observation of NMS is prevented by the onset of parametric instability on the positive detuning side.

6.3.1.2 STEADY-STATE ENTANGLEMENT

Here, we study the steady-state bipartite entanglement of the three possible subsystems, by quantifying it in terms of the logarithmic negativity E_N [122, 123] of bimodal Gaussian states. Now, in order to measure the stationary entanglement between any two modes, we rewrite the linearized equations of motion (14)–(19) in the following compact form [63]:

$$\dot{R}(t)=MR(t)+N(t), \tag{34}$$

where $R(t)=(Q(t),P(t),X_1(t),Y_1(t),X_2(t),Y_2(t))^\tau$ represents the vector of the quadrature fluctuations and $N(t)=(0,W(t),\sqrt{\gamma_{c1}}X_{in1},\sqrt{\gamma_{c1}}Y_{in1},\sqrt{\gamma_{c2}}X_{in2},\sqrt{\gamma_{c2}}Y_{in2})^\tau$ denotes the corresponding vector of noises (the superscript τ denotes the transposition). Furthermore, M is the drift matrix, which is given as:

$$M = \begin{bmatrix} 0 & \omega_m & 0 & 0 & 0 & 0 \\ -\omega_m & -\gamma_m & G_1 & 0 & G_2 & 0 \\ 0 & 0 & -\gamma_{c1}/2 & -\Delta_1 & 0 & \lambda \\ G_1 & 0 & \Delta_1 & -\gamma_{c1}/2 & -\lambda & 0 \\ 0 & 0 & 0 & \lambda & -\gamma_{c2}/2 & -\Delta_2 \\ G_2 & 0 & -\lambda & 0 & \Delta_2 & -\gamma_{c2}/2 \end{bmatrix} \tag{35}$$

The formal solution of eq. (34) can be written as $R(t) = F(t)R(0) + \int_0^t ds F(s)N(t-s)$ where $F(\infty) = 0$. The system reaches a steady state only if it is stable and that can only be possible if all the eigen values of the drift matrix M have negative real parts such that $F(\infty) = 0$. In this situation, the following stability conditions must always be satisfied which can be obtained by applying the Routh-Hurwitz criterion [124, 125]:

$$\begin{aligned} S_1 &= \frac{\gamma_{c1}^2\gamma_{c2}^2\omega_m^2}{16} + \frac{\lambda^2\omega_m^2\gamma_{c1}\gamma_{c2}}{2} - \frac{\omega_m\lambda G_1G_2\gamma_{c1}\gamma_{c2}}{2} + \frac{\gamma_{c1}^2\omega_m^2\Delta_2^2}{4} + \frac{\gamma_{c1}^2\omega_m\Delta_2G_2^2}{4} + \frac{\gamma_{c2}^2\omega_m^2\Delta_1^2}{4} \\ &+ \frac{\gamma_{c2}^2\omega_m\Delta_1G_1^2}{4} + \omega_m^2(\Delta_1^2\Delta_2^2 + \lambda^4 - 2\lambda^2\Delta_1\Delta_2) + \omega_m(G_1^2\Delta_2^2\Delta_1 - G_1^2\lambda^2\Delta_2) \\ &+ \omega_m(2\lambda G_1G_2\Delta_1\Delta_2 - 2G_1G_2\lambda^3 + G_2^2\Delta_1^2\Delta_2 - \Delta_1G_2^2\lambda) > 0, \end{aligned} \tag{36}$$

$$S_2 = (c_5c_4c_3 + c_6c_1c_5 - c_6c_3^2 - c_2c_5^2) > 0, \tag{37}$$

where

$$\begin{aligned} c_1 &= \gamma_m\left[\frac{\gamma_{c1}^2\gamma_{c2}^2}{16} + \frac{\lambda^2\gamma_{c1}\gamma_{c2}}{2} + \frac{\gamma_{c1}^2\Delta_2^2}{4} + \frac{\gamma_{c2}^2\Delta_1^2}{4} + \Delta_1^2\Delta_2^2 + \lambda^4 - 2\lambda^2\Delta_1\Delta_2\right] \\ &+ (\gamma_{c1} + \gamma_{c2})\left(\frac{\omega_m^2\gamma_{c1}\gamma_{c2}}{4} + \lambda^2\omega_m^2 - \omega_m\lambda G_1G_2\right) + \gamma_{c1}(\omega_m^2\Delta_2^2 + \omega_m\Delta_2G_2^2) \\ &+ \gamma_{c2}(\omega_m^2\Delta_1^2 + \omega_m\Delta_1G_1^2), \end{aligned} \tag{38}$$

$$\begin{aligned} c_2 &= \frac{\gamma_{c1}^2\gamma_{c2}^2}{16} + \omega_m^2\left[\frac{\gamma_{c1}^2}{4} + \frac{\gamma_{c2}^2}{4} + \gamma_{c1}\gamma_{c2}\right] + \frac{\gamma_{c1}^2\gamma_{c2}\gamma_m}{4} + \frac{\gamma_{c2}^2\gamma_{c1}\gamma_m}{4} + 2\lambda^2\omega_m^2 - 2\omega_m\lambda G_1G_2 \\ &+ \lambda^2\left[\frac{\gamma_{c1}\gamma_{c2}}{2} + \gamma_m(\gamma_{c1} + \gamma_{c2})\right] + \Delta_2^2\left[\frac{\gamma_{c1}^2}{4} + \omega_m^2 + \gamma_{c1}\gamma_m\right] + \lambda^4 - 2\lambda^2\Delta_1\Delta_2 \\ &+ \Delta_1^2\left[\frac{\gamma_{c2}^2}{4} + \omega_m^2 + \gamma_{c2}\gamma_m\right] + \omega_m(\Delta_2G_2^2 + \Delta_1G_1^2) + \Delta_1^2\Delta_2^2, \end{aligned} \tag{39}$$

$$c_3 = \gamma_{c1}\left[\frac{\gamma_{c2}^2}{4} + \omega_m^2 + \lambda^2 + \Delta_2^2\right] + \gamma_{c2}\left[\frac{\gamma_{c1}^2}{4} + \omega_m^2 + \lambda^2 + \Delta_1^2\right]$$
$$+ \gamma_m\left[\frac{\gamma_{c1}^2}{4} + \frac{\gamma_{c2}^2}{4} + \gamma_{c1}\gamma_{c2} + 2\lambda^2 + \Delta_1^2 + \Delta_2^2\right], \quad (40)$$

$$c_4 = \frac{\gamma_{c1}^2}{4} + \frac{\gamma_{c2}^2}{4} + \gamma_{c1}\gamma_{c2} + \gamma_m(\gamma_{c1} + \gamma_{c2}) + \omega_m^2 + \Delta_1^2 + \Delta_2^2 + 2\lambda^2, \quad (41)$$

$$c_5 = (\gamma_{c1} + \gamma_{c2} + \gamma_m), \quad c_6 = 1. \quad (42)$$

Since the dynamics is linearized and the quantum noises are white in nature, the steady state of the system will be a zero mean Gaussian state. Thus, the system can be completely characterized by a 6×6 correlation matrix $V_{ij} = (< R_i(\infty)R_j(\infty) + R_j(\infty)R_i(\infty) >)/2$. For a stable system, starting from the formal solution of Eq. (34), one arrives at

$$V_{ij}(\infty) = \sum_{k,l} \int_0^\infty ds \int_0^\infty ds' F_{ik}(s)F_{jl}(s')D_{kl}(s-s'). \quad (43)$$

Here, $D_{kl}(s-s') = (< N_k(s)N_l(s') + N_l(s')N_k(s) >)/2$ is the matrix of the stationary noise correlation functions. Moreover, we consider the oscillator with a very high mechanical quality factor $Q = \omega_m / \gamma_m \to \infty$ in order to achieve the mechanical entanglement. In this limit, the quantum Brownian noise becomes delta-correlated [126] such that

$$< W(t)W(t') + W(t')W(t) > \simeq \gamma_m(2n_{th} + 1)\delta(t-t') \quad (44)$$

with bath occupation $n_{th} = \left[\exp\left(\frac{\hbar\omega_m}{2k_B T}\right) - 1\right]^{-1}$ and one recovers a Markovian process. Within this Markovian limit, we finally obtain $D_{kl}(s-s') = D_{kl}\delta(s-s')$, with the diffusion matrix D given as:

$$D = \begin{bmatrix} 0 & 0 & 0 & 0 & 0 & 0 \\ 0 & \gamma_m(2n_{th}+1) & 0 & 0 & 0 & 0 \\ 0 & 0 & \gamma_{c1}/2 & 0 & 0 & 0 \\ 0 & 0 & 0 & \gamma_{c1}/2 & 0 & 0 \\ 0 & 0 & 0 & 0 & \gamma_{c2}/2 & 0 \\ 0 & 0 & 0 & 0 & 0 & \gamma_{c2}/2 \end{bmatrix}, \quad (45)$$

Which is obtained using the definitions of $X_{in1}(t), Y_{in1}(t), X_{in2}(t), Y_{in2}(t)$ and the fact that the five components of N(t) are uncorrelated. As a result, Eq.(43) reduces to $V = \int_0^\infty ds F(s) D F(s)^\tau$, which is equivalent to the following Lyapunov equation

$$MV + VM^\tau \equiv -D, \tag{46}$$

for the correlation matrix in the steady state $[F(\infty) = 0]$. This Lyapunov equation is the linear matrix equation and can be straightforwardly solved for V, which is too cumbersome to be reported here. Now, in order to measure the entanglement between any two modes of the system, one requires to compute E_N, which can be obtained by tracing out the third mode (i.e., removing the rows and columns of V which correspond to the third mode). Thus, the reduce state can then be fully characterized by the 4 × 4 matrix V′ and still remains Gaussian. In the CV case, the logarithmic negativity E_N can be defined as [122, 123]:

$$E_N = \max[0, -\ln 2\mu^-]. \tag{47}$$

Here, with $\mu^- \equiv \frac{1}{\sqrt{2}}\sqrt{A - \sqrt{A^2 - 4\det(V')}}$ which is the smallest symplectic eigenvalue. We have used the 2×2 block form of V' as:

$$V' = \begin{bmatrix} X & Z \\ Z^\tau & Y \end{bmatrix}. \tag{48}$$

The Eq. (47) clearly depicts that the logarithmic negativity is the decreasing function of μ^- which measures how much two Gaussian states are entangled. A Gaussian state is entangled only if $\mu^- < 1/2$ (or $4\det(V') < \sum(V') - 1/4$) which is equivalent to Simon's necessary and sufficient entanglement non-positive partial transpose criterion of the Gaussian states [127]. Now, we study the stationary entanglement in the three possible bipartitions of the system using the logarithmic negativity E_N, where, $E_N^{(1)}$, $E_N^{(2)}$ and $E_N^{(3)}$ denote the logarithmic negativities for the mechanical mode-optical mode a_1, mechanical mode-optical mode a_2 and optical mode a_1-optical mode a_2 entanglements respectively.

In Fig. 6.5 (a), the logarithmic negativities of the three bipartite cases $E_N^{(1)}$ (thin solid line), $E_N^{(2)}$ (dashed line) and $E_N^{(3)}$ (thick solid line) versus normalized effective optical frequency Δ_1/ω_m are plotted for $n_{th} = 20$ (corresponding to a bath temperature T = 0.01K). Figure 6.5 (b) depicts the same plot but at a higher value of bath occupation, $n_{th} = 1250$ (i.e., T = 0.6K). We have also demonstrated the logarithmic negativities for the three bipartite entanglements versus Δ_2/ω_m for n_{th} = 20 (fig.6.5 (c)) and n_{th} = 1250 (fig.6.5 (d)). The stationary entanglement between a single driven optical cavity field mode and a mechanical resonator via radiation pressure has already been examined in Ref. [128]. Our model Hamiltonian includes the two cavity modes with different frequencies, each driven by an intense laser, those are not only separately coupled via pondermotive interaction to the mechanical resonator but are also coupled together via mechanical oscillator. Due to the presence of this second optical cavity mode, we have observed that the steady-state entanglement is now produced within the three subsystems, namely, mechanical mode-optical mode a_1, mechanical mode-optical mode a_2 and optical mode a_1-optical mode a_2. The simultaneous presence of all the three possible bipartite entanglements in the chosen parameter regime for a wide range of effective optical frequencies indicates the strong correlation between the two optical fields and the mechanical oscillator at the steady state. Although, these entanglements vanish at some values of effective optical frequencies. Thus, nonzero entanglements can be achieved by the proper choice of the system parameters. Further note that the presence of nonzero entanglement between the two optical modes is due to the effect of the mechanical motion of the resonator as both the cavity modes are coupled together via mechanical oscillator only. No such entanglement would be possible if the mechanical element is fixed. In addition to this, at a higher temperature T = 0.6K, the qualitative behaviour of all the three possible bipartite entanglements are demonstrated in Figs. 6.5(b) and 6.5(d), which are similar to that of the corresponding Figs. 6.5(a) and 6.5(c), respectively. However, the attainable values of these stationary entanglements are comparatively lower. In spite of the lower values, they are still quite robust against temperature.

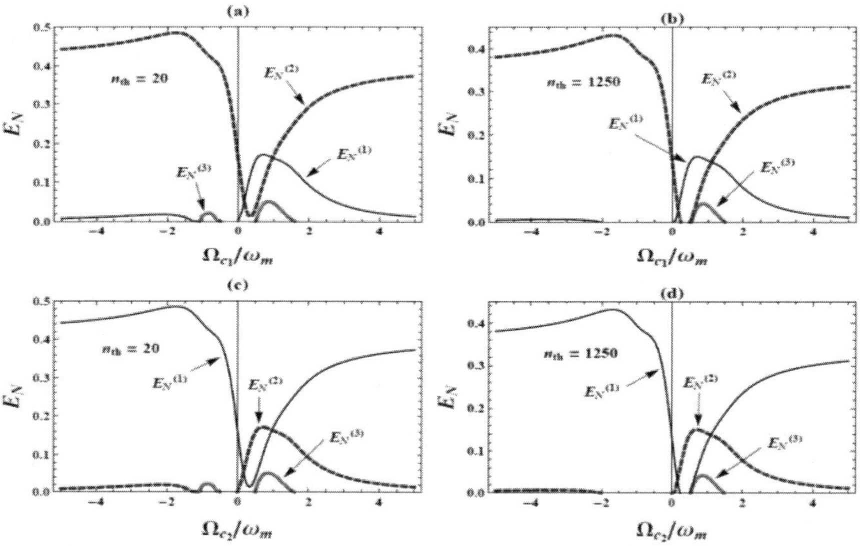

FIGURE 6.5 (a): Plot of logarithmic negativities of the three bipartite cases $E_N^{(1)}$ (thin solid line), $E_N^{(2)}$ (dashed line) and $E_N^{(3)}$ (thick solid line) versus the normalized effective optical frequency (Δ_1/ω_m) for $n_{th} = 20$. We have taken $\Upsilon_m = 10^{-5}\omega_m$, $\Upsilon_{c1} = \omega_m$, $\Upsilon_{c2} = 0.5\omega_m$, $G_1 = 0.1\omega_m$, $G_2 = 0.9\omega_m$, $\lambda = 0.8\omega_m$ and $\Delta_2 = -\omega_m$. (b): Same as in (a) but for $n_{th} = 1250$. (c): Plot of $E_N^{(1)}$ (thin solid line), $E_N^{(2)}$ (dashed line) and $E_N^{(3)}$ (thick solid line) versus Δ_2/ω_m for nth = 20, $\Upsilon_{c1} = 0.5\omega_m$, $\Upsilon_{c2} = \omega_m$, $G_1 = 0.9\omega_m$, $G_2 = 0.1\omega_m$ and $\Delta_1 = -\omega_m$. Other parameters are same as in (a). (d): Same as in (c) but for $n_{th} = 1250$.

This is examined in more detail in Fig. 6.6, where we study the robustness of the steady-state entanglements with respect to reservoir temperature T. As expected, the logarithmic negativities hence the entanglements decrease with the increase in mechanical resonator's environmental temperature. Figs. 6.6 (a) and 6.6(b) further show that the entanglement between the mechanical mode and any of the two optical modes can be selectively made large and more robust against thermal noise by considering an appropriate parameter regime. Thus, one can not only find the optomechanical entanglement but can also selectively optimize and increase it. Furthermore, for T < 2.4K, the simultaneous presence of all the three possible bipartite stationary entanglements again guarantees the strong correlation between the three bosonic modes.

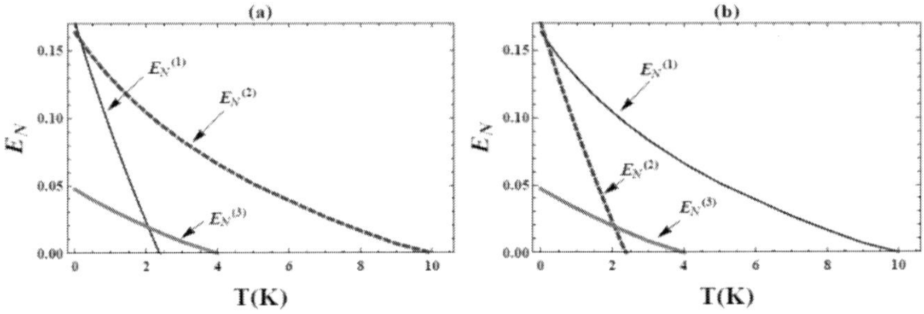

FIGURE 6.6 (a): Plot $E_N^{(1)}$ (thin solid line), $E_N^{(2)}$ (dashed line) and $E_N^{(3)}$ (thick solid line) as a function of bath temperature T for $\omega_m=2\pi\times10^7$Hz, $\Upsilon_m = 10^{-5}\omega_m$, $\Upsilon_{c1}=0.5\omega_m$, $\Upsilon_{c2}=\omega_m$, $G_1 = 0.9\omega_m$, $G_2 = 0.1\omega_m$, $\lambda = 0.8\omega_m$, $\Delta_1=-\omega_m$ and $\Delta_2= \omega_m$. (b): Plot of Plot $E_N^{(1)}$ (thin solid line), $E_N^{(2)}$ (dashed line) and $E_N^{(3)}$ (thick solid line) as a function of T for $\Upsilon_{c1}=\omega_m$, $\Upsilon_{c2}=0.5\omega_m$, $G_1 = 0.1\omega_m$, $G_2 = 0.9\omega_m$, $\Delta_1=\omega_m$ and $\Delta_2=-\omega_m$. Other parameters used are same as in (a).

Now, the experimental prospects for various parameters are illustrated below to get an intuitive picture of the feasibility of the dynamics of the system. The frequency of mechanical oscillator is taken to be $2\pi \times 10^7$Hz with a mechanical quality factor of 10^5in our calculations, which is very close to that of recently performed experiments [5, 6]. The optical cavities of high quality factor can decay with a rate nearly $2\pi\times10$ MHz [129, 130]. The coupling rate of $2\pi\times1$ MHz has been recorded for the optomechanical setups [131]. Such coupling rates can be further increased above 10 MHz by using nanoscales [132] which enhances the local cavity field in such structures. The condition $\gamma_m << \omega_m << k_B T / \hbar$ is always taken into account in various standard optomechanical experiments [93, 133, 134].

6.3.2 PHONON LASER EFFECT AND SUPERRADIANT PHASE TRANSITION IN MAGNETIC CANTILEVER COUPLED TO A BOSE EINSTEIN CONDENSATE

In this section, we propose a phonon laser operating like a two-level optical laser by using a Bose Einstein Condensate of ^{87}Rb atoms coupled to a magnetic cantilever [16]. We also show that the same system can undergo a Dicke-Hepp-Lieb type phase transition [71–74] into a phonon super radiance regime for strong coupling. We also see the importance of atomic two-body interaction in this system dynamics.

6.3.2.1 PHONON LASER EFFECT

Here, we propose a phonon lasing device that operates like a two-level optical laser. As shown in Fig. 6.7(a), it essentially consists of a gas of Bose Einstein Condensate of ^{87}Rb atoms at a distance l_0 above a cantilever resonator with a ferromagnetic tip, which generates a strong magnetic field $G_m = \frac{2\mu_0 \left|\overrightarrow{\mu_m}\right|}{4\pi y_0^4}$ [16]. Here, μ_0 denotes the permeability of free space and $\vec{\mu}_m$ represents the magnetic moment of the ferromagnetic tip. The out-of-plane mechanical oscillations are performed by the magnetic cantilever, which in turn transduces into an oscillatory magnetic field $\vec{B}_r(t)$.

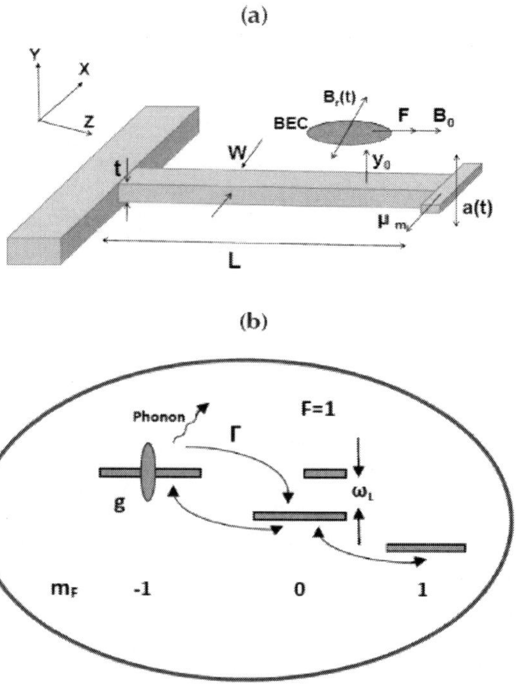

FIGURE 6.7 (a). Schematic representation of the coupling of magnetic cantilever (orange) to the BEC (blue). a(t) represents the out-of-plane mechanical oscillations of the cantilever and y_0 as the distance of BEC from the cantilever. The magnetic cantilever and the atomic spin F are coupled by the oscillatory component of the magnetic field $B_r(t)$. (b): Hyperfine structure of ^{87}Rb BEC atoms showing the emission of phonons due to the transition from the state |F=1|, m_F=–1> to |F=1|, m_F=0>.

There is an interaction between the atomic spin $\vec{F}$ and the magnetic field $\vec{B}_r(t)$ such that $\vec{\mu} = \mu_B g_F \vec{F}$ is the magnetic moment operator. The hyperfine spin levels $|F=1, m_F\rangle$ of ^{87}Rb atoms in the ground state are shown in Fig. 6.7(b). Tunable Larmor frequency $\omega_L(=\mu_B\,|\,g_F\,|\,B_0/\hbar)$ gives the energy splitting between the adjacent m_F levels. The detuning $\delta(=\omega_r-\omega_L)$ can be coherently controlled by the tunability of the Larmor frequency where ω_r is the mechanical frequency of the cantilever mode. There is flipping of spins between the ground state energy level $|g> = |F=1, m_F=0>$ and the excited level $|e> = |F=1, m_F=-1>$ due to atom-phonon coupling near resonance ($\delta \approx 0$). The $|g> \leftrightarrow |e>$ transition can be uncoupled from other m_F levels [16]. This kind of coupling between BEC and magnetic cantilever (i.e., cantilever driving the atoms and atoms driving the cantilever) leads to a kind of positive feedback that occurs in all laser like system. A pumping mechanism must also be introduced to compensate for the loss of condensates and the phonon dissipation in order to observe the phonon laser effect. This can be accomplished by pumping the atoms in the excited state in order to maintain a steady population inversion.

The Hamiltonian describing the coupled dynamics of BEC and magnetic cantilever is given as follows:

$$H_2 = H_{phonon} + H_{atom} + H_{atom-phonon} + H_{atom-atom} \tag{49}$$

where,

$$H_{phonon} = \hbar\omega_r a^\dagger a \tag{50}$$

$$\begin{aligned} H_{atom} &= \int d^3r\psi_g^\dagger(r)\left[\frac{-\hbar^2\nabla^2}{2m} + \hbar\omega_g + V_g(r)\right]\psi_g(r) \\ &+ \int d^3r\psi_e^\dagger(r)\left[\frac{-\hbar^2\nabla^2}{2m} + \hbar\omega_e + V_e(r)\right]\psi_e(r) \end{aligned} \tag{51}$$

$$H_{atom-phonon} = \hbar g\left(a^\dagger + a\right)\left[\int \psi_g^\dagger(r)\psi_e(r)d^3r + \int \psi_e^\dagger(r)\psi_g(r)d^3r\right] \tag{52}$$

$$\begin{aligned} H_{atom-atom} &= \sum_{i=g,e}\left(\frac{2\pi\hbar^2 a_{ii}}{m}\int d^3r\psi_i^\dagger(r)\psi_i^\dagger(r)\psi_i(r)\psi_i(r)\right) \\ &+ \frac{4\pi\hbar^2 a_{ge}}{m}\int d^3r\psi_g^\dagger(r)\psi_g(r)\psi_e^\dagger(r)\psi_e(r) \end{aligned} \tag{53}$$

ψ_g and ψ_e represent the ground state and the excited state wave functions of the condensate respectively; g denotes the atom-phonon coupling constant, which is taken to be real. $V_i(r)$ and ω_i (i=g, e) are the trapping potentials and the energies respectively for the ground state and the excited of the condensate. a_{gg}, a_{ee} and a_{ge} are the s-wave scattering lengths for the ground-ground, excited-excited and ground-excited atomic states respectively where we have taken $a_{ge}=a_{eg}$. m is the mass of a condensate atom. We now consider [129]

$$\psi_g(r,t)=\sqrt{N}b_0(t)\xi_g(r),\ \psi_e(r,t)=\sqrt{N}c_0(t)\xi_e(r), \tag{54}$$

where $\xi_g(r)$ and $\xi_e(r)$ are the single particle wave functions for the ground state and the excited state, respectively. They satisfy the normalization condition $\sum_{i=g,e}\int d^3r\left|\xi_i(r)\right|^2=1$. Also, $b_0(t)$ and $c_0(t)$ are the annihilation operators for the atoms in the ground and the excited states respectively. The Hamiltonian in the second quantized form in terms of the normalized operators $b_0\rightarrow\sqrt{N}b_0$ and $c_0\rightarrow\sqrt{N}c_0$, after ignoring the counter rotating terms, is given as follows:

$$\begin{aligned} H_2 &= \hbar\omega_r a^\dagger a+\hbar\omega_0 b_0^\dagger b_0+\hbar\omega_1 c_0^\dagger c_0+\hbar\left[Gab_0c_0^\dagger+G^*a^\dagger c_0 b_0^\dagger\right] \\ &+\frac{\hbar K_{gg}}{2}b_0^\dagger b_0^\dagger b_0 b_0+\frac{\hbar K_{ee}}{2}c_0^\dagger c_0^\dagger c_0 c_0+\hbar K_{eg}b_0^\dagger b_0 c_0^\dagger c_0 \end{aligned} \tag{55}$$

where,

$$\hbar\omega_0=\int d^3r\,\xi_g^*(r)\left[\frac{-\hbar^2\nabla^2}{2m}+\hbar\omega_g+V_g(r)\right]\xi_g(r) \tag{56}$$

$$\hbar\omega_1=\int d^3r\,\xi_e^*(r)\left[\frac{-\hbar^2\nabla^2}{2m}+\hbar\omega_e+V_{ge}(r)\right]\xi_e(r) \tag{57}$$

$$\hbar K_{gg}=\frac{4\pi\hbar^2 a_{gg}}{m}\int d^3r\left|\xi_g(r)\right|^4 \tag{58}$$

$$\hbar K_{ee}=\frac{4\pi\hbar^2 a_{ee}}{m}\int d^3r\left|\xi_e(r)\right|^4 \tag{59}$$

$$\hbar K_{eg}=\frac{4\pi\hbar^2 a_{eg}}{m}\int d^3r\left|\xi_g(r)\right|^2\left|\xi_e(r)\right|^2 \tag{60}$$

$$\hbar G = \hbar g \int d^3 r \xi_e^* (r) \xi_g (r) \tag{61}$$

6.3.2.1.1 PHONON LASER MEAN FIELD EQUATIONS

Using the Hamiltonian given by Eq. (69), the Quantum-Langevin equations of motion for the phonon operator a and the atomic operators b_0 and c_0 are given as follows:

$$\dot{a} = -i\omega_r a - \frac{\Gamma}{2} a - iG^* c_0 b_0^\dagger, \tag{62}$$

$$\dot{b}_o^\dagger = i\left[\omega_0 b_0^\dagger + aGc_0^\dagger + \frac{K_{gg}}{2}|b_0|^2 b_0^\dagger + K_{eg}|c_0|^2 b_0^\dagger\right] - \frac{\gamma}{2} b_0^\dagger, \tag{63}$$

$$\dot{c}_0 = -i\left[\omega_1 c_0 + aGc_0 + \frac{K_{ee}}{2}|c_0|^2 c_0^\dagger + K_{eg}|b_0|^2 c_0^\dagger\right] - \frac{\gamma}{2} c_0, \tag{64}$$

where Γ and ϒ are the phononic and the atomic damping rates, respectively. Now we rewrite the Eqs. (76)–(78) in the rotating frame of the frequency of the phonon ω_r in terms of the population inversion $\Delta n = \frac{(|c_o|^2 - |b_0|^2)}{N}$ and polarization $p = b_0^\dagger c_0 / N$, which are given as follows:

$$\dot{a} = -\frac{\Gamma}{2} a + G^* Np, \tag{65}$$

$$\dot{p} = -i\Delta\omega p - \frac{\gamma}{2} p + Ga\Delta n, \tag{66}$$

$$\dot{\Delta n} = \frac{\gamma}{2}\left(\Delta n_{eq} - \Delta n\right) - 2\left[G^* a^\dagger p + Gap^*\right]. \tag{67}$$

Here, $\Delta\omega = \omega_L - \omega_r + \frac{K_-}{4} - \left(\frac{K_+}{4} - K_{eg}\right)N\Delta n$, $K_- = K_{gg} - K_{ee}$, $K_+ = K_{gg} + K_{ee}$, $G \rightarrow iG$ and Δn_{eq} is the equilibrium value of Δn. The mean-field steady state solutions of the Eqs. (79)–(81) can be evaluated by factorizing them. It gives the critical number of atoms N_{cr} required to support a continuous wave laser, $N > N_{cr} = \Gamma\left(\Delta\omega^2 + \gamma^2/4\right)\big/\left(|G|^2 \Delta n_{eq}\gamma\right)$. N_{cr} =300/Δn_{eq} for the possible experimentally mentioned values [16]. N_{cr}=1.5×10^3 atoms for Δn_{eq}=0.2 which is reasonable.

6.3.2.1.2 TRANSIENT SOLUTIONS

The coherent population oscillations between an oscillator and a (pseudo) spin, that is, a two-level system are one of the vital predictions of the Jaynes-Cummings model [130]. We can also observe such kind of oscillations in our current system if the effective frequency of the oscillations, $\sqrt{|\Delta n_{eq}|N}G$ is greater than the fastest relaxation of the system γ. We can approximate the minimum number of atoms N_t required to observe this transient phenomena from this condition as $N_t = \gamma^2/(|\Delta n_{eq}|G^2) \approx 1.25\times10^4$. However, instead of energy oscillations, a single pulse can be produced between the atoms and the cantilever if N_t>N>N_{cr}.

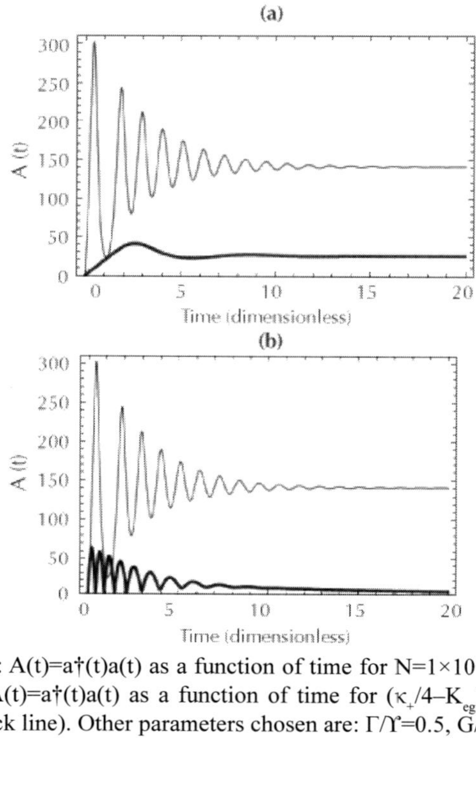

FIGURE 6.8 (a): A(t)=a†(t)a(t) as a function of time for N=1×10^5 (thin line), N=4×10^3 (thick line). (b): A(t)=a†(t)a(t) as a function of time for (κ_+/4–K_{eg})=0 (thin line), (κ_+/4-K_{eg})=2.2×10^{-3} (thick line). Other parameters chosen are: Γ/ϒ=0.5, G/ϒ=0.02, Δn_{eq}=0.2 and ω_L–ω_r=0.

The two transient outputs can be obtained by numerical integration of Eqs. (79)–(81) which are illustrated in Figs. 6.8(a) and (b). Figure 6.8(a) shows the coherent energy oscillations for $N=1\times10^5$ (thin line) and one single pulse for $N=4\times10^3$ (thick line) which takes away a large part of energy at one go. The tails of the output pulses decay as $1/\Gamma$. There is no observation of single pulse for $N<N_{cr}$. The effect of atomic two-body interaction on the transient outputs is depicted in Fig. 6.8(b). It shows that the amplitude of the coherent oscillations decreases while the frequency of oscillations increases with the increase in the value of $(K_+/4-K_{eg})$ for the fixed number of atoms $N=1\times10^5$. Such atom-atom interactions can be modified with the help of Feshbach resonances [131].

6.3.2.2 PHONON SUPERRADIANCE PHASE TRANSITION

We now examine the phonon super radiance for the system considered here. The system consists of a BEC of N two-level atoms having level spacing $\hbar\omega_L$ which can be expressed by a collective spin S=N/2. In the usual Dicke Hamiltonian, the counter rotating terms $a^\dagger S_+$ and aS_- are usually neglected in the weak atom-phonon coupling regime (rotating wave approximation) [71]. Including atom-atom interactions, the Hamiltonian of Eq. (69) in the limit of strong atom-phonon coupling can now be rewritten as:

$$H_2 = \hbar\omega_r a^\dagger a + \hbar(\omega_L + \gamma_0) S_z + \hbar\gamma_1 S_z^2 + \hbar\lambda(a^\dagger + a)(S_+ + S_-), \quad (68)$$

where, $\lambda = G_N/\sqrt{2S}, G_N = G\sqrt{N}, \gamma_0 = (K_{ee} - K_{gg})/4$ and $\gamma_1 = (K_{gg} + K_{ee} - 2K_{eg})/2$.

Also, $S_\pm = S_x \pm iS_y$ and $S_x = \sqrt{2}g_F F_x$ where F_x is the x component of atomic spin [16]. S_+, S_- and S_Z are the collective spin operators of the BEC [16]. The nonlinear term $\hbar\gamma_1 S_z^2$ ($S_z^2 \neq 1$) is supposed to show phase diagrams and rich collective dynamics since Υ_1 can take both the positive and negative values. Hence in order to study the system dynamics, the equations of motion for a, S_Z and S-operators (using the above Hamiltonian) are given as follows:

$$\dot{a} = -i\omega_r a - \kappa a - i\lambda(S_+ + S_-), \quad (69)$$

$$\dot{S}_z = \lambda(a^\dagger + a)(S_- - S_+), \quad (70)$$

$$\dot{S}_- = -i(\omega_L + \gamma_0 + \gamma_1)S_- - 2i\gamma_1 S_- S_z + i2\lambda(a^\dagger + a)S_z, \quad (71)$$

where, the phonon-damping rate $\kappa = \Gamma/2$.

6.3.2.2.1 MEAN FIELD SOLUTIONS AND THE FLUCTUATIONS

We now discuss the long-time, mean-field steady state solutions with $\dot{\vec{S}}=0$ and $\dot{a}=0$. Following Refs. [73, 74], we recognize the two stable states: normal state ($\Downarrow$, represented by N in the phase diagrams), having $S_Z=-N/2$ with all spins pointing down and no phonons, a=0, and inverted state ($\Uparrow$, represented by I in the phase diagrams), having $S_Z=N/2$ with all spins pointing up and no phonons. We now further consider $a=a_1+ia_2$, $\dot{a}=0$, $S_\pm=S_x\pm iS_y$, $\dot{S}_-=0$ and $\dot{S}_Z=0$ such that

$$\left[\kappa+i\omega_r\right]a=-i\lambda S_x, \tag{72}$$

$$\omega_0 S_x=-2\gamma_1 S_z S_x+4\lambda a_1 S_z, \tag{73}$$

$$\left(2\gamma_1 S_z+\omega_0\right)S_y=0. \tag{74}$$

Here, $\omega_0=\omega_L+\gamma_0+\gamma_1$ and Eq. (88) depicts the two classes of solutions. For the first solution $S_y=0$, we observe the usual superradiant phase in the Dicke model with the steady state population difference as:

$S_z=\dfrac{\omega_0\left(\kappa^2+\omega_r^2\right)}{8\lambda^2\omega_r-2\gamma_1\left(\kappa^2+\omega_r^2\right)}$. For the onset of superradiance, the critical coupling strength is obtained by setting $\vec{S}=(0,0,\pm N/2)$. One achieves,

$$\lambda^2 N_\pm=\pm\frac{\left(\omega_0\pm\gamma_1 N\right)\left(\kappa^2+\omega_r^2\right)}{4\omega_r}. \tag{75}$$

We get $\lambda^2 N_\pm=\pm\dfrac{\omega_L\left(\omega_r^2+\kappa^2\right)}{4\omega_r}$ for $\gamma_1=0,\ \gamma_0=0$, which is the standard Dicke model critical point [71]. We further examine that the atomic two-body interactions turn out to be a new and convenient handle to tune the critical point for $\gamma_1\neq 0$. For the second solution $S_Z=-\omega_0/2\gamma_1$, a=0 which turns out to be the normal phase. We now consider the fluctuations of phonon number and the spin S- around the steady state in order to discuss the instability of the normal and inverted phase. Thus we rewrite the operators as $a\to a+\delta a$ and $S_-\to S_-+\delta S_-$, where a=0, $S_-=0$ and $S_Z=\mp N/2$. As a result, the following linearized equations of motion are obtained:

$$\delta\dot{a}=-\left(\kappa+i\omega_r\right)\delta a-i\lambda\left(\delta S_-+\delta S_+\right), \tag{76}$$

$$\delta \dot{S}_{-} = -i\tilde{\omega}_{0\mp}\delta S_{-} \mp i\lambda N\left(\delta a + \delta a^{\dagger}\right), \tag{77}$$

with $\tilde{\omega}_{0\mp} = \omega_0 \mp \gamma_1 N$. We are looking for the solutions having the form $\delta a = Ae^{-i\eta t} + B^{*}e^{i\eta^{*}t}$ and $\delta S_{-} = Ce^{-i\eta t} + D^{*}e^{i\eta^{*}t}$. One can obtain the equations for A, B, C and D by equating the coefficients with the same time dependence. Following Refs. [73, 74], η satisfies the following condition:

$$\left(\eta^2 - \omega_r^2 - \kappa^2\right)\left(\eta^2 - \tilde{\omega}_{0\mp}^{2}\right) \mp 4\lambda^2 N\omega_r\tilde{\omega}_{0\mp} - i2\kappa\eta\left(\eta^2 - \tilde{\omega}_{0\mp}\right) = 0 \tag{78}$$

Equation (92) implies that the boundary between the stable (exponentially decaying) and unstable (exponentially growing) solutions have real solutions for η. The imaginary part of the above equation disappears when $\eta = \tilde{\omega}_{0\mp}$ or $\eta = 0$. $\eta = \tilde{\omega}_{0\mp}$ infers that the real part of Eq. (92) vanishes when $\tilde{\omega}_{0\mp}$=0. This implies the instability of both the normal and inverted states for $\omega_0 = \pm\gamma_1 N$. The real part of Eq. (92) vanishes for η=0 when $\left(\omega_r^2 + \kappa^2\right)\tilde{\omega}_{0\mp}^{2} \mp 4\lambda^2 N\omega_r\tilde{\omega}_{0\mp} = 0$, which gives the same condition as Eq. (89). Hence the onset of superradiant phase transition is accompanied by the instability of both the normal and the inverted phases.

6.3.2.2.2 DYNAMICAL PHASE DIAGRAMS

Now the dynamical phase diagrams corresponding to the dynamics of Eqs. (90, 91) are illustrated in Fig. 6.9. They emerge from the Eigen values, which are given by,

$$\omega_{\mp} = \frac{\pm 2\lambda^2 N}{\omega_0 \mp \gamma_1 N} \pm \sqrt{\frac{4\lambda^4 N^2}{\left(\omega_0 \mp \gamma_1 N\right)^2} - \kappa^2} \tag{79}$$

in this nonequilibrium setting.

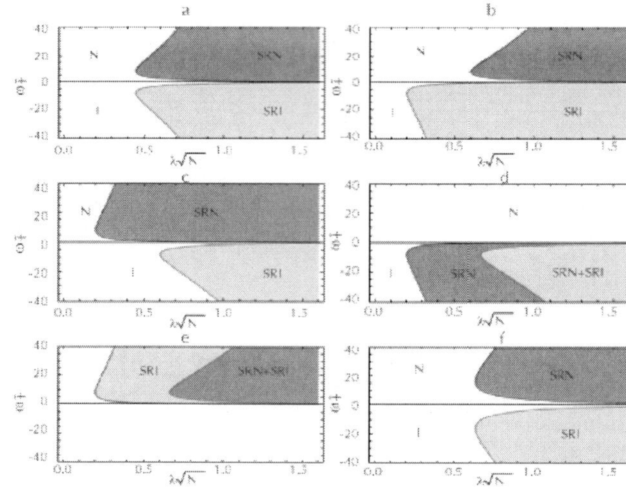

FIGURE 6.9 Plot of dynamical phase diagrams illustrating the various phases (normal phase N, inverted phase I, superradiant phase SRN and superradiant inverted phase SRI). Parameters chosen are (a): Υ_1N=0, κ=8.1, (b): Υ_1N=–0.04, (c): Υ_1N=0.04, (d): Υ_1N=0.06, (e): Υ_1N=–0.06, (f): Υ_1N=0, κ=16.1.

Figure 9(a) depicts the equilibrium phase diagram of the Dicke model for the positive Eigen value ω_- and for Υ_1N=0. It shows a transition from the normal phase (N) at low $\lambda\sqrt{N}$ to the superradiant normal phase (SRN) at higher value of $\lambda\sqrt{N}$. This phase transition arises at $\lambda\sqrt{N} = \dfrac{\omega_L\left(\omega_r^2+\kappa^2\right)}{4\omega_r}$. The critical value of $\lambda\sqrt{N}$ (necessary for superradiance) $\rightarrow$ infinity as $\omega_-\rightarrow$ zero. As shown in Refs. [73, 74], we further notice that this open dynamical system shows the sign of nonequilibrium dynamics for the negative Eigen values ω_+. The normal state ($\Downarrow$) becomes unstable and an inverted state (a=0, S_Z=N/2, represented by I) and superradiant inverted phase (represented by SRI) appears which is a stable state. The inverted state and the normal state are mirror images of each other which can also be depicted from the equations of motion (83)–(85), which have an inversion symmetry for $a \rightarrow a*$, $\vec{S} \rightarrow -\vec{S}$, $\omega_r \rightarrow -\omega_r$ and $\gamma_1 = 0$. However, this inversion symmetry is broken instantly for the finite values of γ_1, which is illustrated in the Figs. 6.9(b) and (c). We can infer from Fig. 6.9(b) that the phase boundary connecting the N and SRN phases recedes to higher values of $\lambda\sqrt{N}$ while the phase boundary between the I and SRI phases shifts towards the lower values of $\lambda\sqrt{N}$ for γ_1N= –0.04. Exactly opposite is the case for γ_1N=0.04, which is reflected in Fig. 6.9(c). Figures 6.9(d) and (e) reflect the coexistence of both the SRN and SRI phases for γ_1N=$\pm$ 0.06. Figure 6.9(f) shows the effect of increasing the phonon-damping rate (κ). It clearly shows that the region of N and I phases increases on increasing the phonon damping rate, that is, SRN and SRI phases get separated out further. In the next section, we described the applications of the above mentioned optomechanical systems.

6.3.3 OPTOMECHANICAL SYSTEMS INVOLVING BEC

In this section, we are describing some of our previous works [17, 132] investigating various cooling techniques in order to cool the optomechanical system involving BEC. The model consists of a Fabry Perot Cavity with one mirror fixed and the other mirror movable. This is known as an optomechanical system, which is the basic model to detect the weak forces such as gravitational forces [133, 134]. The movable mirror can also be treated as a quantum mechanical harmonic oscillator having mass m_m and frequency ω_m. In addition, there is an elongated cigar shaped Bose Einstein Condensates (BEC) of N – two level ^{87}Rb atoms in the |F = 1> state having mass m and transition frequency ω_a of the |F = 1> → |F' = 2> transition of the D_2 line. BEC is coupled to a quantized single standing wave cavity mode with frequency ω_c. It is known that mirror and cavity photons exert radiation pressure on the mirror and results in optomechanical coupling between the mirror and the cavity field.

This system can be described by the optomechanical Hamiltonian in the rotating wave and dipole approximation, which can be written as [12, 17, 132]:

$$H_3 = E_o \Sigma_j b_j^\dagger b_j + J_o(\hbar U_o a^\dagger a + V_{cl})\Sigma_j b_j^\dagger b_j + \frac{U}{2}\Sigma_j b_j^\dagger b_j^\dagger b_j b_j - \hbar\Delta_c a^\dagger a \\ -i\eta\hbar(a - a^\dagger) + \hbar\omega_m a_m^\dagger a_m - \hbar\varepsilon\omega_m a^\dagger a(a_m^\dagger + a_m) \tag{80}$$

where the first term in the Hamiltonian represents the total kinetic energy of the condensate such that E_0 is the onsite kinetic energy of the condensate atoms with b_j representing the annihilation operator for the jth bosonic atom. Second term gives the total potential energy of the atoms with J_0 representing the effective onsite potential energy of the condensate. U_0 is the optical lattice barrier height per photon and V_{cl} is the classical potential. a ($a^\dagger$) is the annihilation (creation) operator of the cavity photon. Third term depicts the two-body atom-atom coupling with U representing the effective onsite atom-atom interaction energy. Fourth term is the energy of the intensity of the light mode with Δ_c as the cavity-pump detuning. Fifth term represents the energy due to external pump laser. Second last term of the Hamiltonian gives the energy of the single vibrational mode of the mechanical oscillator with annihilation (creation) operator a_m ($a_m^\dagger$). Last term gives the energy due to non-linear dispersive coupling between the light field intensity and the position quadrature of the moving mirror.

The dynamics of the optomechanical system is also influenced by the dissipative interaction like photon leakage through the mirror, leakage of BEC atoms from the cavity etc. Also the mirror motion is not only due to the radiation pressure, it also undergoes Brownian motion due to the thermal fluctuations of the external environment.

By solving the pertinent quantum Langevin equations of motion, for photon, phonon and boson operators, obtained by the Hamiltonian, we get the steady state energy for the mechanical oscillator. The dimensionless steady state energy of the mechanical oscillator is given by $\{<Q^2>+<P^2>\}$. $<Q^2>$ and $<P^2>$ represent the position and momentum variances of the mechanical resonator. Here, we present a detailed study to show the possibility of approaching the quantum ground-state of a mechanical oscillator with BEC using different cooling schemes namely Stochastic Cooling feedback scheme, Back-Action cooling scheme and Cold Damping feedback scheme.

6.3.3.1 BACK- ACTION COOLING SCHEME

Using BEC, Back-Action dynamics have been observed in a wide variety of physical systems [15, 16]. Photon shot noise is responsible for the randomness present in the unavoidable stochastic back-action forces, which arises due to the radiation pressure. The coupling between the cavity field and the mechanical oscillator via radiation pressure leads to the self-cooling of the oscillator through dynamical back-action [91, 135, 136]. Generated by the cavity delay, the correlations between the radiation pressure and the Brownian motion of the oscillator will either amplify or cool the system depending upon the laser detuning. Using these effects, a single vibrational mode has been cooled experimentally [5, 6, 11].

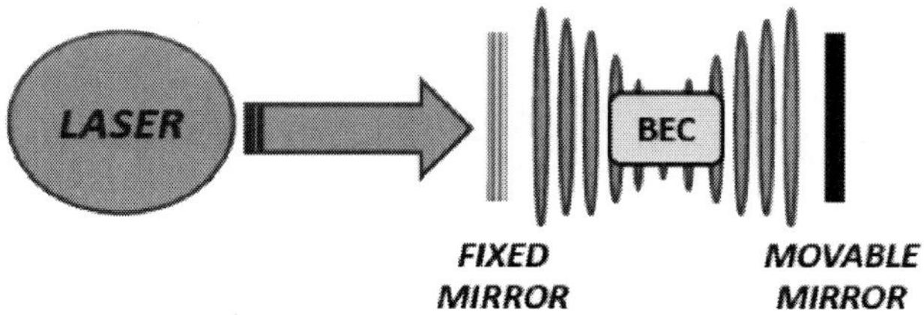

FIGURE 6.10 Schematic representation for Back-Action cooling.

Back-Action cooling scheme helps in achieving the quantum ground state of the mechanical resonator effectively as described below.

Figure 6.11 represents the plot of effective phonon number (n_{eff}) as a function of dimensionless effective detuning (D_d/w_m) of the condensate atoms for the absence of BEC (thick solid line) and for two different values of atomic two-body interaction, U_{eff}=0.8w_m (dashed line) and U_{eff}=w_m (thin solid line). One can observe from the figure that the quantum ground state of the system can be achieved by increasing the condensate two-body interaction such that the minimum value of n_{eff} is nearly 0.52 at Δ_d=–1.13w_m for U_{eff}=w_m. However, we have observed the least value of n_{eff} to be nearly 0.51 at Δ_d=–w_m in the absence of BEC.

As shown in the Fig. 6.11, we have also observed better results in the presence of BEC as compared to that in the absence of BEC for the lower values of effective detuning of condensate atoms. Moreover, the effective temperature of the resonator does not vary significantly by adding BEC to the system. Thus, a low temperature of the movable end mirror can be sustained over a wide range of effective detuning in the presence of BEC.

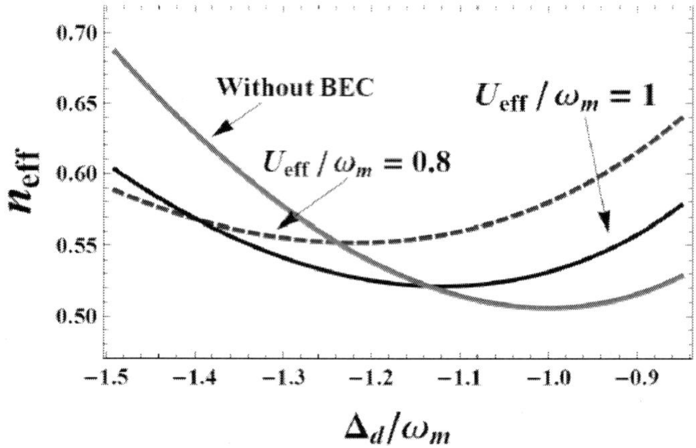

FIGURE 6.11 Plot of effective phonon number (n_{eff}) as a function of dimensionless effective detuning (Δ_d/ω_m) in the absence of BEC (thick solid line) and for different values of two body atom-atom interaction (U_{eff})= $0.8\omega_m$ (dashed line) and $=1\omega_m$ (thin solid line) for Back-Action cooling scheme.

6.3.3.2 FEEDBACK SCHEMES

Now, we describe different feedback cooling schemes such as Stochastic cooling and Cold Damping feedback schemes.

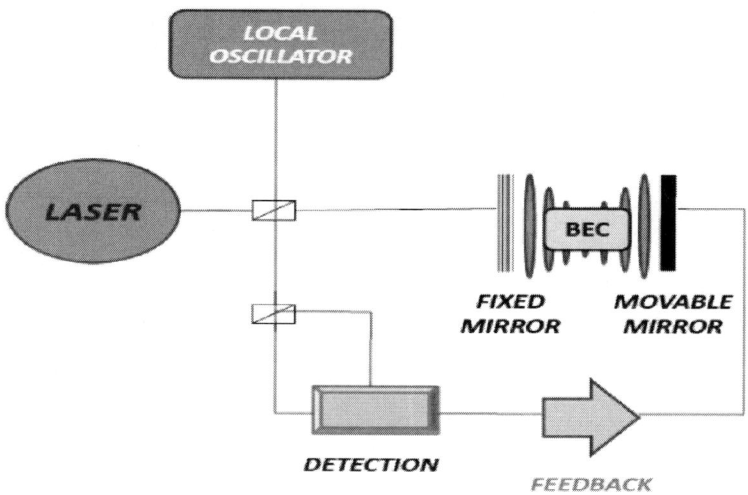

FIGURE 6.12 Schematic representation of the set up involving feedback schemes.

6.3.3.2.1 *STOCHASTIC COOLING FEEDBACK SCHEME*

This feedback technique cools the hybrid optomechanical system by randomly kicking the back of the mirror. Homodyne detection regularly monitors the position of the mirror. Latest experiments have illustrated the efficient cooling of the system using this technique [7, 95, 137]. As suggested [138] even for small frequencies (off resonance), this scheme decreases the thermal noise of the system. In order to detect small forces, the movable mirror can be used as ponderomotive meter [2]. Therefore, an extra term $(-\frac{\hbar}{2}(a_m + a_m^{\dagger})f(t))$ is added in the Hamiltonian of the system for this technique. This additional term represents the effect of classical external force on the displacement of the mirror.

Now, we will describe the effect of Stochastic cooling scheme on the steady state energy of the mechanical oscillator. Figure 6.13(a) depicts the plot of steady state energy as a function of dimensionless mirror-photon coupling G in the absence of BEC (dashed line) and in the presence of BEC (solid line). It clearly shows the decrease in oscillator energy in the presence of BEC for small values of G/ω_m. Figure 6.13(b) illustrates the stationary oscillator energy as a function of G for two different values of atom-atom two body interaction $U_{eff} = 8\omega_m$ (dashed line) and $U_{eff} = 30\omega_m$ (solid line).

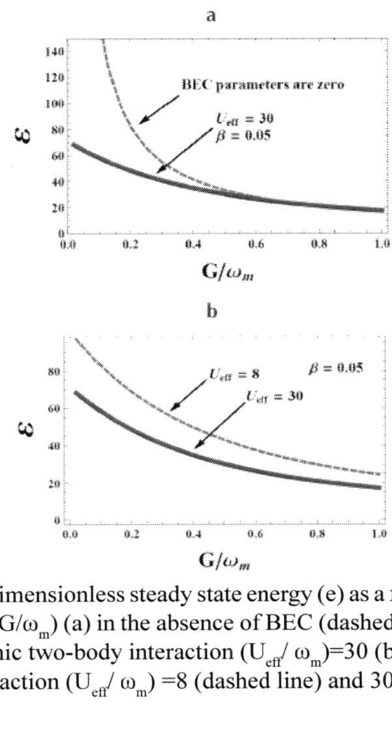

FIGURE 6.13 Plot of dimensionless steady state energy (e) as a function of dimensionless mirror-photon coupling (G/ω_m) (a) in the absence of BEC (dashed line) and the presence of BEC (solid line) for atomic two-body interaction (U_{eff}/ω_m)=30 (b) for two different values of atomic two-body interaction (U_{eff}/ω_m) =8 (dashed line) and 30 (solid line).

Figure 6.13 shows that the quantum ground state of the mechanical oscillator is approachable with the increase in the atomic two-body interaction. It infers that the hybrid system can be cooled in this scheme for the presence of BEC.

6.3.3.2.2 COLD DAMPING FEEDBACK SCHEME

The last cooling technique studied here is the Cold Damping quantum feedback scheme. This technique involves the cooling of oscillating mirror with the help of viscous force applied on the moving mirror through radiation pressure. This viscous force is induced by another intensity modulated laser beam on the back of the mirror [7, 95, 128, 139]. The damping of the system can be enhanced using this scheme without any external effect on the thermal noise of the system [140, 141]. Negative derivative feedback is used in this technique. Phase-sensitive homodyne detection of the output of the optical field is used to determine the displacement of the movable mirror [138, 142]. This feedback scheme helps in approaching the quantum ground state in the bad cavity limit, which can be seen below.

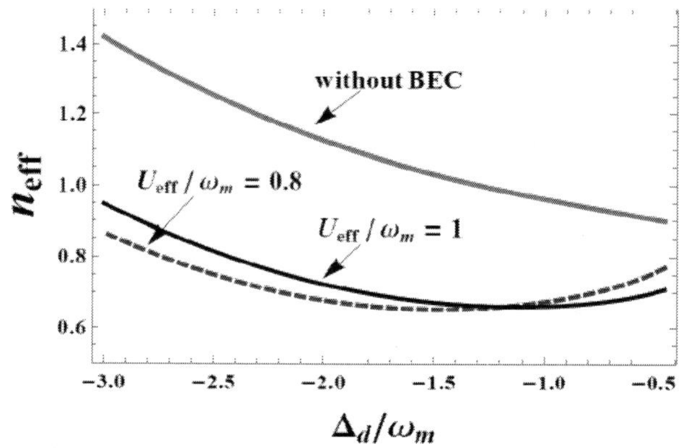

FIGURE 6.14 Plot of effective phonon number (n_{eff}) as a function of dimensionless effective detuning (Δ_d/ω_m) in the absence of BEC (thick solid line) and for different values of two body atom-atom interaction (U_{eff})= $0.8\omega_m$ (dashed line) and $=1\omega_m$ (thin solid line) for Cold Damping feedback scheme.

Figure 6.14 illustrates the plot of effective mean excitation number (n_{eff}) as a function of dimensionless effective detuning (D_d/w_m) of atoms for the absence of BEC (thick solid line) and for two different values of two body atom-atom interaction, U_{eff}=0.8w_m (dashed line) and U_{eff}=w_m (thin solid line). It can be observed from the figure that the value of n_{eff} is decreased considerably by adding BEC to the system. The least value of effective phonon number is 0.658 which is attained for U_{eff}=0.8w_m. It is clearly depicted from the figure that one can approach the quantum ground state cooling of the oscillating mirror by increasing the atomic two-body interaction.

Overall, we have observed that all the above-mentioned techniques help in cooling the mechanical oscillator of the hybrid optomechanical system consisting of BEC. Experimentally, similar system involving BEC coupled to membrane via cavity mode has been studied [54]. It has been observed that membrane dissipation is enhanced with the increase in number of atoms. In the present system, we have seen that the atomic two-body interaction plays a new control parameter in decreasing the oscillator energy. Hence, our result is confirmed from this experiment as U_{eff} is directly proportional to the number of atoms.

6.4 APPLICATIONS

One of the strong applications of the optomechanical systems is quantum information processing [44]. A quantum phase gate can be engineered for photonic or phononic qubits using all optical switch, which is formed on the basis of Kerr type nonlinearity. Quantum memory can also be served by mechanical mode [45]. A quantum interface between atomic, solid-state and optical qubits arises due to the coupling between the mechanical and optical degrees of freedom [46]. To architect a quantum communication and quantum information processing unit, a hybrid design can be made consisting of optical modes and mechanical modes that can be fruitfully used to store and transfer information coherently. Acoustic modes (phonons) in quantum information processing units can store information for an extended period whereas optical modes can transfer information over long distances.

Light (optical photons) is used by the architectures of quantum and classical information processing network for the transmission of information over extended distances, which range from hundreds of meters to hundreds of kilometers [143, 144]. The optical photons have affability to thermal noise due to their high frequency (approx. 200 THz), weak interaction with the environment and large bandwidth of transmission, which stems their utility. Although, there are limitations for phonons (acoustic excitations) in terms of their ability to transmit information farther than a few millimeters and bandwidth. But this information can be stored or delayed for much longer times and this can interact resonantly with RF-microwave electronic systems [145]. This leads to introduce a hybrid phononic–photonic systems comprising of the affirmative nature of photons and phonons. Such hybrid system could be made to perform a wide range of task, which is out of the range of systems consisting purely of phonons or photons. The optomechanical circuitary made by the coherent interfacing of optical and acoustic circuits, forms the building block of the architecture of such hybrid systems.

The acoustic waveguides have delay line, which arises due to the inherent slowness of acoustic waves in comparison to electromagnetic waves(roughly a factor of 105 for waves in silicon). Due to this reason, chip-scale RF-microwave filters are created using electro-acoustic piezoelectric structures. These filters are found in many compact wireless communication devices [146].

Optomechanical cooling of nanomechanical resonator has a variety of application in sensitive measurements such as small masses [147], detection of weak forces [133, 134] and small displacements [148]. The hybrid optomechanical systems explained here can be used as a quantum device to detect weak forces by an adequate choice of spectral range and parameters of the system [17].

6.5 CONCLUSION

In conclusion, we have demonstrated the effects in various nano-optomechanical systems. We have presented that the splitting of normal mode into three modes in a hybrid optomechanical system depicts the possibility of coherently transfer the energy between the three modes of the system. It was also observed that the mechanical mode and the two optical modes of the system are strongly correlated for the temperatures below T = 2.4K. Such a strongly coupled three-mode system showing robust steady-state entanglements against temperature can be exploited for the realization of quantum memories and quantum interfaces within quantum-communication networks. We have also observed that the entanglement between the mechanical mode and any of the two optical modes of the system can be selectively made large and more robust against temperature by the proper choice of the system parameters. In the next generation of quantum communications and quantum information processing units, the hybrid optomechanical system is useful as the photonic modes can exchange energy over long distances and phononic modes can store energy for a longer duration. In the next section of magnetic cantilever coupled to a Bose-Einstein condensate, we have demonstrated that a phonon laser can be designed based on magnetically coupling between mechanical oscillations of a nano scale magnetic cantilever to an ultra-cold atomic cloudin close analogy to a two-level optical laser system. One can switch between transient pulses and solitary pulses by regulating the number of atoms. The atomic two body interaction can also controls the transients. Phonon superradiance can also be produced by using this system. The system can be described by the Dicke type Hamiltonian with large atom-mechanical mode coupling. Some interesting phase diagrams are shown which arises due to non linear term proportional to S_Z^2produced by two-body interaction. In the phase diagrams, we get the regions where the superradiant inverted and superradiant normal phase coexists by appropriately tuning the atomic two-body interaction. In the next section of optomechanical system involving BEC within an optical cavity, we have discussed how the Stochastic cooling, Back-Action cooling and Cold Damping feedback schemes help in cooling the mechanical oscillator to its quantum ground state. In all the schemes, we have found that the cooling of the mirror can be approached by increasing the two body atom-atom interaction. Thus, the condensate two body interaction acts as a new handle to provide a systematic control of the system. By using s-wave scattering length or the number of condensate atoms, this new handle can be altered. We have further noticed that the Cold Damping feedback scheme is more efficient in the bad cavity limit. Back-Action cooling scheme is more relevant in the good cavity limit.

KEYWORDS

- **Bose-Einstein condensate**
- **laser mode squeezing**
- **mean field solution**
- **optical lattice**
- **phase diagram**
- **phonon lasers**

REFERENCES

1. Milburn, G. J., Woodley, M. J., Acta Phys. Slovaca, 61, 483, (2011).
2. Braginski, V. B., YaKhalilli, F., Quantum Measurements (Cambridge University Press, Cambridge, 1992).
3. Caves, C. M., Phys. Rev. Lett. 45, 75, (1980).
4. Hohberger, C., Metzger, K Karrai, Nature 432, (1002). (2004).
5. Gigan, S. et al., Nature 444, 67, (2006).
6. Arcizet, O. et al., Nature 444, 71, (2006).

7. Kleckner, D., Bouwmeester, D., Nature 444, 75, (2006).
8. Favero, I. et al., Appl. Phys. Lett. 90, 104101 (2007).
9. Regal, C., Terfel, J. D., Lehnert, K., Nature Phys. 4, 555, (2008).
10. Carmon, T. et al., Phys. Rev. Lett. 94, 223902 (2005).
11. Schliesser, A. et al., Phys. Rev. Lett. 97, 243905 (2006).
12. Bhattacherjee, A., Phys. Rev. A 80, 043607 (2009).
13. Bhattacherjee, A., J. Phys. B: At. Mol. Opt. Phys. 43, 205301 (2010).
14. Brennecke, F. et al. Science 322, 235, (2008).
15. Murch, K. W. et al. Nature Phys. 4, 561, (2008).
16. Bhattacherjee, A. B., Brandes, T., Canad. J. Phys. Vol. 91, 639, (2013).
17. SonamMahajan, Tarun Kumar, Aranya Bhattacherjee, B., ManMohan, Phys. Rev. A, 87, 013621 (2013).
18. Thompson, J. D. et al., Nature 452, 72, (2008).
19. Marquardt, F. Girvin, S. M., Physics 2, 40, (2009).
20. Purdy, T. P. et al., Phys. Rev. Lett. 105, 133602 (2010).
21. Teufel, J. D. et al., Nature 475, 359, (2011).
22. Chan, J. et al., Nature 478, 89, (2011).
23. Verhagen, E. et al., Nature 482, 63, (2012).
24. Kumar, T., Bhattacherjee, A., ManMohan, Phys. Rev. A. 81, 013835 (2010).
25. O'Connell, A. D., Hofheinz, M., Ansmann, M., Bialczak, R. C., Lenander, M., Lucero, E., Neeley, M., Sank, D., Wang, H., Weides, M., Wenner, J., Martinis, J. M., Cleland, A. N., Nature London 464, 697, (2010).
26. Kittel, C., Phys. Rev. Lett., 6, 449, (1961).
27. Vredevoe, L. A., Silvera, I. F., Solid State Commun., 8, 175, (1970).
28. Wallentowitz, S. et al., Phys. Rev. A., 54, 943, (1996).
29. Lozovik, Y. E., Ovchinnikov, I. V., JETP Lett., 72, 431, (2000).
30. Camps, I. et al., Phys. Rev. B., 64, 125311 (2001).
31. Liu, H. C. et al., Phys. Rev. Lett., 90, 077402 (2003).
32. Kabuss, J., Carmele, A., Brandes, T., Knorr, A., Phys. Rev. Lett. 109, 054301 (2012).
33. Okuyama, R., Eto, M., and Brandes, T., arXiv:1205.6955(2012).
34. Bargatin, I., Roukes, M. L., Phys. Rev. Lett., 91, 138302 (2003).
35. Chudnovsky, E. M., Garanin, D. A., Phys. Rev. Lett., 93, 257205 (2004).
36. Mendonca, J. T., Euro. Phys. Lett., 91, 33001 (2010).
37. Grudinin, I. S. et al., Phys. Rev. Lett., 104, 083901 (2010).
38. Vahala, K. et al., Nature Phys. 5, 682, (2009).
39. Kent, A. J. et al., Phys. Rev. Lett., 96, 215504 (2006).
40. Tucker, E. B., Phys. Rev. Lett., 6, 547, (1961).
41. Hu, P., Phys. Rev. Lett., 44, 417, (1980).
42. Fokker, P. A. et al., Phys. Rev. B., 55, (2925). (1997).
43. Bron, W. E., Grill, W., Phys. Rev. Lett., 40, (1459). (1978).
44. Stannigel, K. et al., Phys. Rev. Lett. 109, 013603 (2012).
45. Chang, D. E. et al., New, J. Phys. 13, 023003 (2011).
46. Stannigel, K. et al., Phys. rev. Lett. 105, 220501 (2010).
47. Parkins, A. S., Kimble, H. J. (1999). Quantum state transfer between motion and light, J. Opt. B: Quantum Semiclass. Opt. 1–496.

48. Massoni, E., Orszag, M. (2000). Phonon–photon translator Opt. Commun. 179–315–21.
49. Orszag, M. (2002). Phononphoton interactions with a trapped ion in a cavity Laser Phys. 12–1054–63.
50. Rodrigues, R. L., Moussa, M. H. Y., Villas-Boas, C. J. (2006). Engineering phonon-photon interactions with a driven trapped ion in a cavity Phys. Rev. A 74–063811.
51. Chang, D. E., Regal, C. A., Papp, S. B., Wilson, D. J., Ye, J., Painter, O., Kimble, H. J., Zoller, P. (2010). Cavity optomechanics using an optically levitated nanosphere Proc. Natl Acad. Sci. USA 107–1005–10.
52. Zhang, J., Peng, K., Braunstein, S. L., Phys. Rev. A 68, 013808 (2003).
53. Eichenfield, M., Chan, J., Camacho, R. M., Vahala, K. J., Painter, O. (2009). Optomechanical crystals Nature 462–78–82.
54. Camerer, S. et al., Phys. Rev. Lett. 107, 223001 (2011).
55. Szirmai, G., D. Nagy and Domokos, P., Phys. Rev. A, 81, 043639 (2010).
56. Hunger, D. et al., Phys. Rev. Lett. 104, 143002 (2010).
57. Chen, B. et al., Phys Rev. A 83, 055803 (2011).
58. Chiara, G. De. et al., Phys. Rev. A 83, 052324 (2011).
59. Steinke, S. K. et al., Phys. Rev. A 84, 023834 (2011).
60. Chen, B. et al., J. Opt. Soc. Am., 28, (2007). (2011).
61. Zhang, K. et al., Phys. Rev. A, 81, 013802 (2010).
62. Singh, S., Bhattacharya, M., Dutta, O., Meystre, P., Phys. Rev. Lett. 101, 263603 (2008).
63. Genes, C., Vitali, D., Tombesi, P., Phys. Rev. A 77, 050307 (2008).
64. Ian, H., Gong, Z. R., Liu, Y., Sun, C. P., Nori, F., Phys. Rev. A 78, 013824 (2008).
65. Geraci, A. A., Kitching, J., Phys. Rev. A 80, 032317 (2009).
66. Hammerer, K., Wallquist, M., Genes, C., Ludwig, M., Marquardt, F., Treutlein, P., Zoller, P., Ye, J., Kimble, H. J., Phys. Rev. Lett. 103, 063005 (2009).
67. Meiser, D., and Meystre, P., Phys. Rev. A., 73, 033417 (2006).
68. Hunger, D. et al., ComptesRendus Physique, 12, 871, (2011).
69. Eichenfield, M., Camacho, R., Chan, J., Vahala, K. J., Painter, O. (2009). A picogram- and nanometerscale photonic-crystal optomechanical cavity Nature 459–550–5.
70. Roels, J., De Vlaminck, I., Lagae, L., Maes, B., Van Thourhout, D., Baets, R. (2009). Tunable optical forces between nanophotonic waveguides Nat. Nanotechnol. 4–510–3.
71. Emary, C., Brandes, T., Phys. Rev. E., 67, 066203 (2003).
72. Hepp, K., Lieb, E. H., Phys. Rev. A 8, (2517). (1973).
73. Bhaseen, M. J. et al., Phys. Rev A, 85, 013817 (2012).
74. Bhaseen, M. J. et al., Phys. Rev. Letts., 105, 043001 (2010).
75. Kepler, J., Letter to Galileo Galilei (1610).
76. Kepler, J., De CometisLibelliTres (1619).
77. Nichols, E. F., Hull, G. F., About Radiation pressure, Ann. Phys. 12, 225, (1903).
78. Lebedev, P., Untersuchungen ¨uber die Druckkr¨afte des Lichtes, Ann. Phys. 6, 433, (1901).
79. Braginsky, V. B., Eksp. Teor. Fiz. 53, (1434). (1967).
80. Braginsky, V. B. Vorontsov, Y. I., Sov. Phys. Usp. 17, 644, (1975).

81. Meystre, P., E. M.Wright, McCullen, J. D., Vignes, E., J. Opt. Soc. Am. B 2, (1830). (1985).
82. Jacobs, K., Tombesi, P., Collett, M. J., D. F.Walls, Phys. Rev. A 49, (1961). (1994).
83. Pinard, M., Fabre, C., Heidmann, A., Phys. Rev. A 51, 2443, (1995).
84. Fabre, C., Pinard, M., Bourzeix, S., Heidmann, A., Giacobino, E., Reynaud, S., Phys. Rev. A 49, 1337, (1994).
85. Mancini, S., Tombesi, P., Phys. Rev. A 49, (4055). (1994).
86. Mancini, S., Vitali, D., Tombesi, P., Phys. Rev. Lett. 80, 688, (1998).
87. Mancini, S., V. I. Man'ko, Tombesi, P., Phys. Rev. A 55, (3042). (1997).
88. Bose, S., Jacobs, K., Knight, P. L., Phys. Rev. A 56, (4175). (1997).
89. Marshall, W., Simon, C., Penrose, R., Bouwmeester, D., Phys. Rev. Lett. 91, 130401 (2003).
90. Gr¨oblacher, S., Hammerer, K., Vanner, M. R., Aspelmeyer, M., Nature 460, 724, (2009).
91. Corbitt, T., Chen, Y., Innerhofer, E., H. M¨uller-Ebhardt, Ottaway, D., Rehbein, H., Sigg, D., Whitcomb, S., Wipf, C., Mavalvala, N., Phys. Rev. Lett. 98, 150802 (2007).
92. Kippenberg, T., Rokhsari, H., Carmon, T., Scherer, A., Vahala, K., Phys. Rev. Lett. 95, 033901 (2005).
93. Tittonen, I., Breitenbach, G., Kalkbrenner, T., T. M¨uller, Conradt, R., Schiller, S., Steinsland, E., Blanc, N., N. F. de Rooij, Phys. Rev. A 59, (1038). (1999).
94. Briant, T., Cohadon, P., Pinard, M., Heidmann, A., Eur. Phys. J. D 22, 131, (2003).
95. Arcizet, O., P.-Cohadon, F., Briant, T., Pinard, M., Heidmann, A., J.-Mackowski, M., Michel, C., Pinard, L., O. Francais, Rousseau, L., Phys. Rev. Lett. 97, 133601 (2006).
96. Kippenberg, T. J., Vahala, K. J., Opt. Express 15, 17172 (2007).
97. Aspelmeyer, M. and Schwab, K. C., New, J. Phys. 10, 095001 (2008).
98. Favero, I., Karrai, K., Nature Photonics. 3, 201, (2009).
99. Genes, C., Mari, A., Vitali, D., Tombesi, P., Adv. At. Mol. Opt. Phys. 57, 33, (2009).
100. Aspelmeyer, M., S. Gr¨oblacher, Hammerer, K., Kiesel, N., J. Opt. Soc. Am. B 27, A189 (2010).
101. Wineland, D. J., Itano, W. M., Phys. Rev. A 20, (1521). (1979).
102. Grimm, R., Weidemuller, M., Ovchinnikov, Y. B. Adv. At. Mol. Opt. Phys. 42, 95–170 (2000).
103. Jessen, P. S., Deutsch, I. H. Optical lattices: Advances in Atomic, Molecular and Optical Physics, 37, (1996).
104. Bose, S., Z. Phys., 26, 178, (1924).
105. Einstein, A., SitzungsberKgl. Press. Akad. Wiss., 261(1925).
106. Pais, A., "Subtle is the lord …" (1982). Oxford University Press: Oxford.
107. London, F., Nature, 141, 643, (1938).
108. Tisza, L., Nature, 141, 913, (1938).
109. Chu, S., Rev. Mod. Phys., 70, (3): 685–706 (1998).
110. Claude, N. Cohen Tannoudji, Rev. Mod. Phys., 70, (3): 707–719, (1998).
111. Phillips, W. D., Rev. Mod. Phys., 70, (3): 721–741 (1998).
112. Anderson, M. H., J. Ensher, R., Mathews, M. R., Wiemann, C. E., Cornell, E. A. (1995). Science 269, 198.
113. Devis, K. B., Mews, M. O., Andrews, M. R., N. J. Van Druten, Durfee, D. S., Kurn, D. M. Ketterle, W., (1995). Phy. Rev. Lett. 75, 3969.

114. Bardley, C. C., Sackett, C. A., Tollet, J. J., Hulet, R. G., (1997). Phy. Rev. Lett., 78, 985.
115. Bardley, C. C., Sackett, C. A., Tollet, J. J., Hulet, R. G., (1995). Phy. Rev. Lett., 75, 1687.
116. Leggett, J., (2001). Rev.Mod. Phys., 73(2), 307.
117. Safavi-Naeini, A. H., Painter, O., (2011). New J. Phys. 13, 013017
118. Max Ludwig et al., (2012). Phys. Rew. Lett.109, 063601.
119. Thompson, J. D. et al., (2008). Nature 452, 900.
120. Walls, D. and Milburn, G. J., Quantum Optics (Springer, Berlin, 1995).
121. Giovannetti, V., Tombesi, P., Vitali, D., (2001).Phys. Rev. A 63, 023812.
122. Tanaka, Y., Asano, T. Noda, S., (2008). J. Lightwave Technol. 26, 1532.
123. Notomi, M., Kuramochi, E., Taniyama, H., (2008). Opt. Express 16, 11095.
124. Chan, J. et al., (2012). Appl. Phys. Lett. 101, 081115
125. Robinson, J. T. et al., (2005). Phys. Rev. Lett. 95, 143901.
126. Schliesser, A. et al., (2008). Nat. Phys. 4, 415.
127. Hadjar, Y. et al., (1999). Europhys. Lett. 47, 545.
128. Cohadon, P. F., Heidmann, A., Pinard, M., (1999). Phys. Rev. Lett. 83, 3174.
129. Steel, M. J., Collett, M. J., (1998). Phys. Rev. A, 57, 2920.
130. Jaynes, E. T. and Cummings, F. W., (1963). Proc. IEEE 51, 89.
131. Bose-Einstein Condensation in Dilute Gases, C. J. Pethick and Smith, H., Cambridge University Press (Cambridge, UK) 2002.
132. Mahajan, S., Neha Aggarwal, Aranya B. Bhattacherjee, ManMohan, (2013). Journal of Physics B, Vol. 46, 085301 [arXiv:1302.0339].
133. Braginsky, V., Manukin, A., (Cambridge University Press 1977).
134. A. Abramovici et al., Science 256, 325, (1992).
135. Braginsky, V. B., Strigin, S. E., Vyatchanin, S. P., Phys. Lett. A 287, 331, (2001).
136. Sankey, J. C. et al., Nature Physics 6, 707, (2010).
137. Degan, C. L., Poggio, M., Mamin, H. J., and Rugar, D., Phys. Rev. Lett. 99, 250601 (2007).
138. Vitali, D., Mancini, S., Ribichini, L., Tombesi, P., (2002). Phys. Rev. A. 65, 063803.
139. Poggio, M. et al., Phys. Rev. Lett. 99, 017201 (2007).
140. Milatz, J., M. W., J. J. van Zolingen, Physica (Amsterdam) 19, 181, (1953).
141. Grassia, F. et al., Eur. Phys. J. D 8, 101 (2000).
142. Genes, C. et al., Phys. Rev. A 77, 033804 (2008).
143. Agrawal, G. P. (2002). Fiber-Optic Communication Systems 3rd edn (New York: Wiley-Interscience).
144. Duan, L.-M., Lukin, M. D., Cirac, J. I., Zoller, P. (2001). Long-distance quantum communication with atomic ensembles and linear optics Nature 414–413–8.
145. Pozar, D. M. (2004). Microwave Engineering 3rd edn (New York: Wiley).
146. Lakin, K. M., Kline, G. R. Mc Carron, K. T. (1995). Development of miniature filters for wireless applications IEEE Trans. on Microw. Theory Techn. 43, 2933–2939.
147. Jensen, K., Kim, K., Zettl, A., (2008). Nature Nanotechnol. 3, 533.
148. Latlaye, M. D. et al., (2004). Science 304, 74.

CHAPTER 7

MAGNETIC ANISOTROPY IN CASE STUDIES

MARILENA FERBINTEANU and FANICA CIMPOESU

CONTENTS

ABSTRACT

In this chapter, we present a survey on a hot keyword of a nowadays-molecular magnetism, the magnetic anisotropy. This phenomenon, appearing by the interplay of spin and orbital magnetic moments is a key ingredient for obtaining systems behaving as magnets at molecular and nano-scale systems. The effect is exhibited by several d-type transition metal ions and most of the f-type, lanthanide systems, contained in coordination compounds and extended systems. Given the local nature of the magnetic anisotropy, it is tuned by environment features, termed and modeled in the frame of ligand field theory, with its specific premises for d and f sites. In the last years we had pioneering contributions in this field, taking newly synthesized systems as case studies, proposing advanced methodologies of analysis, modeling and prediction. We presented the so-called state-specific magnetization surfaces as a powerful and intuitive tool for revealing the poles of molecular magnets. Particularly interesting achievements were recorded in the field of f and d-f complexes, acknowledged to be very difficult due to the electronic structure particularities, such as nonaufbau configuration and weak interaction with ligands and neighbor magnetic sites. Taking simple and idealized examples, for the sake of clarity, combined with certain excerpts from more complicated realistic systems, we showed how the causal factors of the magnetic anisotropy, tuned by ligand field effects can be understood, advancing toward to the desiderata of rational property engineering, achieved investing equal interests in experimental and theoretical aspects.

7.1 BRIEF SURVEY OF THE MOLECULAR MAGNETISM

The magnetic anisotropy is a running paradigm of the nowadays molecular magnetism, stirring interest both from academic point of view, since it has intricate mechanisms, and also for application purposes, its understanding serving to design systems behaving as magnets at molecular and nano-scale levels [1]. The modern molecular magnetism originated about two decades ago, following the fresh perspective brought in the milieu by research articles and monographic works of scientists like Kahn, Gatteschi, Miller and others [2], who imported in the deals of coordination chemistry subjects that previously were merely the area of solid state physics [3]. The clue of this enterprise was worthy and fruitful, since the realm of coordination compounds, with their structural variety, allowed to identify at the molecular scale interactions acting also in bulk magnetic materials, such as oxides, enabling a more detailed focus on basic mechanisms due to the metal ions and their interaction with the immediate environment. Such an aim fitted naturally the investigation interests and working methodologies of coordination chemists, the quest for causal relationships bringing also the theoretical approach into the equation. Being parallel, in the timeline of recent history, with the increased availability of electronic structure methods on modern computers, the molecular magnetism was born with built-in interest for the electron structure underlying aspects, either correlating experimental data with computational counterparts, or at least invoking qualitatively orbital factors of the investigated interactions. Many electron structure aspects of molecular magnetism can be undertaken in relatively simple manner by means of Density Functional Theory (DFT) [4, 5] and user-friendly computer codes, even though the full account is realized through the so-called many-configuration methods, such as Complete Active Space Self Consistent Field (CASSCF) [6].

The first stage of the molecular magnetism focused on exchange coupling [7, 8], namely the quantum effects that govern the interaction between neighbor metal ions, most often mediated by bridges offered by the ligands that form the skeleton of a polynuclear coordination compound or of the extended structures. The exchange interaction is equated with the help of the so-called Heisenberg Hamiltonian (also cited with the acronym HDvV, from the complete list of contributors, Heisenberg, Dirac, van Vleck) [9]. This is ascribed as a sum over the closest pairs of interacting metal ions, *A–B*,

$$\hat{H}_{\mathrm{HDvV}} = -2\sum_{AB} J_{AB}\,\hat{S}_A \cdot \hat{S}_B \qquad (1)$$

where J_{AB} are the coupling constants, while $\hat{S}_A$ and $\hat{S}_B$ the spin operators, the approach being limited to sites without orbital degeneracy and reduced implication of one-center spin-orbit effects. The equation is not empirical, being, in fact, much more general and powerful in conceptual respects than the routine and frequent use in magneto-chemical modeling may let to guess, namely, an effective way to rewrite basic premises of Valence Bond theory [10, 11], the very first quantum model of molecular electron structure.

A plethora of complexes with different nuclearities, various topologies and electron count on metal ion sites, were synthesized and analyzed in the key of magnetic properties due to the exchange coupling effects [7, 12, 13]. The most often encountered situation is the antiferromagnetic coupling, which corresponds to negative parameters, $J_{AB} < 0$, since this is in line with the general trend of establishing a bond by antiparallel correlation of the electron spins. Even though the interaction between metal ion centers is noncovalent, the exchange effects work partly similarly to the Valence Bond picture that assimilates spin pairing to the energy stabilization trend. More rare, but desirable from the point of view of constructing systems with large spins, suitable for magnets, is the ferromagnetic coupling, with $J_{AB} > 0$ [14]. Fortunately, the situation of ferrimagnetism, namely a coupling of antiferro-type, $J_{AB} < 0$, but with unequal spins, $S_A \neq S_B$, leads to systems with net spin in groundstate, suited for making the whole cluster or lattice a possible candidate for application as magnetic material or as academic case study, in this key. In fact, the most useful magnetic materials, such as magnetite itself [15], are based on a ferrimagnetic ordering in the bulk.

Must emphasize the phenomenological parametric nature of the Heisenberg Hamiltonian, since the J_{AB} is not a pure exchange integral, containing also overlap effects [16]. Thus, for the case when the A and B centers have, each, one orbital only, a and b, respectively, the $J_{AB} \equiv J_{ab}$ parameter can be written as a pure exchange integral, K_{ab}, with positive sign, and an overlap dependent part, in terms of S_{ab}, with negative contribution [17]:

$$2J_{ab} = 2K_{ab} - U_{ab}S_{ab}^2, \tag{2}$$

where U_{ab} is the inter-center Coulomb repulsion term. For sites with many unpaired electrons, n_A and n_B, respectively, the total exchange parameter results as an average over the various J_{ab} orbital coupling channels:

$$J_{AB} = \frac{1}{n_A n_B} \sum_{a \in A} \sum_{b \in B} J_{ab} \tag{3}$$

From the elements of Eq. (2), one may understand the frequent encounter of antiferomagnetic coupling as due to the overlap part, the pure exchange having smaller absolute value and decaying rapidly with the inter-center distance. The occurrence of ferromagnetism appears in rather special situations where the overlap vanishes by symmetry, [16] or there is the case of real small chemical involvement, as may happen with lanthanide ions, where the f-type orbitals are contracted inside of the atomic body. In fact, the propensity for ferromagnetism in the d-f complexes (systems with d-type transition metal ions and lanthanides) is explained by more intricate mechanisms, as we analyzed previously, with focus on Cu-Gd complexes [18], but with reasons pointing to larger generality. In a recent work we established also the prerequisites of the d-f antiferromagnetic situation [19].

The rationalization of exchange coupling, as function of various orbital overlap conjunctures, electron count and topological effects, done picking-up case studies from the zoo of coordination compounds, formed the first stage of molecular magnetism.

7.2 THE MICROSCOPIC SOURCE OF MAGNETIC ANISOTROPY AND ITS KEY ROLE IN DESIGNING MOLECULAR AND NANO-SIZED MAGNETS

The large overall spin of a molecular or a supramolecular system is not the single condition to have a material behaving like a magnet and the engineering of exchange coupling is not the sole leverage that should be tuned in this purpose. Therefore, after solving the basic principles of the exchange coupling, the paradigm of molecular magnetism shifted, bringing into focus a subtle effect, the magnetic anisotropy. In an early stage it was regarded as a sort of *deux ex machina*, conceptually acknowledged, but having few technical clues to it, being primarily an object from the physicist's world, impenetrable to simple chemical intuition. In the recent years, the studies aiming to conquer this territory have become increasingly numerous and insightful [20]. We had our own contribution of rationales in this quest [21, 22], offering new keys for this major keyword, by approaching several relevant prototypic case studies and proposing methodological advances in the analysis.

In the previous section we introduced the spin of transition metal ions as origin of magnetism. The electronic spin is, in fact, the smallest magnet, as the panel (a) of Fig. 7.1 suggests. However, the spin behaves isotropicaly in a three-dimensional (3D) space, where the molecules are located. Heuristically, this can be figured recalling the quantum and relativistic nature of the spin, as a degree of freedom apart from the 3D space itself. If we place a probe magnetic field around a spin carrier, the expectation value of the magnetic moment will lead to components aligned with the vector of the applied field. Therefore, the polar map of the magnetic moment, as function of field orientation will be, simply, a sphere, namely isotropic. Then, we cannot rely on spin only to build a magnet in a molecule, since we cannot fix a stable axis with north and south poles. In order to accomplish such desiderata we need another source of magnetic moment, which originates from objects residing in the 3D space itself. At quantum level, this is the orbital momentum, which can be assimilated to the magnetism generated by a coil of electric current. The orbital degeneracy offers the possibility to move the electrons between states with different density patterns, which appears, in heuristic key, similar to the possibility to orbit a coil. This analogy is pictured in the panel (b) of the Fig. 7.1.

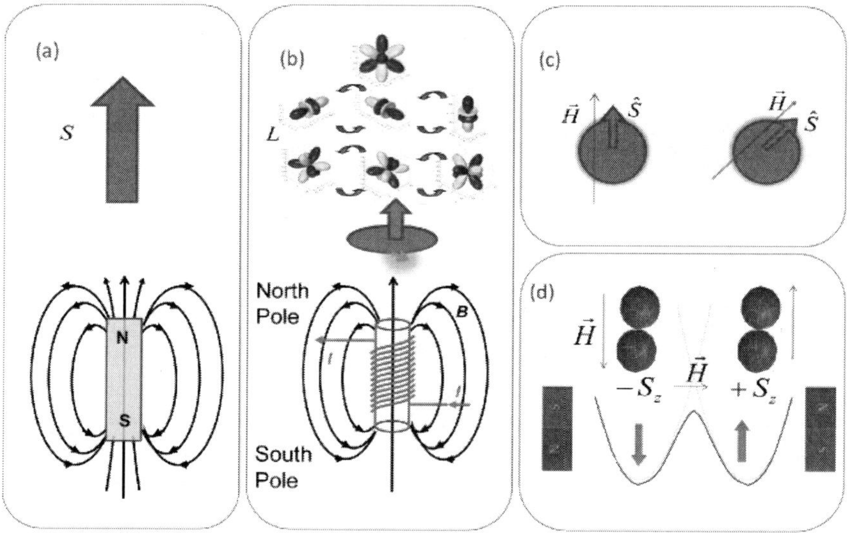

FIGURE 7.1 Synopsis of the basic ingredients of the atomic and molecular magnetism: (a) the spin magnetic moment of the electron is the "smallest permanent magnet"; (b) the orbital magnetism, related to the free movement of electrons between equivalent (degenerate) orbitals is, figuratively, the equivalent of the magnetic field induced by a coil of current; (c) the pure spin is isotropic- the magnetization follows freely the field and the magnetization surface is a sphere (d) the magnetic anisotropy (nonspherical, e.g., two-lobe, magnetization polar diagrams) and the energy barrier between opposed spin projections states are the mandatory premises of building systems behaving as magnets.

Despite of customary image that the electrons are orbiting always the nucleus, as in the Bohr historic model, the case is that the electron trapped in a single (nondegenerated) orbital cannot change the electronic cloud distribution, having no possibility to generate something similar to an electric current. Such analogies should be taken with reserve, but the fact is that the orbital momentum is a degree of freedom located in the 3D space that can serve as leverage to impose a preferred axis of magnetization, by its interaction with the spin momentum and the environment established by the ligands. The orbital degeneracy is an expression of high spherical symmetry of the free ion. The orbital and spin moments interact trough the so-called spin-orbit coupling that has relativistic origin. Phenomenologically, it is similar to the spin-spin interaction ascribed by Eq. (1), except that now the active components are the determined by the total spin quantum number, *S,* and the orbital one, *L*, if rely to the spectral term (many-electron) perspective,

$$\hat{H}_{SO} = \lambda\, \hat{L} \cdot \hat{S}\,, \tag{4a}$$

or by elementary spins with s=1/2 and the orbital l quantum number of the electronic shell,

$$\hat{H}_{SO} = \zeta \sum_i \hat{l}(i) \cdot \hat{s}(i)\,, \tag{4b}$$

if rely to the one-electron description. In the first case, Eq. (4a), the momentum is composed by the vector coupling of the total quantum numbers characterizing a spectral term, *L* and *S*, the resultant *J* taking values between $|L\text{-}S|$ and $|L+S|$. In the second case, Eq. (4b), the spin-orbit Hamiltonian is obtained handling the $j_z=l_z+s_z$ projections of the relativistic orbital sets described by the $j = l\pm1/2$ quantum numbers for the electrons occupying a given l-type shell (i.e., l=2 for transition metal ions and l=3 for lanthanides). Both approaches are equivalent, considering that the *L* and *S* are obtained subsequently from sets of respective l_z and s_z projections. The sum in the Eq. (4b) runs on the electrons, *i*, that occupy a given configuration of the shell characterized by the quantum number l and playing the α or β spin projections, described by the s=1/2 spin quantum number. The relationship between the orbital spin-orbit coupling and the overall one is $\lambda=\zeta/2S$ for configurations before half-filling and $\lambda = -\zeta/2S$ for situations with more than $2l+1$ electrons. We recalled the above elementary issues since the spin-orbit is the essential effective force of the magnetic anisotropy, the proper introduction being worthy.

The spherical symmetry that sustains the algebra behind the *L* and *S* quantum numbers and their coupling via spin-orbit operator is affected when the metal ions are immersed in low symmetry environments from coordination compounds or in the texture of extended systems. The low symmetry destroys the degrees of freedom that are generating the orbital magnetic moment and also decouples the spin-orbit interaction. This happens to a higher extent in d-type transition metal compounds, while the f-type ones are retaining a good part of atomic-like behavior, since the f shell is rather interior, both as radial extension and in the energy scale, being shielded from environment effects. If the lowering of symmetry do not destroys completely the manifestations of the spin-orbit coupling, then, one may arrive to the picture described in the following, aiming to reveal the microscopic prerequisites for having the permanent magnet alike behavior. Even if the *L*, *S* and *J* are no longer good quantum numbers, may happen that some of their projections can still be used as effective descriptors of lowest states. Thus, in the case of d-type transition metal ions, the ligand field (the name for environment effects) overrides the scale of spin-orbit coupling, removing the description by J quantum numbers, and, in certain circumstances, the $2S+1$ degeneracy of the spin, letting as ground levels a couple of degenerate or quasi-degenerate states assignable to the $\pm S_z$ opposite projections. In the case of f-type metal ions, the spin-orbit is stronger than the ligand field effects, but the J multiplets are yet sufficiently perturbed to determine a couple of $\pm J_z$ degenerate of quasi-degenerate levels. Because in the formation of the $\pm S_z$ or $\pm J_z$ doublets there is a reminiscent trace of orbital magnetic moments, the states acquire a firm orientation of the magnetic moments, as if, in the classical analogy, the orientation of a solenoid is fixed in the space. Having a preferred space direction containing the magnetization of groundstate pair of levels is like having an axis analog to the north-south poles in a magnet.

After premises as the above-described ones, another effect is necessary to finally have a molecular magnet: a reaction coordinate and a barrier between the $+S_z$ and $-S_z$ or the $+J_z$ and $-J_z$ couples. If such a barrier does not exist, then, a rapid random flipping of the opposite magnetization vectors will occur and the system cannot be yet a magnet, the average of the moment being null. Customarily, the opposite projections and the barrier between them are often represented by the double well pattern [23] depicted in the panel (c) of Fig. 7.1. In spite of the widespread representation of such a picture in the introductory part of research or review papers dedicated to the magnetic anisotropy, the fact is that the clear definition of the reaction coordinate that carries the given profile is, to the best of our knowledge, not yet well understood. The height of the barrier is often associated to the total gap in the spectrum of S_z states of a given $2S+1$ multiplet in the case of transition metal ion systems. The coordinate of such a representation cannot be simply the magnetic field, since in that case no minima can result, to have a double well representation. It possibly implies small distortions of the molecular frame, accompanying the magnetic state of the system [23], but the complete subtleties are not clear, being the next battlefield in the extension of the molecular magnetism towards new territories of concepts and application challenges.

A molecule fulfilling the mentioned structural prerequisites of: (i) high molecular spin resulted from ferro- or ferri-magnetic coupling of the contained coordination spheres: (ii) magnetic anisotropy determined by the local environment features and electron count: (iii) a sizeable barrier for the relaxation of magnetization, realized by supra-molecular cooperativity, is called

Single Molecule Magnet (SMM) [24]. Systems extended in one dimension, that is, coordination polymer chains having SMM alike behavior, are called Single Chain Magnets (SCM) [21, 25]. A single anisotropic center cannot provoke alone the SMM phenomenon, the assistance of neighbor spins being necessary. However, a special class of systems named Single Ion Magnets (SIM) [26] achieves a SMM behavior even in mono-nuclear compounds. In this case, the companion magnetic moment is the nuclear spin, if the isotope distribution contains appropriate species. In relative recent works we dealt with the synthesis and thorough analysis of prototypic cases of SMM and SCM systems [21, 22], abstracting in the following several details revealed about the microscopic mechanisms to be considered when aim for magnets incorporated in molecules and nano-structures. A SMM can be thought as the analog of the magnetization domain known from the solid state magnetism. In bulk magnets, the borders of the magnetization domains are established by small structural defects, while the SMM is a much more regular edifice, defined by the chemical structure of the given complex.

The magnetic anisotropy is a *sine qua non*, necessary, ingredient for having the magnet behavior at microscopic level. Its origin is local, from inside the metal ion sites, the exchange coupling between centers being needed to consolidate a collective order that cooperatively determines the hysteretic loops of the magnetization recorded in the bulk magnets. We discarded from this discussion another source of anisotropy, due to the trough-space dipolar interaction between spins [27], since, having small magnitude on energy scale (fractions of reciprocal centimeters), is regarded as less relevant.

7.3 THE MAGNETIC ANISOTROPY IN d-TYPE TRANSITION METAL IONS

The magnetic anisotropy in d-shell based complexes is easily tractable with the help of the so-called Zero Field Splitting (ZFS) Hamiltonian:

$$\hat{H}_{ZFS} = D\cdot\left(\hat{S}_z^2 - \frac{1}{3}S(S+1)\right) + E\cdot\left(\hat{S}_x^2 - \hat{S}_y^2\right), \tag{5}$$

where D and E are effective parameters resulted from the interplay of ligand field and spin-orbit effects. As matter of principle, the ZFS can take place in any ion having $S > 1/2$, provide that the reminiscent perturbation effects from the spin-orbit coupling are sufficient to yield sizeable D and E parameters, the Mn(III) compounds being the most prominent candidates [28]. Other systems, such as Cr(III), Fe(II), Ni(II) complexes show this kind of behavior, [2] but the most efficient carrier of anisotropy remains the d^4 configuration. As suggests the name, the ZFS removes the $2S+1$ degeneracy of a spin multiplet in the absence of the magnetic field. The D component acts in tetragonal and lower symmetry point groups (being null in octahedral and cubic cases), the part driven by the E parameter appearing only in low symmetries (rhombic). For Mn(III) systems there is a simple correlation with the stereochemistry of octahedral coordination, having $D < 0$ for elongated octahedra, and $D > 0$ for the compressed ones, as consequence of different orbital ordering in the ligand field scheme (d-type orbitals of the complex) and in the diagram of spectral terms, as sketched in the Fig. 7.2.

For a clearer insight, beyond the qualitative issues usually exposed about the ZFS effect, we will advance in the following a quantitative modeling based on the ligand field and spin-orbit parameterization. The ligand field (LF) Hamiltonian for tetragonal symmetry is [29]:

$$\hat{H}_{LF}^{d}(D_{4h}) = 14Dq\sqrt{\pi}\left[Y_{4,0}(\theta,\phi)+\sqrt{\frac{5}{14}}\left(Y_{4,4}(\theta,\phi)+Y_{4,-4}(\theta,\phi)\right)\right]-14Dt\sqrt{\pi}\ Y_{4,0}(\theta,\phi)-14Ds\sqrt{\frac{\pi}{5}}\ Y_{2,0}(\theta,\phi), \quad (6)$$

where Y_{lm} are Spherical Harmonics which, for application perspectives, can be perceived as trigonometrical functions that can be found tabulated in many textbooks devoted to the atom structure or general chemical theory. Generally, a parameter in the front of a Y_{lm} Spherical Harmonics component varies approximately in a $1/R^{l+1}$ dependence on the metal-ligand distance [29].

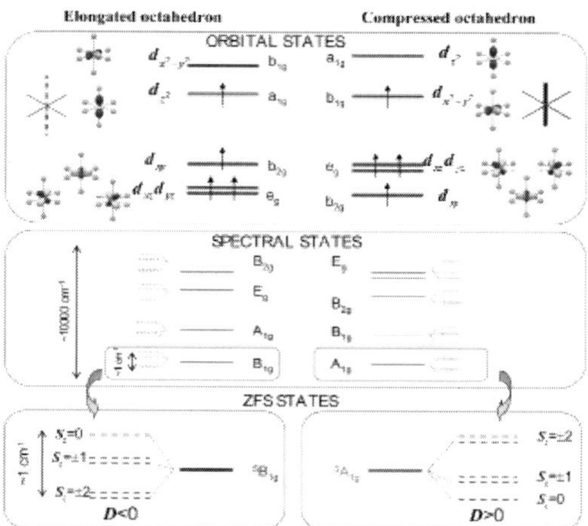

FIGURE 7.2 Scheme of ligand field (orbital diagram in the upper panel and spectral terms in the middle one) and ZFS levels (lower panel) occurring in elongated vs. compressed octahedral geometries of d^4 metal ion complexes (elongated – left side, compressed – right side).

Having a $[MA_4B_2]$ complex with A ligands in equatorial plane and B in the axial positions, the Dt parameter is proportional to the $1/R_A^5 - 1/R_B^5$ difference and Ds to the $1/R_A^3 - 1/R_B^3$ form. It turns that Ds and Dt are negative in the compressed octahedron, with $R_B < R_A$, and positive for the reverse case, $R_B > R_A$. The Ds and Dt vanish at pure octahedron, letting the Hamiltonian as sole function of the celebrated $10Dq$ parameter, (Dq being, literally, one tenth from the $10Dq$). With a total Hamiltonian taken as sum of the ligand field from Eq. (6) and the spin-orbit from Eq. (4), one may set a model that consists in solving a 25×25 matrix, the dimension resulting from the $(2S+1)(2L+1)$ multiplicity of microstates with L=2 and S=2 quantum numbers corresponding to the ^{5}D ground term of the Mn(III) free ion.

In a recent work [30], about a compressed Mn(III) complex we retrieved the following parameter values: $10Dq$ = 21146 cm^{-1}, Ds = −1565 cm^{-1}, Dt = −324 cm^{-1}, by means of a state-of-the-art TD-DFT (Time Dependent DFT) calculations that allowed a simple approach to the simulation of electron spectra and excited states. We will take these values as orientation for setting a numeric experiment with $10Dq$ = 20000 cm^{-1} and the tetragonal parameters tuned continuously, Ds from −1500 to 1500 cm^{-1} and Dt from −300 to 300 cm^{-1}, to simulate the change from compressed to elongated pattern, passing through the regular octahedron. The results are shown in Fig. 7.3. The panel (a) shows the overall spectral term variation. Note the fact, specific to the d^4 configuration, that the spectral term diagram is the reverse of the ligand field orbital scheme, since it can be assimilated to a hole in a half-filled d shell. The hole formally performs the reverse transitions, from upper levels toward the lower ones, yielding the reverse energy scheme. Thus, while in octahedron the ligand field d-type orbitals show the well-know sequence of t_{2g} and e.g. sets, the levels of the D-type term are ordered reversely, E_g and T_{2g}. In the D_{4h} symmetry the octahedral E_g parent splits into the A_{1g} and B_{1g} nondegenerate terms, the first one being groundstate in the compressed octahedron and vice-versa. The well known mutation of magnetic anisotropy described by the sign change, D>0 in compressed case and D<0 in the elongated one, is the consequence of level crossing trough the high symmetry point, the B_{1g} carrying the positive and A_{1g} the negative D values.

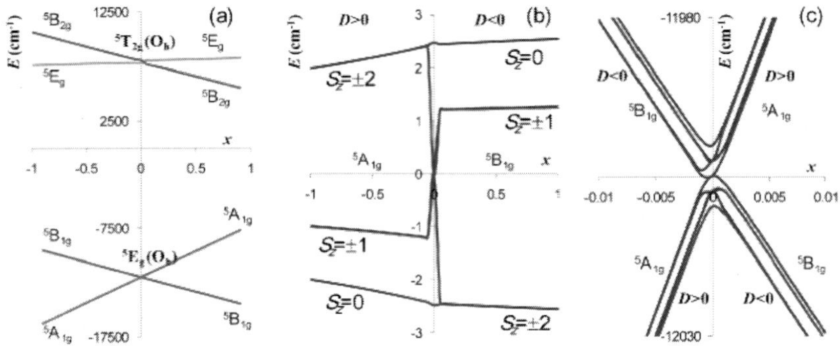

FIGURE 7.3 Energy diagrams resulted from the modeling of a d^4 system, tuning the axial field parameters as follows: Ds=1500x Dt=300x (in cm^{-1}), where x is the scaling factor, between −1 and 1, represented in abscissas. The octahedral parameter and the spin-orbit coupling are kept fixed at 10Dq=20,000 cm^{-1} and z=360 cm^{-1}. The x<0 left side corresponds to compressed octahedron while the x>0 to the elongated one. (a) The spectral terms are in the labeling of D_{4h} point group for $x\neq0$ and O_h at x=0. (b) The ZFS splitting computed as combined effect of ligand field and spin-orbit parameter. The panel (b) corresponds to the magnification of previous diagram to the energy scale of few cm^{-1}, with the shift in zero for the barycenter of the lowest spin multiplet. The discontinuity at x=0 is the consequence of crossing between $^5A_{1g}$ and $^5B_{1g}$ states. (c) Magnification of the crossing point around x=0 illustrating the ZFS patterns of both $^5A_{1g}$ and $^5B_{1g}$ states.

The panels (b) and (c) from Fig. 7.3 are showing details, probably lesser known, about the microscopic origin of the D parameter. Thus, from the 7.3(b) diagram one may see that the D is not continuously tuned trough zero, passing, as one may tacitly believe, if consider the two opposable cases of compressed vs. elongated octahedra and the supposed magnetic isotropy of the regular octahedron. In fact, there is a sudden change due to the mentioned level crossing, the $|D|$ absolute value remaining approximately constant during the tuning of axial ligand field. This feature is revealed because the spin-orbit coupling was not turned off during the numeric experiment, its manifestation resulting in small, but persistent, split of the spin multiplets on each electronic level. The presented modeling is a bit richer than the simple phenomenology of effective ZFS Hamiltonian from Eq. (5). The Fig. 7.3(c) shows a close-up to the crossing event of the $^5A_{1g}$ and $^5B_{1g}$ states, within spin orbit coupling, or, in other words, the spin-orbit split of the ^5E_g octahedral degenerate state. As a matter of fact, due to the so-called E×e Jahn-Teller effect [31], that removes the degeneracy by distortion, a Mn(III) system is not expected to resist in pure octahedral form. The picture from panel 3(c) clarifies that the discontinuity of the diagram from panel 3(b) is just a matter of graphical representation, by resetting the zero of the energy scale, not a physical event or a computational accident. The Fig. 7.4 shows a couple of mono-nuclear Mn(III) units having the compressed and elongated axial patterns. The compressed system, $[Mn^{III}(3\text{-MeO-sal-N}(1,5,9,13))]NO_3$, is realized with a hexadentate Schiff base ligand, 3-MeO-sal-N(1,5,9,13), obtained from N, N'-bis(3-aminopropyl)-1,3-propanediamine and 2-hydroxy-3-methoxy benzaldehyde [30]. The computed ZFS parameters, (using the BP86 DFT functional and VTZ basis set in Orca program) [32] are $D = -1.09$ cm^{-1} and $E = -0.01$ cm^{-1}, in line with the above discussed phenomenology.

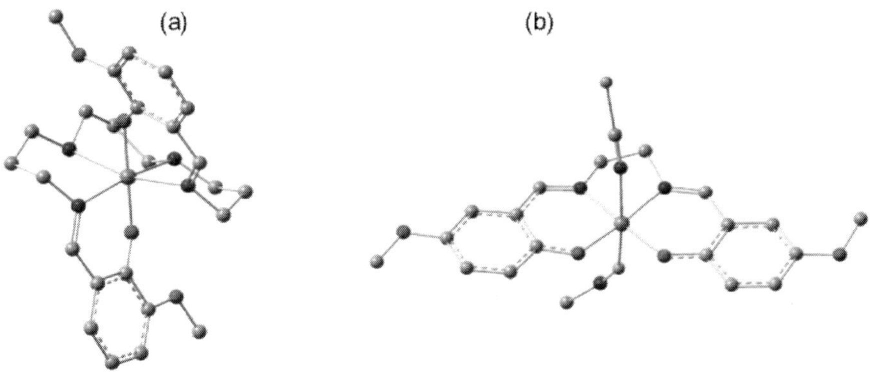

FIGURE 7.4 Examples of tetragonally compressed vs. elongated Mn(III) complexes: (a) The six-coordinated compressed high-spin Mn(III) Schiff base mono-nuclear complex, $[Mn^{III}(3\text{-Meo-sal-N}(1,5,9,13))]NO_3$. (b) The Mn(5-MeOsalen)]$^+$ moiety (completed with axial NC-Fe bridges), representing a case of elongated Mn(III) complex, cut from the zigzag chain $[(Tp)Fe(CN)_3Mn(5\text{-MeOsalen})\cdot 2CH_3OH]_n$. The hydrogen atoms are omitted.

Must point that the compressed octahedral species are rare [33, 34], the vast majority of Mn(III) complexes showing an elongated pattern. The unusual stereochemistry is due to the strain features of the encapsulating hexadentate ligand, enforcing the axial compression. The Mn(III) complexes with ligands belonging to the same class show other unusual magnetic property, namely the spin conversion phenomenon, the feature being a subtle balance of effects due to peripheral substituents of the Schiff base ligand and supra-molecular interaction of the mono-nuclear complex units [34]. The elongated example is embedded in the extended system [35] obtained from the assembling of [Mn(5-MeOsalen)$(H_2O)]^+$ and $[(Tp)Fe(CN)_3]^-$ leading to the one-dimensional (1D) zigzag chain

$[(Tp)Fe(CN)_3Mn(5\text{-}MeOsalen)\cdot 2CH_3OH]_n$ where, Tp^- = hydrotris (pyrazolyl) borate, 5-MeO-salen^{2-} = *N*, *N'*-ethylenebis(5-methoxysalicylideneiminate). The computed ZFS parameters, in the same setting as discussed for the above example, are in agreement with expectations for an elongated system, D= 3.87 cm^{-1} and E = 0.11 cm^{-1}.

We continue the analysis introducing a methodology proposed by us in recent papers [19, 22d,e,f], that displays the issue of magnetic anisotropy in a very picturesque and intuitive manner. The technical route consists in adding to the Hamiltonian containing ligand field and spin-orbit ingredients the dependence on the magnetic field, in the form of Zeeman Hamiltonian:

$$\hat{H}_{\text{Zeeman}} = -\sum_i \left(g_L \mu_B \hat{l}(i) \cdot \vec{B} + g_S \mu_B \hat{s}(i) \cdot \vec{B} \right), \quad (7)$$

where μ_B is the Bohr magneton, g_L=1 and g_S=2 are orbital and spin gyromagnetic factors and $\vec{B}$ the outer field.

Running the orientation of the magnetic field $\vec{B} = B \cdot \{\cos(\phi)\sin(\theta), \sin(\phi)\sin(\theta), \cos(\theta)\}$ on a mesh of (θ, φ) polar angles resembling the grid of latitude and longitude representations on the globe, we resolve at every point the eigenvalues, $E_i(\theta, \phi, B)$, of the total Hamiltonian $\hat{H} = \hat{H}_{\text{LF}} + \hat{H}_{\text{SO}} + \hat{H}_{\text{Zeeman}}$, and take the following derivative for any desired state, labeled i:

$$M_i(\theta, \phi, B) = \frac{dE_i(\theta, \phi, B)}{dB} \quad (8)$$

This magnitude has the meaning of response of the selected state to a perturbation with outer magnetic field, being called by us, state-specific magnetization [19]. The polar diagram of the $M_i(\theta, \varphi, B)$ function, namely the surface drawn joining the points having the radial $|M|$ extension at a given (θ, φ) set of polar angles, is a very clear image of the magnetic anisotropy. In Fig. 7.5, we illustrate the situation of the isotropic S=2 spin multiplet. At D=0 or, in fact, in the case of no spin-orbit perturbation, ζ=0, the response of each state is isotropic, the magnetization of the states following the field, with a magnitude simply equal to the product of S_z spin projection and Landé spin factor, M=$|g_S S_z|$.

In the case of anisotropy, the state responds to the spin perturbation with the maximal value $|g_S S_z|$ only in specific points, defining the so-called easy magnetization axis, and, progressively, with lower extents, for field orientations departed from this preferred direction. In a plane perpendicular to the easy axis there is no response to the magnetic field, the polar surface being contracted to a nodal point. In this way, the 3D polar representation of Eq. (8) takes the aspect of lobes, as seen in Figs. 7.6 and 7.7. The state with S_z=0 gives no response to magnetic field. The compressed octahedron has a nonmagnetic groundstate, in principle this case being not favorable for constructing magnetic devices. However, the compressed octahedron is interesting from chemical point of view by the rare occurrence of such species and the manifestation of an unusual phenomenon: the quintet-triplet spin transition, as our works revealed [34]. The elongated octahedron is perfectly suited for yielding the Single Molecule Magnet (SMM) manifestations, the $\pm S_z$ projections representing the different orientations of the magnetization along an axis perpendicular to the equatorial plane of the complex [21].

As pointed previously, the magnetic anisotropy must interplay with other effects, such as the exchange coupling, in order to yield the aimed magnetic ordering. The results are very sensitive to the details of interaction, as is illustrated comparing the systems displayed in the Fig. 7.8. Thus, a heterometallic chain [21a] was assembled from Mn^{III} salen-type dimeric units and hexacyanoferrate(III), see Fig. 7.8(a), showing two kinds of exchange interactions, Mn•••Fe via –CN– bridge and Mn•••Mn via bis-phenolate bridge. Both couplings are ferromagnetic and the whole system behaves as a Single Chain Magnet (SCM). The significant zero-field splitting on the Mn^{III} complex offers the anisotropy required for magnet-type behavior.

The assembling of the same cation with the complex anion $[(Tp)Fe(CN)_3]^-$, where Tp^- = hydrotris(pyrazolyl) borate, leads to the zigzag chain shown in Fig. 7.8(b) [35a]. This system shows less usual pattern of the recorded magnetic properties as function of the applied field, which can be characterized as metamagnetism, being non magnetic at null field and small temperatures, but acquiring, progressively, a magnetization when these two external variables are raised.

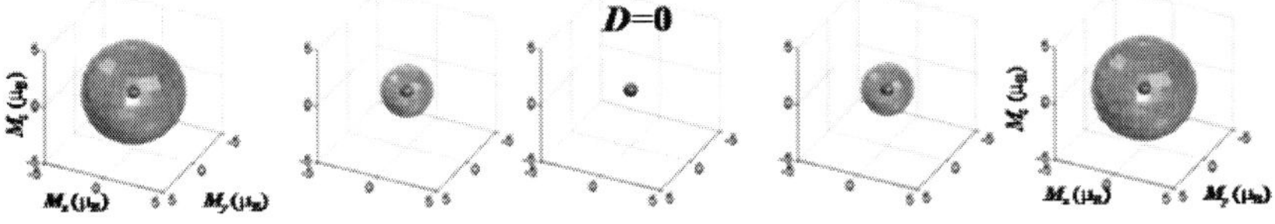

FIGURE 7.5 Polar diagrams for state-specific magnetization functions for the isotropic spin quintet components, D=0, E=0. The magnetization functions are spheres with radius $|gS_z|$, in Bohr magnetons (μ_B).

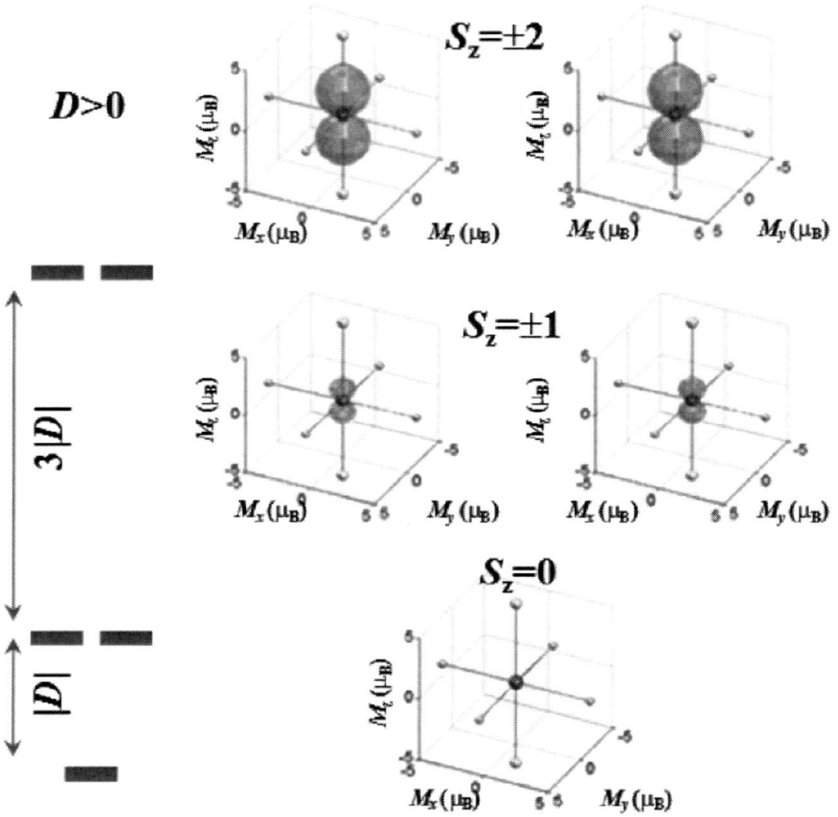

FIGURE 7.6 Polar diagrams for state-specific magnetization functions of the spin quintet components in the circumstances of magnetic anisotropy with $D>0$ (compressed octahedron). The axial lobes have $|gS_z|$ maximal extension, in Bohr magnetons (μ_B).

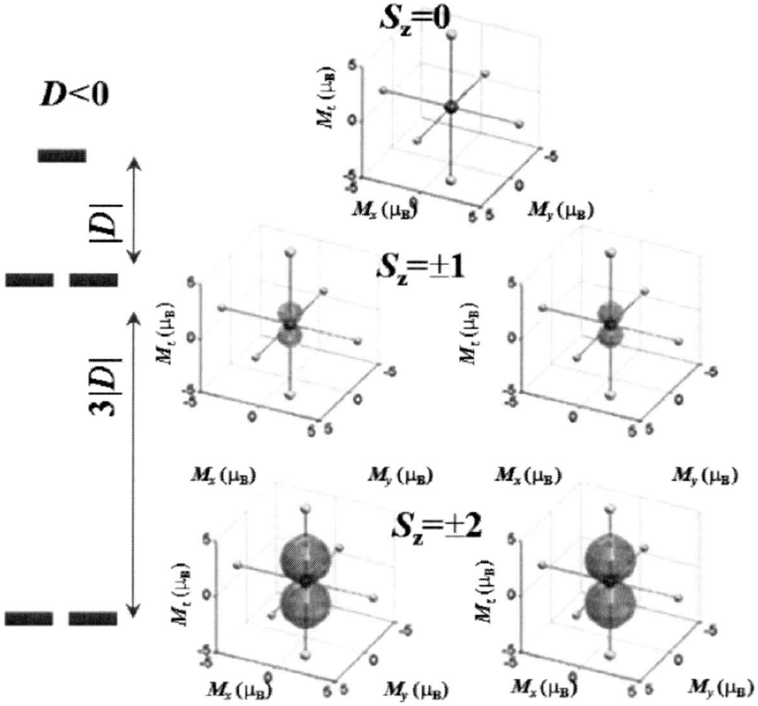

FIGURE 7.7 Polar diagrams for state-specific magnetization functions of the spin quintet components in the circumstances of magnetic anisotropy with $D<0$ (elongated octahedron). The axial lobes have $|gS_z|$ maximal extension, in Bohr magnetons (μ_B).

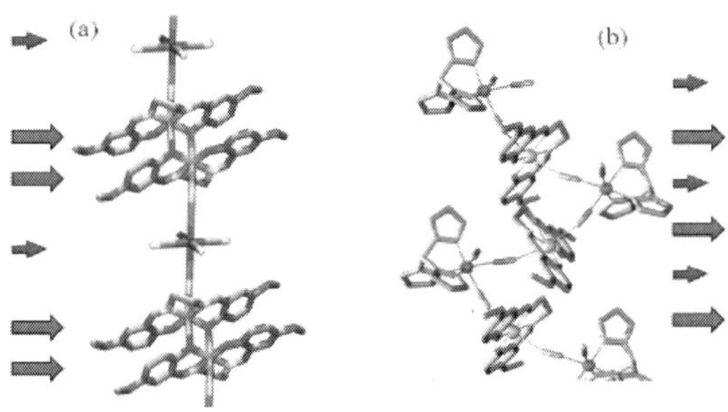

FIGURE 7.8 Extended structures obtained by the supra-molecular assembling of $[Mn(MeOSalen)]^+$ building blocks with different counterions, exhibiting differentiated magnetic properties: (a) a Single Chain Magnet system obtained with $[FeCN_6]_2$- bridges [21a] (b) a system resulted from coupling with pyrazolyl-borate cyanoferrate, $[(Tp)Fe(CN)_3]^-$, with metamagnetic behavior [35a].

The system (a) has a {$Mn^{III}Mn^{III}Fe^{III}$} repeating pattern with Mn^{III} dimeric sequences, while the (b) has a {$Mn^{III}Fe^{III}$} elementary unit. The (b) case is the same system discussed previously as example containing a Mn(III) elongated complex unit. In spite of the fact that both systems show extended ferromagnetic coupling and the same type of anisotropy on the Mn^{III} sites, their magnetic properties are quite different. This leads to the idea (substantiated by more advanced analyzes) that the (Mn^{III}-Mn^{III}) sequences are the key units which, behaving themselves as SMMs, are inducing the SCM character of the whole 1D system. The separation between the Mn^{III} units in the [Mn(MeOsalen)] [(Tp)Fe(CN)$_3$] compound precluded a similar effect in this system. The above tableau of differentiated magnetic properties, as consequence of modified coordination and supramolecular features, offers a taste for the challenge of designing materials with desired magnetic properties, by understanding the magnetic properties of the building blocks and the factors driving their assembling. At the same time, even small, the supra-molecular effects, the interchain interaction in our examples, exert a crucial role on the magnetic properties at nano-scale.

The onset of ferromagnetic ordering in the first system is assisted by a sort of ferro-type coupling between chains also, while the metamagnetism of the second case is due to an opposite antiferro-like long range interaction. In order to simulate such effects we will call the Ising model, which can be used in certain circumstances to mimic a combined effect of the exchange coupling and anisotropy. In a very evasive manner, one may say that, if the ZFS Hamiltonian leads to the split in elements characterized as S_z projections, then, the exchange may be termed as a product of such elements instead of the scalar product of complete spin operators, as ascribed in the Heisenberg Hamiltonian (see Eq. (1)). Therefore, the Ising Hamiltonian consists in the simple products $-J_{AB}S_z^A S_z^B$, the Hamiltonian being resolved already as energies, since the operators became numbers. However, this sort of approximation is too extreme and can be used only on limited sequences of states. To be distinguished from the customary use of Ising interaction between ion centers, considering here the interaction between two chains, we will denote the coupling parameter by zJ. In fact, we take the interchain Ising coupling as a resonant interaction model, based on the assumption that states having equal energies in the two subsystems are interacting more than any arbitrary couple of levels. Thus, taking equivalent states, i, from the individual chains, we get a gap equated by $2|zJ|(S_z^i)^2$, as sketched in the middle panel of the Fig. 7.9.

Note that the conceived states are collective ones on each chain, not the site-localized projections. However, one may guess that the resonant interaction is ultimately due to genuine Ising terms between relatively close magnetic centers from neighbor chains. Even assigning different splitting on all the i chain states, the key one is those of the groundstate, so that we can work with an unique zJ parameter, dedicated to the lowest levels. In the spirit of the Ising formalism, there are four states grouped in two degenerate pairs, one originating from the $\{(+S_z^i,+S_z^i),\ (-S_z^i,-S_z^i)\}$ set, and another from the $\{(+S_z^i,-S_z^i),\ (-S_z^i,+S_z^i)\}$ couple. The first couple can be characterized as the magnetic doublet, having the $-zJ\cdot S_z^2$ energy and the $\pm 2S_z^i$ projections. The other doublet has the $zJ\cdot S_z^2$ energy and null total projection, the opposite spin components being summed in a null magnetic moment.

The field has effect only on the first couple, and if the ordering is determined by an antiferro-like $zJ<0$ parameter, it can tune a crossing of the groundstate levels, leading to a phase transition. This induces the metamagnetism described for the case of [Mn(MeOSalen)][(Tp)Fe(CN)$_3$] system, having the mechanism sketched in the right side of the Fig. 7.9. The opposite case with $zJ>0$ shows a magnetic supra-molecular groundstate and an overall extended ferromagnetic ordering, from 1D to 2D-3D, that determines the SCM manifestation. In fact, the SCM is not a pure 1D effect, being tuned critically by the weak interaction between chains.

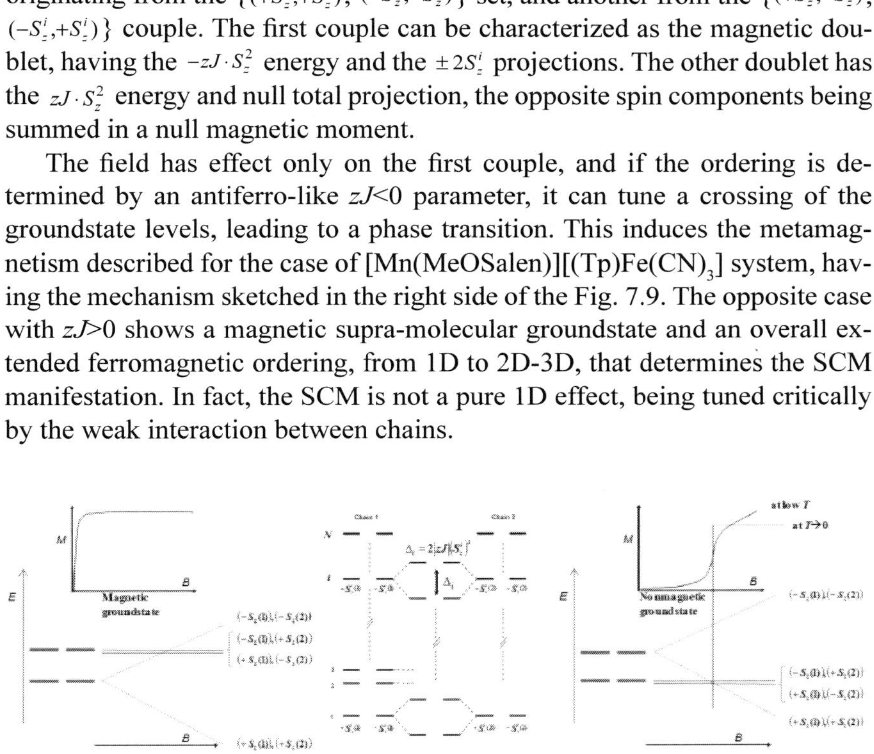

FIGURE 7.9 Scheme of Ising-like resonant interactions (middle diagram), the sign of parameters determining the onset of full magnetic ordering ($zJ>0$, left side) and the SCM effective behavior, or the metamagnetic result ($zJ<0$, left side).

7.4 THE ELECTRONIC STRUCTURE AND PROPERTIES OF THE f-TYPE SYSTEMS

The chemistry of lanthanide complexes, mostly in the Ln(III) oxidation states, forms a distinct part of the nowadays molecular magnetism [36, 37]. In the vast majority of cases the lanthanide ions are carriers of strong magnetic anisotropy, due to the fact that the f shell is weakly perturbed by the outer environment, retaining many of free atom features for the ions embedded in complexes and lattices. Then, in quasi-spherical symmetry, a good part of the orbital moment and spin-orbit coupling survives, determining the strong anisotropy. The f shell do not contributes to the molecular bonding energy, as our analyzes revealed [38], but determine the optical and magnetic properties of lanthanide-based materials. In fact, the electronic structure of lanthanide systems is quite tricky, demanding nonstandard procedures and special control of the calculation, given the non-*aufbau* nature of the f shell [18]. Namely, as is suggested in the Fig. 7.10, the f shell is contracted both in the radial scale, being inner to the atomic body and shielded by the outer shells of the Ln(III) and also in the overall energy scheme, the energies of f–type orbitals in a complex being lower than those of many occupied ligand-type orbitals, or even than those of the d-shells carrying other unpaired electrons, in the case of the d-f complexes. In the panel (a) of Fig. 7.10 we shown the radial distribution of the population carried by the f shell (the curve with sharp peak) against the total population of the Gd(III) atom (the steep increasing curve). The dashed components show the decomposition into three Slater-type primitives. One observes the announced fact that the maximum of radial wave functions of the f shell is placed below the atomic radii, making it not suited for bonding effects.

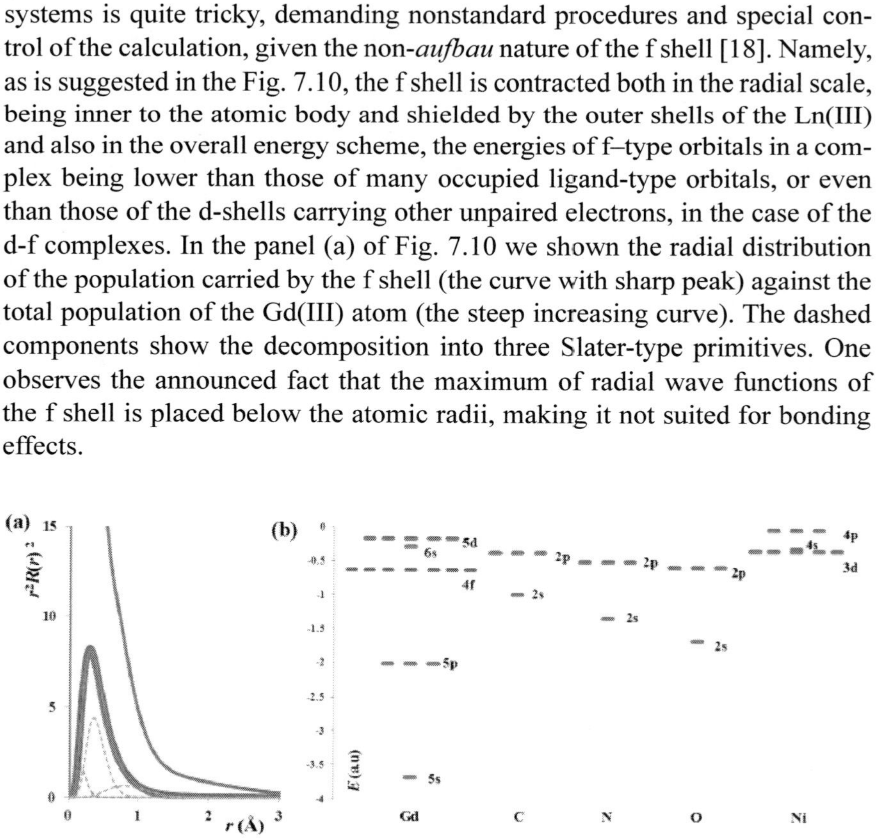

FIGURE 9.10 Atomic data illustrating special features of the f shell: (a) the density supported by the f radial function, represented against the total density carried by other squared radial functions of the occupied orbitals appears contracted inside the atomic body; (b) the atomic calculations show that the energy level of the f orbitals is lower than tose of 2p or 3d shells, the partly occupied f shells being in non-*aufbau* situation with respect of ligand orbitals or other d-type paramagnetic centers in heterometallic d-f complexes.

In the panel (b) of Fig. 9.10 is shown, also at the level of atomic calculations, that the energy of the f shell is placed below those of p-type orbitals of the atoms constituting the ligands (O and N as donors, C as part of overall molecular skeleton), or below those of the d-type shells of transition metals, determining the non-*aufbau* situation of the f and d-f complexes (namely, having unpaired electrons in the f shell, below the occupied levels of the other atoms in molecule).

We pointed in the introductory part that many works on modern molecular magnetism were accompanied by electronic structure calculations and theoretical considerations. However, for long time the modeling was confined to the d-type transition metal ion systems, the lanthanide complexes remaining uncharted area for molecular theory, even though in the solid-state physics special procedures, like so-called DFT+U, are able to account the particularities of lanthanide ions in lattices, by plane-wave procedures [39]. We claim pioneering works, doing the very first multiconfiguration calculations on realistic f and d-f complexes [18], presented as case studies for various magnetic properties implying exchange coupling and magnetic anisotropy [22].

We devised the methodology of assembling the wave function of the complex from the molecular orbital data of previously prepared fragments: ligands and d-type complexes (where is the case) plus naked Ln(III) ions. The orbital functions of the whole assembly were prepared having the combination coefficients of the atoms in fragments as the blocks of the initial matrix, placing zero-blocks for the interfragment areas of the complete LCAO matrix (linear combination of atomic orbitals). Then, we start directly the multiconfiguration calculation with the preliminary assembled wavefunctions. This nonstandard procedure accounts for the physical truth of weakly interacting nature of the lanthanide systems.

Very probably, the absence of molecular calculations on lanthanide complexes prior to our methodological breakthrough was due to the systematic negative results obtained when standard approaches were followed, namely either limiting to DFT attempts or making the multi-configuration approaches tributary to Hartree-Fock starting orbitals. However, single determinant procedures, DFT or Hartree-Fock, diverge wildly in the case of a non-*aufbau* configuration, since most of the existing codes cannot tackle such a situation, being based on *aufbau* assumption, in the design of the iteration procedures.

Even with codes explicitly enabling the control of individual occupation numbers of molecular orbitals, the calculation of large complexes, at realistic scale and with low symmetry, is quite problematic, while in the case of small systems with high symmetry, the DFT approach can be successful [42b, c].

As our practice of treating the experiment and theory on equal footing proved [22], making the considered systems case studies for causal structure property relationships, the *ab initio* calculations are a powerful complement to the interpretation of experimental data and the way to advance towards the engineering of special properties of molecular and nanosystems.

7.5 THE WEAK MAGNETIC ANISOTROPY IN THE LANTHANIDE COMPLEXES WITH f^7 CONFIGURATION

As pointed previously, the lanthanide ions are intrinsic carriers of magnetic anisotropy, due to large expectation values for the orbital momentum and strong spin-orbit coupling. This is true for most of the cases, except the ions with f^7 configuration, as the gadolinium(III), where there is no orbital component of magnetism, having a single half-filled electron configuration. However, in the case of f^7 configuration there is a ZFS effect, which even small, is significant as the basis of important applications, such as agent of Magnetic Resonance Imaging (MRI) [40] in tomography procedures. The effect results from to the mixing of the ground spin octet, originating from the ^8S atomic term, with the excited sextet levels, related with the ^6D, ^6F, ^6G atomic terms [41].

The ZFS effect of Gd(III) complexes is accounted by the general Hamiltonian ascribed in the Eq. (5) the matrix representation having a particular structure that generates doubly degenerate couples, because consists in two identical blocks, for positive and negative projections. Thus, the full 8×8 matrix of the spin octet is factorized in two 4×4 blocks:

$$\mathbf{H}_{ZFS}(\mathrm{f}^7) = \begin{pmatrix} \mathbf{h}_{ZFS}(\mathrm{f}^7) & \mathbf{0} \\ \mathbf{0} & \mathbf{h}_{ZFS}(\mathrm{f}^7) \end{pmatrix}, \tag{9a}$$

where

$$\mathbf{h}_{ZFS}(\mathrm{f}^7) = \begin{pmatrix} 7D & \sqrt{21}E & 0 & 0 \\ \sqrt{21}E & -3D & 2\sqrt{15}E & 0 \\ 0 & 2\sqrt{15}E & -5D & 3\sqrt{5}E \\ 0 & 0 & 3\sqrt{5}E & D \end{pmatrix}, \tag{9b}$$

while **0** denotes 4×4 blocks with null elements. The entries of the $\mathbf{h}_{ZFS}(\mathrm{f}^7)$ subblock rows and columns are the {+7/2, +3/2, −1/2, −5/2} or the {−7/2, −3/2, +1/2, +5/2} series of projections. The matrix shows this pattern because the $E \cdot (S_x^2 - S_y^2)$ part of ZFS determines nondiagonal elements between the $\Delta S_z = \pm 2$

spin projections. When a Zeeman Hamiltonian is added, the states related by $\Delta S_z = \pm 1$ are determining nondiagonal couples, removing the factorization of **H** in independent **h** blocks, since these are interconnected by the $S_z = \pm 1/2$ elements.

A picturesque view of the small anisotropy driven by the ZFS parameters in the case of the f^7 configuration is offered in the synopsis from the Fig. 7.11 employing the above defined state-specific magnetization functions (see Eq. (8)), taken as polar diagrams from the derivatives of a ZFS+Zeeman Hamiltonian. In the trivial isotropic case, with both ZFS parameters null, we have the case of degenerate octet with spin projections from the S=7/2 multiplet, the magnetization functions being spheres with the $|gS_z|$ radius.

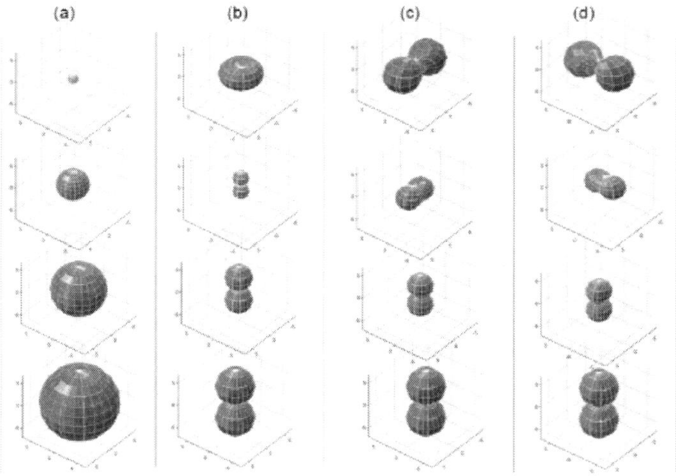

FIGURE 7.11 Magnetization surfaces as function of ZFS parameters for the states of a generic Gd(III) case: (a) The isotropic case with D=0 and E=0, when the magnetization surfaces are spheres with the radius $|gS_z|$; (b) the axial symmetry with $D\neq 0$ and E=0) the components corresponding from bottom to top, to the S_z={±7/2, ±5/2, ±3/2, ±1/2} series; (c) the general $D\neq 0$, $E\neq 0$ case with $D/E > 0$; (d) the general $D\neq 0$, $E\neq 0$ case for $D/E < 0$.

Since the degenerate spin states can be arbitrarily mixed, any perturbation by the magnetic field can be formulated as a rotation of the z axis along the outer field direction. Then, any derivative with respect of the field can be taken as the equivalent of the $\varepsilon = -g\mu_B S_z B_z$ with respect of B_z, namely $M_z = g\mu_B S_z$, explaining the spherical pattern. In the D-only case, the magnetization along the z axis can be simply understood, being given by the $M_z = -d\varepsilon / dB_z = g\mu_B S_z$

formula, the maximal extension of the magnetization state-specific functions in this direction being $g\mu_B S_z$, as seen in the column (b) of Fig. 7.11. The two-lobe pattern observed for the $S_z = \pm 7/2, \pm 5/2, \pm 3/2$ states can be described in this way: the x and y field components are represented only in the nondiagonal elements of the Zeeman Hamiltonian and, at infinitesimal values, do not contribute to the derivatives of the energies because the difference of related diagonal elements is larger than the perturbation, resulting in a nodal xy plane for the magnetization functions. The situation is a bit different for the $S_z = \pm 1/2$ couple, where the equal diagonal energies allow an arbitrary mixing of these eigenvectors, bringing the field in x and y components into the new diagonal elements. This allows the response to a perturbation in the xy plane to follow continuously the field, resulting in a torus shape of the magnetization for the $S_z = \{-1/2, +1/2\}$ pair. The shapes obtained in the general case, with nonnull D and E parameters, cannot be explained in easy manner, but one may retain the biaxial nature of this situation. The magnetization functions show lobes along the z axis and along the x when $D/E>0$ or y, when $D/E<0$, as seen in the columns (c) and (d) of the Fig. 7.11. The sign change in E does not affect the energies of the ZFS spectrum.

Recently, we realized a thorough analysis of the ZFS and exchange effects in two systems containing Gd(III) sites, a dinuclear, $[Fe_{LS}^{III}$(bpca)(μ-bpca)$Gd(NO_3)_4]$ and a coordination polymer $[Fe_{LS}^{II}$(μ-bpca)$_2Gd(NO_3)_2(H_2O)]NO_3$, where Hbpca=bis(2-pyridilcarbonyl)amine [19]. The magnitude of the ZFS split is small, described by $D = 0.047\ cm^{-1}$ and $|E| = 0.004\ cm^{-1}$ for the first system, while $D = 0.063\ cm^{-1}$ and $|E| = 0.005\ cm^{-1}$ for the second one. A non-trivial thing was the identification of the anisotropy axes with respect of the molecular plane, the situation corresponding to the biaxial pattern illustrated in the Fig. 7.11(c) and (d) panels.

7.6 THE STRONG MAGNETIC ANISOTROPY IN THE LANTHANIDE COMPLEXES WITH QUASI-DEGENERATE ORBITAL GROUNDSTATES OF THE F IONS

In the general case, for other configurations than the previously discussed f^7 half filled shell, the magnetic anisotropy cannot be treated in the simple form of the ZFS effect. A key of the successful treatment is the good account of the ligand field effects on the f shell [42] that determines the corresponding split of orbital multiplets, followed by the inclusion of spin-orbit coupling. Alternatively, considering that the spin-orbit acted in advance of the ligand field, one have to consider the split of the *J*-type multiplets as consequence of environment factors.

In order to make such aspects transparent, we will consider in the following very simple model systems, investigating the ligand field as causal factor of the magnetic anisotropy effects. Constructing another perspective to the work presented by Baldovi et al. [43] we will aim analyzing eight-coordination units, $\{LnL_8\}$, in several symmetries: cube (O_h), elongated and compressed tetragonal prisms (D_{4h}), tetragonal antiprism (D_{4d}) and triangular dodecahedron (D_{2d}). All these geometries can be presented in the unique general form from Table 7.1 giving the polar coordinates as function of the α, β, γ angles, as described in the following. For the sake of simplicity we take the same metal-ligand bond length, *R*, for all the coordinating atoms, the Cartesian coordinates $\{x, y, z\} = \{R\sin(\theta)\cos(\varphi), R\sin(\theta)\sin(\varphi), R\cos(\theta)\}$ being obtained with the polar angles presented in the Table 7.1.

The $\arccos(+1/\sqrt{3}) \sim 54.73°$ value corresponds to the half of the angle known to represent the regular tetrahedron (109.47°). This is also the LML angle made, trough the metal ion center, between ligands situated at opposed corners in a square face of a regular cube. The cube can be presented as superposition of two tetrahedra, mutually rotated by 90°. For α=β=γ=0 we have the case of cube and O_h symmetry.

TABLE 7.1 Polar coordinates for donor atom positions in idealized $\{LnL_8\}$ coordination Units. Tuning the α, β, γ angles, the molecular geometry can be continuously varied, specific values corresponding to special polyhedra, with high and intermediate symmetry (O_h, D_{4h}, D_{4d}, D_{2d}).

L	θ	φ
1	$\arccos(+1/\sqrt{3})+\alpha$	0
2	$\arccos(+1/\sqrt{3})+\beta$	π/2
3	$\arccos(+1/\sqrt{3})+\alpha$	π
4	$\arccos(+1/\sqrt{3})+\beta$	3π/2
5	$\arccos(-1/\sqrt{3})-\alpha$	g
6	$\arccos(-1/\sqrt{3})-\beta$	γ+π/2

TABLE 7.1 *(Continued)*

L	θ	φ
7	$\arccos\left(-1/\sqrt{3}\right)-\alpha$	γ+π
8	$\arccos\left(-1/\sqrt{3}\right)-\beta$	γ+3π/2

For α=β≠0 and γ=0 there is the case of tetragonal prism, elongated when α<0 and compressed when α>0, both situations showing D_{4h} symmetry. For any case with α = β and γ = π/4 we have the D_{4d} antiprism, the sign and magnitude of α not altering the symmetry of the complex. At arbitrary γ angles, nonnull and different from multiples of π/4, the system adopts S_8 symmetry. In addition, we will consider the case when α = –β and γ = π/2. This is a distorted cube, called triangular dodecahedron, made by interlocking a compressed tetrahedron, for example, the subsystem depending on α>0, and an elongated one, with α<0, the system having the D_{2d} symmetry.

The general Hamiltonian for the f shell can be expressed as follows [44]:

$$\hat{H}_{LF}^{f} = \sum_{k=2,4,6}\sqrt{\frac{4\pi}{2k+1}}\left(B_k^0 Y_{k,q}(\theta,\phi)+\sum_{q=1}^{2k+1} B_k^q Z_{k,q}(\theta,\phi)+i\sum_{q=1}^{2k+1} B_k^{-q} Z_{k,-q}(\theta,\phi)\right), \quad (10.a)$$

where the Z terms are combination of Spherical Harmonics,

$$Z_{k,q}(\theta,\varphi)=\left(Y_{k,q}(\theta,\varphi)+(-1)^q Y_{k,-q}(\theta,\varphi)\right), \quad (10.b)$$

$$Z_{k,-q}(\theta,\varphi)=\left(Y_{k,q}(\theta,\varphi)-(-1)^q Y_{k,-q}(\theta,\varphi)\right), \quad (10.c)$$

arranged to have real B_k^q parameters. In general, for an asymmetric case, there are 27 parameters, the count being obtained by summing the $2k$+1 elements of the k = 2, 4, 6 sets. In high and intermediate symmetries, the number of effective parameters is drastically reduced to few nonnull ones. The B_k^q can be regarded as effective parameters themselves, or can be dichotomized in further components. Even though is definitely acknowledged that the ligand field parameters cannot be conceived in the initial electrostatic meaning of crystal field theory, it is conventionally convenient to rely on this formalism. Then, from each kind of ligand L the B_k^q term includes contributions proportional to formal $eq_L\left\langle r^k\right\rangle/R_{ML}^{k+1}$ quantities, where eq_L is the positive amount assigned to the repeal between negative charges of the electron in the f shell and the cloud of the ligand, $\langle r^k \rangle$ is an expectation value determined by the radial shape of the f shell and R_{ML} the metal-ligand distance. In fact, these components are not taken distinctly, the whole block $eq_L \langle r^k \rangle / R_{ML}^{k+1}$ being conceived as a parameter [45] including, in the spirit of effective theories, many contributions, such as orbital stabilization and exchange, not only the electrostatic part, being nonnull for neutral ligands. When we deal with a single sort of ligands, characterized by an unique bond-length, $R_{ML} = R$, as in our idealized eight-coordinated systems, the above quantity can be taken as overall factor in B_k^q formulas, in front of a part determined by the geometry factors, as will be seen in the following. All the systems will be described using only three radial parameters, $eq_L \langle r^2 \rangle / R^3$, $eq_L \langle r^4 \rangle / R^5$ and $eq_L \langle r^6 \rangle / R^7$, the O_h symmetry case retaining only the last two.

The Hamiltonian terms for the square prism (compressed or elongated, including regular cube as particular case) are expressed as follows, as function of the α angle accounting the deviation from the regular cube:

$$B_2^0(\mathrm{D_{4h}}) = \frac{eq_L \langle r^2 \rangle}{R^3} \left(-4\sqrt{2}\sin(2\alpha) - 2\cos(2\alpha) + 2 \right) \tag{11.a}$$

$$B_4^0(\mathrm{D_{4h}}) = \frac{eq_L \langle r^4 \rangle}{R^5} \left(\frac{35\sqrt{2}}{18}\sin(4\alpha) - \frac{245}{72}\cos(4\alpha) - \frac{5\sqrt{2}}{3}\sin(2\alpha) - \frac{5}{6}\cos(2\alpha) + \frac{9}{8} \right) \tag{11.b}$$

$$B_4^4(\mathrm{D_{4h}}) = \frac{eq_L \langle r^4 \rangle}{R^5} \left(\frac{\sqrt{35}}{18}\sin(4\alpha) - \frac{7\sqrt{70}}{144}\cos(4\alpha) + \frac{\sqrt{35}}{3}\sin(2\alpha) + \frac{\sqrt{70}}{12}\cos(2\alpha) + \frac{3\sqrt{70}}{16} \right) \tag{11.c}$$

$$\begin{aligned} B_6^0(\mathrm{D_{4h}}) = \frac{eq_L \langle r^6 \rangle}{R^7} \Bigg(& \frac{385\sqrt{2}}{288}\sin(6\alpha) + \frac{1771}{576}\cos(6\alpha) + \frac{7\sqrt{2}}{8}\sin(4\alpha) - \frac{49}{32}\cos(4\alpha) \\ & - \frac{35\sqrt{2}}{32}\sin(2\alpha) - \frac{35}{64}\cos(2\alpha) + \frac{25}{32} \Bigg) \end{aligned} \tag{11.d}$$

$$\begin{aligned} B_6^4(\mathrm{D_{4h}}) = \frac{eq_L \langle r^6 \rangle}{R^7} \Bigg(& \frac{55\sqrt{7}}{288}\sin(6\alpha) + \frac{253\sqrt{14}}{1152}\cos(6\alpha) - \frac{13\sqrt{7}}{24}\sin(4\alpha) + \frac{91\sqrt{14}}{192}\cos(4\alpha) \\ & - \frac{5\sqrt{7}}{32}\sin(2\alpha) - \frac{5\sqrt{14}}{128}\cos(2\alpha) + \frac{15\sqrt{14}}{64} \Bigg) \end{aligned} \tag{11.e}$$

At α=0 the above formulas represent the O_h cubic case, the values from the parentheses as described above as trigonometric expansions becoming respectively, 0, $-28/9$, $2\sqrt{70}/9$, 16/9, $8\sqrt{14}/9$, for the $B_2^0(\mathrm{O_h})$, $B_4^0(\mathrm{O_h})$, $B_4^4(\mathrm{O_h})$,

$B_6^0(\mathrm{O_h})$, $B_6^4(\mathrm{O_h})$, series of ligand field terms. One notes that the second-order term, $B_2^0(\mathrm{O_h})$, disappears. The B_k^0 terms determine the diagonal of the 7×7 matrix of ligand field in the basis of f-type orbitals, taken as $Y_{3,m}$ ($m = -3$ to 3) Spherical Harmonics, while the B_k^4 terms (k=4, 6) create nondiagonal elements between m and m' components having the $|m-m'|=4$ orbital projection indices.

A rather interesting situation is recorded for the D_{4d} antiprism, where the B_k^0 terms are the same like in the prism case, while the other components are vanishing:

$$B_2^0(\mathrm{D_{4d}}) = B_2^0(\mathrm{D_{4h}}) \tag{12.a}$$

$$B_4^0(\mathrm{D_{4d}}) = B_4^0(\mathrm{D_{4h}}) \tag{12.b}$$

$$B_4^4(\mathrm{D_{4d}}) = 0 \tag{12.c}$$

$$B_6^0(\mathrm{D_{4d}}) = B_6^0(\mathrm{D_{4h}}) \tag{12.d}$$

$$B_6^4(\mathrm{D_{4d}}) = 0 \tag{12.e}$$

Since only the B_k^0 -type terms are acting, the ligand field matrix will be already diagonal. The coincidence of a part of ligand field terms from D_{4d} with the D_{4h} ones and the vanishing of the other ones seems somewhat surprising. It can be heuristically explained pointing the so-called holohedrization effect that acts inside of the ligand field models, both in the d and f shell cases. Namely, confining to a Hamiltonian based on a single shell, irrespective the parity of the orbital set, for example, even (g) in the case of d and odd (u) in the case of f atomic orbitals, the constructed matrix will behave symmetrically, since the products between rows and columns are even: $u \otimes u = g \otimes g = g$. Therefore, in the frame of a ligand field model, the asymmetry along a given axis cannot be accounted, or, in other words, the asymmetric potential along that axes is perceived as the symmetric average. In a prism we have four *trans*-L-M-L axes connecting ligands from upper and lower faces, related by inversion operations. In the antiprism there are no such axes, but the holohedrization makes the field exerted by one face equivalent to one from a virtual prism having half of the perturbation power, like four *trans*(L/2)-M-(L/2) axes, since the perturbation effect from a M-L bond is smeared equally in the opposite L-M direction. Therefore, the antiprism acts as two prisms mutually rotated by 45°, having each half of the coordination power of the prism. On the diagonal, this pattern leads to the same amount like the prism, the vanishing of the terms that lead to nondiagonal ligand field elements being due to the cancelation of angular coefficients coming from the formal two-prism assembly of the anti-prism. For the so-called triangular dodecahedron we will confine to the case when it can be described as realized by opposite distortions, elongation and compression, of the two tetrahedra that are making a cube. The coordinates, are expressed by $+\alpha$ distortion in the azimuth angle for half of the ligands and $-\alpha$ for the complementary set. In this case, the ligand field terms are:

$$B_2^0(\mathrm{D_{2d}}) = \frac{eq_L\left\langle r^2\right\rangle}{R^3}\left(-2\cos(2\alpha)+2\right) \tag{13.a}$$

$$B_4^0(\mathrm{D_{2d}}) = \frac{eq_L\left\langle r^4\right\rangle}{R^5}\left(-\frac{245}{72}\cos(4\alpha)-\frac{5}{6}\cos(2\alpha)+\frac{9}{8}\right) \tag{13.b}$$

$$B_4^4(\mathrm{D_{2d}}) = \frac{eq_L\left\langle r^4\right\rangle}{R^5}\left(-\frac{7\sqrt{70}}{144}\cos(4\alpha)+\frac{\sqrt{70}}{12}\cos(2\alpha)+\frac{3\sqrt{70}}{16}\right) \tag{13.c}$$

$$B_6^0(\mathrm{D_{2d}}) = \frac{eq_L\left\langle r^6\right\rangle}{R^7}\left(\frac{1771}{576}\cos(6\alpha)-\frac{49}{32}\cos(4\alpha)-\frac{35}{64}\cos(2\alpha)+\frac{25}{32}\right) \tag{13.d}$$

$$B_6^4(\mathrm{D_{2d}}) = \frac{eq_L\left\langle r^6\right\rangle}{R^7}\left(\frac{253\sqrt{14}}{1152}\cos(6\alpha)+\frac{91\sqrt{14}}{192}\cos(4\alpha)-\frac{5\sqrt{14}}{128}\cos(2\alpha)+\frac{15\sqrt{14}}{64}\right) \tag{13.e}$$

Here, one observes another simple relationship with the Eq. (11) dedicated to the square prism. Namely, the $B_k^q(\mathrm{D_{2d}})$ are obtained from the $B_k^q(\mathrm{D_{4h}})$ ones by picking only the terms in cosines. This can be understood, again, recalling the holohedrization effect. Namely, the described triangular dodecahedron is made of two tetrahedra with opposite distortions, $+\alpha$ vs. $-\alpha$. Each tetrahedron acts as a prism of half coordination power, obtained by smearing the tetrahedron via de inversion center. To be distinguished from the antiprism, the two prisms are not mutually rotated, being distinguished only by their opposite shape, compressed vs. elongated. In this case, because of the different sign of angular coordinate, the terms in $\sin(\pm m\alpha)$ cancel each other, having conserved the $\cos(\pm m\alpha)$ ones, insensitive to the sign change.

The above unitary description of geometry and related ligand field parameterization for a series of octa-coordination systems offers a clear perspective on the aimed problem. Even though such things are relatively accessible exercises in ligand field modeling, and previous analyzes focused on magnetic properties of systems with such topologies with the help of a numerical analysis were reported,[43] to the best of our knowledge, this is a rather unprecedented analytical outline for the selected series of the $\{LnL_8\}$ complexes.

We aim in this section to illustrate how the environment determines the magnetic anisotropy of $\{LnL_8\}$ complexes, trough the ligand field factors. In order to find a set of ligand field parameters, to do the desired numeric experiments, we will use as source the *ab initio* calculations on very idealized systems. Namely, we will mimic the ligands taking clusters of eight neon noble gas atoms, at R=2.3 Å distance from central ion. Usually, the lanthanide-ligand bond lengths are in the range 2.5–2.7 Å, but we enforced a slightly shorter bond, to rise the perturbation effect, compensating the lower propensity for interaction of the noble gas atoms.

TABLE 7.2 Computed and fitted ligand field levels for several $[TbNe_8]^{3+}$ model complexes with geometries tuned according to the coordinates mentioned in Table 7.1.

	Cube (O_h) and Prisms (D_{4h})			**Antiprisms (D_{4d})**			**Cube (O_h) and Triangular Dodecahedron (D_{2d})**		
	irrep.	**calc.**	**fit**	**irrep.**	**calc.**	**fit**	**irrep.**	**calc.**	**fit**
α=0	T_{1u}	–106.0	–104.9	E_1	–89.1	–95.5	T_{1u}	–106.0	–104.9
	T_{2u}	–82.0	–79.6	E_3	–81.9	–89.1	T_{2u}	–82.0	–79.6
	A_{2u}	564.2	553.7	B_2	–75.0	–79.6	A_{2u}	564.2	553.7
				E_2	208.4	224.4			
α=–5°	E_{1u}	–216.6	–222.6	E_3	–209.7	–215.7	A_1	–112.0	–112.4
	B_{1u}	–85.5	–78.3	B_2	–55.5	–70.6	E_1	–87.8	–85.5
	A_{2u}	–64.6	–70.6	E_1	23.9	28.0	B_2	–82.2	–80.6
	E_{1u}	21.1	34.9	E_2	213.6	222.9	E_1	–80.2	–80.1
	B_{2u}	541.1	524.2				A_2	530.3	524.1
	E_{1u}	–191.5	–192.1	E_1	–184.3	–187.1	A_1	–112.0	–112.4
	B_{1u}	–139.9	–146.5	B_2	–96.1	–90.5	E_1	–87.8	–85.5
α=+5°	A_{2u}	–94.0	–90.5	E_3	64.5	43.6	B_2	–82.2	–80.6
	E_{1u}	40.0	48.6	E_2	167.9	188.8	E_1	–80.2	–80.1
	B_{2u}	536.8	524.0				A_2	530.3	524.1

All energies are in cm^{-1}. The α=–5° corresponds to elongated prism and antiprism, while α=+5° to the compressed ones. The triangular dodecahedra with α=±5° are equivalent each to other and α=0 coincident to the regular cube.

Since the f shell do not contributes to the bonding and the perturbation is not purely electrostatic, the quantum exchange effects between atomic bodies, usually termed as Pauli interaction [46], essentially nonbonding, carrying a good part of perturbation, the artificial choice of noble gas as surrogate for ligands does not impinge upon the desired goal. Obtaining a total gap of several hundreds of reciprocal centimeters is a sufficient outcome of the imposed environment, this fact being satisfactorily accomplished, as seen from Table 7.2. We avoided taking other, apparently more natural choices, such as fluoride or oxygen anions, since this leads to highly negatively charged complexes, determining more artificial and less controllable consequences in the distribution of the total electronic cloud. Avoiding such effects, or the supplementary adjusting tricks, for example, placing positive point charges to compensate the undesired excess of negative charge, we confined to the neutral ligands, the positive charge of the complex being more convenient in computational respects and even as physical meaning.

A convenient choice for obtaining the ligand field parameters is the case of Tb(III) complexes, where the split of the ^7F term can be considered as parallel to the ligand field diagram itself. Must point that such a handling of the spectral terms, in the frame of multiconfiguration calculations, is more correct, as principle, than the use of orbital energies. In spite of the current belief that the orbital levels are rendering the ligand field diagrams, there is no rigorous reason, even though in certain circumstances a relative correlation may exist. The ligand field parameters should be, nominally, one-electron quantities, while the orbital energies show only effective one-electron nature, incorporating averages of two-electron effects. The two-electron part is very tributary to conventions intrinsic to the computational frame, being quite different in Hartree-Fock, DFT or multiconfiguration methods. In fact, the orbitals should be regarded merely as pieces of construction for density or active space, rather than fully meaningful objects in themselves. In the case of Tb(III) complexes, with f^8 configuration, the lower terms results from the seven possibilities to run the β electron (or, in other words, the doubly occupied orbital) over the f-type orbitals. Doing a CASSCF(7,8) calculation (i.e. an active space with 8 electrons in 7 orbitals) under state-average conditions (in order to preserve the unity of the orbital space over all the configurations, in the spirit of ligand field theory) we obtained for the different topologies of $[TbNe_8]^{3+}$ complexes the results cumulated in the Table 7.2. Even though a certain variation of the ligand field parameters is expected along the lanthanide series, the changes are relatively small, within few percents, due to so-called lanthanide contraction effects [47], not essential in semiquantitative respects, so that Tb(III) complexes are considered to provide reasonable information for congener systems too. The calculations were done with the GAMESS code [48].

Since all the systems presented in Table 7.2 are conceived as having unique bond length, then, there are only three ligand field parameters, with the following values: $eq_L\left\langle r^2\right\rangle/R^3$ = 360.2 cm^{-1}, $eq_L\left\langle r^4\right\rangle/R^5$ = 299.3 cm^{-1} and $eq_L\left\langle r^6\right\rangle/R^7$ = 216.3 cm^{-1}. It is quite remarkable that the few parameters accounted a relatively good match for all the couples of computed-fitted values, at different geometries, certifying the acceptable validity of superposition principle staying behind the ligand field approach.

The ligand field modeling of the d complexes is simpler than those of lanthanides, being in many circumstances invoked in qualitative respects. On the other hand, the above examples showed that the f systems are also tractable, the modern deals of molecular magnetism provoking a resurrection [26] of the old art of f-type ligand field modeling, practiced until now, and mostly in the early ages before the boom of computational chemistry, in rather small scientific communities. In spite of relative algebraic difficulty, the ligand field of lanthanides benefits from certain shortcuts, like the Stevens equivalent operator techniques, that make easier the application to specific problems, such as spectroscopy and magnetism [49]. Thus, the ligand field matrix accounts only for the one-electron part, the complete modeling demanding the use of Slater-Condon parameters for the two-electron interactions (exchange and Coulomb) and the spin-orbit coupling. However, in the frame of Stevens equivalent operator treatment [49], one finds a shortcut that do not needs the explicit use of Slater-Condon and spin-orbit parameters. This is because, considering that ligand field part is smaller than the other terms of the Hamiltonian, one may use the predefined basis set obtained in the free atom, after the action of two-electron and spin-orbit interactions. Thus, placing certain factors in the front of B_k^q parameters, one may pass from the basis of $Y_{3,m}\equiv|3,m\rangle$ orbital quantum numbers of the f shell directly to the $|J,J_z\rangle$ basis set of the ground multiplet described by the J quantum number of a given ion. There are series of {α_{Ln}, $\beta_{\mathrm{, Ln}}$, $\gamma_{\mathrm{, Ln}}$ } coefficients for the groundstate multiplets of each Ln ion, that transform the Hamiltonian parameters as follows, $B_2^q\rightarrow\alpha_{Ln}B_2^q$, $B_4^q\rightarrow\beta_{Ln}B_4^q$ and $B_6^q\rightarrow\gamma_{Ln}B_6^q$. Such coefficient sets are {–1/99, 2/16335, −1/891891} for the J=6 multiplet of Tb(III), {–2/315, −8/135135, 4/3864861} for the J=15/2 multiplet of Dy(III), {–1/450, −1/30030, −5/3864861} for the J=8 multiplet of Ho(III) and {4/1575, 2/45045, 8/3864861} for the J=15/2 multiplet of Er(III) [49]. At the same time, the Zeeman Hamiltonian from Eq. (7) can be replaced with the simpler form $-g_J\mu_B\hat{J}\cdot\vec{B}$. The magnetization for each desired state of the lowest J multiplets can be obtained working a matrix $(2J+1)\times(2J+1)$ dimension containing the Stevens-type operators and the Zeeman term. The Figs. 7.12 and 7.13 are showing the versatility of the spectral levels and the corresponding patterns of magnetic anisotropy for all the states. The Fig. 7.12 combines the systems with geometry close to the cube (prisms and triangular dodecahedron), while the Fig. 7.13 is reserved to antiprisms. One may note that in the first series only few states are carrying magnetic anisotropy. Particularly, only the elongated tetragonal prism shows this manifestation in groundstate, the other systems showing visible lobes in the magnetization functions of higher states. The lobes should be considered in relative comparison, each to other, not insisting here in the quantitative aspects.

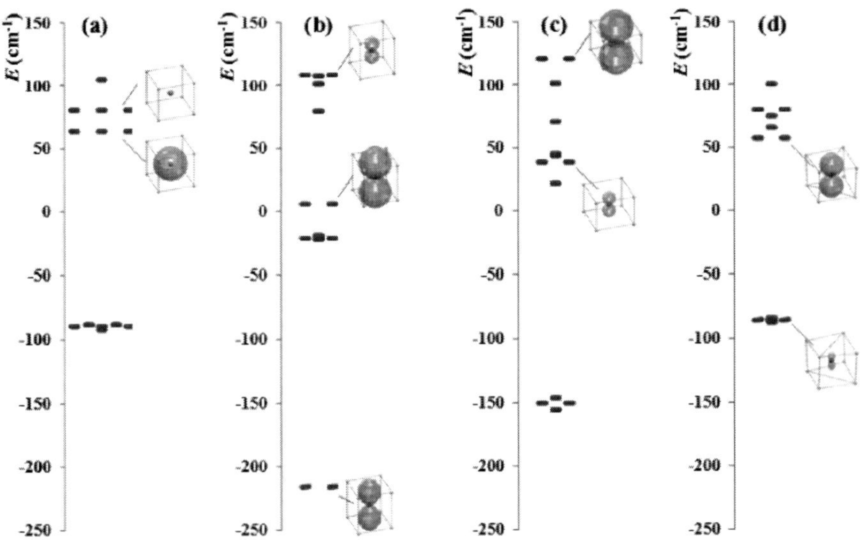

FIGURE 7.12 Ligand field split of the states resulted from the J=6 multiplet of Tb(III) and polar diagrams of the magnetization functions for states with nonnull magnetic response in several $[TbNe_8]^{3+}$ model complexes with geometries related to the cube: (a) regular cube (O_h), (b) elongated prism (D_{4h}), (c) compressed prism (D_{4h}), (d) triangular dodecahedron (D_{2d}).

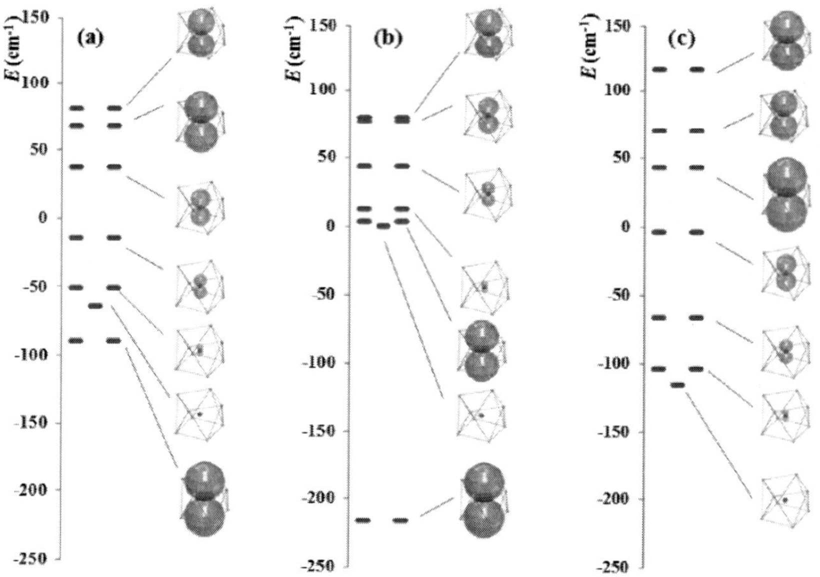

FIGURE 7.13 Ligand field split of the states resulted from the J=6 multiplet of Tb(III) and polar diagrams of the magnetization functions in several $[TbNe_8]^{3+}$ square antiprisms: (a) derived from regular cube, (b) elongated, (c) compressed. All the systems show D_{4d} symmetry.

To be distinguished from the cases in the Fig. 7.12, the antiprism series from Fig. 7.13 shows a very rich content in magnetic anisotropy for all the states, except one element in each system, assignable to J_z=0. The other states are arranged in doublets related to the $\pm J_z$ couples. All three antiprisms, obtained with α=0, −5, +5 degrees from geometries defined in Table 7.1, follow the D_{4d} point group. In spite of relatively small differences in Cartesian coordinates and visual aspect of the coordination polyhedra, the spectral and anisotropy patterns are varying strongly among the three systems. All the states show axial anisotropy, along the fourfold symmetry axis, the maximal extension of the lobes being precisely the $|g_J J_z|$, in Bohr magnetons (μ_B). In fact, this case is relatively simple, because the Hamiltonian is diagonal also in the $|J,J_z\rangle$ basis, for the same reasons, explained previously, that determine the diagonal pattern in the $|3,m\rangle$ basis of f orbitals, having an operator based on the B_k^0 terms only (see Eq. 12). Thus, the field dependence along the diagonal is simply written as $-\mu_B g_J J_z B_z$. Taking the derivative along the z axis one obtains the $\mu_B g_J J_z$ value, explaining the above discussed maximal extensions. The dependence on the field along the x and y axes is contained in nondiagonal elements (Zeeman perturbation has matrix elements between J_z and $J_z \pm 1$ microstates). At small fields, the nondiagonal perturbation is smaller than the gaps due to the ligand field on the diagonal, having therefore no response in the xy plane, explaining the nodal plane that determines the two-lobe aspect. For the J=6 groundstate of Tb(III) we have the g_J=3/2 ideal value. The maximal extension of groundsates in the panels (a) and (b) of Fig. 7.13 is $9\mu_B$, corresponding to the J_z=±6 projections. The magnetic moment of the excited states varies nonmonotonously, as can be observed from the size of magnetization lobes assigned along the spectra. The case from panel (c) corresponds to the J_z=0 singlet as groundstate, having no magnetic response. Returning to the systems from Fig. 7.12, the lesser occurrence of magnetic response along the spectral sequence can be explained by the more advanced mixture due to symmetry operations, transmuting the elements from diagonal, that in principle can contribute to the derivative with respect of the field, in nondiagonal positions, from where their contribution is extinguished.

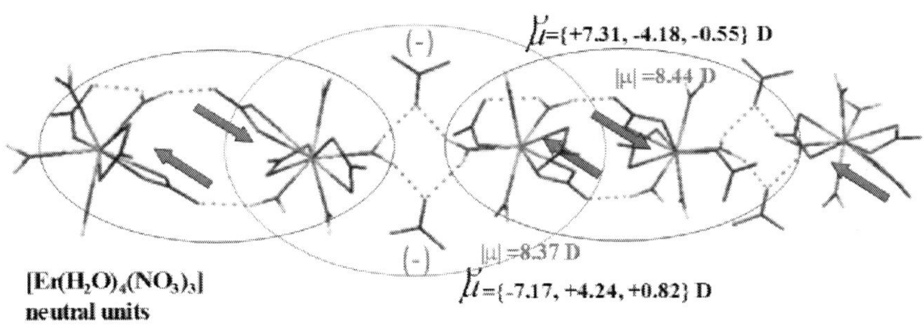

FIGURE 7.14 The chain formed by hydrogen bonding and electric dipole coupling of the $[Er(NO_3)_3(H_2O)_4]$ complex units [22b].

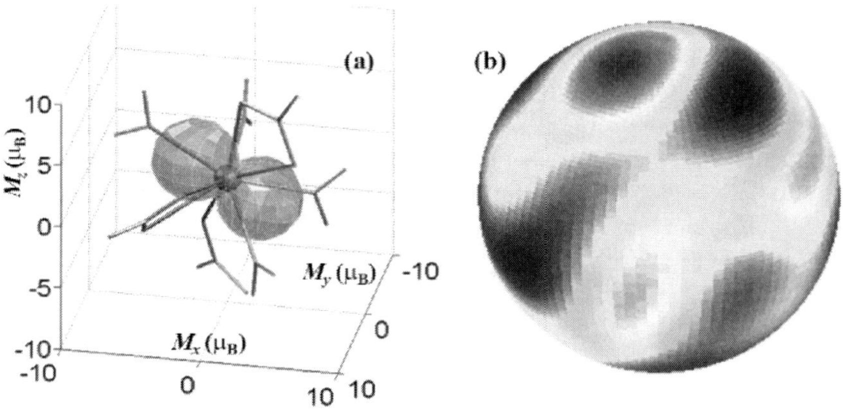

FIGURE 7.15 Synopsis of magneto-structural analysis of a system based on $[Er(NO_3)_3(H_2O)_4]$ units [22b]. (a) The magnetic anisotropy of the groundstate level obtained equating the ligand field parameters; (b) The map of ligand field potential obtained by detailed analysis of the Tb hypothetical analog. Note the approximate coincidence between the axis of magnetic anisotropy and higher ligand field potential poles.

In the following, we will offer a brief survey on the f units of the system with the $[Fe(bpca)_2]$ $[Er(NO_3)_3(H_2O)_4]$ $\times NO_3 \times H_2O$ formula [22b]. The complex d cations form a distinct sublattice with the nitrate counterion, while the f neutral units are assembled by forces specific to supramolecular chemistry in chains, by hydrogen bonds and alternating ordering of their dipoles. In fact, the nitrate anions are interposed also in sequence of the neutral f complex units. The image of the discussed chain is presented in the Fig. 7.14. The $[Er(NO_3)_3(H_2O)_4]$ units are rather strongly polar, with a moment of about 8.4 Debye, since the negative ions are concentrated in a half of the coordination sphere, the neutral aqua-ligands occupying the opposite part. This relatively simple system is suitable for a glance in the ligand field and magnetic anisotropy issues, and their causal relationship, in a low symmetry situation. For the reasons motivated previously, we considered the Tb(III) congener for the aim of ligand field analysis. Given the lack of the symmetry, the procedure is a bit more complicated, implying nonstandard operations. The CASSCF(7,8) calculations yield a series of orbital energies, {0.0, 50.5, 163.1, 257.6, 358.0, 423.7, 530.7} in cm^{-1}, which represent the ligand field split of the ^7F term, and a set of canonical orbitals which are almost pure f functions, except that these look mutually mixed, as compared to their standard shapes, known usually from the axial symmetry definition. Since the ligand field matrix is defined with respect of standard orbitals, for example, in the $\{y(3x^2\text{-}y^2), xyz, yz^2, z^3, xz^2,$

$z(x^2-y^2)$, $x(x^2-3y^2)\}$ sequence, we applied to the canonical orbitals a transformation bringing the set to the maximal resemblance to the above mentioned atomic orbitals. Since an unitary transformation inside the active space of a CASSCF procedure does not change the eigenvalues, the new eigenvectors obtained repeating the configuration-interaction calculation with modified orbitals are used to perform a back-transformation leading to the *ab initio* equivalent of the ligand field matrix. Fitting the computed matrix to its algebraic content resulted from the general Eq. (10), we obtain the full set of B_k^q parameters, with $k = 2, 4, 6$ and q running from $-k$ to k, as follows: B_2^q :{29.8, 112.2, 78.7, −141.6, 102.6} cm^{-1}, B_4^q: {19.2, −55.5, −399.4, −338.9, 102.5, −175.9, −59.3, 110.3, 153.7} cm^{-1} and B_6^q: {66.3, 186.4, 37.8, −61.3, −122.7, −207.7, 379.5, −208.0, 206.1, −199.1, −135.5, 43.9, 243.1, 541.2, 70.2, 7.5} cm^{-1}. One observes that, to be distinguished from the previously discussed idealized systems with high or intermediate symmetries, where the number of active parameters was reduced to few, here we face the full set of twenty-seven B_k^q 's, since the symmetry is low and ligands different each from other, in spite of their separation in only two different chemical classes (aqua and nitrate). Since the parameters themselves are not very intuitive, we offered in Fig. 7.15(b), as picturesque insight, the intensity map of the ligand field represented on a sphere, considering the same orientation of the molecular skeleton as shown in panel 15(a). This is, literally, the Eq. (10) worked with the above-mentioned list of parameters and represented as function of polar θ, φ angles, drawing in red-brown-yellow the heights of the potential and blue the negative areas. With the help of Stevens equivalent operators, we used the ligand field parameters to obtain the split of the lowest multiplet of Er(III), J=15/2. Taking the corresponding derivatives with respect of the magnetic field and running its orientation around the central ion, we obtained the map of the magnetic anisotropy of the groundstate, as illustrated in the Fig. 7.15(a). The revealing of the easy magnetization axis is a nontrivial thing, however not an unreachable goal, having illustrated here the technical route to such desiderata.

7.7 CONCLUSION

In this material we tried to show, using mostly simple examples, with just a touch of more complex cases picked up from our recent works, the large picture of causal relationships whose mastering is the prerequisite for understanding and engineering the magnetic anisotropy in molecules and materials. We hold the advent of pioneering contributions in the field, constructed by relevant case studies from new syntheses and original methodologies of analysis and prediction [19, 21, 22]. Our procedure for identifying the easy axes of magnetization for ground or excited states can be called a path to the discovery of the poles in molecular magnets. The rational advance, via structure-property correlations, is the key in which the fundamental research and academic insight contribute to questions coming from the field of applied sciences. We advocated also for placing on equal footing the theory and experiment, defending the market of ideas and debates from the point of hegemonic prevalence of one side or another. The applied chemistry of any sort would look terribly cryptic and limited without the help of basic theoretical constructs such as electrons, orbitals, states, while, in turn, the theoretical concepts were born, as we well known, from puzzles started in experiments. The numerous achievements of the nowadays magneto-chemistry, having the single-molecule, single-chain or single-ion magnets (SMMs, SCMs and SIMs) as spectacular objects and the magnetic anisotropy as key property, allow us already to dream about new revolutionary technologies, such as spintronics and quantum computing [50], replacing or complementing the electronic devices. However, we are yet on the learning curve, the academic perspective serving to design proper engineering recipes for the further new technologies. We hope that the present work is a good building block in this field, where interesting constructions are yet raising or even just sketched as blueprints.

ACKNOWLEDGMENT

This work was supported by the CNCS-UEFISCDI PCCE 9/2010 grant, Romania.

KEYWORDS

- **f-type nanosystems**
- **lanthanide compounds**
- **ligand field theory**
- **molecular magnetism**
- **transitional metal chemistry**
- **weak and strong magnetic anysotropy**

REFERENCES

1. Gatteschi, D., Sessoli, R., Villain, J. *Molecular Nanomagnets (Mesoscopic Physics and Nanotechnology)*, Oxford University Press: Oxford, 2006.
2. (a) Kahn, O. (1993). *Molecular Magnetism*, Wiley-VCH, Weinheim, (b) Coronado, E., Delhaès, P., Gatteschi, D., Miller, J. S. *Molecular Magnetism: From Molecular Assemblies to the Devices*, Springer: Heidelberg, 1996. (c) Miller, J. S., Drillon, M. (eds), *Magnetism: Molecules to Materials*, Wiley-VCH: Weinheim, 2001.
3. Goodenough, J. B. *Magnetism and the Chemical Bond,* Interscience: New York, 1963.
4. Koch, W., Holthausen, M. C. *A Chemist's Guide to Density Functional Theory*, Wiley-VCH: Berlin, 2001.
5. (a) Daul, C. A., Ciofini, I., Bencini, A. (2002). In: *Reviews of Modern Quantum Chemistry*, part II; Sen, K. D. Ed.; World Scientific: Singapore, p. 1247. (b) Noodleman, L., Peng, C. Y., Case, D. A., Mouesca, J. M. (1995). *Coord. Chem. Rev.* 144, 199–244.
6. (a) Jensen, F. *Introduction to Quantum Chemistry*, Wiley: New York, 2002. (b) Roos, B. O. The Complete Active Space Self-Consistent Field Method and its Applications in Electronic Structure Calculations. In *Advances in Chemical Physics: Ab Initio Methods in Quantum Chemistry Part 2*; Lawley, K. P., Ed.; John Wiley & Sons, Inc.: Hoboken, NJ, 2007; Vol. 69.
7. Willett, R. D., Gatteschi, D., Kahn, O. *Magneto-structural correlations in exchange coupled systems*, Reidel Publishing Co.: Hingham, MA, 1985.
8. (a) Ruiz, E., Alemany, P., Alvarez, S., Cano, J. (1997). *Inorg. Chem.* 36, 3683–3688. (b) Desplanches, C., Ruiz, E., Rodriguez–Fortea, A., Alvarez, S. *J. Am. Chem. Soc.* (2002). 124, 5198–5205. (c) Triki, S., Gomez–Garcia, C. J., Ruiz, E., Sala-Pala, J. (2005). *Inorg. Chem.* 44, 5501–5508.
9. (a) Heisenberg, W. (1928). *Z. Phys.*, 49, 619–636. (b) van Vleck, J. H., Sherman, A. *Rev. Mod. Phys.* (1935). 7, 167–228. (c) Anderson, P. W. (1959). *Phys. Rev.* 115, 2–13.
10. (a) Heitler, W., London, F. (1927). *Z. Phys.* 44, 455–472. (b) Shaik, S., Hiberty, P. C. *A Chemist's Guide to Valence Bond Theory*, John Wiley & Sons Inc.: New Jersey, 2008. (c) Cooper, D. L., Ed. *Valence Bond Theory*; Elsevier Science: Amsterdam, 2001.
11. (a) Raimondi, M., Cooper, D. L. (1999). *Top. Curr. Chem.* 203, 105–120. (b) Cooper, D. L., Thorsteinsson, T; Gerratt, J. *Adv. Quantum Chem.* (1999). 32, 51–67. (c) Thorsteinsson, T., Cooper, D. L. (1998). *Int. J. Quant. Chem.* 70, (4–5), 637–650.
12. Verdaguer, M., Bleuzen, A., Marvaud, V., Vaissermann, J., Seuleiman, M., Desplanches, C., Scuiller, A., Train, C., Garde, R., Gelly, G., Lomenech, C., Rosenman, I., Veillet, P., Cartier, C., Villain, F. (1999). *Coord. Chem. Rev.* 190–192, 1023–1047.
13. (a) Novoa, J. J., Deumal, M., Jornet–Somoza, J. (2011). *Chem. Soc. Rev.* 40, 3182–3212. (b) Li, L., Turnbull, M. M., Landee, C. P., Jornet, J., Deumal, M., Novoa, J. J., Wikaira, J. L. (2007). *Inorg. Chem.* 46, 11254–11265. (c) Her, J.–H.; Stephens, P. W., Ribas–Ariño, J., Novoa, J. J., Shum, W. W;. Miller, J. S. (2009). *Inorg. Chem.* 48, 3296–3307.
14. (a) Kahn, O., Galy, J., Journaux, Y., Morgenstern–Badarau, I. (1982). *J. Am. Chem. Soc.* 104, 2165–2176. (b) Kahn, O., Prins, R., Reedjik, J., Thompson, J. S. (1987). *Inorg. Chem.* 26, 3557–3561. (c) Benelli, C., Dei, A., Gatteschi, D., Pardi, L. (1988). *Inorg. Chem.* 27, 2831–2836.

15. (a) Néel, L. (1948). *Annales de Physique* (Paris) 3, 137–198. b) Anderson, P. W. (1956). *Phys. Rev.* 102, 1008–1013. (c) Spaldin, N. A. *Magnetic materials: fundamentals and applications*, Cambridge University Press.: Cambridge, 2010.
16. (a) Hay, P. J., Thibeault, J. C., Hoffmann, R. (1975). *J. Am. Chem. Soc.* 97, 4884–4899. (b) Girerd, J. J.,Journaux, Y., Kahn, O. *Chem. Phys. Lett.* (1981). 82, 534–537.
17. (a) Ruiz, E., Rodriguez–Fortea, A., Alvarez, S., Verdaguer, M. (2005). *Chem. Eur. J.* 11, 2135–2144. (b) Mouesca, J.–M. (2000). *J Chem Phys* 113, 10505–10511.
18. Paulovic, J., Cimpoesu, F., Ferbinteanu, M., Hirao, K. (2004). *J. Am. Chem. Soc.* 126, 3321–3331.
19. Ferbinteanu, M., Cimpoesu, F., Gîrțu, M. A., Enachescu, C., Tanase, S. (2012). *Inorg. Chem.* 51(1), 40–50.
20. (a) Sessoli, R., Powell, A. K. *Coord. Chem. Rev.* (2009). 253, 2328– 2341. (b) Cucinotta, G., Perfetti, M., Luzon, J., Etienne, M., Car, P.–E.; Caneschi, A., Calvez, G., Bernot, K., Sessoli, R. (2012). *Angew. Chem. Int. Ed.* 51, (1606). –1610. (c) Ruamps, R., Maurice, R., Batchelor, L., Boggio–Pasqua, M., Guillot, R., Barra, A.–L.; Liu, J., Bendeif, E.–E.; Pillet, S., Hill, S., Mallah, T., Guihéry, N. (2013). *J. Am. Chem. Soc.* 135, (8), 3017–3026.
21. (a) Ferbinteanu, M., Miyasaka, H., Wernsdorfer, W., Nakata, K., Sugiura, K., Yamashita, M., Coulon, C., Clérac, R. (2005). *J. Am. Chem. Soc.* 127, 3090. (b) Cimpoesu, F. ; Ferbinteanu, M., Frecus, B. Gîrțu, M. A., *Polyhedron*, (2009). 28, 2039–2043.(c) Oprea, C. I., Cimpoesu, F., Panait, P., Ferbinteanu, M., Girtu, M. A. (2011). *Theor. Chem. Acc.* 129(6), 847–857.
22. (a) Ferbinteanu, M., Kajiwara, T., Choi, K.–Y.; Nojiri, H., Nakamoto, A., Kojima, N., Cimpoesu, F., Fujimura, Y., Takaishi, S., Yamashita, M. (2006). *J. Am. Chem. Soc.* 128, 9008–9009. (b) Ferbinteanu, T. Kajiwara, F. Cimpoesu, K. Katagari, M. Yamashita, *Polyhedron* (2007). 26, 2069–2073. (c) Ferbinteanu, M., Cimpoesu, F., Kajiwara, T., Yamashita, M. *Solid State Sciences*, (2009). 11,760–765. (d) Tanase, S., Ferbinteanu, M., Cimpoesu, F., *Inorg. Chem.* (2011). 50(19), 9678–9687. (e) Ferbinteanu, M., Cimpoesu, F., Gîrtu, M. A., Enachescu, C., Tanase, S. *Inorg. Chem.* (2012). 51, 40–50. (f) Cimpoesu, F., Dahan, S., Ladeira, S., Ferbinteanu, M., Costes, J.–P. (2012). *Inorg. Chem.* 51, 11279–11293.
23. Aubin, S. M. J., Sun, Z., Eppley, H. J., Rumberger, E. M., Guzei, I. A., Folting, K., Gantzel, P. K., Rheingold, A. I., Christou, G., Hendrickson, D. N. (2001). *Inorg. Chem.* 40, 2127–2146.
24. (a) Christou, G., Gatteschi, D., Hendrickson, D. N., Sessoli, R. *MRS Bull.* (2000). 66–71. (b) Gateschi, D., Sessoli, R. (2003). *Angew. Chem. Int. Ed.* 42, 268–297.
25. (a) Caneschi, A., Gateschi, D., Lalioti, C. S., Sessoli, R., Venturi, G., Vindigni, A., Rettori, A., Pini, M. G., Novak, M. A. (2001). *Angew. Chem., Int. Ed.* 40, 1760–1763. (b) Lescouëzec, R., Vaissermann, J., Ruiz–Pérez, C., Lloret, F., Carrasco, R., Julve, M., Verdaguer, M., Dromzée, Y., Gatteschi, D., Wernsdorfer, W. (2003). *Angew. Chem. Int. Ed.* 42, 1483–1486.
26. (a) Ishikawa, N., Sugita, M., Ishikawa, T., Koshihara, S. Y., Kaizu, Y. (2003). *J. Am. Chem. Soc.* 125, 8694−8695. (b) Gomez–Coca, S., Cremades, E., Aliaga–Alcalde, N., Ruiz, E. (2013). *J. Am. Chem. Soc.* 135, (18), 7010–7018.
27. Maretti, L., Ferbinteanu, M., Cimpoesu, F., Islam, S. M., Ohba, Y., Kajiwara, T., Yamashita, M., Yamauchi, S. (2007). *Inorg. Chem.* 46, 660–669.

28. Boca, R. (2004). *Coord. Chem. Rev.* 248, 757–815.
29. (a) Figgis, B. N., Hitchman, M. A. *Ligand Field Theory and its Applications*, Wiley-VCH: New York, 2000. (b) Solomon, E. I., Lever, A. B. P. (eds.), *Inorganic Electronic Structure and Spectroscopy*, Wiley & Sons: New York, 1999.
30. Wang, S., He, W.-R.; Ferbinteanu, M., Li, Y.–H.; Huang, W. (2013). *Polyhedron* 52, 1199–1205.
31. (a) Bersuker, I. B. (2001). *Chem. Rev.* 101, 1067–1114. (b) Bersuker, I. B. (2013). *Chem. Rev. 113*, 1351–1390.
32. (a) Neese, F. (2012). The ORCA program system. In WIREs Comput Mol. Sci. 2, 73–78. (b) Neese, F. *J. Am. Chem. Soc.* (2006). 128, 10213–10222. (c) Schmitt, S., Jost, P., van Wüllen, C. (2011). *J. Chem. Phys.* 134, 194113–194124.
33. (a) Morgan, G. G., Murnaghan, K. D., Müller–Bunz, H., McKee, V., Harding, C. J. (2006). Angew. Chem. Int. Ed. 45, 7192–7195. (b) Pandurangan, K., Gildea, B., Murray, C., Harding, C. J., Müller–Bunz, H., Morgan, G. G. *Chem. Eur. J.* (2012). 18, 2021–2029. (c) Scheifele, Q., Piplinger, C., Neese, F., Weihe, H., Barra, A.–L.; Juranyi, F., Podlesnyak, A., Tregenna–Piggott, P. L. W. (2008). *Inorg. Chem.* 47, 439–447.
34. Wang, S., Ferbinteanu, M., Marinescu, C., Dobrinescu, A., Ling, Q. D., Huang, W. (2010). *Inorg. Chem.* 49, 9839–9851.
35. (a) Wang, S., Ferbinteanu, M. ; Yamashita, M. (2007). *Inorg. Chem.,* 46, 610–612; (b) Wang, S., Ferbinteanu, M. ; Yamashita, M. (2008). *Solid State Sciences*, 10, 915–920.
36. Cotton, S. *Lanthanide and actinide chemistry*, John Wiley & Sons: New York, 2006.
37. (a) Benelli, C., Gatteschi, D. (2002). *Chem. Rev.* 102, 2369–2388. (b) Sakamoto, M., Manseki, K., Okawa, H. *Coord. Chem. Rev.* (2001). 219–221, 379–414. (c) Sorace, L., Benelli, C., Gatteschi, D. (2011). *Chem. Soc. Rev.* 40, 3092–3104.
38. Ferbinteanu, M., Zaharia, A., Gîrţu, M. A., Cimpoesu, F. (2010). *Cent. Eur. J. Chem. 8*, 519–529.
39. (a) Liechtenstein, A. I., Anisimov,V. I., Zaane, (1995). *J. Phys. Rev. B* 52, R5467–R5470. (b) Dudarev, S. L., Botton, G. A., Savrasov, S. Y., Humphreys; C. J., Sutton, A. P. *Phys. Rev. B* (1998). 57, 1505–1509. (c) Czyzyk, M. T., Sawatzky, G. A. *Phys. Rev. B* (1994). 49, 14211–14288.
40. Caravan, P., Ellison, J. J., McMurry, T. J., Lauffer, R. B. (1999). *Chem. Rev.* 99, 2293–2352.
41. (a) Wybourne, B. G. *Phys. Rev.* (1966). 148, 317–327. (b) Newman, D. J., Urban, W. (1975). *Adv. Phys.* 24, 793–843.
42. (a) Ishikawa, N., Sugita, M., Okubo, T., Tanaka, N., lino, T., Kaizu, Y. (2003). *Inorg. Chem.* 42, 2440–2446. (b) Zbiri, M., Daul, C., Wesolowski, T. A. *J. Chem. Theor. Comput.* (2006). 2, 1106–1111. (c) Atanasov, M., Daul, C., Güdel, H. U., Wesolowski, T. A., Zbiri, M. (2005). *Inorg. Chem. 44*, 2954–2963.
43. Baldovi, J. J., Cardona–Serra, S., Clemente–Juan, J. M., Corronado, E., Gaita–Arino, A., Palii, (2012). A. *Inorg. Chem.* 51, 12565–12574.
44. (a) Newman, D. J., Ng, B. K. C. *Crystal Field Handbook*, Cambridge University Press: Cambridge, 2000. (b) Edvardsson, S., Åberg, D. *Comp. Phys. Commun.* (2001). 133, 396–406.
45. Stevens, K. W. H. (1952). *Proc. Phys. Soc.* A *65*, 209–215.

46. (a) Gritsenko, O. V., Schipper, P. R.T.; Baerends, E. J. *Phys. Rev. A* (1998). 57, 3450. (b) Dykstra, C. E., Frenking, G., Kim, K. S., Scuseria, G. E. (2005). *Theory and Applications of Computational Chemistry*, Elsevier, B. V.: Amsterdam, pp. 291.
47. Cotton, F. A., Wilkinson, F. G. (1999). *Advanced Inorganic Chemistry*, 6th edition, John Wiley: New York, p. 1108.
48. Schmidt, M. W., Baldridge, K. K., Boatz, J. A., Elbert, S. T., Gordon, M. S., Jensen, J. H., Koseki, S., Matsunaga, N., Nguyen, K. A., Su, S. J., Windus, T. L., Dupuis, M., Montgomery, J. A. (1993). *J. Comput. Chem.* 14, 1347–1363.
49. (a) Abragam, A., Bleaney, B. E*lectron Paramagnetic Resonance*, Clarendon Press: Oxford, 1970. (b) Lueken, H. *Magnetochemie*, B. G. Teubner: Stuttgart, 1999.
50. (a) Troiani, F., Affronte, M. (2011). *Chem. Soc. Rev.* 40, 3119–3129. (b) Lehmann, J., Gaita–Ariño, A., Coronado, E., Loss, D. *Nature Nanotechnol.* (2007). 2, 312–317. (c) Bogani, L., Wernsdorfer, W. (2008). *Nature Materials* 7, 179–186.

CHAPTER 8

FUNCTIONAL SUPRAMOLECULAR SYSTEMS CONTROLLED BY LIGHT

TOMMASO AVELLINI MASSIMO BARONCINI ENRICO MARCHI MONICA SEMERARO, and MARGHERITA VENTURI

CONTENTS

ABSTRACT

Molecules can be used as building blocks for the assembly of multicomponent (supramolecular) structures exhibiting novel and complex functions that derive from the cooperation of simpler functions performed by each component. These systems, if suitably designed, behave as devices and machines of nanometric size. In analogy to their macroscopic counterparts, molecular devices and machines need energy to operate and signal to communicate with the operator. Light provides an answer to this dual requirement: photons are indeed both quanta of energy and elements of information (by means of the spectroscopic techniques). In this chapter we describe a few recent examples developed in our laboratory with the aim to show that, in the frame of research on supramolecular photochemistry, the design and construction of nanoscale devices and machines capable of performing useful light-induced functions can indeed be attempted. They are (a) molecular machines capable of undergoing light-controlled unidirectional linear movements, (b) dendrimers in which predetermined energy transfer processes occur, (c) quantum dots that behave as luminescent chemosensors, and (d) assemblies of molecules that upon light inputs and/or outputs can perform logic functions. These systems represent some small, although interesting, advancements in exploiting the peculiar properties of light and its interaction with matter to obtain useful functions.

8.1 INTRODUCTION

The interaction between light and matter lies at the heart of the most important processes of life [1]. Light is made of photons, which are exploited by natural systems as both quanta of energy and elements of information. All the natural phenomena related to the interaction between light and matter and the great number of applications of photochemistry in science and technology can ultimately be traced back to these two aspects of light. Living examples of this double-faced nature of light are provided by the two most important photochemical processes taking place in the biological world: photosynthesis and vision.

A variety of functions can also be obtained from the interaction between light and matter in artificial systems [2]. The type and utility of such functions depend on the degree of complexity and organization of the chemical systems that receive and process the photons. Indeed, understanding the interaction between light and molecules, together with the progress in chemical synthesis, has led to the point where one can conceive artificial multicomponent systems capable of using light as an energy supply (to sustain energy-expensive functions and/or to induce extensive conformational changes) or as an input signal (to be processed and/or stored). The construction and the working mechanisms of these systems are based on the concepts of supramolecular chemistry [3] and photochemistry [4].

Supramolecular chemistry is a highly interdisciplinary field that has developed at an astonishingly fast rate during the last three decades. In a historical perspective, as pointed out by Jean-Marie Lehn, supramolecular chemistry originated from Paul Ehrlich's receptor idea, Alfred Werner's coordination chemistry, and Emil Fischer's lock-and-key image. It was only after 1970, however, that fundamental concepts such as molecular recognition, preorganization, self-assembly and self-organization were introduced in chemistry. Supramolecular chemistry then began to emerge as a well-defined discipline and was consecrated by the award of the Nobel Prize in Chemistry to Charles Pedersen, Donald Cram, and Jean-Marie Lehn in 1987 [3].

Supramolecular chemistry, according to its most popular definition, is "the chemistry beyond the molecule, bearing on organized entities of higher complexity that result from the association of two or more chemical species held together by intermolecular forces." As the field developed, it became evident that a definition strictly based on the nature of the bond that links the components would be limiting. Many scientists, therefore, started to distinguish between what is molecular and what is supramolecular based on the degree of intercomponent interactions. In a general sense, one can say that with supramolecular chemistry there has been a shift in focus from molecules to molecular assemblies or multicomponent structures driven by the emerging of new functions.

On the basis of this consideration, about 20 years ago, in the frame of research on supramolecular chemistry, the idea began to arise [3, 5, 6] that the concept of macroscopic device and machine can be transferred to the molecular level.

In other words, molecules might be used as building blocks for the assembly of multicomponent structures exhibiting novel and complex functions that derive from the cooperation of simpler functions performed by each component. This strategy, encouraged by a better understanding of biomolecular devices and machines, has been implemented on a wide variety of chemical systems, leading to highly interesting results. As a matter of fact, the molecular bottom-up construction of nanoscale devices and machines has become one of the most stimulating challenges of nanoscience.

In analogy to their macroscopic counterparts, molecular devices and machines need energy to operate and signal to communicate with the operator. Light provides an answer to this dual requirement. Indeed, a great number of molecular devices and machines are powered by light-induced processes, and light can also be useful to "read" the state of the system and thus to control and monitor its operation. In this regards, light energy possesses a number of further advantages: (i) the amount of energy conferred to a chemical system by using photons can be carefully controlled by the wavelength and intensity of the exciting light, in relation to the absorption spectrum of the targeted species, and (ii) such an energy can be transmitted to molecules without physically connecting them to the source (no 'wiring' is necessary), the only requirement being the transparency of the matrix at the excitation wavelength.

In this chapter a few recent examples of light-controlled molecular devices and machines developed in our laboratory are described. They present some small, although interesting, advancements in exploiting the peculiar properties of light and its interaction with matter to obtain useful functions.

8.2 UNIDIRECTIONAL LINEAR MOVEMENTS IN SUPRAMOLECULAR SYSTEMS

The control of motion on the molecular scale is of fundamental importance for living organisms [7], and one of the most fascinating challenges in nanoscience [8, 9]. Artificial molecular machines have been realized on the laboratory scale [10–12]. and utilization of such systems to construct responsive materials [13] and surfaces [14], control catalytic processes [15], and develop test structures for information storage devices [16] and drug delivery [17] has been investigated. Nevertheless, the construction of synthetic nanoscale motors capable of showing directionally controlled linear or rotary movements still pose a considerable challenge to chemists [18]. Although a few examples of artificial molecular rotary motors [15, 19] and DNA-based linear motors [20] have been described, only one prototype of a fully synthetic linear motor molecule is available [21]. Moreover, the use of such systems to perform the tasks that natural molecular motors do [22] – particularly, active transport of substrates over long distances or across membranes – remains a very difficult endeavor, further complicated by the fact that most currently available synthetic molecular motors are based on sophisticated chemical structures and/or operation procedures [10–12, 18].

In this context, the development of (supra)molecular systems that exhibit directionally controlled relative motions of their components, based on a minimalist design, and activated by convenient inputs is of the highest importance. Particularly suitable species for the construction of linear molecular machines are pseudorotaxanes, supramolecular complexes minimally composed of an axle-like molecule surrounded by a macrocycle, and rotaxanes, structurally similar to pseudorotaxanes except for the presence of bulky groups at the extremities of the axle to prevent dethreading and render these systems kinetically inert [23, 24]. The operation of most pseudorotaxane and rotaxane machines is based on classical switching processes between thermodynamically stable states [10, 11, 25]. It has become clear, however, that functional molecular motors will only be realized if the reaction rates between states can also be controlled, thus enabling the implementation of ratchet-type mechanisms [26, 27]. Therefore, the ability of adjusting the threading-dethreading kinetics by modulating the corresponding energy barriers through external stimulation is an important goal.

8.2.1 LIGHT-CONTROLLED UNIDIRECTIONAL TRANSIT OF A MOLECULAR AXLE THROUGH A MACROCYCLE

Herein we describe the operation of a simple supramolecular assembly in which a molecular axle passes unidirectionally through the cavity of a molecular wheel in response to photochemical and chemical stimulation [28]. A system of this kind constitutes a first step towards the construction of an artificial molecular pump; it can also lead to the realization of molecular linear motors based on rotaxanes and rotary motors based on catenanes.

The working strategy of the system composed of a molecular wheel and a nonsymmetric molecular axle is shown in Fig. 8.1. The axle molecule is made of three different functional units: (i) a passive pseudostopper (D), (ii) a central recognition site (S) for the wheel, (iii) a bistable *trans-cis* photoswitchable unit (P) at the other end. In acetonitrile solution, for kinetic reasons, the wheel threads through the axle exclusively from the side of the photoactive gate in its starting *trans* configuration (*E*) (Fig. 8.1a), affording a pseudorotaxane in which the molecular wheel encircles the recognition site S. Light irradiation converts the *E*-P end group into the bulkier *cis* form *(Z)*, a process which also destabilize the supramolecular complex (Fig. 8.1b). Therefore, a dethreading of the wheel is expected, which occurs by slippage of the wheel through the D moiety of the axle (Fig. 8.1c). The system is brought back to its initial state by photochemical or thermal conversion of the *Z*-P gate back to the *E* configuration (Fig. 8.1d). Overall, the photoinduced directionally controlled transit of the axle through the wheel would be obtained according to a flashing energy ratchet mechanism [26, 27].

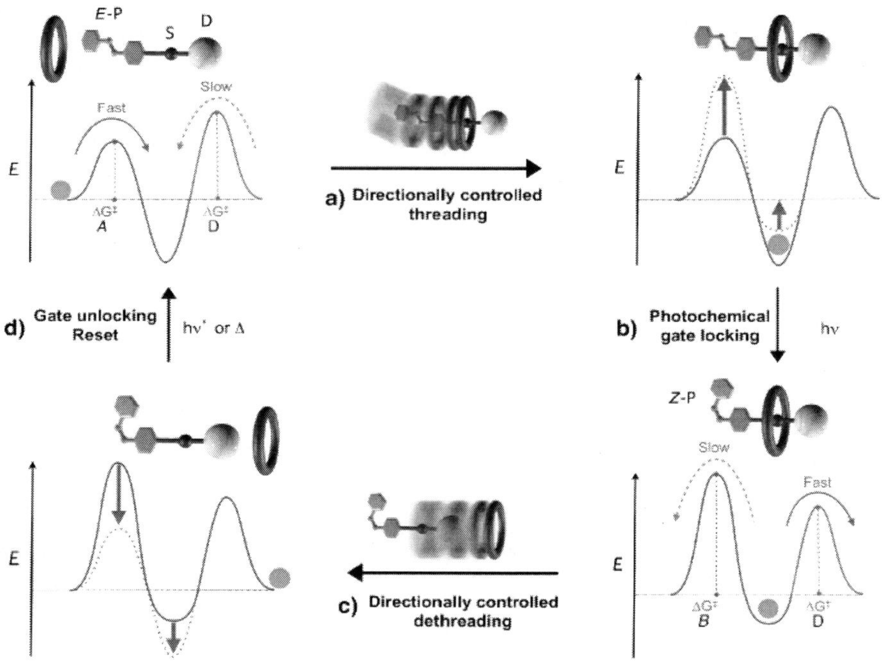

FIGURE 8.1 Strategy for the photoinduced unidirectional transit of a nonsymmetric axle through a molecular wheel. Simplified potential energy curves (free energy versus wheel-axle distance) for the states shown, describing the operation of the system in terms of a flashing ratchet mechanism.

Two basic requirements are needed for this strategy to work: (i) the kinetic barriers for the slippage of the wheel through the axle end groups [29] should follow the $\Delta G^{\#}(E\text{-P}) < \Delta G^{\#}(\text{D}) < \Delta G^{\#}(Z\text{-P})$ order, and (ii) the wheel should form a more stable pseudorotaxane when the axle has the phtoswitchable end group in its *E* configuration compared to the *Z* one. It is also important that the differences in the kinetic and stability constants are sufficiently large, and that the photochemical interconversion of the P gate between its *E* and *Z* forms is fast, efficient, and reversible.

In a previous investigation [30] it was studied that the formation of the supramolecular complexes (Fig. 8.2) between the dibenzo[24]crown-8 wheel (**DB24**) and axle *EE*-**1**H$^+$, composed of a dialkylammonium recognition site with two azobenzene end units. The obtained results showed that the threading-dethreading rate constants are slowed down by four orders of magnitude when the *E*-azobenzene end units are photoisomerized to the *Z* form, practically transforming the complex from a pseudorotaxane into a rotaxane; moreover, the stability constant drops by a factor of two [31].

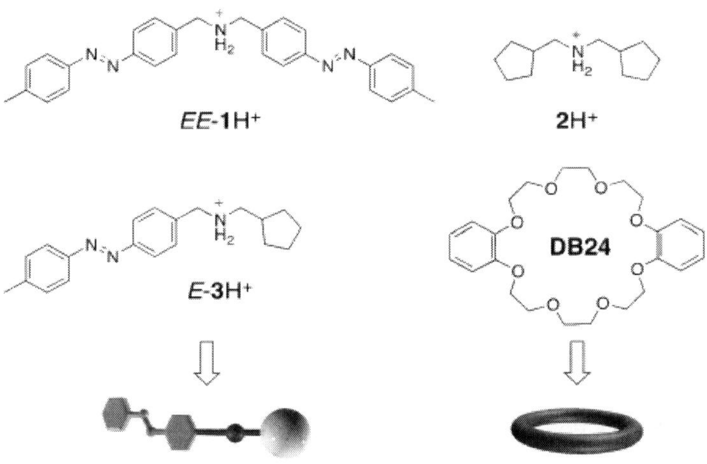

FIGURE 8.2 Structure formulas and cartoon representation of the examined axle and wheel components.

In Ref. [32], it was also reported that the bis(cyclopentylmethyl)ammonium ion **2**H$^+$ is complexed by **DB24** to form a pseudorotaxane, with threading and dethreading rate constants that fall in between those observed for *EE*-**1**H$^+$ and *ZZ*-**1**H$^+$ with the same wheel [30]. Hence, the strategy shown in Fig. 8.1 can be implemented with a nonsymmetric axle such as *E*-**3**H$^+$ (Fig. 8.2).

^{1}H NMR Spectroscopic titration experiments showed that in acetonitrile the wheel **DB24** threads through *E*-**3**H$^+$ exclusively from the *E*-azobenzene terminus. Irradiation of *E*-**3**H$^+$ with UV light affords *Z*-**3**H$^+$ in an almost quantitative way. The increased bulkiness of the azobenzene end group upon photoisomerization forces *Z*-**3**H$^+$ to thread **DB24** through its methylcyclopentyl terminus. It is noteworthy that the $E \rightarrow Z$ photoisomerization of the azobenzene end group of **3**H$^+$ takes place efficiently also when it is surrounded by **DB24**. Therefore, we can kinetically control the threading-dethreading side of **3**H$^+$ by photoadjusting the steric hindrance of its azobenzene end group.

In contrast with the results found for the [**DB24**$\supset$*EE*-**1**H]$^+$ and [**DB24**$\supset$*ZZ*-**1**H]$^+$ pseudorotaxanes, the stability constants of [**DB24**$\supset$*E*-**3**H]$^+$ and [**DB24**$\supset$*Z*-

3H]$^+$ are identical within errors. Therefore, the dethreading of *Z*-**3**H$^+$ from the wheel cannot be caused by the same photochemical stimulus that triggers the azobenzene *E*→*Z* isomerization. Because deprotonation of the ammonium recognition site of *Z*-**3**H$^+$ with a base causes the fast dethreading from **DB24**, thereby neutralizing the stoppering ability of the *Z*-azobenzene unit, K$^+$ ions were used, as competitive guests for **DB24** [33], in order to promote the disassembly of the complexes. The addition of two equivalents of KPF_6 causes the complete dethreading of both [**DB24**⊃*E*-**3**H]$^+$ and [**DB24**⊃*Z*-**3**H]$^+$; however, while the K$^+$-induced disassembly of the former complex is fast, the latter one exhibits a dethreading half-life of 51 min. This finding indicates that the chemically induced disassembly of *Z*-**3**H$^+$ and **DB24** takes place exclusively by slippage of the wheel through the methylcyclopentyl unit of the axle.

The results of an experiment that illustrates the directional transit of the axle through the wheel are summarized in Fig. 8.3 (i) *E*-**3**H$^+$ pierces **DB24** with its *E*-azobenzene side to form the pseudorotaxane [**DB24**⊃*E*-**3**H]$^+$, which equilibrates fast with its free components; (ii) irradiation in the near UV converts quantitatively [**DB24**⊃*E*-**3**H]$^+$ into [**DB24**⊃*Z*-**3**H]$^+$, characterized by much slower assembly disassembly kinetics; (iii) the successive addition of K$^+$ ions promotes the dethreading of *Z*-**3**H$^+$ from **DB24** by the passage of the methylcyclopentyl moiety through the cavity of the wheel. It should be noted that equilibration of the [**DB24**⊃*Z*-**3**H]$^+$ complex with its separated components, that would cause the loss of the information on the threading direction of *E*-**3**H$^+$, is much slower than the time required for the activation of the dethreading stimulus (addition of K$^+$). Therefore, after the threading event the system is 'locked' by photoisomerization, and the successive addition of potassium ions causes dethreading in the same direction along which threading of *E*-**3**H$^+$ has initially occurred. The starting species *E*-**3**H$^+$ can be fully regenerated by thermal *Z*→*E* isomerization and sequestration of K$^+$ by an excess of 18-crown-6 affords the reassembly of [**DB24**⊃*E*-**3**H]$^+$ and the full reset of the system.

This supramolecular system, however, if it would be incorporated in a compartmentalized structure (e.g., embedded in the membrane of a vesicle), could not be used to 'pump' the molecular axle and generate a transmembrane chemical potential because the wheel component has two identical faces. Despite this deficiency the described system is characterized by a minimalist design, facile synthesis, convenient switching and reversibility: all these features constitute essential requirements for real world applications.

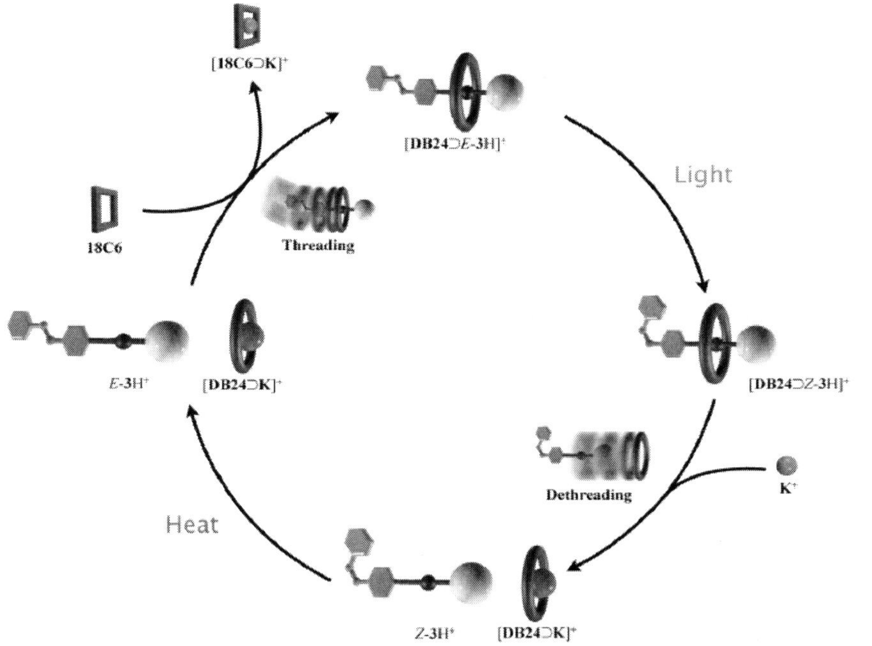

FIGURE 8.3 Representation of the photochemically and chemically controlled transit of **DB24** through **3**H^+.

In the next section a strategy similar to the one just described is applied to supramolecular assemblies based on three-dimensional nonsymmetric macrocycles whose length can approach the thickness of a bilayer membrane [34], and in which face-selective threading can be realized.

8.2.2 SOLVENT- AND LIGHT-CONTROLLED UNIDIRECTIONAL TRANSIT OF A MOLECULAR AXLE THROUGH A NONSYMMETRIC MACROCYCLE

An increase in the structural complexity of the self-assembled pseudorotaxane/rotaxane systems can be obtained by using as a wheel component the macrocycle tris(phenylureido)calix[6]arene [35] derivative **4** (Fig. 8.4). Because of the nonsymmetric nature of this wheel it is possible to selectively thread suitable axles from the "upper" or "lower" rim of the macrocycle leading to "up" and "down" oriented isomers (Fig. 8.5). It is known [36] that in apolar media wheel **4** is able to be threaded exclusively from the upper rim by axles derived from 4,4′-bipyridinium salts [37]. This behavior can be explained by the peculiar chemical and structural features of compound **4** as a host, which are (i) a π-donor cavity that, because of its width, can include the positively charged bipyridinium unit of the axle, but not together with its counter anions, (ii) three efficient hydrogen-bonding donor ureidic groups at the upper rim that, by complexing the counter anions of the axle, can assist the insertion of the cationic portion of the latter, and (iii) three methoxy groups at the lower rim that, in apolar media, are oriented towards the interior of the cavity [38], thereby hindering the access of the guest from this direction. The use of more polar solvents has a profound effect on these interactions. In fact, the solvent polarity, by changing the extent of ion pairing of the axle and decreasing the pivoting role of the three ureidic groups of the host affects both the concentration of the active guest available in solution and the binding ability of the wheel.

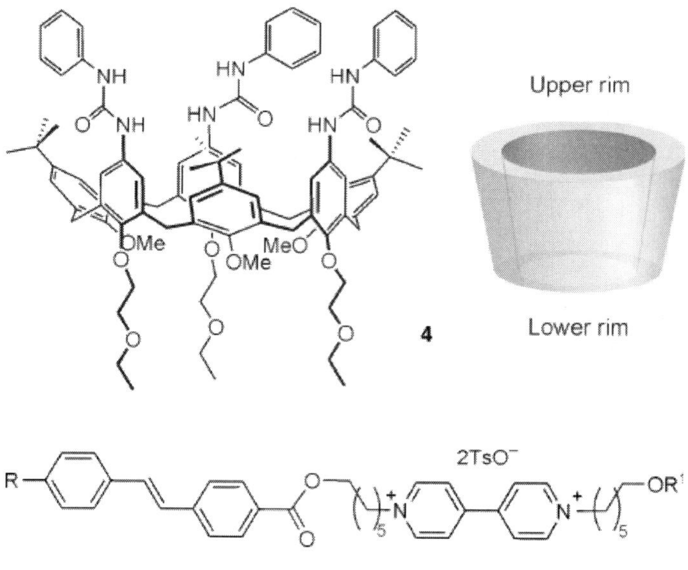

5 R = R¹ = H **7** R = H; R¹ = Ph_2CHCO
6 R = ᵗBu; R¹ = H **8** R = ᵗBu; R¹ = Ph_2CHCO

FIGURE 8.4 Structure formulas and cartoon representation of the examined axles and wheel components.

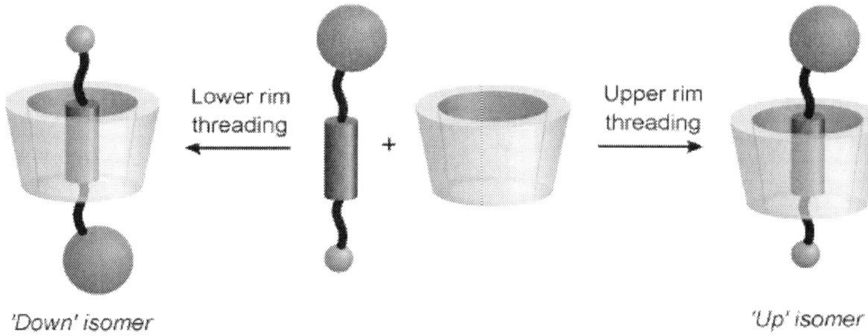

FIGURE 8.5 Formation of "up" and "down" oriented pseudorotaxane isomers by self-assembly of the nonsymmetric wheel and axle components.

To obtain pseudorotaxane systems based on wheel **4** capable of undergoing unidirectional threading/dethreading motion, the axles **5** and **6** (Fig. 8.4) were employed [39]. They are composed by a central electron-acceptor 4,4′-bipyridinium unit functionalized with a hexanol chain at one side, and a stilbene (**5**) or a *t*Bu-substituted stilbene (**6**) at the other side. The terminal OH group has been selected because it can be involved in stoppering reactions, while the stilbene and the *t*Bu-substituted stilbene head groups have been employed because they are not too bulky to prevent their slippage [40, 41] through wheel **4**, but large enough to enable a kinetic control of the threading/dethreading motions of the wheel. NMR spectra of the equilibrium mixture obtained after mixing wheel **4** with an excess of axle **5** or **6** in C_6D_6 at room temperature prove that only one pseudorotaxane type complex does form predominantly, confirming that both axles enter the wheel through the calixarene upper rim with their OH terminus to yield the oriented species $P[\mathbf{4}\supset\mathbf{5}]_{up}$ and $P[\mathbf{4}\supset\mathbf{6}]_{up}$ (Fig. 8.6). This orientational control during the formation of the pseudorotaxane indicates that the wheel promotes the threading of the axles from the upper rim and that the latter components access the macrocycle through the less bulkier OH terminus in a process that is kinetically controlled by the different size of the end groups of the axles. When the OH end group of pseudorotaxanes $P[\mathbf{4}\supset\mathbf{5}]_{up}$ and $P[\mathbf{4}\supset\mathbf{6}]_{up}$ is replaced by bulky diphenylacetyl moieties, the pseudorotaxanes are converted to rotaxanes $R[\mathbf{4}\supset\mathbf{7}]_{up}$ and $R[\mathbf{4}\supset\mathbf{8}]_{up}$ (Fig. 8.6). These species exhibit a rotaxane-like behavior because one end of their axle is stoppered by the presence of the bulky diphenylacetyl moiety while dethreading from the side carrying the stilbene-type unit is greatly slowed down by the steric hindrance of the latter.

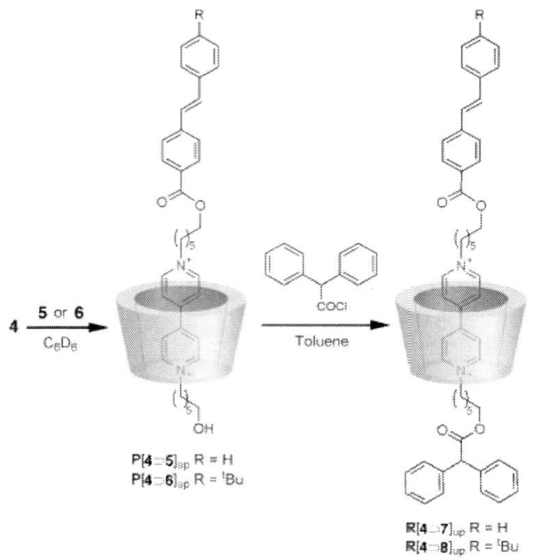

FIGURE 8.6 Self-assembly of P[**4**⊃**5**]$_{up}$ and P[**4**⊃**6**]$_{up}$, and synthesis of R[**4**⊃**7**]$_{up}$ and R[**4**⊃**8**]$_{up}$.

In polar solvents the multiple interactions that stabilize compounds R[**4**⊃**7**]$_{up}$ and R[**4**⊃**8**]$_{up}$ are weakened and, therefore, dissolution of compounds R[**4**⊃**7**]$_{up}$ and R[**4**⊃**8**]$_{up}$ in such kind of solvents induces axle dethreading. Because of the presence of the diphenylacetyl stopper at one end of the axles, the dethreading occurs through the slippage of the sufficiently slim stilbene-type unit from the lower rim of the wheel. The dethreading rate in DMSO for R[**4**⊃**8**]$_{up}$ is two order of magnitude lower than that of R[**4**⊃**7**]$_{up}$ due to the higher hampering effect of the *t*Bu-substituted stilbene present in axle **8** compared with the unsubstituted axle **7**.

In these rotaxane-like systems the dethreading rate can also be photocontrolled upon UV light-induced *E–Z* isomerization of the stilbene end group of the axles. For both R[**4**⊃**7**]$_{up}$ and R[**4**⊃**8**]$_{up}$ compounds at the photostationary state about 70% of the stilbene units are converted from the *E* to the *Z* isomer. In the case of R[**4**⊃**7**]$_{up}$ the higher hampering effect of the *Z* isomer compared with the *E* one results in a much more difficult slippage of this unit through the lower rim of the wheel once it is dissolved in a polar solvent. It is interesting to notice that the *E–Z* photoisomerization affects the dethreading rate constant more than the incorporation of the *tert*-butyl group on the stilbene unit. The rate constant observed upon isomerization is indeed about one order of magnitude slower than that observed in the case of the *E*-isomer of the substituted stilbene.

In the case of R[**4**⊃**8**]$_{up}$ the photoisomerized compound does not undergo dethreading in polar solvents at all. The *Z* isomer of the *t*Bu-stilbene is too bulky to pass through the lower rim of the wheel, and R[**4**⊃**8**]$_{up}$ in its *Z* configuration behaves as a real rotaxane also in highly polar solvents.

According to the sequence of transformations described above and as schematized in Fig. 8.7, the unidirectional transit of the nonsymmetric molecular axles **5** and **6** through the nonsymmetric molecular wheel **4** is achieved. The strategy is based on the use of appropriately designed molecular components, an essential feature of which is their nonsymmetric structure, and exploits the following steps: (i) in apolar solvents axle **5** or **6** threads into wheel **4** from its upper rim leading to an oriented pseudorotaxane structure in which the OH group is positioned at the lower rim of the wheel; this threading mode is favored because of the small hampering effect of the OH group substantially lower than that of the stilbene moiety (Fig. 8.7a); (ii) by a stoppering reaction that introduces a bulky diphenylacetyl moiety, the pseudorotaxane is converted in a rotaxane-like species (Fig. 8.7b); and (iii) replacement of the apolar solvent with a polar one weakens the interactions that stabilize the assembled structure and induces the axle dethreading from the lower rim of the wheel (Fig. 8.7c), that is, in the same direction of the axle threading.

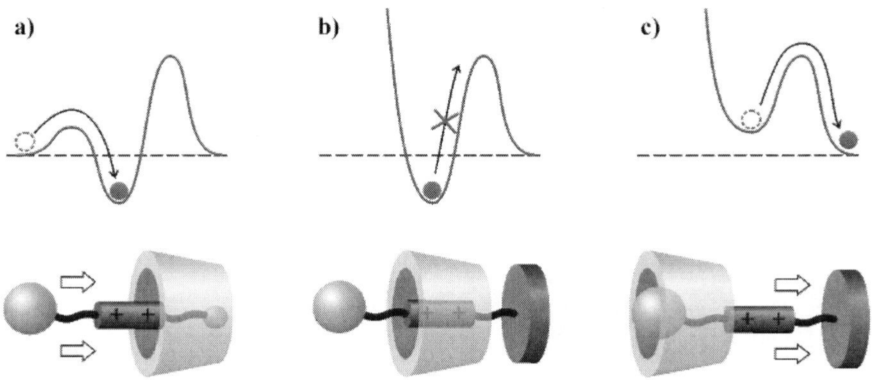

FIGURE 8.7 Simplified potential energy curves representing the unidirectional transit of the axle through the wheel. The horizontal coordinate of the diagrams represents the axle–wheel distance when they approach one another along the direction and with the orientation shown in the cartoons. (a) Threading of the axle through the upper rim of the wheel in apolar solvents. (b) The stoppering reaction that converts the pseudorotaxanes into rotaxane-like species. (c) Dethreading of the axle from the lower rim of the wheel in polar solvents.

It is also important to stress the essential role played by the stilbene unit incorporated at one end of the axle. It indeed enables: (i) to achieve the unidirectional transit of the axle through the wheel, because its dimensions are not too big to prevent slippage through wheel, but big enough to induce a kinetic control of the axle threading/dethreading processes; and (ii) to tune the dethreading rate because of the possibility to modify its hampering effect upon the use of stilbene unit substituted with relatively bulky groups, or, more interesting, upon photoisomerization.

8.3 CONTROLLED ENERGY TRANSFER IN DENDRIMERS

Dendrimers constitute a class of multibranched molecules that exhibit a defined structure and a high degree of order, but also a high degree of complexity [42, 43]. From a topological viewpoint, dendrimers contain three different regions: core, branches, and surface (Fig. 8.8).

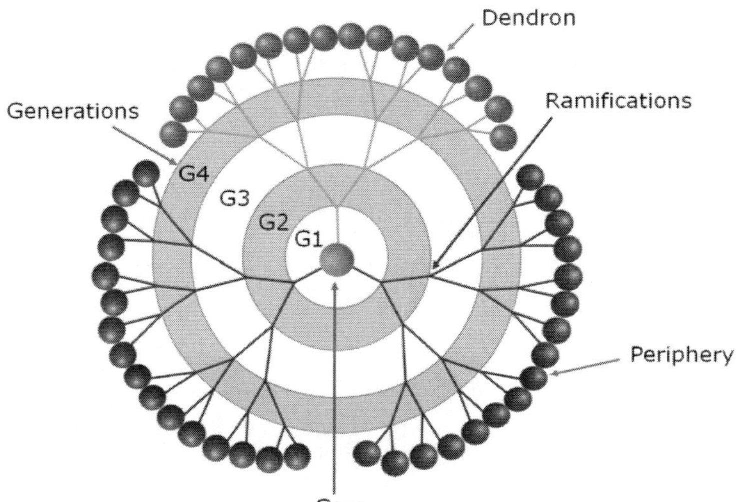

FIGURE 8.8 Schematic representation of a dendrimer.

A most important feature of dendrimer chemistry is the possibility to insert selected chemical units in predetermined sites of the dendritic architecture. Moreover, thanks to their three-dimensional structure, internal dynamic cavities are present, where ions or molecules can be hosted. It is thus possible to construct large nanoobjects capable of performing complex functionality that derives from the integration of the specific properties of the constituent moieties.

Dendrimers containing photochromic [44, 45] and/or luminescent [46] units and/or performing as ligands of metal ions [47, 48] have been extensively investigated in the past decade. Because of their proximity, the various functional groups of a dendrimer can lead to interesting photophysical properties [49], such as, (i) interactions of the units in the ground and/or in the excited states (possible formation of excimers and exciplexes in luminescent dendrimers), (ii) quenching of dendrimer luminescence by external or internal species via energy or electron transfer processes, and (iii) sensitization of luminescent metal ions, or dyes encapsulated by the dendrimer.

In a dendritic structure containing luminescent units at the periphery the formation of excimers and exciplexes is favored because of the proximity of the interacting molecules (Fig. 8.9). In such a case as many as three different types of luminescence can be observed, namely "monomer", exciplex, and excimer emissions. Compared with the "monomer" emission, the emission of an excimer or exciplex is always displaced to lower energy and usually corresponds to a broad and rather weak band.

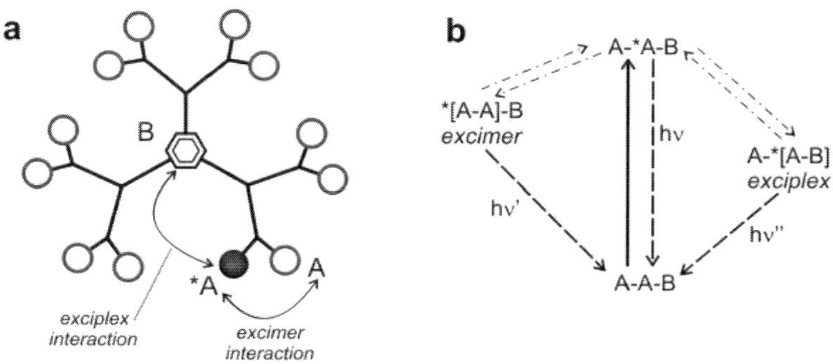

FIGURE 8.9 Schematic representation of (a) excimer and exciplex formation in a dendrimer and (b) an energy level diagram showing the three different emissions that can be observed.

As previously mentioned dendrimers are ideal scaffolds to organize many active units in a restricted space according to a predetermined pattern. Indeed, dendrimers have been extensively investigated in the last few years both from a fundamental viewpoint [50] and for a variety of applications, including (i) sensing with signal amplification [51], (ii) quenching and sensitization processes [52], and (iii) light harvesting [53] (Fig. 8.10).

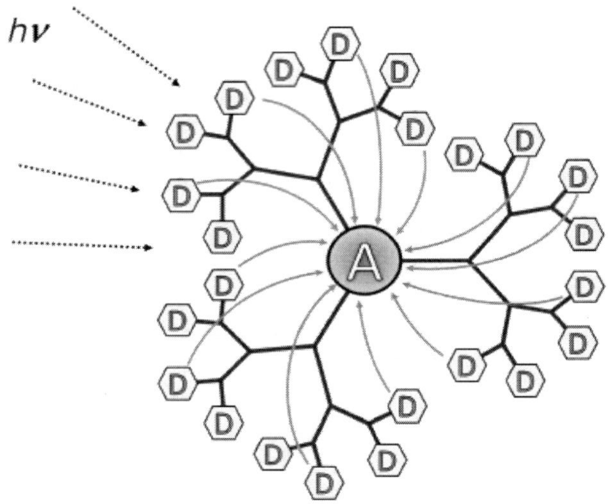

FIGURE 8.10 Schematic representation of a light harvesting process in a dendritic structure from peripheral donor chromophores **D** toward the acceptor core **A**.

An antenna for light harvesting is a multicomponent system organized in terms of energy, space and time in which many chromophoric molecular units absorb the incident light and then channel the excitation energy to a common acceptor component [54, 55]. This requirement implies the occurrence of a sequence of energy transfer steps along predetermined directions. In order to have a high light-harvesting efficiency, each energy transfer step must successfully compete with the intrinsic decay of the excited state as well as with other excited state deactivation processes (e.g., electron transfer, exciplex and excimer formation). Clearly, the quantum yield of the sensitized emission of the acceptor component cannot be larger than the quantum yield of the emission obtained upon direct excitation of the chromophore.

Here we report on two specific examples of light harvesting supramolecular systems containing dendrimers in which it is possible to tune the efficiency of the energy transfer process using metal ions.

8.3.1 ENERGY TRANSFER IN A SELF ASSEMBLED TRIAD

One of the most extensively investigated ligands in coordination chemistry is 1,4,8,11-tetraazacyclotetradecane (cyclam). Both cyclam and its 1,4,8,11-tetramethyl derivative in aqueous solution can be mono- and di-protonated and can coordinate metal ions such as CO_2^+, Ni^{2+}, Cu^{2+}, Zn^{2+}, Cd^{2+}, Hg^{2+} and lanthanide ions with very large stability constants [56]. Furthermore, cyclam and its derivatives have been studied for several medical applications [57] like carrier of metal ions in antitumor, imaging contrast agents, and anti-HIV agents.

Dendrimer **9** (Fig. 8.11, left), consisting of a cyclam core appended with 12 dimethoxybenzene and 16 naphthyl units, was extensively studied [58]. In CH_3CN/CH_2Cl_2 (1:1 *v/v*) solution the absorption spectrum is dominated by the naphthalene bands and the dendrimer exhibits three types of emissions, assigned to naphthyl localized excited state (λ_{max} = 337 nm), naphthyl excimer (λ_{max} ~ 390 nm), and naphthyl-amine exciplex (λ_{max} = 480 nm) [58a]. Extensive investigations have also been performed to study the interaction of dendrimer **9** with metal ions [58b, c, d, e]. Coordination of Zn^{2+}, a metal ion with d^{10} electronic configuration that is difficult to oxidize and reduce, leads to complexes that do not exhibit electronic excited states at low energy. Nevertheless, Zn^{2+} coordination by dendrimer **9** causes strong changes in the dendrimer emission spectrum: engagement of the nitrogen lone pairs in metal ion coordination prevents indeed exciplex formation, with a resulting increase of the naphthyl fluorescence. Such a fluorescent signal is quite suitable for monitoring the formation of the complex in dendrimer/metal titration experiments. Surprisingly, at low Zn^{2+} concentration, dendrimer **9** gives rise to a complex with 2:1 dendrimer/metal stoichiometry and a high formation constant (logβ > 13) as evidenced by both fluorescence and ^{1}H-NMR titrations. The unexpected $[Zn(\mathbf{9})_2]^{2+}$ species shows that the dendrimer branches do not hinder, but in fact favor coordination of cyclam to Zn^{2+} with respect to coordination of solvent molecules or counter ions. Furthermore, the two cyclam cores, to account for the coordination number ($\leq$ 6) of Zn^{2+}, are likely forced to adopt a structure in which not all the four nitrogen atoms are available for Zn^{2+} coordination, thereby favoring a 2:1 stoichiometry. A possible configuration of the $[Zn(\mathbf{9})_2]^{2+}$ species in which the branches of the two coordinated dendrimers do not interact but impose to the cyclam core a very specific coordination structure is reported in the right part of Fig. 8.11.

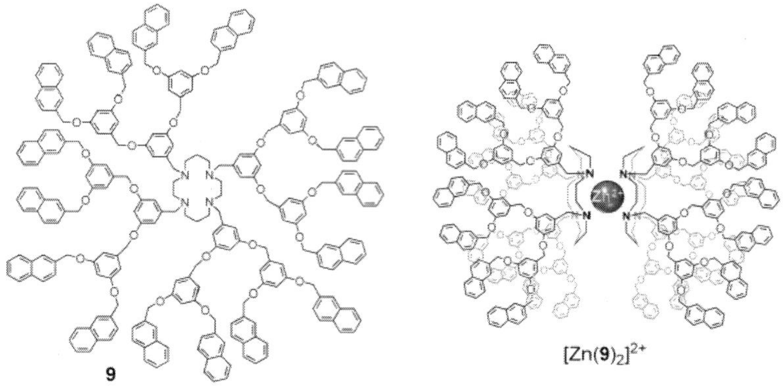

FIGURE 8.11 (left) Structure formula of dendrimer **9** consisting of a 1,4,8,11-tetraazacyclotetradecane (cyclam) core appended with 12 dimethoxybenzene and 16 naphthyl units. (right) Schematic representation of the $[Zn(\mathbf{9})_2]^{2+}$ species in which the dendrimer branches are extending outward.

In $[Zn(\mathbf{9})_2]^{2+}$ species, a single Zn^{2+} ion is able to enhance the luminescence of all the 32 naphthyl units present in the two dendritic structure. This result shows that dendrimers can be profitably used as supramolecular fluorescent sensors for metal ions. In this case, the sensor fluorescence is switched on upon metal ion coordination, and low Zn^{2+} concentrations (ca. 1 μM) can be easily detected.

When a solution of dendrimer **9** is titrated with lanthanide ions different results are obtained. Lanthanide metal ions exhibit long-lived and line-like luminescence, but direct excitation of lanthanide metal ions is inefficient because of the forbidden nature of their electronic transitions. Therefore, coordinating organic or inorganic chromophores are usually exploited to sensitize the luminescence of the ions (antenna effect). In the case of the interaction of dendrimer **9** with lanthanide ions (M^{3+}) such as Nd^{3+}, Eu^{3+}, Gd^{3+}, Tb^{3+}, Dy^{3+} [58b, c, d], a 1:3 (metal/dendrimer) complex ($\log\beta_{1:3} = 20.3$) is formed at low metal ion concentration as demonstrated by fluorescence and NMR titration. It is likely that in the $[M(\mathbf{9})_3]^{3+}$ complex not all the 12 nitrogen atoms of the three cyclam cores are engaged in metal ion coordination. However, upon metal coordination the exciplex emission band completely disappears. Clearly, as it is also shown by NMR results, the presence of the +3 ion is "felt" by all the nitrogen atoms of the three cyclam moieties, thereby raising the energy of the exciplex excited state above that of the naphthyl-based one. For all the lanthanide complexes of **9** no sensitized emission of the lanthanide ion was observed. Therefore, energy transfer from either the S_1 or T_1 excited state of the naphthyl units of **9** to the lanthanide ion is inefficient. By contrast, efficient energy transfer from naphthalene-like chromophores to Eu^{3+} has been reported in the case in which naphthalene is linked through an amide or carboxylate bond to the lanthanide [59]. Apparently, the nature of the first coordination sphere plays an important role concerning energy transfer efficiency.

To overcome the lack of sensitization of the lanthanide emissions for the complex with dendrimer **9** (see above), a supramolecular approach using a Ru^{2+} complex has been pursued [60].

Complexes of Ru^{2+} containing 2,2′-bipyridine (bpy) and cyanide ligands, such as $[Ru(bpy)_2(CN)_2]$ and $[Ru(bpy)(CN)_4]_2$, are particularly interesting because they are luminescent and can play the role of ligands giving rise to supercomplexes [61]. In particular $[Ru(bpy)_2(CN)_2]$ has low-energy absorption bands and a luminescence band in the visible region which are related to metal-to-ligand (bpy) charge-transfer (MLCT) excited states [61c]. Titration of a CH_3CN/CH_2Cl_2 (1:1 *v/v*) solution of $[Ru(bpy)_2(CN)_2]$ with Nd^{+3} causes changes in the absorption spectrum of the $[Ru(bpy)_2(CN)_2]$ and quenching of its emission with the concomitant appearance of the sensitized Nd^{+3} emission [60]; this finding demonstrates the ability of $[Ru(bpy)_2(CN)_2]$ to complex the lanthanide metal ion.

Titration of a 1:1 mixture of dendrimer **9** and $[Ru(bpy)_2(CN)_2]$ in CH_3CN/CH_2Cl_2 (1:1 *v/v*) with $Nd(CF_3SO_3)_3$ brings about changes in the absorption and emission spectra. The lowest energy absorption band due to the Ru^{2+} complex is blue-shifted, as already observed in the titration of $[Ru(bpy)_2(CN)_2]$ with the same lanthanide ion in the absence of dendrimer. Upon excitation at 260 nm, where most of the light is absorbed by dendrimer **9**, the intensity of the naphthyl monomer emission at 337 nm does not show a monotonous increase, as instead observed in the absence of the $[Ru(bpy)_2(CN)_2]$ complex: it reaches a maximum at 0.5 equivalent and then decreases up to about 1.0 equivalent of Nd^{3+} to rise again for higher metal ion concentration. The emission intensity at 1.0 equivalent is lower than that observed in the absence of $[Ru(bpy)_2(CN)_2]$. These results show that the three-component system $\{$**9**•Nd•$[Ru(bpy)_2(CN)_2]\}^{3+}$ (Fig. 8.12) is formed in which the dendrimer emission is quenched. This three-component system can be disassembled by addition of an excess of each component, or of cyclam.

The main photophysical processes of the $\{$**9**•Nd•$[Ru(bpy)_2(CN)_2]\}^{3+}$ adduct are summarized in Fig. 8.13, which shows the energy levels of the three components. In the two-component dendrimer-Nd^{3+} system, energy transfer from either the lowest singlet (S_1) or triplet (T_1) excited state of the naphthyl units of the dendrimer to the lanthanide ion does not occur.

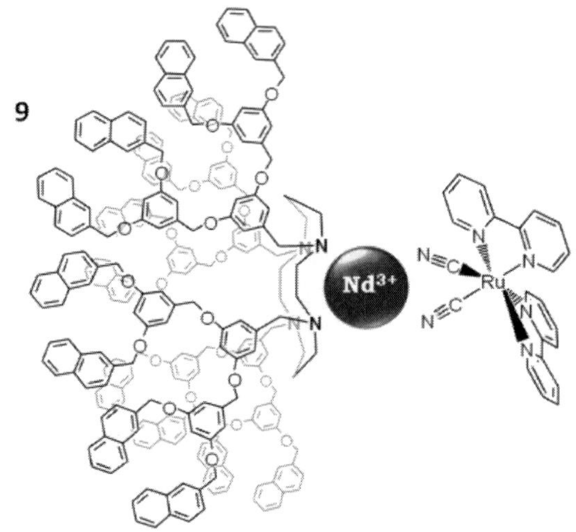

FIGURE 8.12 Schematic representation of the self-assembled {**9**•Nd•[Ru(bpy)$_2$(CN)$_2$]}$^{3+}$.

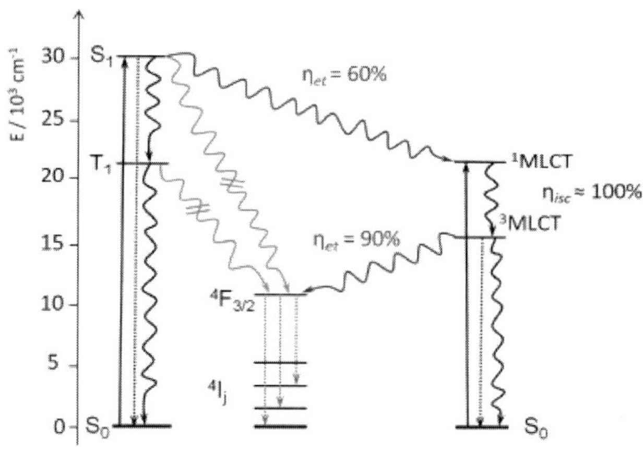

FIGURE 8.13 Energy level diagram showing the excited states involved in the main photophysical processes (absorption: solid lines; radiative deactivation: dotted lines; nonradiative deactivation processes: wavy lines) of the {**9**•Nd•[Ru(bpy)$_2$(CN)$_2$]}$^{3+}$ three-component system. Naphthyl excimer energy level has been omitted.

Therefore, it can be concluded that in the three-component system sensitization of the Nd^{3+} emission upon dendrimer excitation is mediated by the $[Ru(bpy)_2(CN)_2]$ component. Comparison between the emission quantum yield of $[Ru(bpy)_2(CN)_2]$ upon excitation at 260 nm (dendrimer absorption) and 450 nm ($[Ru(bpy)_2(CN)_2]$ absorption) enabled to estimate that the energy transfer efficiency from the S_1 excited state of the naphthyl groups to the ^{1}MLCT excited state of $[Ru(bpy)_2(CN)_2]$ is about 60% (Fig. 8.13). The energy transfer efficiency from the ^{3}MLCT excited state of $[Ru(bpy)_2(CN)_2]$ to Nd^{3+} can be assumed to be equal to the efficiency of the quenching of the $[Ru(bpy)_2(CN)_2]$ emission (ca 90%). Note that quenching by electron transfer can be ruled out because of the Nd^{3+} redox properties. Furthermore, in the adduct no evidence of energy transfer from the naphthyl-localized T_1 excited state of the dendrimer to the lowest ^{3}MLCT state of $[Ru(bpy)_2(CN)_2]$ has been found since no change in the T_1 lifetime at 77 K has been observed.

The three components of the self-assembled structure have complementary properties, so that new functions emerge from their assembly. Dendrimer **9** has a very high molar absorption coefficient in the UV spectral region, but it is unable to sensitize the emission of a Nd^{3+} ion placed in its cyclam core. On the other hand the $[Ru(bpy)_2(CN)_2]$ complex can coordinate and sensitize the emission of Nd^{3+} ions. Self-assembly of the three species leads to a quite unusual Nd^{3+} complex which exploits a dendrimer and a Ru^{2+} complex as ligands. Such a system behaves therefore as an antenna that can harvest from UV to VIS light absorbed by both the Ru^{2+} complex and the dendrimer and emit in the NIR region with line-like bands. It is also important to notice that this supramolecular antenna can be assembled or disassembled changing the concentration of the lanthanide ion and in principle the emission wavelength can be tuned by replacing Nd^{3+} with other lanthanide ions possessing low-lying excited states.

8.3.2 ENERGY TRANSFER CONTROLLED BY METAL IONS

As said before dendrimers are ideal scaffolds to organize multiple active units in a nanoobject. Dendrimers containing photochromic [44, 45], or luminescent [46] units or performing as ligands of metal ions [47, 48] have been extensively investigated in the last decade, but there is only one example were all the three above mentioned functions are simultaneously present [62].

Dendrimers **10-*t*** and **11-*t*** (Fig. 8.14) contain two cyclam coordinating units linked by a photoswitching azobenzene moiety and are functionalized at the periphery with six or 12 light-harvesting (naphthalene) chromophores, respectively. Azobenzene has been chosen because it undergoes an efficient and fully reversible photoisomerization reaction from the *trans* to the *cis* form [63]. For this reason, it has been extensively used to construct photoswitchable devices [55].

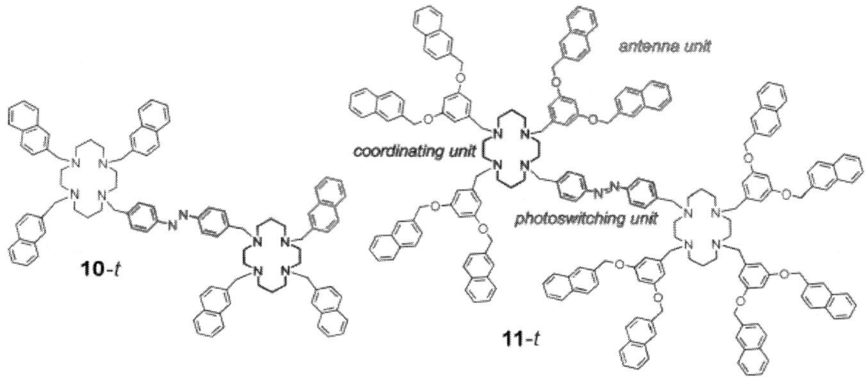

FIGURE 8.14 Structures of dendrimers **10-*t*** and **11-*t*** in *trans* (*t*) conformation in which the three different functional units are evidenced.

The absorption spectra of **10-*t*** and **11-*t*** in CH_3CN/CH_2Cl_2 (1:1 *v/v*) show the characteristic bands of both naphthalene at 275 nm and azobenzene at 336 (ππ* transition) and 450 nm (nπ* transition); therefore the lowest energy excited state is localized on the azobenzene moiety. By excitation at 275 nm, where most of the light is absorbed by the naphthalene chromophores, **10-*t*** and **11-*t*** exhibit a very weak luminescence with emission quantum yields that are much lower than those previously observed for cyclam-cored dendrimers [58d]. This result suggests a quenching of the naphthyl emission by the azobenzene unit (see Fig. 8.15). In agreement with the behavior previously reported for the two dendrimers containing only a cyclam core [58a] dendrimers **10-*t*** and **11-*t*** exhibit two different emissions: a band with maximum at 335 nm, assigned to a naphthyl localized excited states, and a shoulder at 470 or 400 nm for dendrimers **10-*t*** and **11-*t***, respectively. These shoulders can be assigned to the naphthyl excimer (λ_{max} ~ 390 nm) and naphthyl-amine exciplex (λ_{max} = 470 nm, as a result of the interaction of an excited naphthalene with an amine group of the cyclam unit) emissions. The emission spectrum of **11-*t*** does not show the exciplex emission, probably because the energy transfer to the azobenzene core (see Fig. 8.15) is faster than exciplex formation, while in **10-*t*** the close proximity of naphthalene and cyclam nitrogen atoms enables exciplex formation in competition with energy transfer to the azobenzene core.

Upon irradiation of the two dendrimers at 365 nm, where only the azobenzene absorbs, spectral changes ascribed to the *trans* → *cis* isomerization of azobenzene are observed with a decrease of the $\pi\pi^*$ band at 336 nm and an increase of the $n\pi^*$ band. The *cis*-azobenzene species can be converted back to the *trans* isomer by irradiation at 436 nm in the $n\pi^*$ band. The value of the photochemical *trans* → *cis* isomerization quantum yield ($\Phi_{t\rightarrow c}$) and the percentage of *trans* isomer present at the photostationary state obtained for **10-*t*** and **11-*t*** (0.15 and 5%, 0.09 and 7%, respectively) are very similar to those obtained for an azobenzene model compound (0.15 and 5%). In **11-*t*** the lower value of $\Phi_{t\rightarrow c}$ and a somewhat larger fraction of the *trans* isomer are indicative of a higher stability of the *trans* isomer compared to the *cis* one with respect to the model compound.

In order to elucidate the interaction between naphthalene and azobenzene units in the dendritic structures and test the antenna abilities of these systems, photoisomerization of azobenzene has been investigated upon selective excitation of the naphthalene at 275 nm, where more than 95% of the light is absorbed by the naphthalene both in the *trans* and *cis* isomers. The results show that azobenzene isomerization takes place, demonstrating that energy transfer from the naphthalene to the *trans* isomer of the azobenzene core occurs with an efficiency (η_{ET}) of 20% and 40% for **10-*t*** and **11-*t***, respectively.

Upon titration of a CH_3CN/CH_2Cl_2 (1:1 *v/v*) solution of **11-*t*** with $Zn(CF_3SO_3)_2$ the absorption spectrum shows a blue shift and an increase in intensity of the $\pi\pi^*$ band of azobenzene together with a decrease of absorbance at 260 nm. The plot of the normalized absorption changes at 304 nm versus the added amount of Zn^{2+} is quite linear and exhibits a plateau at ca. 2.3 equivalents of Zn^{2+} per dendrimer: this finding demonstrates that up to two metal ions can be coordinated, that is, one per cyclam unit, with high association constants ($K_1 = 7 \times 10^7\ M^{-1}$, $K_2 = 1 \times 10^5\ M^{-1}$ for 1:1 and 2:1 metal/dendrimer stoichiometry, respectively). In the 1:1 complex only one of the two cyclam units is linked to Zn^{2+} because the *trans*-azobenzene unit keeps the two cyclam quite far apart (Fig. 8.15a, left). Titration with $Zn(CF_3SO_3)_2$ has been also performed on the solution obtained at the photostationary state upon irradiation of **11-*t*** at 365 nm. In this case the absorbance changes at 304 nm reach a plateau at ca. 1.4 equivalents of Zn^{2+} ions per dendrimer, instead of 2.3 equivalents as obtained for the *trans* isomer. These results can be rationalized on the basis of a structural rearrangement; in the *cis* isomer the two cyclam units are indeed much closer, so that one Zn^{2+} ion can be coordinated by both of them (Fig. 8.15a, right) with a higher association constant ($K_1 = 1 \times 10^8\ M^{-1}$), evidencing a positive effect of the second cyclam unit on the stability constant. Qualitatively similar results are obtained in the case of **10-*t*** and **10-*c*** upon titration with $Zn(CF_3SO_3)_2$.

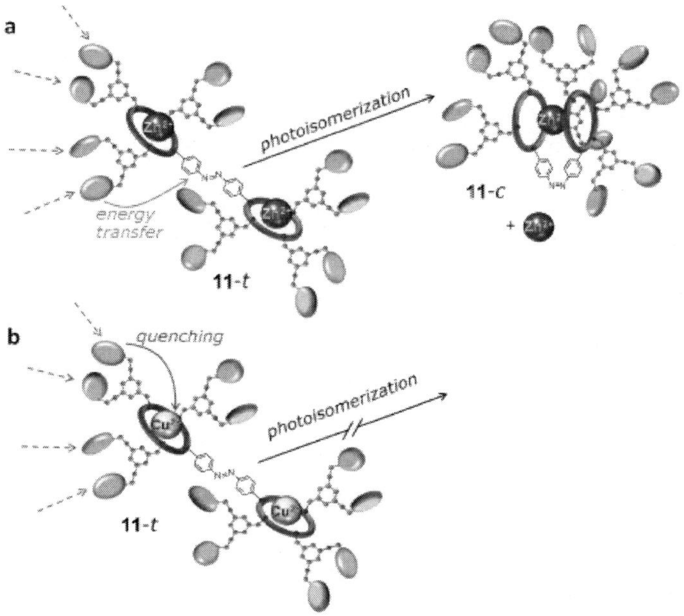

FIGURE 8.15 Schematic representation of the functions performed by dendrimer **11**: different coordination ability exhibited by the cyclam moieties (empty rings) of **11-*t*** (left) and **11-*c*** (right), light-harvesting capability of the naphthalene units (ovals) and sensitized photoisomerization of the azobenzene core.

In order to investigate the effect of coordinated metal ions on the energy transfer and isomerization abilities of the system, photochemical experiments on the metal complexes of both dendrimers have been performed upon addition of an excess of Zn^{2+} ions. The presence of two metal ions per dendrimer disfavors the *trans* → *cis* photoisomerization carried out by irradiating at 365 nm, as demonstrated by the lower value of $\Phi_{t \to c}$ and higher molar fraction of *trans* isomer present at the photostationary state (0.07 and 10%, 0.06 and 13% for $[Zn_2(\mathbf{10}\text{-}t)]^{4+}$ and $[Zn_2(\mathbf{11}\text{-}t)]^{4+}$, respectively).

Regarding the isomerization sensitized by the naphthalene excitation (antenna effect) a strong increase in the efficiency of energy transfer from naphthalene to azobenzene is observed (Fig. 8.15a), consistent with the fact that coordination of Zn^{2+} to the cyclam units prevents excimer and exciplex formation.

Dendrimers **10-*t*** and **11-*t*** have been also titrated with $Cu(CF_3SO_3)_2$; in such a case a new absorption band at 330 nm appears, which can be assigned to a ligand-to-metal charge transfer (LMCT) transition [64]. As expected, the cyclam moiety is able to coordinate both Zn^{2+} and Cu^{2+} ions, but profound differences are observed in the photophysical and photochemical properties. Indeed, in the dendrimer-Cu^{2+} complexes no sensitized isomerization takes place upon excitation of the naphthalene units and a lower isomerization quantum yield is obtained (Fig. 8.15b) upon direct excitation of the azobenzene core (0.03 and 0.06 for $[Cu_2(\mathbf{10}\text{-}t)]^{4+}$ and $[Cu_2(\mathbf{11}\text{-}t)]^{4+}$, respectively). These results are consistent with the quenching of the naphthalene luminescence and, to some extent, with the reduced occurence of the azobenzene isomerization by energy/electron transfer to the coordinated Cu^{2+} ions.

In conclusion dendrimers **10-*t*** and **11-*t***, containing simultaneously photochromic (azobenzene), luminescent (naphthalene) and metal coordinating (cyclam) units, can perform three different functions: light-harvesting, metal coordination and photoswitching. Because of their proximity, the various functional groups of the dendrimers interact and the azobenzene unit enables to control the distance between the two cyclam moieties upon light stimulation leading to different coordination properties for the *cis* and *trans* isomers. Furthermore, in this photocontrolled metal coordinating tweezers the coordinated metal ion is not only a spectator, but it can also switch ON/OFF the light-harvesting and photoisomerization functions. In this way metal coordination and photosensitized isomerization can control each other: coordination of Zn^{2+} ions enables 100% sensitization of azobenzene by the light-harvesting antenna function, whereas coordination of Cu^{2+} ions prevents this effect.

8.4 FUNCTIONALIZED QUANTUM DOTS FOR CHEMICAL SENSING

Chemosensors are molecules able to bind selectively and reversibly the analyte of interest with concomitant change in one or more properties of the system (e.g., redox potentials, absorption or fluorescence spectra) [65]. Luminescent chemosensors are a particular class of chemosensors in which the association of an analyte causes a change in the luminescence properties.

Multicomponent supramolecular systems are particularly suitable to act as luminescent chemosensors [66]. Their design requires the presence of three units which play different roles: the luminophore, which emits light upon photoexcitation; a receptor, which influences the luminophore emission upon binding an analyte; and a spacer, which modulates the communication between the fluorophore and the receptor [65]. The interaction between the luminophore and the receptor is typically based on either electron- or energy-transfer processes [67]. The association of the analyte changes this interaction causing the establishment or the suppression of a quenching path for the luminophore with the concomitant decrease or enhancement of its emission, respectively.

In literature many examples of luminescent chemosensors based on this supramolecular approach are reported; they are luminescent sensors for cation [68], anions [69], and molecules [70] and also self-assembled fluorescence chemosensors [71].

Luminescent semiconductor nanocrystals (quantum dots, QDs) are emerging nanostructured materials with unique characteristics that are not present in their bulk parent materials [72]. Quantum dots "core", based on CdSe, CdS, CdTe, InAs and InP semiconducting material, have tunable size-dependent broad absorption bands with high extinction coefficients, and size-dependent Gaussian emission profiles [73]. In particular CdSe-based nanocrystals exhibit a remarkable resistance to chemical- and photo-degradation [72c]. Epitaxial growth of another semiconductor material of wider band gap (e.g., ZnS) onto QD "core" (e.g., CdSe), affords the formation of the so called "core shell" nanocrystals, (e.g., CdSe-ZnS) [74, 75]. The shell confers stability against photobleaching [76], preserving the luminescence properties of the core [77].

Owing to their peculiar spectroscopic characteristics, QDs could find application in many fields ranging from solar cells [78] and LED fabrication [79] to medicine (e.g., imaging) [80] and sensing [77, 81, 82].

The development of luminescence chemosensors bearing QDs is based on the operating principles of conventional fluorescent chemosensors: [65, 67] semiconductor nanocrystals are conjugated with molecules able to sense specific analytes (receptors) and the communication between the two subunits can occur via electron or energy transfer (or both) [77].

Because the electron transfer from excited quantum dots to a redox active species (and viceversa) can efficiently quench the QD's luminescence [83, 84], this process can be used to modulate the luminescence of the nanoparticles. Therefore using the molecular recognition to activate or deactivate an electron transfer mechanism between QDs and the receptors, a significant change in the luminescence intensity can be obtained. For example the maltose binding protein (MBP) and its recognition property can be used to switch the luminescence of conjugated quantum dots via electron transfer [85].

MBP was engineered with a thiol group, able to bind the QD's surface, and a ruthenium complex. When maltose is not bound, the ruthenium complex is close enough to the surface of the nanocrystal to transfer an electron to the quantum dot upon excitation thereby quenching its emission. Following the binding of a maltose molecule, MBP undergoes a conformational change that moves the ruthenium complex far enough from the nanocrystal's surface. As a consequence the electron transfer quenching pathway is suppressed with an increase in the QD's luminescence quantum yield: the binding of maltose (analyte) is therefore transduced in a significant luminescence enhancement.

Semiconductor nanocrystals can also undergo energy transfer to one or more acceptors [81]. Energy transfer involving QDs was first observed in closed-packet film of nanoparticles of different diameter. Small and big (3.8 nm and 6.2 nm, respectively) CdSe core were used. Upon excitation on the first absorption band of the smaller particle, energy transfer from the small QDs (high energy) to the big ones (low energy) has been clearly evidenced by steady state and time resolved fluorescence measurements [86].

Quantum dots are usually employed as energy donor to an acceptor. For an efficient energy transfer the QD and the acceptor need to be closed and the emission bands of the nanoparticle must overlap with the absorption bands of the acceptor. Molecular recognition can be used to change the separation between the nanoparticles and the energy acceptor and to modify the overlap of their spectra. Supramolecular association of a substrate to a receptor can indeed be exploited to modulate the efficiency of energy transfer and QD's emission quantum yield: the presence of an analyte can be indeed transduced into a detectable luminescence signal [87]. As an example, QD-squaraine dye nanohybrid was successfully used as ratiometric pH sensor based on FRET (Fluorescence Resonance Energy Transfer) from nanoparticle to the dye. The energy transfer is modulated by the environment because the absorption profile of the squaraine is a function of pH: at low pH the energy transfer is more efficient and the emission of the squaraine is predominant; conversely at basic pH the spectral overlap is smaller, the energy transfer less efficient and the overall emission is mainly due to the nanocrystals [88].

In the following sections two examples of chemosensors based on quantum dots are reported.

8.4.1 A RATIOMETRIC OXYGEN SENSOR

Luminescent ratiometric chemosensors represent a suitable way for the determination of an analyte, avoiding problems concerning local concentration differences of the probe compounds [89]. In this context determination of oxygen concentration is particularly important because it is required for a variety of applications [90], such as in food packaging manipulated in modified atmosphere [91].

The preparation of an oxygen sensor based on CdSe-ZnS core shell QDs with pyrenyl units covalently attached on the surface of the nanocrystal has been recently reported [92]. Pyrene is a well-known fluorophore the luminescence of which is strongly quenched by oxygen [93], while QD's luminescence is not quenched by dissolved oxygen [94]. Emissions coming from both moieties are fully independent and no energy or electron transfer takes place upon photoexcitation. This latter observation and the difference in the oxygen quenching feature enable the ratiometric response for the nanohybrid.

CdSe-ZnS nanocrystals (QD) capped with trioctyl phosphine oxide (TOPO) were reacted with the pyrene ligand (PYI) bearing an imidazole group as anchoring point for the QD. From a comparison of the absorption data of the separated components and the nanohybrid (**QD-PYI**), 55 PYI units per QD were estimated. The luminescence spectrum of the **QD-PYI** exhibits similar shape and λ_{max} compared to those obtained for QD and PYI units, but the quantum yield of both the units is substantially lower. FRET process from PYI to QD can be excluded on the basis of the lifetime that does not show any shortening, and the excitation spectra. Because the quantum yield of PYI (0.018) is very similar to the one obtained for pyrene in hexane (0.022), the observed decrease of the quantum yield can be assessed to the non polar environment experimented by PYI on the nanohybrid; it is indeed surrounded by the long alkyl chains of the TOPO ligands. Concerning the decrease of the QD's emission quantum yield (from 0.080 to 0.020) it can be explained by the replacement of TOPO molecules with the PYI units with a consequent less surface passivation.

In the presence of oxygen the following results have been obtained: (i) the luminescence spectrum of **QD-PYI** is influenced by the O_2 concentration; (ii) the oxygen-quenching of the pyrene fluorescence ($k_q = 1.1 \times 10^{10}$ $M^{-1} s^{-1}$) can be interpreted in terms of Stern-Volmer equation and is dynamic in nature; (iii) the CdSe luminescence is unaffected by the O_2 concentration so that it can be used as internal reference to normalize the pyrene fluorescence. On the basis of such data the optical signal of the sensor is calculated as the ratio between the QD emission (632 nm) and the PYI emission (398 nm). The schematic representation of this ratiometric O_2 sensor is reported in Fig. 8.16.

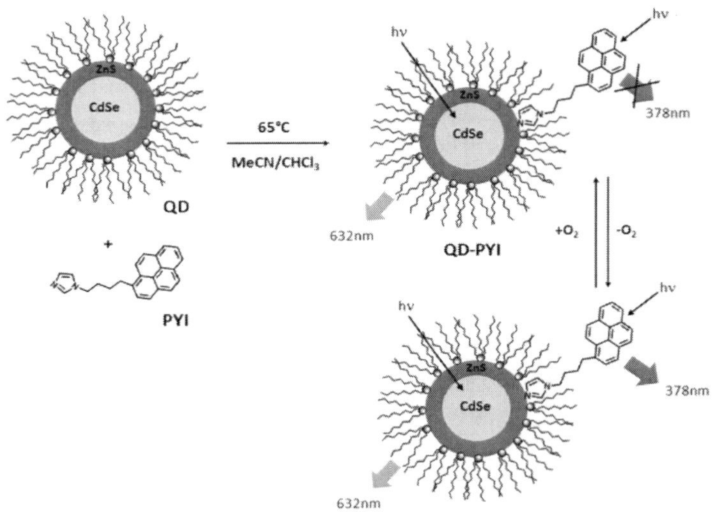

FIGURE 8.16 Synthesis and schematic representation of the **QD-PYI** system as ratiometric oxygen sensor.

8.4.2 QD-DOPAMINE NANOASSEMBLY AS LUMINESCENT pH SENSOR

Determination of proton concentration is a demanding analysis for a wide range of applications [70], however, only a few examples of luminescent pH sensors based on semiconductor nanocrystals have been reported.

In the cases in which the nanocrystals are coupled with pH-sensible oxazine [95], squaraine [88], and fluorescein isothiocyanate [96], an energy transfer mechanism between the molecule and the nanocrystal accounts for the luminescent response of the nanoassemblies.

When dopamine is used as pH-sensible molecule [97, 98] to modify the surface of the nanocrystals an electron transfer mechanism takes place. Dopamine, a natural compound involved in different biological activities [99], can be converted between two forms, the quinone and the hydroquinone. Quinone molecules behave as good electron acceptors in biological and abiotic formats [100, 101], while hydroquinone is a poor electron acceptor. Reduced dopamine (cathecol), as other hydroquinones, undergoes to autoxidation and oxidation by O_2 generating a dopamine in quinone form. This electron-proton transfer system exhibits slow kinetics whose rate constants in water depend linearly on pH: [100, 102] the rates of oxidation to quinone increase as pH increases from acidic to basic [100, 103].

A **QD-dopamine** nanohybrid based on the self-assembly of the dopamine unit, covalently attached to an histidine-rich peptide, on the QD's surface has been recently reported. The pH-dependent interconversion between catechol and quinone was exploited to modulate the electron transfer quenching mechanism in **QD-dopamine** assembly by changing the external pH. Upon photoexcitation an electron is transferred from the conduction-band of the quantum dot to the lowest unoccupied molecular orbital of the quinone acceptor, resulting in luminescence quenching. The pH dependent concentration of the oxidized dopamine in **QD-dopamine** assembly determines the extent of the electron transfer and consequently the quenching of the QD emission. At acid pH the concentration of the quinone form of dopamine is small and consequently the QD quenching is marginal. As pH increases the cathecol form of dopamine undergoes a progressive oxidation to quinone, with concomitant increasing in QD quenching. This nanohybrid was successfully used to sense the intracellular pH.

Very recently it has been reported on a **QD-dopamine** system with a more controllable structure, where the dopamine group is covalently attached at the end of hydrophilic QD ligands [104]. The CdSe-ZnS core shell quantum dots used in this work were coated *via* cap-exchange with a mixture of DHLA-PEG_{750}-OCH_3 (inert) and DHLA-PEG_{600}-NH_2 (functional) ligands. They provide a robust hydrophilic coating characterized by little to no changes in the absorption and luminescence properties in the 4–12 pH range. The use of functional amine ligand (DHLA-PEG_{600}-NH_2) enables the postsynthetic coupling of QD with dopamine bearing an isothiocianate group, affording the **QD-dopamine** nanohybrid in which the dopamine is covalently attached to the semiconductor nanocrystals (Fig. 8.17). The reaction leads to the formation of an isothiourea bond, which is pH independent. Using this method different hybrids were prepared in which the number of dopamine per QD was controllably changed by varying the percentage of functional ligands during the cap-exchange step. This coupling strategy that provides the direct attachment of the functional unit on the ligand shell, without interference concerning the inorganic QD's surface, avoids any kind of pH-induced alterations in the optical properties of the QDs. Moreover using this strategy the average distance between the nanocrystal center and the redox active dopamine units is maintained.

In such kind of assembly, the effect of pH on the luminescence properties of QDs was tested. Increasing the pH from 4 to 10, a progressive loss in QD luminescence was observed according to the increasing amount of dopamine in the quinone form, which quenches the QD's luminescence. Time resolved fluorescence analysis shows a shortening in the luminescence lifetime. In a reverse configuration where the **QD-dopamine** assembly was first dissolved in basic buffer and then the pH was reduced to acidic values, the luminescence of QDs undergoes a progressive recovery (Fig. 8.18).

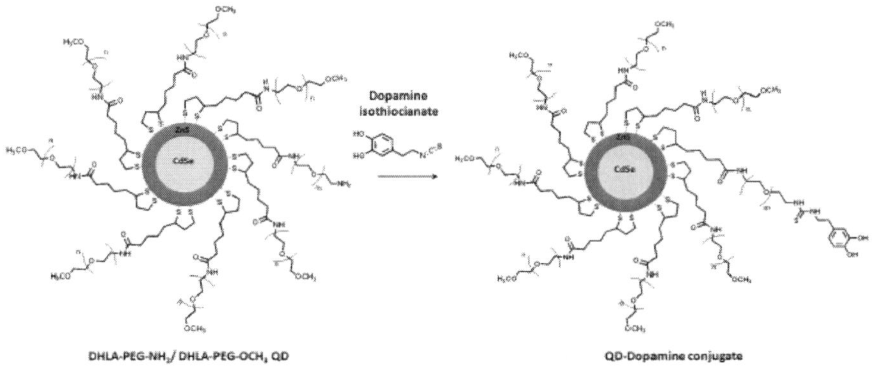

FIGURE 8.17 General scheme of the **QD-dopamine** conjugation strategy.

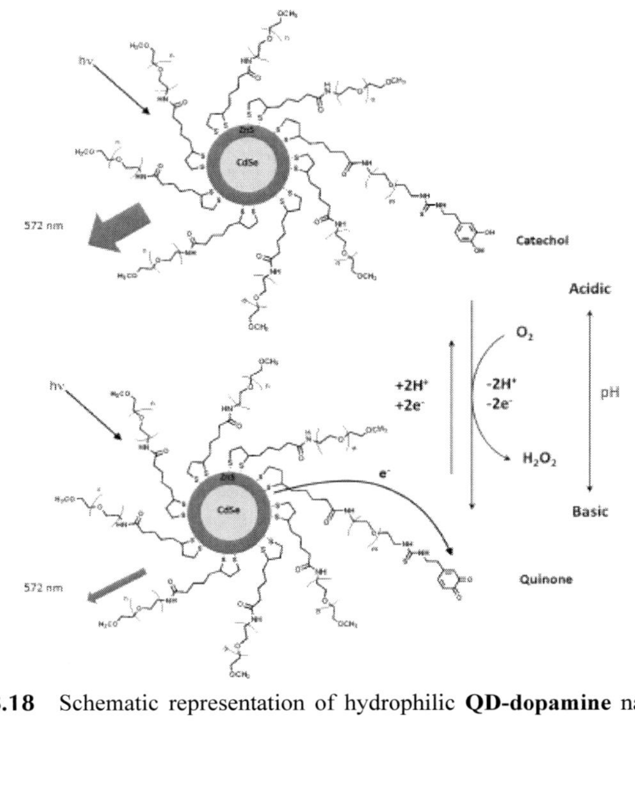

FIGURE 8.18 Schematic representation of hydrophilic **QD-dopamine** nanohybrid as pH sensor.

The key role of the oxygen was also investigated. Oxygen-free dispersions of *QD-dopamine* at different pH were compared to dispersions of the same hybrid in air-equilibrated water. At basic pH for the first set of dispersions a slight quenching was observed, while a pronounced quenching was observed for the second set of dispersion. These experiments, together with absorption changes of dopamine molecule in oxygen free and air equilibrated water, prove that oxygen plays a crucial role in the pH-dependent oxidation of dopamine and the resulting quenching of the QD emission.

In order to evaluate the quenching mechanism, transient absorption spectroscopy analysis on conjugated and unconjugated *QD-dopamine* dispersed in pH 4 and pH 10 buffers were performed. Intraband transitions of both hole and electron of the photoexcited *QD-dopamine* hybrid change on moving from pH 4 to pH 10. Changes for unconjugated QDs are negligible and similar to the conjugated system at pH 4. Specifically, time resolved analysis shows a faster relaxation of both electron and hole in the presence of dopamine at pH 10, proving that the charge transfer quenching mechanism involves both charge carriers. These observations have been attributed to a combination of three processes that occur simultaneously: (i) electron transfer from a photoexcited QD to the lowest unoccupied molecular orbital of the quinone; (ii) electron transfer from hyroquinone dopamine to the valence band of the QD; (iii) weak electron transfer from hydroquinone dopamine to unexcited QD. Increasing the concentration of quinone favors higher electron transfer from QD to the proximal oxidized dopamine, while oxidation of the reduced dopamine promotes easier electron transfer to the valence band of the QD. A scheme of the proposed relaxation mechanism is reported in Fig. 8.19.

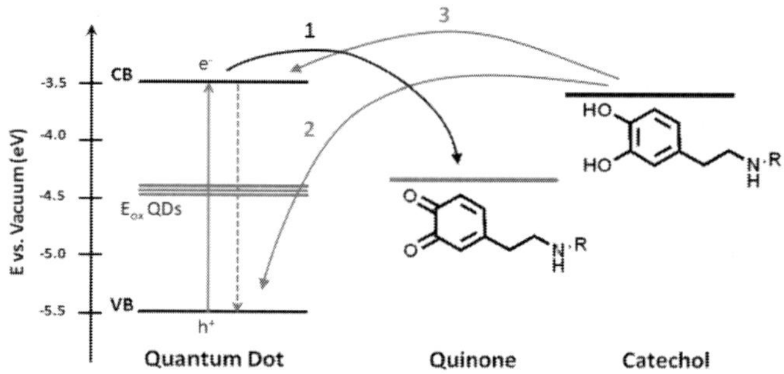

FIGURE 8.19 Schematic representation of the proposed relaxation mechanisms for the electron and hole transfer: (1) electron transfer from photoexcited QD conduction band to quinone; (2) electron transfer from dopamine catechol to valence band of photoexcited QD; (3) weak electron transfer from catechol dopamine to unexcited QD.

8.5 MOLECULAR LOGIC

Molecular level systems can respond to external stimulation (such as a pulse of light, an electric potential or an encounter with another molecule) by changing some physical or chemical properties that is reaching a new stable or metastable state different from the initial one. In this sense, any chemical system that can be reversibly switched between two different states by means of a chemical or physical stimulus, can be taken as a basis for storing information, that is, for memory purposes [105]. In fact the initial state (state "0") of the system can "be written" by using a suitable input and its switched form (state "1") can "be read", because its properties differ from those of the initial state, and "erased" by using an opposite stimulus to go back to the state "0" (Fig. 8.20).

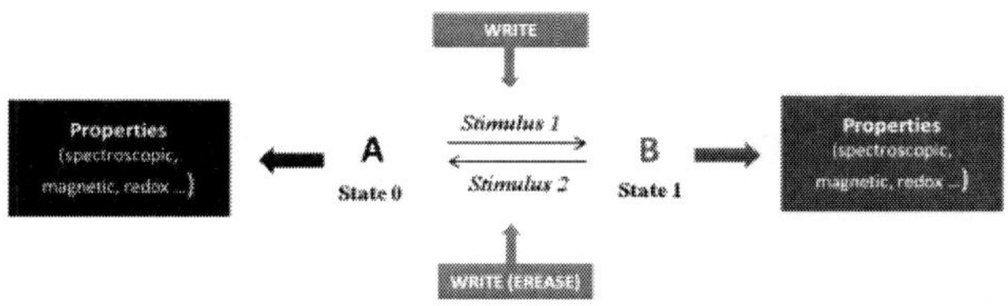

FIGURE 8.20 Bistable molecular or supramolecular systems reversibly switched between two states (A and B) by external stimuli can store information for memory purposes.

Any chemical or physical change of the system (e.g., conformational change, red-ox change, disassembling, etc.), that is caused by an external stimulation and that accompanies its switching, can be viewed as a transformation of inputs into outputs, and thus can be used to process information. In more complex systems, like multistate-multifunctional systems, switching can be performed between more than two states by means of different and independent stimuli. This possibility can be exploited for obtaining memories that are permanent unless they are erased on purpose or for performing logic operations. Therefore, research in supramolecular chemistry provides a wide range of possibilities to develop devices capable of processing information at molecular-level, similarly to what happens in living organisms, in which information is elaborated, transported and stored using "soft" molecular or ionic "substrates" [106]. Indeed, several recent achievements have been obtained in the field of molecular logic combining molecular chemionics, photonics, and electronics.

It is straightforward that molecular-level switches can operate as logic gates performing binary operation. A logic gate is a device that transforms one or more inputs into an output according to relationships described in the so-called truth table that is specific of each type of logic gate. The logic behavior of a molecular switch is based on the assignment of appropriate threshold values to the input and output signals so that they can assume only two values (e.g., 0/1, Yes/Not, True/False, On/Off) according to binary logic. The inputs and outputs do not need to share the same logic convention and different logic conventions applied to the same molecular device can lead to different logic functions, similar to semiconductor devices [105].

Even if input/output molecular-level processes are very common, their logic aspects have been recognized only towards the end of '80. The first proposal to execute logic operations at the molecular level was made in 1988 [107], but only after the experimental investigation of simple logic operations performed by molecular switches [108], the field of molecular logic starts to get a footing. During the last few years, several kinds of molecular-level logic functions have been reported and discussed. Although any kind of input/output signal can be used to perform logic operation with molecular switches, as matter of fact most of molecular logic gates investigated so far are operated by chemical inputs and give optical output signals (absorption or luminescence). While presenting some limitations, chemical input signals provide a high versatility because a wide range of molecules and ions can be employed. Light outputs as fluorescence, besides having several advantages relative to the controlled switch of the system, can be detected easily even at the single molecule limit.

A critical issue of molecular logic gates is the connection of basic elements to create complex circuits, which is indeed a key feature of electronic logic gates owing to full input/output homogeneity. The construction of molecular logic networks, however, can take advantage of functional integration that can be achieved within one molecule by rational chemical design, rather than relying on intensive physical connection of elementary gates [109]. Such a high functional integration is possible thanks to the possibility of defining different logic operations for the same gate. This important feature of molecular logic gates, called reconfigurability [110], can be achieved by changing the type of input/output signals or by using a separate switching stimulus. For instance the logic expression for molecular logic gates with optical output signals and significant input-induced spectral shifts, can be conveniently reconfigured by monitoring the output at a different wavelength. In a wavelength-reconfigurable logic gate, multiple wavelengths (outputs) can be observed at the same time, that is such gate can perform simultaneously different logic functions on a given set of inputs. Such a property is called superposition [111] and cannot occur with electronic systems since it is a consequence of the multichannel nature of light with whom only molecular logic gates can operate.

The molecular-level information processing can potentially be used to develop "molecular computers" [113d] much smaller and much powerful than the current silicon-based ones. Clearly the way to reach this ambitious objective is very long and such idea seems too futuristic at the present time. Nevertheless, artificial molecular-level devices developed so far and capable of performing quite complex logic functions, could be used for specific applications in areas such as diagnostics, medicine, and material science, where problems need to be addressed in places (for example inside a cell) that are out of reach for the "top down" techniques currently available to the microelectronics industry [111].

In this section, we described some molecular logic gates based on synthetic molecular and supramolecular (multicomponent) systems that operate in solution environment, and are investigated at the ensemble level.

8.5.1 BIDIRECTIONAL HALF SUBTRACTOR AND REVERSIBLE LOGIC DEVICE

In 1997, it was studied the first example of a molecular XOR (eXclusive OR) gate based on the controlled assembly and disassembly of a host-guest pseudorotaxane-type species in solution [112]. The paper received a considerable attention [113] because the XOR gate is of particular importance as it compares the digital state of two signals. At that time, however, the field of molecular logic was still in its infancy and the potentialities of the system were not fully explored. Very recently, an improved version of such a system has been reported in which its UV-visible spectroscopic properties in response to acid/base stimulation in solution have been investigated in detail, showing that it can operate as a bidirectional half subtractor and it exhibits logic reversibility [114].

The molecular components of the system are the electron-rich wheel **12** and the axle-type molecule $\mathbf{13}^{2+}$, containing an electron-deficient diazapyrenium (DAP) unit (Fig. 8.21).

In organic solution both the molecular components exhibit intense and structured absorption spectra and are strongly fluorescent (Fig. 8.22).

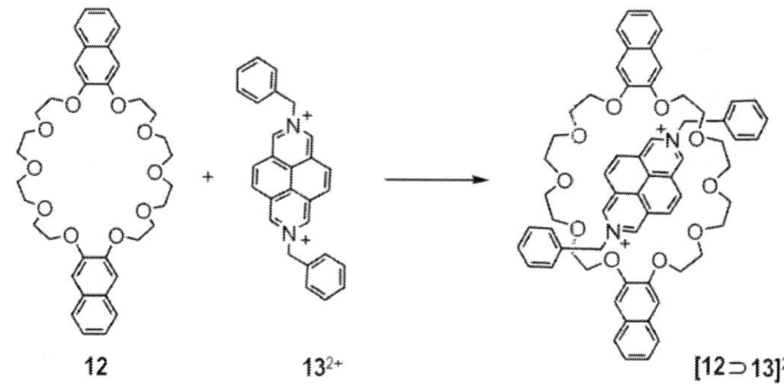

FIGURE 8.21 The self-assembly of compounds **12** and $\mathbf{13}^{2+}$ to yield the pseudorotaxane $[\mathbf{12}\supset\mathbf{13}]^{2+}$.

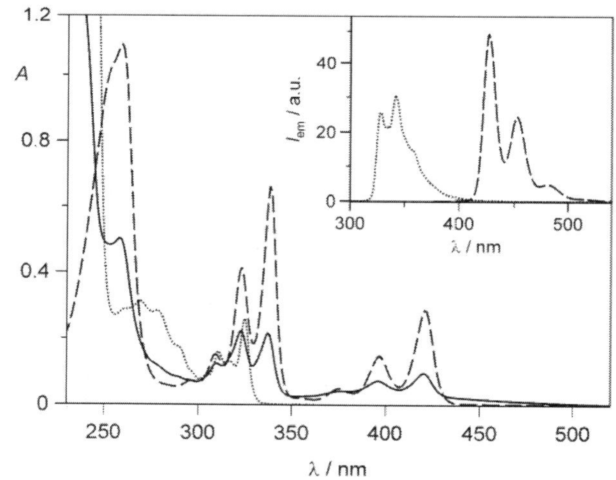

FIGURE 8.22 Absorption and luminescence (inset; λ_{ex} = 264 nm) spectra of **12** (dotted line) and $\mathbf{13}^{2+}$ (dashed line) at a concentration of 18 μM in CH_2Cl_2/CH_3CN (9:1 *v/v*) at room temperature. The full line represents the absorption spectrum of a 1:1 mixture of **12** and $\mathbf{13}^{2+}$ under the same conditions.

The pseudorotaxane $[\mathbf{12}\supset\mathbf{13}]^{2+}$ (Figs. 8.21 and 8.22), coming from the self-assembly of the molecular components in CH_2Cl_2/CH_3CN (9:1 *v/v*), is stabilized by a charge-transfer (CT) interaction. Complexation is signaled by as many as three different optical channels: (i) appearance of a yellow color because of the presence of a CT absorption tail in the visible region; (ii) disappearance of the blue-green fluorescence of $\mathbf{13}^{2+}$ (λ_{max} = 428 nm), and (iii) disappearance of the UV fluorescence of **12** with λ_{max} = 340 nm. The fluorescent signals of the two molecular components are quenched because of the presence of the lower lying CT state of the complex.

The XOR function is performed by the pseudorotaxane [**12**⊃**13**]$^{2+}$, using acid (CF_3SO_3H) and base (n-Bu_3N) as the inputs and choosing the fluorescence of **12** at 340 nm as the output. The working mechanism of this system is illustrated schematically in Fig. 8.23 and Table 8.1. As mentioned above, in the absence of the two inputs the fluorescence of **12** is quenched in the pseudorotaxane (output {0}). When the n-Bu_3N input (B in Fig. 8.23) alone is applied (30 equivalents with respect to the pseudorotaxane), the pseudorotaxane dethreads because of the formation of a stronger CT interaction between the amine and **13**$^{2+}$. Under such conditions, **12** is free and its fluorescence is no longer quenched (amine input {1}, output {1}). The fluorescence typical of free **13**$^{2+}$ (λ_{max} = 428 nm) is still quenched and a luminescence band with λ_{max} = 666 nm, arizing from the complex between **13**$^{2+}$ and the amine, is observed. Application of the H^+ input (30 equivalents with respect to the pseudorotaxane) causes protonation of **12** and, again, dethreading of the pseudorotaxane occurs with the concomitant restoration of the fluorescence of free **13**$^{2+}$ (λ_{max} = 428 nm). Since protonation of **12** (presumably at the aliphatic ether oxygens) does not perturb its emission compared to the neutral form, activation of the H^+ input switches on the output at 340 nm. Therefore the output achieves logic state {1} in the two situations in which exclusively one of the two inputs is present. However, when both inputs are applied, acid-base neutralization results; as a consequence the pseudorotaxane remains intact and the 340 nm emission is quenched (output {0}).

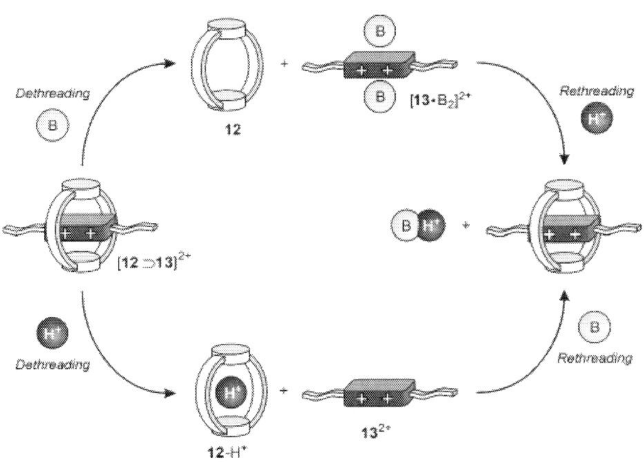

FIGURE 8.23 Working mechanism of the logic system based on compounds **12** and **13**$^{2+}$ whose formulas are shown in Fig. 8.21.

Interestingly, wavelength reconfiguration of the luminescence readout affords double Inhibit (INH) behavior (Table 8.1). If the output is monitored at 428 nm (fluorescence of free **13**$^{2+}$), the system behaves as a INH gate with tris-*n*-butylamine as the disabling input. By monitoring the emission of the CT complex between **13**$^{2+}$ and *n*-Bu_3N (λ_{max} = 666 nm) a INH gate disabled by a proton input is obtained. The superposition of the XOR and two complementary INH functions enables Boolean subtraction of two one-bit digits, x and y, in either order ($x - y$ and $y - x$), thereby mimicking the behavior of a bidirectional half subtractor. This logic feature corresponds to the physical inversion of the input channel that represents the minuend and the subtrahend data in a conventional half subtractor. It should be noted that this molecular ensemble can exhibit such advanced Boolean functionalities because of the high degree of logic integration, which in turn is made possible because of the peculiar – and for some aspects unusual – physicochemical properties of the system.

TABLE 8.1 Truth Table of the Logic Operations Performed by the Chemical Ensemble shown in Fig. 8.23. Experimental conditions are those described in Fig. 8.22.

In_1 (amine)[a]	In_2 (acid)[b]	Out_1 (340 nm)[c, d]	Out_2 (428 nm)[c, d]	Out_3 (666 nm)[c, e]	Out_1' (666 nm)[c, e]
0	0	0 *(23)*	0 *(10)*	0 *(1.0)*	0 *(23)*
1	0	1 *(35)*	0 *(3.0)*	1 *(94)*	1 *(35)*
0	1	1 *(36)*	1 *(62)*	0 *(2.0)*	1 *(36)*
1	1	0 *(27)*	0 *(18)*	0 *(2.0)*	0 *(27)*
		↓	↓	↓	↓
		XOR	**INH(In_1)**	**INH(In_2)**	**OR**

[a]Binary state {1} corresponds to addition of 30 equivalents of *n*-Bu_3N.
[b]Binary state {1} corresponds to addition of 30 equivalents of CF_3SO_3H.
[c]The values in parentheses indicate the experimental fluorescence intensity values at the wavelength indicated, in arbitrary units, upon excitation at 264 nm.
[d]Binary **states** determined by applying a threshold value I_{em} = 30 a.u.
[e]Binary **states** determined by applying a threshold value I_{em} = 25 a.u.

The truth table displayed in Table 8.1 shows that two different input strings, namely {0,0} and {1,1}, produce the same combination of Out_1, Out_2, and Out_3, namely {0,0,0}. Such a behavior, which occurs for most molecular logic gates and circuits reported so far [12, 115, 116], determines a loss of information upon performing the operation as distinct input strings can no longer be distinguished on the basis of the output state [117]. Logic gates that erase information in their operation are said to be irreversible. It has been showed that, because of the increase of the entropic content of the system, irreversible logic computations dissipate $kT\ln 2$ Joules of heat energy for each bit of information lost [118]. In contrast, reversible logic operations do not erase information and therefore they do not generate heat as a result of entropy increase. The study of molecular systems exhibiting logically reversible behavior is interesting for basic science reasons [119] and, in a perspective, also because heat dissipation is a main issue for the construction of ultra-miniaturized information-processing devices [120].

The present system would become logically reversible if a signal could be found that enables to distinguish between the {0,0} (no inputs added) and {1,1} (amine and acid added together in stoichiometric amounts) states. A careful analysis of the luminescence spectra obtained upon sequential addition of acid and base reveals that the signals corresponding to Out_1 (340 nm) and Out_2 (428 nm) do not go back to the initial values upon reset, because the triflate anions generated by input annihilation compete with **12** for the $\mathbf{13}^{2+}$ guest, thus diminishing the apparent stability of the pseudorotaxane [121]. Such an interference hampers the reversibility of the threading-dethreading process but enables to distinguish the {0,0} from the {1,1} state. For example, if a new threshold is applied to Out_1 (see Out_1' in Table 8.1) an OR response can be obtained, complementing the set of the already discussed XOR and INH functions and allowing reversible logic behavior. Therefore, in this case chemical irreversibility of the processes caused by the application of the inputs translates into logic reversibility of the Boolean operations performed [114].

8.5.2 A SIMPLE UNIMOLECULAR MULTIPLEXER-DEMULTIPLEXER

An important function in information technology is signal multiplexing/demultiplexing. A 2:1 multiplexer (MUX) is a circuit with two data inputs, one address input, and one output. The MUX selects the binary state from one of the data inputs and directs it to the output; the selected input depends on the binary state of the address input (Fig. 8.24a). Conversely, a 1:2 demultiplexer (DEMUX) is a circuit that possesses one data input, one address input, and two outputs. The demultiplexer routes the data input to one of the output lines; the selected output is determined by the binary state of the address input (Fig. 8.24b). Hence, a multiplexer allows the encoding of multiple data streams into a single data line for transmission, and a demultiplexer can decode such entangled data streams from the received single signal. The logic circuits corresponding to a 2:1 multiplexer and 1:2 demultiplexer are shown in Figs. 8.24c and 8.24d, respectively.

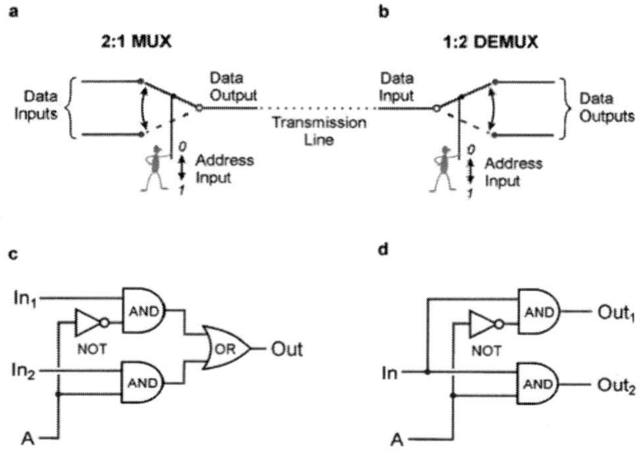

FIGURE 8.24 Operation scheme (top) and equivalent logic circuit (bottom) corresponding to a 2:1 multiplexer (a and c) and a 1:2 demultiplexer (b and d).

Molecules that can mimick the function of a 2:1 multiplexer [122] or a 1:2 demultiplexer [123, 124] have been reported in the past few years. However, these systems either rely on carefully designed multicomponent species [122, 123] and coupling to an external optical device [123], or imply a dependence of the data input on the binary state of the address input [124]. It has been shown that the reversible acid-base switching of the absorption and photoluminescence properties of a fluorophore as simple as 8-methoxyquinoline (**14** in Fig. 8.25) in solution can form the basis for molecular 2:1 multiplexing and 1:2 demultiplexing with a clear-cut digital response [125].

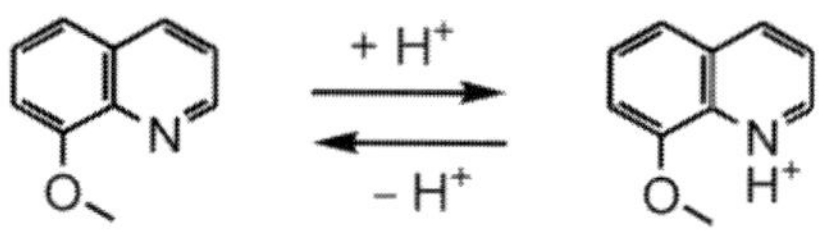

14 **14-H⁺**

FIGURE 8.25 The acid-base-controlled switching between 8-methoxyquinoline **14** and its protonated form **14-H**$^+$.

In CH_3CN solution, **14** shows an absorption band with λ_{max} = 301 nm and an intense fluorescence band with λ_{max} = 388 nm (Fig. 8.26a, b). The addition of one equivalent of triflic acid (CF_3SO_3H) to **14** affords the protonated form **14-H**$^+$ (Fig. 8.25), whose absorption and fluorescence spectra are markedly different from those of **14**. Specifically, the absorption band of **14** (270–320 nm) decreases substantially, and a new band (λ_{max} = 358 nm) is observed in a spectral region where **14** does not absorb. Moreover the fluorescence band of **14** (λ_{max} = 388 nm) is replaced by a weaker emission band with λ_{max} = 500 nm (Fig. 8.26c, d). The nonprotonated form can be regenerated on addition of one equivalent of tris-*n*-butylamine (*n*-Bu_3N) to **14**-H^+. The acid-base-controlled switching between **14** and **14**-H^+ is fully reversible and can be repeated many times on the same solution without appreciable losses in the absorption and fluorescence spectra. This peculiar spectroscopic behavior and the chemical reversibility of the acid-base switching can be used to obtain both 2:1 MUX and 1:2 DEMUX functions, using proton concentration as the address input (A), and excitation and emission optical signals as the data inputs and outputs, respectively.

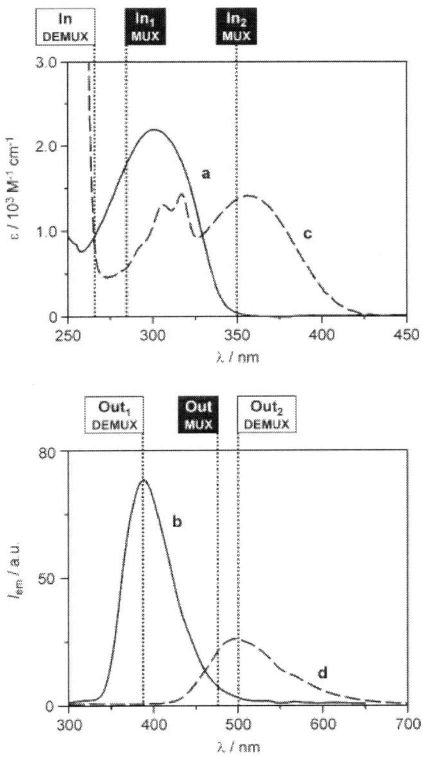

FIGURE 8.26 Absorption (top) and fluorescence (λ_{ex} = 262 nm, bottom) spectra of **14** (a and b) and **14**-H^+ (c and d). The wavelengths of the inputs and outputs signals relevant for the MUX/DEMUX binary functions are indicated. Conditions: CH_3CN solution, 15 μM, room temperature.

In the case of the 2:1 multiplexer, the data inputs are represented by the incident light intensity at 285 nm (In_1) and 350 nm (In_2); these wavelengths are chosen in order to afford selective excitation of **14** and **14**-H^+, respectively (Fig. 8.26a, c). The output (Out) is coded for by the fluorescence intensity at 474 nm, a wavelength at which both **14** and **14**-H^+ fluoresce (Fig. 8.25b, d). The fluorescence output levels measured for a CH_3CN solution of **14** under the conditions corresponding to the eight combinations of the binary data and address inputs are listed in Table 8.2. A threshold value can be easily identified in order to assign binary output states such that the truth table corresponds to that of a 2:1 multiplexer. If the address input is {0} (no H^+ added, **14** is present), the output reports the state of data input In_1, whereas if the address input is {1} (one equivalent of H^+ added, **14**-H^+ is present), the output mirrors the state of data input In_2.

TABLE 8.2 Truth Table of the 2:1 Multiplexer Function based on Compound **14**. Experimental Conditions are those Described in Fig. 8.26.

Data inputs		**Address input**	**Data output**
In_1 (285 nm)[a]	**In_2** (350 nm)[a]	**A** (acid)[b]	**Out** (474 nm)[c]
0	0	0	0 *(0)*
1	0	0	1 *(21)*
0	1	0	0 *(5.1)*
1	1	0	1 *(26)*
0	0	1	0 *(0)*
1	0	1	0 *(9.2)*
0	1	1	1 *(21)*
1	1	1	1 *(30)*

[a]Binary state {1} corresponds to irradiation with the excitation lamp of the spectrofluorimeter; state {0} corresponds to no excitation (lamp off).
[b]Binary state {1} is obtained upon addition of one equivalent of CF_3SO_3H.
[c]The values in parentheses indicate the experimental fluorescence intensity values in arbitrary units; the corresponding binary states are determined by applying a threshold value I_{em} = 15 a.u.

The system can be straightforwardly reconfigured to behave as a 1:2 demultiplexer by changing the optical input and output channels. In this case, the data input (In) is coded for by the incident light intensity at 262 nm, which is an isosbestic point for **14** and **14**-H^+ (Fig. 8.26a, c). The two output signals are represented by the fluorescence intensity at 388 nm (λ_{max} for **14**, Out_1) and

500 nm (λ_{max} for **14**-H$^+$, Out_2), respectively (Fig. 8.26b, d). Table 8.3 shows the fluorescence output levels measured for a CH_3CN solution of **14** under the conditions corresponding to the four combinations of the binary data and address inputs. On fixing an appropriate threshold for the fluorescence output, the truth table of the 1:2 demultiplexer is obtained. If the address input is {0} (**14** is present), the binary state of the data input is transmitted to Out_1; conversely, if the address input is {1} (**14**-H$^+$ is present), the binary data input is transmitted to Out_2.

TABLE 8.3 Truth Table of the 1:2 Demultiplexer Function based on Compound **14**. Experimental Conditions are those Described in Fig. 8.26.

Data input	**Address input**	**Data outputs**	
In (262 nm)[a]	**A** (acid)[b]	**Out$_1$** (388 nm)[c]	**Out$_2$** (500 nm)[c]
0	0	0 *(0)*	0 *(0)*
1	0	1 *(71)*	0 *(0.5)*
0	1	0 *(0)*	0 *(0)*
1	1	0 *(2.1)*	1 *(21)*

[a]Binary state {1} corresponds to irradiation with the excitation lamp of the spectrofluorimeter; state {0} corresponds to no excitation (lamp off).
[b]Binary state {1} is obtained upon addition of one equivalent of CF_3SO_3H.
[c]The values in parentheses indicate the experimental fluorescence intensity values in arbitrary units; the corresponding binary states are determined by applying a threshold value I_{em} = 10 a.u.

Interestingly, the acid-base switching of the absorption and fluorescence bands observed for **14** in solution can also be performed if the molecule is embedded in a polystyrene thin films and exposed to solutions of CF_3COOH and *n*-Bu_3N [125]. More importantly, this study demonstrates that functions achieved by circuits whose logic design requires the interconnection of several basic elements can be implemented with simple molecules [126], taking advantage of logic reconfiguration.

8.5.3 MEMORY EFFECT IN A BISTABLE MECHANICAL MOLECULAR SWITCH

A bistable rotaxanes in which the wheel component can be switched between two different co-conformations in response to external stimuli, constitute an artificial molecular switch. The operation of bistable molecular switches is based on classical switching processes between thermodynamically stable states. It has become clear that the achievement of useful logic functions for the development of molecular memories will only be possible if the rates of the mechanical movement between such states can also be controlled.

Herein, a [2]rotaxane [127] (Fig. 8.27), which operates as a bistable memory element under kinetic control is described. In this mechanically interlocked system, (i) the relative mechanical movements of its wheel and axle components are controlled by redox stimuli so that it can be thermodynamically switched between two states, and (ii) the energy barriers between these two states can be controlled kinetically by photochemical inputs.

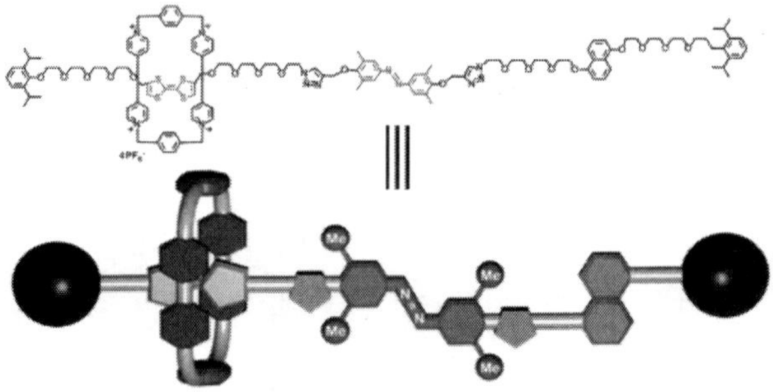

FIGURE 8.27 Structure formula of [2]rotaxane $\mathbf{15}^{4+}$.

The functional units of the [2]rotaxane $\mathbf{15}^{4+}$ are: (i) the π-electron-deficient cyclobis(paraquat-*p*-phenylene) wheel ($CBPQT^{4+}$); (ii) the π-electron-donating recognition sites of the dumbbell component, constituted by a tetrathiafulvalene (TTF) unit and a 1,5-dioxynaphthalene (DNP) unit; and (iii) a photoactive 3,5,3′,5′-tetramethylazobenzene (TMeAB) moiety, located in between the TTF and DNP units, which can be reversibly and efficiently switched between its *trans* and *cis* configurations by photochemical stimuli. Since the TTF unit is more π-electron-rich than the DNP one, the $CBPQT^{4+}$ wheel prefers to reside on the TTF unit rather than on the DNP unit in the ground state co-conformation (GSCC) of $\mathbf{15}^{4+}$. Upon chemical or electrochemical oxidation of the TTF unit to its radical cation ($TTF^{\cdot+}$) form, the $CBPQT^{4+}$ shuttles to the DNP recognition site on account of the Coulombic repulsion caused by the oxidized TTF unit ($TTF^{\cdot+}$ radical cation). Upon reduction of the $TTF^{\cdot+}$ unit to its neutral state, the $CBPQT^{4+}$ wheel resides on the DNP recognition site (metastable state co-conformation, MSCC) for some time before its relaxation to the GSCC is complete. The lifetime of the MSCC can be controlled (Fig. 8.28) by isomerization of the TMeAB unit from its *trans* to *cis* configuration, a process, which brings about a large geometrical change capable of affecting substantially the free-energy barrier for the shuttling of the $CBPQT^{4+}$ along the axle component [128]. The azobenzene unit in its *cis* isomer poses indeed a much larger steric hindrance to the shuttling of the wheel than does a *trans* azobenzene unit.

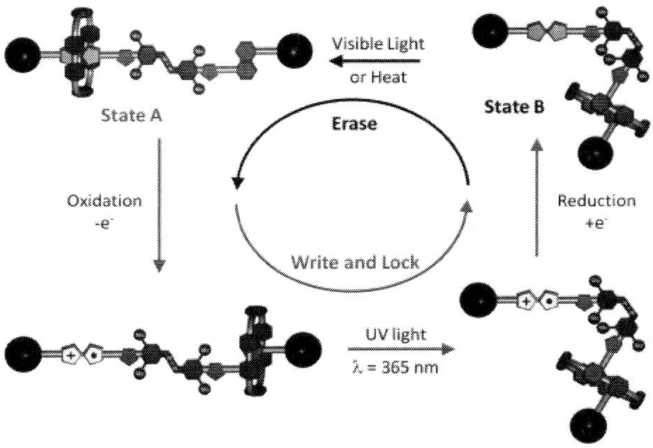

FIGURE 8.28 Chemically and photochemically triggered memory switching cycle of the [2]rotaxane $\mathbf{15}^{4+}$.

The redox- and light-induced switching of $\mathbf{15}^{4+}$ and its corresponding axle **16** were monitored by both steady-state and time-resolved UV/Vis absorption spectroscopies [127]. The results confirmed the possibility of trapping the shuttle in the MSCC by photochemically closing the azobenzene gate when the $CBPQT^{4+}$ resides on the DNP secondary recognition site.

Specifically, the chemical oxidation causes the shuttling of $CBPQT^{4+}$ from the $TTF^{\cdot+}$ to the DNP recognition site, then light irradiation converts the TMeAB unit from the *trans* to the *cis* configuration (gate closed). Subsequent chemical reduction regenerates the neutral TTF unit with the $CBPQT^{4+}$ still residing on the DNP unit and successive thermal *cis* → *trans* isomerization opens the gate and enables the repositioning of the $CBPQT^{4+}$ onto the TTF primary recognition site, thereby affording (Fig. 8.28) full reset of the system.

The results of such a "write-lock-erase" experiment can be summarized as follows (Fig. 8.29): (i) the data can be written on the rotaxane by an oxidation stimulus, and locked by UV light irradiation; (ii) after the writing session, the oxidized species can be reduced back to the original form without losing the written data for a remarkably longer time, compared to most thermodynamically controlled molecular switches, in which the MSCCs have short lifetimes; (iii) the data remain stored for a few hours in the dark at room temperature until the thermal opening of the azobenzene gate occurs. It is also important to note that light irradiation not only locks the data previously recorded by oxidation, but also protects the nonoxidized rotaxanes from accidental writing. These properties have positive implications for the use of such molecules in engineered test devices [129, 130].

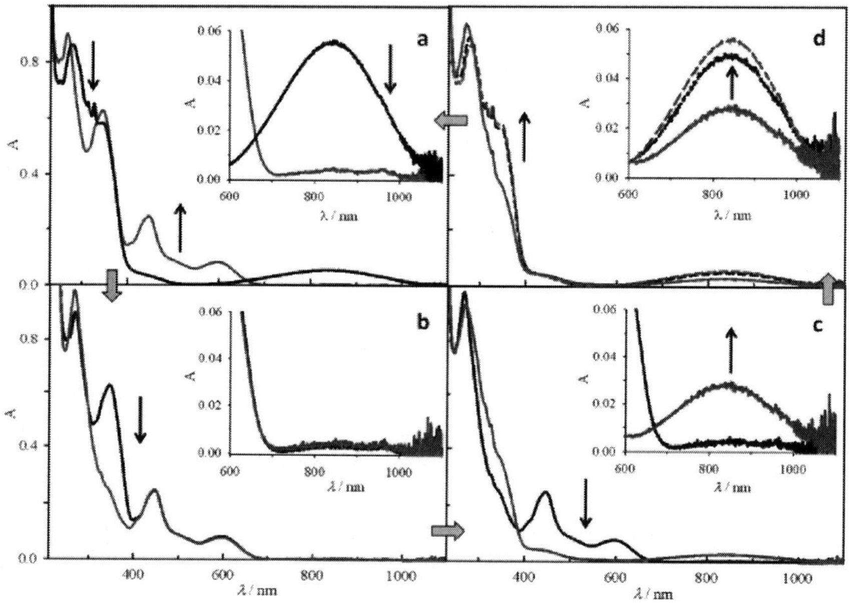

FIGURE 8.29 Changes in the absorption spectrum observed in CH_3CN at 295 K for a 19 μM solution of *trans*-**15**$^{4+}$ resulting from the following sequence of operations: (a) oxidation with up to 1 equivalent of $Fe(ClO_4)_3$ to obtain *trans*-**15**$^{5+}$ (from black to gray traces); (b) exhaustive irradiation at 365 nm to obtain *cis*-**15**$^{5+}$ (from black to gray traces); (c) reduction with 1 equivalent of $Me_{10}Fc$ to obtain *cis*-**15**$^{4+}$ (from black to gray traces); (d) equilibration in the dark to obtain *trans*-**15**$^{4+}$(from gray to dashed black traces). The absorption bands of the $Me_{10}Fc^+$ cation, generated in the reaction between **15**$^{5+}$ and $Me_{10}Fc$, were arithmetically subtracted from the spectra shown in (c) (gray trace) and (d).

8.6 CONCLUSIONS

In this chapter, we described few recent examples developed in our laboratory with the aim to show that one of the most interesting aspects of supramolecular (multicomponent) systems is their interaction with light and that, in the frame of research on supramolecular chemistry and photochemistry, the design and construction of nanoscale devices and machines capable of performing useful light-induced functions can be attempted. The systems here discussed, as the most multicomponent systems developed so far, are studied in solution, that is, they operate in an incoherent fashion and without the control of their spatial position. Although the solution studies are fundamental to understand their operation mechanisms, before such systems can find applications, they have to be interfaced with the macroscopic world. It is indeed necessary that they are ordered and organized to behave coherently and to be addressed in space. Promising possibilities are deposition on surfaces, incorporation into polymers, organization at interfaces, or immobilization into membranes or porous materials. Recent achievements in this direction [131] let one optimistically hope that useful devices and machines based on functional supramolecular systems could be developed in a not too distant future. We foresee for them various applications that range from energy conversion to sensoring and catalysis. Furthermore, as research in this area is progressing, two interesting kinds of nonconventional applications of such systems are also emerging: (i) their behavior can be exploited for processing information at the molecular level and, in the long run, for the construction of chemical computers; (ii) their mechanical features can be utilized to transport nanoobjects, to gate molecular-level channels, and to develop nanorobotics.

Apart from these foreseeable applications related to the development of nanotechnology, researches on photochemical molecular devices and machines are important not only to increase the basic understanding of photoinduced reactions and to develop reliable theoretical models, but also to stimulate the ingenuity of chemists, thereby instilling new ideas and life into the old science of chemistry.

Finally it is important to point out that all the new devices and machines capable of performing useful light-induced functions created by the progress are of the highest importance in current times. It has become clear, indeed, that products and services in the years ahead, including those of a nanotechnology-based industry, will have to exploit renewable energy sources, especially solar energy.

KEYWORDS

- **light harvesting antenna systems**
- **memory effects**
- **molecular devices and machines**
- **photoisomerization**
- **pseudorotaxanes and rotaxanes**
- **supramolecular chemistry**

REFERENCES

1. Hader, D.-P., Tevini, M. *General Photobiology*, Pergamon: Oxford, 1987.
2. Nalwa, H. S. Ed. (2003). *Handbook of Photochemistry and Photobiology*, American Scientific Publishers: Stevenson Ranch, vols. 1–4.
3. Lehn, J.-M. (1988). Supramolecular chemistry − Scope and perspectives molecules, supermolecules, and molecular devices (Nobel Lecture), *Angew. Chem. Int. Ed. Engl.* 27, 89–112.
4. Balzani, V. Ed. *Supramolecular Photochemistry*, Reidel: Dordrecht, 1987.
5. Joachim, C., Launay, J. P. (1984). Sur la possibilité d'un traitement moléculaire *Nouv. J. Chem.*, 8, 723–728.
6. Balzani, V., Moggi, L., Scandola, F. (1987). Towards a supramolecular photochemistry: assembly of molecular components to obtain photochemical molecular devices in *Supramolecular Photochemistry*, Balzani, V. Ed.; Reidel: Dordrecht, pp. 1–28.
7. Mann, S. (2008). Life as a nanoscale phenomenon *Angew. Chem. Int. Ed.* 47, 5306–5320.
8. Jones, R. A. L. *Soft Machines, Nanotechnology and Life*, Oxford University Press: New York, 2004.
9. For recent outstanding examples see: (a) Ye, A., Kumar, A. S., Saha, S., Takami, T., Huang, T. J., Stoddart, J. F., Weiss, P. S. (2010). Changing stations in single bistable rotaxane molecules under electrochemical control *ACS Nano* 4, 3697–3701; (b) Lussis, P., Svaldo-Lanero, T., Bertocco, A., Fustin, C.-A.; Leigh, D. A., Duwez, A.-S. A single synthetic small molecule that generates force against a load *Nat. Nanotech.* (2011). 6, 553–557; (c) Kudernac, T., Ruangsupapichat, N., Parschau, M., Macia, B., Katsonis, N., Harutyunyan, S. R., Ernst, K.-H.; Feringa, B. L. (2011). Electrically driven directional motion of a four-wheeled molecule on a metal surface *Nature* 479, 208–211.
10. Pease, A. R., Jeppesen, J. O., Stoddart, J. F., Luo, Y., Collier, C. P., Heath, J. R. (2001). Switching devices based on interlocked molecules *Acc. Chem. Res.* 34, 433–444.
11. Kay, E. R., Leigh, D. A., Zerbetto, F. (2007). Synthetic molecular motors and mechanical machines *Angew. Chem. Int. Ed.* 46, 72–191.
12. Balzani, V., Credi, A., Venturi, M. *Molecular Devices and Machines – Concepts and Perspectives for the Nano World*, Wiley-VCH: Weinheim, 2008.

13. (a) van Delden, R. A., Koumura, N., Harada, N., Feringa, B. L. (2002). Unidirectional rotary motion in a liquid crystalline environment: color tuning by a molecular motor *Proc. Natl. Acad. Sci.* 99, 4945–4949; (b) Pijper, D. L., Feringa, B. L. Molecular transmission: controlling the twist sense of a helical polymer with a single light-driven molecular motor *Angew. Chem. Int. Ed.* (2007). 46, 3693–3696; (c) Fang, L., Hmadeh, M., Wu, J., Olson, M. A., Spruell, J. M., Trabolsi, A., Yang, Y.-W.; Elhabiri, M., Albrecht-Gary, A. M., Stoddart, J. F. Acid–base actuation of [*c*2]daisy chains, *J. Am. Chem. Soc.* (2009). 131, 7126–7134; (d) Deng, H., Olson, M. A., Stoddart, J. F., Yaghi, O. M. Robust dynamics *Nat. Chem.* (2010). 2, 439–443; (e) Mercer, D. J., Vukotic, V. N. S., Loeb, J. (2011). Linking [2]rotaxane wheels to create a new type of metal organic rotaxane framework *Chem. Commun.* 47, 896–898.
14. Berna, J., Leigh, D. A., Lubomska, M., Mendoza, S. M., Perez, E. M., Rudolf, P., Teobaldi, G., Zerbetto, F. (2005). Macroscopic transport by synthetic molecular machines *Nat. Mater.* 4, 704–710; (b) Harada, A., Kobayashi, R., Takashima, Y., Hashidzume, A., Yamaguchi, H. . (2011). Macroscopic self-assembly through molecular recognition *Nat. Chem*3, 34–37.
15. Wang, J. B., Feringa, B. L. (2011). Dynamic control of chiral space in a catalytic asymmetric reaction using a molecular motor *Science* 331, 1429–1432.
16. J Green, J. E., Choi, J. W., Boukai, A., Bunimovich, Y., Johnston-Halperin, E., DeIonno, E., Luo, Y., Sheriff, B. A., Xu, K., Shin, Y. S., Tseng, H.-R.; Stoddart, J. F., Heath, J. R. (2007). A 160-kilobit molecular electronic memory patterned at (10^{11}). bits per square centimetre *Nature* 445, 414–417.
17. Ambrogio, M. W., Thomas, C. R., Zhao, Y.-L.; Zink, J. I., Stoddart, J. F. (2011). Mechanized silica nanoparticles: a new frontier in theranostic nanomedicine *Acc. Chem. Res.* 44, 903–913.
18. von Delius, M., Leigh, D. A. Walking molecules *Chem. Soc. Rev.* (2011). 40, 3656–3676; (b) Coskun, A., Banaszak, M., Astumian, R. D., Stoddart, J. F. Great expectations: can artificial molecular machines deliver on their promise? *Chem. Soc. Rev.* (2012). 41, 19–30.
19. Haberhauer, G., Kallweit, C. A bridged azobenzene derivative as a reversible, light-induced chirality switch *Angew. Chem. Int. Ed.* (2010). 49, 2418–2421; (b) Kelly, T. R., De Silva, H., Silva, R. A. Unidirectional rotary motion in a molecular system *Nature* (1999). 401, 150–152; (c) Hernández, J. V., Kay, E. R., Leigh, D. A. A reversible synthetic rotary molecular motor *Science* (2004). 306, 1532–1537; (d) Canary, J. W. (2009). Redox-triggered chiroptical molecular switches *Chem. Soc. Rev.* 38, 747–756.
20. Sherman, W. B., Seeman, N. C. (2004). A Precisely controlled DNA biped walking device *Nano Lett.* 4, 1203–1207; (b) Simmel, F. C. Processive motion of bipedal DNA walkers *ChemPhysChem* (2009). 10, 2593–2597; (c) Bath, J., Turberfield, A. J. DNA nanomachines *Nat. Nanotechnol.* (2007). 2, 275–284; (d) Special issue on DNA-based nanoarchitectures and nanomachines *Org. Biomol. Chem.* (2006). 4, 3369–3540.
21. von Delius, M., Geertsema, E. M., Leigh, D. A. (2010). A synthetic small molecule that can walk down a track *Nat. Chem.* 2, 96–101.
22. Schliwa, M. Ed. *Molecular Motors*, Wiley-VCH: Weinheim, 2002.
23. Sauvage, J.-P.; Dietrich-Buchecker, C. Eds. *Molecular Catenanes, Rotaxanes and Knots: A Journey Through the World of Molecular Topology*, Wiley-VCH: Weinheim, 1999.

24. Amabilino, D. B., Stoddart, J. F. (1995). Interlocked and intertwined structures and superstructures *Chem. Rev.* 95, 2725–2828.
25. Browne, W. R., Feringa, B. L. (2006). Making molecular machines work *Nature Nanotech.* 1, 25–35.
26. Astumian, R. D. (2007). Design principles for Brownian molecular machines: how to swim in molasses and walk in a hurricane *Phys. Chem. Chem. Phys.* 9, 5067–5083.
27. Chatterjee, M. N., Kay, E. R., Leigh, D. A. (2006). Beyond switches: ratcheting a particle energetically uphill with a compartmentalized molecular machine, *J. Am. Chem. Soc.* 128, 4058–4073; (b) Serreli, V., Lee, C.-F.; Kay, E. R., Leigh, D. A. **A molecular information ratchet** *Nature* (2007). 445, 523–527; (c) Alvarez-Pérez, M., Goldup, S. M., Leigh, D. A., Slawin, A. M. Z. (2008). A chemically-driven molecular information ratchet, *J. Am. Chem. Soc.* 130, 1836–1838.
28. Baroncini, M., Silvi, S., Venturi, M., Credi, A. (2012). Photoactivated directionally controlled transit of a non-symmetric molecular axle through a macrocycle *Angew. Chem. Int. Ed.* 51, 4223–4226.
29. Asakawa, M., Ashton, P. R., Ballardini, R., Balzani, V., Belohradsky, M., Gandolfi, M. T., Kocian, O., Prodi, L., Raymo, F. M., Stoddart, J. F., Venturi, M. (1997). The slipping approach to self-assembling [*n*]rotaxanes, *J. Am. Chem. Soc.* 119, 302–310.
30. Baroncini, M., Silvi, S., Venturi, M., Credi, A. (2010). Reversible photoswitching of rotaxane character and interplay of thermodynamic stability and kinetic lability in a self-assembling ring–axle molecular system *Chem. Eur. J.* 16, 11580–11587.
31. For other pseudorotaxane-type systems in which the threading-dethreading kinetics can be photocontrolled, see: (a) Hirose, K., Shiba, Y., Ishibashi, K., Doi, Y., Tobe, Y. (2008). An anthracene-based photochromic macrocycle as a key ring component to switch a frequency of threading motion *Chem. Eur. J.* 14, 981–986; (b) Tokunaga, Y., Akasaka, K., Hashimoto, N., Yamanaka, S., Hisada, K., Shimomura, Y., Kakuchi, S. Using photoresponsive end-closing and end-opening reactions for the synthesis and disassembly of [2]rotaxanes: implications for dynamic covalent chemistry, *J. Org. Chem.* (2009). 74, 2374–2379; (c) Ogoshi, T., Yamafuji, D., Aoki, T., Yamagishi, T. (2011). Photoreversible transformation between seconds and hours time-scales: threading of pillar[5]arene onto the azobenzene-end of a viologen derivative, *J. Org. Chem.* 76, 9497–9503.
32. Ashton, P. R., Campbell, P. J., Chrystal, E. J. T., Glink, P. T., Menzer, S., Philp, D., Spencer, N., Stoddart, J. F., Tasker, P. A., Williams, D. J. (1995). Dialkylammonium ion/crown ether complexes: the forerunners of a new family of interlocked molecules *Angew. Chem. Int. Ed.* 34, 1865–1869.
33. Takeda, Y., Kudo, Y., Fujiwara, S. (1985). Thermodynamic study for complexation reactions of dibenzo-24-crown-8 with alkali metal ions in acetonitrile *Bull. Chem. Soc. Jpn.* 58, 1315–1316.
34. Arduini, A., Credi, A., Faimani, G., Massera, C., Pochini, A., Secchi, A., Semeraro, M., Silvi, S., Ugozzoli, F. (2008). Self-assembly of a double calix[6]arene pseudorotaxane in oriented channels *Chem. Eur. J.* 14, 98–106.
35. Arduini, A., Bussolati, R., Credi, A., Faimani, G., Garaudee, S., Pochini, A., Secchi, A., Semeraro, M., Silvi, S., Venturi, M. (2009). Towards controlling the threading direction of a calix[6]arene wheel by using nonsymmetric axles *Chem. Eur. J.* 15, 3230–3242.

36. Credi, A., Dumas, S., Silvi, S., Venturi, M., Arduini, A., Pochini, A., Secchi, A. (2004). Viologen-calix[6]arene pseudorotaxanes. Ion-pair recognition and threading/dethreading molecular motions, *J. Org. Chem.* 69, 5881–5887.
37. Arduini, A., Ciesa, F., Fragassi, M., Pochini, A., Secchi, A. (2005). Selective synthesis of two constitutionally isomeric oriented calix[6]arene-based rotaxanes *Angew. Chem. Int. Ed.* 44, 278–281.
38. van Duynhoven, J. P. M., Janssen, R. G., Verboom, W., Franken, S. M., Casnati, A., Pochini, A., Ungaro, R., De Mendoza, J., Nieto, P. M., Prados, P., Reinhoudt, D. N. (1994). Control of calix[6]arene conformations by self-inclusion of 1,3,5-tri-O-alkyl substituents: synthesis and NMR studies, *J. Am. Chem. Soc.* 116, 5814–5822.
39. Arduini, A., Bussolati, R., Credi, A., Monaco, S., Secchi, A., Silvi, S., Venturi, M. (2012). Solvent- and light-controlled unidirectional transit of a nonsymmetric molecular axle through a nonsymmetric molecular wheel *Chem. Eur. J.* 18, 16203–16213.
40. Ashton, P. R., Baxter, I., Fyfe, M. C. T., Raymo, F. M., Spencer, N., Stoddart, J. F., White, A. J. P., Williams, D. J. (1998). Rotaxane or pseudorotaxane? That is the question! *J. Am. Chem. Soc.* 120, 2297–2307.
41. Affeld, A., Hubner, G. M., Seel, C., Schalley, C. A. (2001). Rotaxane or pseudorotaxane? Effects of small structural variations on the deslipping kinetics of rotaxanes with stopper groups of intermediate size *Eur. J. Org. Chem.* 15, 2877–2890.
42. Newkome, G. R., Vögtle, F. *Dendrimers and Dendrons*; Wiley-VCH: Weinheim, 2001; (b) *Dendrimers and Other Dendritic Polymers*; Fréchet, J. M. J., Tomalia, D. A. Eds.; John Wiley & Sons: Chichester, 2001; (c) Vögtle, F., Richardt, G., Werner, N. *Dendrimer Chemistry*; Wiley-VCH: Weinheim, 2009.
43. For some recent reviews, see: (a) Gregory, F., Kakkar, A. K. (2010). "Click" methodologies: efficient, simple and greener routes to design dendrimers *Chem. Soc. Rev.* 39, 1536–1544; (b) Wilms, D., Stiriba, S. E., Frey, H. Hyperbranched polyglycerols: from the controlled synthesis of biocompatible polyether polyols to multipurpose applications *Acc. Chem. Res.* (2010). 43,129–141; (c) Astruc, D., Boisselier, E., Ornelas, C. Dendrimers designed for functions: from physical, photophysical, and supramolecular properties to applications in sensing, catalysis, molecular electronics, photonics, and nanomedicine *Chem. Rev.* (2010). 110, 1857–1959; (d) Jang, W. D., Selim, K. M. K; Lee, C. H., Kang, I. K. Bioinspired application of dendrimers: from bio-mimicry to biomedical applications *Progr. Polym. Sci.* (2009). 34, 1–23; (e) Astruc, D., Ornelas, C., Ruiz, J. Metallocenyl dendrimers and their applications in molecular electronics, sensing, and catalysis *Acc. Chem. Res.* (2008). 41, 841–856; (f) Special Issue on Dendrimers, In *New, J. Chem.*; Majoral, J. P. Ed.; 2007; vol. 31; (g) Gingras, M., Raimundo, J. M., Chabre, Y. M. (2007). Cleavable dendrimers *Angew. Chem. Int. Ed.* 46, 1010–1017.
44. Deloncle, R., Caminade, A.-M. (2010). Stimuli-responsive dendritic structures: the case of light-driven azobenzene-containing dendrimers and dendrons, *J. Photochem. Photobiol. C* 11, 25–45.
45. For examples of dendrimers with azobenzene units, see, (a) Nguyen, T.-T.-T.; Turp, D., Wang, D., Nolscher, B., Laquai, F., Müllen, K. (2011). A fluorescent, shape-persistent dendritic host with photoswitchable guest encapsulation and intramolecular energy transfer, *J. Am. Chem. Soc.* 133, 11194–11204; (b) del Barrio, J., Tejedor, R. M., Chinelatto, L. S., Sanchez, C., Pinol, M., Oriol, L. Photocontrol of the supramolecular

chirality imposed by stereocenters in liquid crystalline azodendrimers *Chem. Mater.* (2010). 22, 1714–1723; (c) Li, Z., Wu, W., Li, Q., Yu, G., Xiao, L., Liu, Y., Ye, C., Qin, J., Li, Z. High-generation second-order nonlinear optical (NLO) dendrimers: convenient synthesis by click chemistry and the increasing trend of NLO effects *Angew. Chem. Int. Ed.* (2010). 49, 2763–2767; (d) Puntoriero, F., Ceroni, P., Balzani, V., Bergamini, G., Vögtle, F. (2007). Photoswitchable dendritic hosts: a dendrimer with peripheral azobenzene groups, *J. Am. Chem. Soc.* 129, 10714–10719.

46. Restani, R. B., Morgado, P. I., Ribeiro, M. P., Correia, I. J., Aguiar-Ricardo, A., Bonifácio, V. D. B. (2012). Biocompatible polyurea dendrimers with pH-dependent fluorescence *Angew. Chem. Int. Ed.* 51, 5162–5165; (b) Bozdemir, O. A., Erbas-Cakmak, S., Ekiz, O. O., Dana, A., Akkaya, E. U. Towards unimolecular luminescent solar concentrators: Bodipy-based dendritic energy-transfer cascade with panchromatic absorption and monochromatized emission *Angew. Chem. Int. Ed.* (2011). 50, 10907–10912; (c) Uetomo, A., Kozaki, M., Suzuki, S., Yamanaka, K., Ito, O., Okada, K. Efficient light-harvesting antenna with a multi-porphyrin cascade, *J. Am. Chem. Soc.* (2011). 133, 13276–13279; (d) Bergamini, G., Ceroni, P., Fabbrizi, P., Cicchi, S. A multichromophoric dendrimer: from synthesis to energy up-conversion in a rigid matrix *Chem. Commun.* (2011). 47, 12780–12782; (e) Kuroda, D. G., Singh, C. P., Peng, Z., Kleiman, V. D. (2009). Mapping excited-state dynamics by coherent control of a dendrimer's photoemission efficiency *Science* 326, 263–267.

47. Shellaiah, M., Rajan, Y. C., Lin, H.-C. (2012). Synthesis of novel triarylamine-based dendrimers with N^4,N^6-dibutyl-1,3,5-triazine-4,6-diamine probe for electron/energy transfers in H-bonded donor–acceptor–donor triads and as efficient Cu^{2+} sensors, *J. Mater. Chem.* 22, 8976–8987; (b) Yang, J., Lee, S., Lee, H., Lee, J., Kim, H. K., Lee, S. U., Sohn, D. (2011). Metal ion coordination with an asymmetric fan-shaped dendrimer at the air–water interface *Langmuir* 27, 8898–8904; (c) Ochi, Y., Fujii, A., Nakajima, R., Yamamoto, K. (2010). Stepwise radial complexation of triphenylmethyliums on a phenylazomethine dendrimer for organic−metal hybrid assembly *Macromolecules* 43, 6570–6576; (d) Branchi, B., Ceroni, P., Balzani, V., Klaerner, F.-G.; Vögtle, F. (2010). A light-harvesting antenna resulting from the self-assembly of five luminescent components: a dendrimer, two clips, and two lanthanide ions *Chem. Eur. J.* 16, 6048–6055.

48. For cyclam dendrimers, see: Bergamini, G., Marchi, E., Ceroni, P. (2011). Metal ion complexes of cyclam-cored dendrimers for molecular photonics *Coord. Chem. Rev.* 255, 2458–2468.

49. Hwang, S.-H.; Shreiner, C. D., Moorefield, C. N., Newkome, G. R. (2007). Recent progress and applications for metallodendrimers *New, J. Chem.* 31, 1192–1217; (b) Ceroni, P., Bergamini, G., Marchioni, F., Balzani, V. (2005). Luminescence as a tool to investigate dendrimer properties *Prog. Polym. Sci.* 30, 453–473.

50. Wöll, D., Uji-I, H.; Schnitzler, T., Hotta, J., Dedecker, P., Herrmann, A., De Schryver, F. C., Müllen, K., Hofkens, J. (2008). Radical polymerization tracked by single molecule spectroscopy *Angew. Chem. Int. Ed.* 47, 783–787; (b) Uji-i, H., Melnikov, S. M., Deres, A., Bergamini, G., De Schryver, F. C., Herrmann, A., Müllen, K., Enderlein, J., Hofkens, J. Visualizing spatial and temporal heterogeneity of single molecule rotational diffusion in a glassy polymer by defocused wide-field imaging *Polymer* (2006). 47, 2511–2518; (c) De Schryver, F. C., Vosch, T., Cotlet, M., Van der Auweraer, M.,

Müllen, K., Hofkens, J. (2005). Energy dissipation in multichromophoric single dendrimers *Acc. Chem. Res.* 38, 514–522.
51. Li, W. S., Aida, T. (2009). Dendrimer porphyrins and phthalocyanines *Chem. Rev.* 109, 6047–6076; (b) Kaifer, A. E. (2007). Electron transfer and molecular recognition in metallocene-containing dendrimers *Eur. J. Inorg. Chem.* 5015–5027.
52. Puntoriero, F., Bergamini, G., Ceroni, P., Balzani, V., Vögtle, F. (2008). A fluorescent guest encapsulated by a photoreactive azobenzene dendrimer *New, J. Chem.* 32, 401–406; (b) Ramakrishna, G., Bhaskar, A., Bauerle, P., Goodson III, T. Oligothiophene dendrimers as new building blocks for optical applications, *J. Phys. Chem. A* (2008). 112, 2018–2026; (c) Cao, D., Dobis, S., Gao, C., Hillmann, S., Meier, H. (2007). Optical switching and antenna effect of dendrimers with an anthracene core *Chem. Eur. J.* 13,9317–9323.
53. Balzani, V., Bergamini. G.; Ceroni, P., Marchi, E. (2011). Designing light harvesting antennas by luminescent dendrimers *New, J. Chem.* 35, 1944–1954.
54. Balzani, V., Ceroni, P., Maestri, M., Vicinelli, V. (2003). Light-harvesting dendrimers *Curr. Opin. Chem. Biol.* 7, 657–665.
55. Balzani, V., Credi, A., Venturi, M. (2008). *Molecular Devices and Machines – Concepts and Perspectives for the Nano World*, Wiley-VCH: Weinheim, pp. 132–173.
56. Lukeš, I., Kotek, J., Vojtíšek, P., Hermann, P. (2001). Complexes of tetraazacycles bearing methylphosphinic/phosphonic acid pendant arms with copper(II), zinc(II) and lanthanides(III). A comparison with their acetic acid analogues *Coord. Chem. Rev.* 216–217, 287–312; (b) Fabbrizzi, L., Licchelli, M., Pallavicini, P., Sacchi, D. (2001). Supramolecular functions related to the redox activity of transition metals *Supramol. Chem.* 13, 569–582.
57. Liang, X., Sadler, P. J. (2004). Cyclam complexes and their applications in medicine *Chem. Soc. Rev.* 33, 246–266.
58. Bergamini, G., Ceroni, P., Maestri, M., Lee, S.-K.; van Heyst, J., Vögtle, F. (2007). Cyclam cored luminescent dendrimers as ligands for Co(II), Ni(II) and Cu(II) ions *Inorg. Chim. Acta* 360, 1043–1051; (b) Saudan, C., Ceroni, P., Vicinelli, V., Maestri, M., Balzani, V., Gorka, M., Lee, S.-K.; van Heyst, J., Vögtle, F. Cyclam-based dendrimers as ligands for lanthanide ions *Dalton Trans.* (2004). 1597–1600; (c) Saudan, C., Balzani, V., Gorka, M., Lee, S.-K.; van Heyst, J., Maestri, M., Ceroni, P., Vicinelli, V., Vögtle, F. Dendrimers as ligands: an investigation into the stability and kinetics of Zn^{2+} complexation by dendrimers with 1,4,8,11-tetraazacyclotetradecane (cyclam) core *Chem. Eur. J.* (2004). 10, 899–905; (d) Saudan, C., Balzani, V., Gorka, M., Lee, S.-K.; Maestri, M., Vicinelli, V. Dendrimers as ligands. Formation of a 2:1 luminescent complex between a dendrimer with a 1,4,8,11-tetraazacyclotetradecane (cyclam) core and Zn^{2+} *J. Am. Chem. Soc.* (2003). 125, 4424–4425; (e) Saudan, C., Balzani, V., Ceroni, P., Gorka, M., Maestri, M., Vicinelli, V., Vögtle, F. (2003). Dendrimers with a cyclam core. Absorption spectra, multiple luminescence, and effect of protonation *Tetrahedron* 59, 3845–3852.
59. Beeby, A., Parker, D., Williams, J. A. G. (1996). Photochemical investigations of functionalised 1,4,7,10-tetraazacyclododecane ligands incorporating naphthyl chromophores, *J. Chem. Soc., Perkin Trans. 2,* 1565–1579; (b) Tung, C.-H.; Wu, L.-Z. Enhancement of intramolecular excimer formation, photodimerization and energy transfer of naphthalene end-labelled poly(ethyleneglycol) oligomers via complexa-

tion of alkali-metal and lanthanide cations, *J. Chem. Soc., Faraday Trans.* (1996). 92, 1381–1385; (c) Parker, D., Williams, J. A. G. (1995). Luminescence behaviour of cadmium, lead, zinc, copper, nickel and lanthanide complexes of octadentate macrocyclic ligands bearing naphthyl chromophores, *J. Chem. Soc. Perkin Trans. 2,* 1305–1314.

60. Giansante, C., Ceroni, P., Balzani, V., Vögtle, F. (2008). Self-assembly of a light-harvesting antenna formed by a dendrimer, a Ru^{II} complex, and a Nd^{III} ion *Angew. Chem. Int. Ed.* 47, 5422–5425.
61. Lazarides, T., Easun, T. L., Veyne-Marti, C., Alsindi, W. Z., George, M. W., Deppermann, N., Hunter, C. A., Adams, H., Ward, M. D. (2007). Structural and photophysical properties of adducts of $[Ru(bipy)(CN)_4]^{2-}$ with different metal cations: metallochromism and its use in switching photoinduced energy transfer, *J. Am. Chem. Soc.* 129, 4014–4027; (b) Bernhardt, P. V., Bozoglian, F., Font-Bardia, M., Martinez, M., Meacham, A. P., Sienra, B., Solans, X. The influence of ligand substitution at the electron donor center in molecular cyano-bridged mixed-valent Co^{III}/Fe^{II} and Co^{III}/Ru^{II} complexes *Eur. J. Inorg. Chem.* (2007). 5270–5276; (c) Ward, M. D. $[Ru(bipy)(CN)_4]^{2-}$ and its derivatives: photophysical properties and its use in photoactive supramolecular assemblies *Coord. Chem. Rev.* (2006). 250, 3128–3141; (d) Balzani, V., Sabbatini, N., Scandola, F. (1986).“Second-sphere” photochemistry and photophysics of coordination compounds *Chem. Rev.* 86, 319–337.
62. Marchi, E., Baroncini, M., Bergamini, G., Van Heyst, J., Vögtle, F., Ceroni, P. (2012). Photoswitchable metal coordinating tweezers operated by light-harvesting dendrimers, *J. Am. Chem. Soc.* 134, 15277–15280.
63. For a recent review, see: Bandara, H. M. D., Burdette, S. C. (2012). Photoisomerization in different classes of azobenzene *Chem. Soc. Rev.* 41, 1809–1825.
64. Bergamini, G., Ceroni, P., Balzani, V., Cornelissen, L., van Heyst, J., Lee, S.-K.; Vögtle, F. (2005). Dendrimers based on a bis-cyclam core as fluorescence sensors for metal ions, *J. Mater. Chem.* 15, 2959–2964.
65. Prodi, L., Bolletta, F., Montalti, M., Zaccheroni, N. (2000). Luminescent chemosensors for transition metal ions *Coord. Chem. Rev.* 205, 59–83.
66. Balzani, V., Credi, A., Venturi, M. (2008). Processing energy and signals by molecular and supramolecular systems *Chem. Eur. J.* 14, 26–39.
67. de Silva, A. P., Fox, D. B., Huxley, A. J. M., Moody, T. S. (2000). Combining luminescence, coordination and electron transfer for signalling purposes *Coord. Chem. Rev.* 205, 41–57
68. Valeur, B., Leray, I. (2000). Design principles of fluorescent molecular sensors for cation recognition *Coord. Chem. Rev.* 205, 3–40.
69. Fabbrizzi, L., Licchelli, M., Rabaioli, G., Taglietti, A. (2000). The design of luminescent sensors for anions and ionisable analytes *Coord. Chem. Rev.* 205, 85–108.
70. Parker, D. (2000). Luminescent lanthanide sensors for pH, pO_2 and selected anions *Coord. Chem. Rev.* 205, 109–130.
71. Mancin, F., Rampazzo, E., Tecilla, P., Tonellato, U. (2006). Self-assembled fluorescent chemosensors *Chem. Eur. J.* 12, 1844–1854.
72. Brus, L. (1986). Electronic wave functions in semiconductor clusters: experiment and theory, *J. Phys. Chem.* 90, 2555–2560; (b) Alivisatos, A. P. (1996). Semiconductor clusters, nanocrystals, and quantum dots *Science* 271, 933–937; (c) Murray, C. B., Kagan, C. R., Bawendi, M. G. (2000). Synthesis and characterization of monodisperse

nanocrystals and close-packed nanocrystal assemblies *Annu. Rev. Mater. Sci.* 30, 545–610; (d) Talapin, D. V., Lee, J.-S.; Kovalenko, M. V., Shevchenko, E. V. (2010). Prospects of colloidal nanocrystals for electronic and optoelectronic applications *Chem. Rev.* 110, 389–458.

73. Murray, C. B., Noms, D. J., Bawendi, M. G. (1993). Synthesis and characterization of nearly monodisperse CdE (E=S, Se, Te) semiconductor nanocrystallites, *J. Am. Chem. Soc.* 115, 8706–8715; (b) Micic, O. I., Curtis, C. J., Jones, K. M., Sprague, J. R., Nozik, A. J. (1994). Synthesis and characterization of InP quantum dots, *J. Phys. Chem.* 98, 4966–4969; (c) Guzelian, A. A., Banin, U., Kadavanich, A. V., Peng, X., Alivisatos, A. P. (1996). Colloidal chemical synthesis and characterization of InAs nanocrystal quantum dots *Appl. Phys. Lett.* 69, 1432–1434; (d) Peng, Z. A., Peng, X. (2001). Formation of high-quality CdTe, CdSe, and CdS nanocrystals using CdO as precursor, *J. Am. Chem. Soc.* 123, 183–184.
74. Hines, M. A., Guyot-Sionnest, P. (1996). Synthesis and characterization of strongly luminescing ZnS-capped CdSe nanocrystals, *J. Phys. Chem.* 100, 468–471.
75. Dabbousi, B. O., Mikulec, F. V; Heine, J. R., Mattoussi, H., Ober, R., Jensen, K. F., Bawendi, M. G. (1997). (CdSe)ZnS core-shell quantum dots: synthesis and characterization of a size series of highly luminescent nanocrystallites, *J. Phys. Chem. B* 101, 9463–9475.
76. Jaiswal, J. K., Mattoussi, H., Mauro, J. M., Simon, S. M. (2003). Long-term multiple color imaging of live cells using quantum dot bioconjugates *Nat. Biotechnol.* 21, 47–51; (b) Chen, Y., Vela, J., Htoon, H., Casson, J. L., Werder, D. J., Bussian, D. a; Klimov, V. I., Hollingsworth, J. A. (2008). "Giant" multishell CdSe nanocrystal quantum dots with suppressed blinking, *J. Am. Chem. Soc.* 130, 5026–5027; (c) Hollingsworth, J. A. (2013). Heterostructuring nanocrystal quantum dots toward intentional suppression of blinking and Auger recombination *Chem. Mater.* 25, 1153–1154.
77. Raymo, F. M., Yildiz, I. (2007). Luminescent chemosensors based on semiconductor quantum dots *PhysChemChemPhys* 9, 2036–2043.
78. Kamat, P. V; Tvrdy, K., Baker, D. R., Radich, J. G. (2010). Beyond photovoltaics: semiconductor nanoarchitectures for liquid-junction solar cells *Chem. Rev.* 110, 6664–6688.
79. Shirasaki, Y., Supran, G. J., Bawendi, M. G., Bulović, V. (2013). Emergence of colloidal quantum-dot light-emitting technologies *Nat. Photonics* 7, 13–23.
80. Alivisatos, P. (2004). The use of nanocrystals in biological detection *Nat. Biotechnol.* 22, 47–52; (b) Medintz, I. L., Uyeda, H. T., Goldman, E. R., Mattoussi, H. (2005). Quantum dot bioconjugates for imaging, labelling and sensing *Nat. Mater.* 4, 435–446; (c) Michalet, X., Pinaud, F. F., Bentolila, L. a; Tsay, J. M., Doose, S., Li, J. J., Sundaresan, G., Wu, a, M., Gambhir, S. S., Weiss, S. (2005). Quantum dots for live cells, in vivo imaging, and diagnostics *Science* 307, 538–544.
81. Medintz, I. L., Mattoussi, H. (2009). Quantum dot-based resonance energy transfer and its growing application in biology *PhysChemChemPhys* 11, 17–45.
82. Credi, A. (2012). Quantum dot–molecule hybrids: a paradigm for light-responsive nanodevices *New, J. Chem.* 36, 1925–1930.
83. El-Sayed, M. A. (2004). Small is different: shape-, size-, and composition-dependent properties of some colloidal semiconductor nanocrystals *Acc. Chem. Res.* 37, 326–333.

84. Burda, C., Chen, X., Narayanan, R., El-Sayed, M. A. (2005). Chemistry and properties of nanocrystals of different shapes *Chem. Rev.* 105, 1025–1102.
85. Sandros, M. G., Gao, D., Benson, D. E. (2005). A modular nanoparticle-based system for reagentless small molecule biosensing, *J. Am. Chem. Soc.* 127, 12198–12199.
86. Kagan, C., Murray, C., Nirmal, M., Bawendi, M. (1996). Electronic energy transfer in CdSe quantum dot solids *Phys. Rev. Lett.* 76, 1517–1520.
87. Medintz, I. L., Clapp, A. R., Mattoussi, H., Goldman, E. R., Fisher, B., Mauro, J. M. (2003). Self-assembled nanoscale biosensors based on quantum dot FRET donors *Nat. Mater.* 2, 630–638.
88. Snee, P. T., Somers, R. C., Nair, G., Zimmer, J. P., Bawendi, M. G., Nocera, D. G. (2006). A ratiometric CdSe/ZnS nanocrystal pH sensor, *J. Am. Chem. Soc.* 128, 13320–13321.
89. Guliyev, R., Coskun, A., Akkaya, E. U. (2009). Design strategies for ratiometric chemosensors: modulation of excitation energy transfer at the energy donor site, *J. Am. Chem. Soc.* 131, 9007–9013.
90. Stetter, J. R., Li, J. (2008). Amperometric gas sensors - A review *Chem. Rev.* 108, 352–366.
91. Mills, A. (2005). Oxygen indicators and intelligent inks for packaging food *Chem. Soc. Rev.* 34, 1003–1011.
92. Amelia, M., Lavie-Cambot, A., McClenaghan, N. D., Credi, A. (2011). A ratiometric luminescent oxygen sensor based on a chemically functionalized quantum dot *Chem. Comm.* 47, 325–327.
93. Montalti, M., Credi, A., Prodi, L., Gandolfi, M. T. *Handbook of Photochemistry - Third Edition*; CRC Press: Boca Raton, FL (USA), 2006.
94. Walker, G. W., Sundar, V. C., Rudzinski, C. M., Wun, A. W., Bawendi, M. G., Nocera, D. G. (2003). Quantum-dot optical temperature probes *Appl. Phys. Lett.* 83, 3555–3557.
95. Tomasulo, M., Yildiz, I., Raymo, F. M. (2006). pH-sensitive quantum dots, *J. Phys. Chem. B* 110, 3853–3855.
96. Jin, T., Sasaki, A., Kinjo, M., Miyazaki, J. (2010). A quantum dot-based ratiometric pH sensor *Chem. Comm.* 46, 2408–2410.
97. Clarke, S. J., Hollmann, C. A., Zhang, Z., Suffern, D., Bradforth, S. E., Dimitrijevic, N. M., Minarik, W. G., Nadeau, J. L. (2006). Photophysics of dopamine-modified quantum dots and effects on biological systems *Nat. Mater.* 5, 409–417.
98. Medintz, I. L., Stewart, M. H., Trammell, S. A., Susumu, K., Delehanty, J. B., Mei, B. C., Melinger, J. S., Blanco-canosa, J. B., Dawson, P. E., Mattoussi, H. (2010). Quantum-dot/dopamine bioconjugates function as redox coupled assemblies for in vitro and intracellular pH sensing *Nat. Mater.* 9, 676–684.
99. Usiello, A., Baik, J. H., Rougé-Pont, F., Picetti, R., Dierich, A., LeMeur, M., Piazza, P. V; Borrelli, E. (2000). Distinct functions of the two isoforms of dopamine D2 receptors *Nature* 408, 199–203.
100. Laviron, E. (1984). Electrochemical reactions with protonations at equilibrium: Part, X. The kinetics of the parabenzoquinone hydroquinone couple on a platinum-electrode, *J. Electroanal. Chem.* 164, 213–227.

101. Wraight, C. A. (2004). Proton and electron transfer in the acceptor quinone complex of photosynthetic reaction centers from Rhodobacter sphaeroides *Front. Biosci.* 9, 309–337.
102. Finklea, H. O. (2001). Theory of coupled electron-proton transfer with potential-dependent transfer coefficients for redox couples attached to electrodes, *J. Phys. Chem. B* 105, 8685–8693.
103. Klegeris, A., Korkina, L. G., Greenfield, S. A. (1995). Autoxidation of dopamine: a comparison of luminescent and spectrophotometric detecion in basic solution *Free Radic. Biol. Med.* 18, 215–222.
104. Ji, X., Palui, G., Avellini, T., Na, H. Bin; Yi, C., Knappenberger, K. L., Mattoussi, H. (2012). On the pH-dependent quenching of quantum dot photoluminescence by redox active dopamine, *J. Am. Chem. Soc.* 134, 6006–6017.
105. Feringa, B. L. Ed. *Molecular Switches*, 2nd Ed.; Wiley-VCH: Weinheim, 2008.
106. Goodsell, D. S. *Bionanotechnology-Lessons from Nature*, John Wiley & Sons, Inc.: Hoboken, 2004.
107. Aviram, A. (1988). Molecules for memory, logic, and amplification, *J. Am. Chem. Soc.* 110, 5687–5692.
108. de Silva, A. P., Gunaratne, H. Q. N. ; McCoy, C. P. (1993). A molecular photoionic AND gate based on fluorescent signalling *Nature* 364, 42–44.
109. de Silva, A. P. (2005). Molecular computation: molecular logic gets loaded *Nat. Mater.* 4, 15–16.
110. de Silva, A. P., McClenaghan, N. D. (2002). Simultaneously multiply-configurable or superposed molecular logic systems composed of ICT (Internal Charge Transfer) chromophores and fluorophores integrated with one- or two-ion receptors *Chem. Eur. J.* 8, 4935–4945; (b) de Silva, A. P., McClenaghan, N. D. Molecular-scale logic gates *Chem. Eur. J.* (2004). 10, 574–586.
111. Uchiyama, S., Ieai, K; de Silva, A. P. (2008). Multiplexing sensory molecules map protons near micellar membranes *Angew. Chem. Int. Ed.* 47, 4667–4669.
112. Balzani, V; Credi, A., Langford, S. J., Stoddart, J. F. (1997). Logic operations at the molecular level. An XOR gate based on a molecular machine, *J. Am. Chem. Soc.* 119, 2679–2681.
113. Ball, P. (1997). Take it to the limit …- The superwhizzy computers of the future will only be born if we can make a whole transistor out of a molecule. But right now we're struggling to make just one wire, says Philip Ball *New Sci.* 155(2093), 32; (b) Leigh, D. A., Murphy, A. Molecular tailoring: the made-to-measure properties of rotaxanes *Chem. Ind.* (1999). 1, March (5), 178–183; (c) Freemantle, M. Molecular computation. Two logic gates, working in parallel, perform simple arithmetic operations *Chem. Eng. News* (2000). 78(18), 12; (d) Ball, P. Chemistry meets computing *Nature* (2000). 406, 118–120.
114. Semeraro, M; Credi, A. (2010). Multistable self-assembling system with three distinct luminescence outputs: prototype of a bidirectional half subtractor and reversible logic device, *J. Phys. Chem. C* 114, 3209–3214.
115. Credi, A. (2011). The research front on molecular logic *Aust. J. Chem.* (2010). 63(2), 145; (b) Special issue on Molecular and biomolecular information processing systems, Katz, E. Ed.; *Isr. J. Chem.* 51(1).

116. (a) Raymo, F. M. (2002). Digital processing and communication with molecular switches *Adv. Mater.* 14, 401–414; (b) Steinitz, D., Remacle, F., Levine, R. D. On spectroscopy, control, and molecular information processing *ChemPhysChem* (2002). 3, 43–51; (c) Balzani, V; Credi, A., Venturi, M. Molecular logic circuits *ChemPhysChem* (2003). 4, 49–59; (d) Gust, D., Moore, T. A., Moore, A. L. Molecular switches controlled by light *Chem. Commun.* (2006). 8, 1169–1178; (e) de Silva, A. P., Uchiyama, S. Molecular logic and computing *Nature Nanotech.* (2007). 2, 399–410; (f) Pischel, U. Chemical approaches to molecular logic elements for addition and subtraction *Angew. Chem. Int. Ed.* (2007). 46, 4026–4040; (g) Szaciłowski, K. Digital information processing in molecular systems *Chem. Rev.* (2008). 108, 3481–3548; (h) Willner, I., Shlyahovsky, B., Zayats, M., Willner, B. DNAzymes for sensing, nanobiotechnology and logic gate applications *Chem. Soc. Rev.* (2008). 37, 1153–1165; (i) Wagner, N., Ashkenasy, G. Systems chemistry: logic gates, arithmetic units, and network motifs in small networks *Chem. Eur. J.* (2009). 15, 1765–1775; (j) Benenson, Y. Biocomputers: from test tubes to live cells *Mol. Biosys.* (2009). 5, 675–685; (k) Wenger, O. Long-range electron transfer in artificial systems with d^6 and d^8 metal photosensitizers *Coord. Chem. Rev.* (2009). 253, 1439–1457; (l) Andréasson, J., Pischel, U. Smart molecules at work—mimicking advanced logic operations *Chem. Soc. Rev.* (2010). 39, 174–188; (m) Katz, E., Privman, V. Enzyme-based logic systems for information processing *Chem. Soc. Rev.* (2010). 39, 1835–1857; (n) Amelia, M., Zou, L., Credi, A. Signal processing with multicomponent systems based on metal complexes *Coord. Chem. Rev.* (2010). 254, 2267–2280.
117. Guo, Z., Zhao,V; Zhu, W., Huang, X., Xie, Y., Tian, H. (2008). Intramolecular charge-transfer process based on dicyanomethylene-4*H*-pyran derivative: an integrated operation of half-subtractor and comparator, *J. Phys. Chem. C* 112, 7047–7053.
118. Landauer, R. (1961). Irreversibility and heat generation in the computing process *IBM J. Res. Dev.* 5, 183–191.
119. Remon, P., Ferreira, R., Montenegro, J-M.; Suau, R., Perez-Inestrosa, E., Pischel, U. (2009). Reversible molecular logic: a photophysical example of a Feynman gate *ChemPhysChem* 10, 2004–2007.
120. *The International Technology Roadmap for Semiconductors (ITRS)*, (2009). edition and (2010). update www.itrs.net/reports.html.
121. Clemente-Leon, M., Pasquini, C., Hebbe-Viton, V., Lacour, J., Dalla Cort, A., Credi, A. (2006). Ion-pairing effects in the self-assembly of a fluorescent pseudorotaxane *Eur. J. Org. Chem.* 105–112.
122. Andréasson, J., Straight, S. D., Bandyopadhyay, S., Mitchell, R. H., Moore, T. A., Moore, A. L., Gust, D. (2007). Molecular 2:1 digital multiplexer *Angew. Chem. Int. Ed.* 46, 958–961.
123. Andréasson, J; Straight, S. D., Bandyopadhyay, S., Mitchell, R. H., Moore, T. A., Moore, A. L., Gust, D. (2007). A molecule-based 1:2 digital demultiplexer, *J. Phys. Chem. C* 111, 14274–14278.
124. Perez-Inestrosa, E., Montenegro, J.-M.; Collado, D., Suau, R. (2008). A molecular 1:2 demultiplexer *Chem. Commun.* 1085–1087.
125. Amelia, M., Baroncini, M., Credi, A. (2008). A simple unimolecular multiplexer/demultiplexer *Angew. Chem. Int. Ed.* 47, 6240–6243.

126. Langford, S. J., Yann, T. (2003). Molecular logic: a half-subtractor based on tetraphenylporphyrin, *J. Am. Chem. Soc.* 125, 11198–11199; (b) Szaciłowski, K. (2004). Molecular logic gates based on pentacyanoferrate complexes: from simple gates to three-dimensional logic systems *Chem. Eur. J.* 10, 2520–2528; (c) Margulies, D., Melman, G., Shanzer, A. (2005). Fluorescein as a model molecular calculator with reset capability *Nature Mater.*, 4, 768–771; (d) Margulies, D., Melman, G., Shanzer, A. (2006). A molecular full-adder and full-subtractor, an additional step toward a moleculator, *J. Am. Chem. Soc.* 128, 4865–4871.
127. Avellini, T., Li, H., Coskun, A., Barin, G., Trabolsi, A., Basuray, A. N., Dey, S. K., Credi, A., Silvi, S., Stoddart, J. F., Venturi, M. (2012). Photoinduced memory effect in a redox controllable bistable mechanical molecular switch. *Angew. Chem. Int. Ed.* 51, 1611–1615.
128. Coskun, A., Friedman, D. C., Li, H., Patel, K., Khatib, H. A., Stoddart, J. F. (2009). A light-gated STOP–GO molecular shuttle, *J. Am. Chem. Soc.* 131, 2493–2495.
129. Liu, Y., Flood, A. H., Bonvallett, P. A., Vignon, S. A., Northrop, B. H., Tseng, H.-R.; Jeppesen, J. O., Huang, T. J., Brough, B., Baller, M., Magonov, S., Solares, S. D., Goddard, W. A., Ho, C. M., Stoddart, J. F. (2005). Linear artificial molecular muscles, *J. Am. Chem. Soc.* 127, 9745–9759.
130. van der Molen, S. J., Liljeroth, P. (2010). Charge transport through molecular switches, *J. Phys. Condens. Matter* 22, 133001.
131. Gale, P. A., Steed, J. W. Eds. (2012). *Supramolecular Chemistry: from Molecules to Nanomaterials*, John Wiley & Sons: Hoboken, vols. 1–8.

CHAPTER 9

NOVEL PARALLEL STACKING INTERACTIONS OF AROMATIC MOLECULES

DUŠAN P. MALENOV and SNEŽANA D. ZARIĆ

CONTENTS

ABSTRACT

Properties of nanosystems containing aromatic rings can depend on noncovalent interactions of aromatic rings. This review is about novel parallel alignment interactions between aromatic rings and water and stacking parallel interactions between aromatic rings at large horizontal displacements. Novel parallel stacking interactions of aromatic rings show that aromatic rings have capability to form parallel interactions with substantial energy at very large parallel displacements (offsets), out of the ring, even beyond the C-H bond region, where two rings almost do not overlap. Interactions of aromatic rings at large offsets were neglected until recently. Recognizing these interactions in nanosystems can help to understand and predict properties of the systems.

9.1 INTRODUCTION

Aromatic molecules are very important in various systems, including nanosystems, since nanosystems very often contain aromatic rings [1–3]. Properties of those nanosystems can depend on noncovalent interactions of aromatic rings. Aromatic molecules can form various noncovalent interactions; π-system of aromatic molecules can be involved in interactions, however, hydrogen atoms of aromatic ring or substituents of the ring can also form interactions. Aromatic molecules can interact with other aromatic molecules; however, they can also interact with nonaromatic molecules [4]. Well-known interactions of aromatic molecules are aromatic/aromatic stacking interactions, XH/π interactions (or weak hydrogen bonds, where X can be O, N, C, S), cation-π interactions, anion-π interactions, and C-H/O interactions [4].

Stacking noncovalent interactions, where aromatic molecules are parallel, have been extensively studied for several decades [4]. This review will cover novel, recently described stacking interactions of aromatic molecules. It was recently pointed that nonaromatic molecules can form stacking interactions similar to aromatic molecules, as well as that aromatic molecules can form parallel interactions at very large offset (parallel displacement) values. Number of studies showed that various nonaromatic molecules can form stacking interactions with aromatic molecules [5–19]. Among these molecules is water molecule; whole water molecule or one of OH bonds can lay parallel to the aromatic ring [18, 19]. Some recent studies also showed that stacking interaction energies between aromatic molecules at large horizontal displacements are significant [20, 21].

This review is about novel parallel alignment interactions between aromatic rings and water and stacking parallel interactions between aromatic rings at large horizontal displacements.

9.2 PARALLEL ALIGNMENT INTERACTIONS OF WATER WITH AROMATIC RINGS

9.2.1 WATER/BENZENE PARALLEL ALIGNMENT INTERACTIONS

Cambridge Structural Database (CSD, November 2009 release, version 5.31) was searched for intermolecular contacts involving water molecules and C_{6-} aromatic rings with mutual parallel orientation [18, 19]. Contacts characterized by the distance R_1 shorter than 4.0 Å, offset r_1 shorter than 3.5 Å, absolute ΔR_1 values shorter than 0.1 Å ($\Delta R_1 = R_1 - R_O$), and negative Δr_1 values ($\Delta r_1 = r_1 - r_O$) (Fig. 9.1) were analyzed.

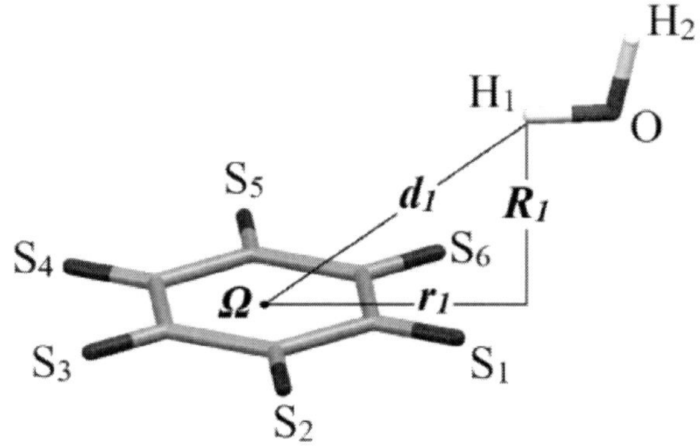

FIGURE 9.1 Geometric parameters and atom labeling used for the description of parallel alignment interactions of water and benzene. Ω is the centroid of aromatic ring; d_1, d_2 and d_O are the distances between Ω and atoms H_1, H_2 and O, respectively; R_1, R_2 and R_O are the normal distances from the ring plane to H_1, H_2 and O atoms, respectively; horizontal displacements (offsets) r_1, r_2 and r_O are the distances from centroid Ω to the projections of H_1, H_2 and O atom positions, respectively, on the plane of the ring; S_1 to S_6 is any atom or group.

The contacts found in this search were separated into two groups, with the whole water molecule parallel to the aromatic ring plane (set A) and with one bond, O-H_1, parallel to the ring plane (set B). Contacts belonging to set A are characterized by absolute ΔR_2 value shorter than 0.1 Å ($\Delta R_2 = R_2 - R_O$). A total of 787 contacts were found by performing the search, of which 71 belong to set A and 716 belong to set B.

The correlation of normal distances R_1 and offset values r_1 (Fig. 9.2) shows that most of the normal distances are longer than 3.0 Å, which is similar to distances typical for stacking interactions of aromatic groups (3.3–3.8 Å) [22–27]. In contacts with water molecules above the aromatic ring, all normal distances are longer than 3.0 Å; however, for offset values longer than 1.5 Å normal distances can be below 3.0 Å. Since the offset value of aromatic carbon atoms is ~1.4 Å, it can be noted that normal distances shorter than 3.0 Å are in C-H bond region, but mostly beyond that region (Fig. 9.2). Hence, the largest portion of parallel alignment water/aromatic interactions are at large horizontal displacements (r_1>2.5 Å, Fig. 9.2).

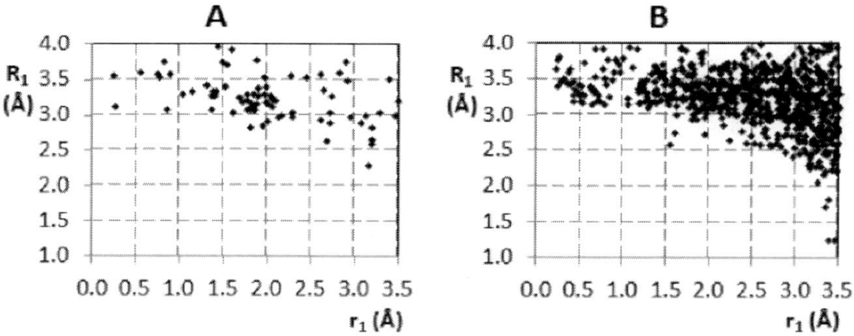

FIGURE 9.2 Correlation of normal distance R_1 and offset r_1 in contacts of the sets A and B.

Calculations of interaction energies at different values of normal distances R for various offset values r were performed at MP2/cc-pVTZ level with BSSE correction [28] in Gaussian03 program [29] for six model systems (Fig. 9.3). CCSD(T) energies at basis set limit for the most stable geometries within each model system were also estimated by performing the method of Helgaker (Table 9.1) [30].

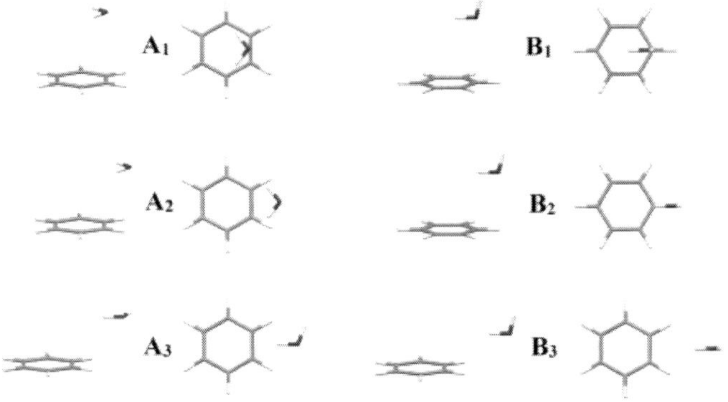

FIGURE 9.3 Two views of model systems for water/benzene parallel alignment interactions at various horizontal displacements.

TABLE 9.1 Estimated CCSD(T)$_{limit}$ Interaction Energies (in kcal/mol) and Normal Distances for Model-systems Depicting Benzene/water Parallel Alignment Interactions (Fig. 9.3) r is the Offset Value of Oxygen Atom of Water

Geometry	Energy	R_1 (Å)	r (Å)
A_1	–0.90	3.6	1.2
A_2	–1.45	3.1	2.5
A_3	–1.73	2.5	2.9
B_1	–0.78	3.4	1.4
B_2	–1.60	3.0	2.5
B_3	–2.45	2.3	3.7

The calculated CCSD(T) energies at basis set limit for model systems A_3 and B_3 are significant, −1.73 kcal/mol and −2.45 kcal/mol, respectively (Table 9.1). The results show that parallel alignment interactions are strongest at large offsets, out of the aromatic ring and out of the C-H bond region (Fig. 9.3). Interactions in the C-H bond region are weaker (–1.45 and −1.60 kcal/mol for model systems A_2 and B_2, respectively, Table 9.1), while parallel alignment interactions above aromatic ring are the weakest (–0.90 and −0.78 kcal/mol for model systems A_1 and B_1, respectively, Table 9.1). Parallel alignment interactions at large horizontal displacements (model systems A_3 and B_3, Fig. 9.3) are somewhat weaker than water/benzene OH/π interactions

(–3.21 kcal/mol) [18]; however, they are significantly stronger than water/benzene CH/O interactions (–1.41 kcal/mol) [31]. The energy of the interaction in model system B_3 is comparable to energy of benzene/benzene stacking interaction in slipped-parallel geometry (–2.74 kcal/mol) [27]. It is interesting that the interaction energies for model systems A and B at even larger horizontal displacements of closer hydrogen (r_1= 3.5 Å) are also substantially strong, −1.34 kcal/mol and −1.49 kcal/mol, respectively (Table 9.2), which is comparable to energy of CH/O interaction.

The analysis of various components of the interaction energy for model systems A and B shows that the correlation energy is significant component of interaction energy (Table 2), and even larger than electrostatic component at shorter (1.5 Å) and longer (3.5 Å) offset values. Electrostatic energy is somewhat larger than correlation energy for the most stable geometries of model systems A_3 and B_3 (offset values of 2.9 Å and 2.6 Å, respectively). Both correlation and electrostatic components are larger for model systems A and B in their most stable geometries than in geometries at shorter and longer offset values (Table 2).

TABLE 9.2 Geometric Data and Interaction Energies (in kcal/mol) for Water/benzene Model Systems A_3 and B_3 (Fig. 9.3)

System	r_1 (Å)	R_1 (Å)	$\Delta E_{CCSD(T)limit}$	ΔE_{MP2} [a]	ΔE_{HF} [a]	ΔE_{CORR} [b]	ΔE_{EL}	ΔE_{EX} [c]
	1.5	3.6	–0.89	–0.86	–0.21	–0.65	–0.48	0.27
A	2.9	2.5	–1.73	–1.58	–0.57	–1.01	–1.21	0.64
	3.5	2.0	–1.34	–1.23	–0.40	–0.83	–0.47	0.07
	1.5	2.9	–1.68	–1.65	–0.17	–1.50	–1.19	1.02
B	2.6	2.3	–2.45	–2.32	–0.73	–1.59	–1.67	0.94
	3.5	1.8	–1.49	–1.55	–0.61	–0.94	–0.51	–0.10

[a]cc-pVQZ **basis** set; [b] $\Delta E_{CORR} = \Delta E_{MP2} - \Delta E_{HF}$; [c] $\Delta E_{EX} = \Delta E_{HF} - \Delta E_{EL}$

The data found in crystal structures (Fig. 9.2) and by calculations (Table 9.1) are in very good agreement. By increasing offset values in model systems used for calculations (Fig. 9.3), the normal distances were decreasing (Tables 9.1 and 9.2). The same trend was observed in crystal structures (Fig. 9.2). Also, the number of interactions in crystal structures at large horizontal displacements, which are the strongest parallel alignment interactions (Table 9.2), is higher than the number of interactions in C-H bond region, and even higher than the number of interactions in the aromatic ring region, which are the weakest (Fig. 9.2).

By performing the CSD search on water/benzene OH/π interactions, using not very restrictive criteria (distance between centroid of the ring Ω and water hydrogen atom ≤ 3.5 Å; angle Ω-H-O ≥ 110°; angle between H-O bond and the normal to the ring ≤ 30°) [32–34], a total of 365 contacts were found [19]. In spite of being stronger, OH/π interactions occur less frequently than parallel alignment interactions in crystal structures. The reason for this lower occurrence is the fact that one O-H bond is forming OH/π interaction and is therefore disabled in forming other interactions, most notably classical hydrogen bonds. Both O-H bonds of water molecule that forms parallel alignment interaction are able to form additional interactions with surrounding molecules (Fig. 9.4), which leads to additional stabilization and more frequent occurrence of these interactions in crystal structures. Also, by forming parallel alignment interaction with water molecule at large horizontal displacement, the aromatic ring can also form additional interactions through its π-system (Fig. 9.4), which is, in addition to significant interaction energy, the reason of more frequent occurrence of these interactions in comparison to interactions at smaller offsets.

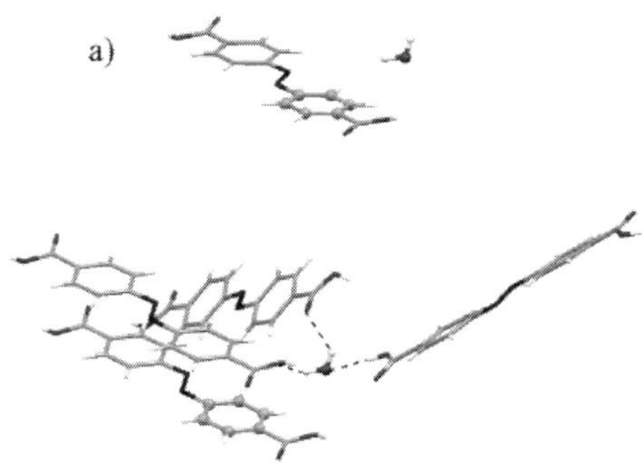

FIGURE 9.4 Parallel alignment interaction at large horizontal displacement (r_1 = 2.83 Å) of water molecule with disubstituted aromatic ring in crystal structure SENRUZ (4,4-azodibenzoic acid dihydrate) (a); all atoms of water molecule form additional hydrogen bonds with atoms of surrounding molecules of 4,4-azodibenzoic acid, while aromatic ring forms stacking interaction with another aromatic ring of 4,4-azodibenzoic acid (b).

9.2.2 WATER/PYRIDINE PARALLEL ALIGNMENT INTERACTIONS

Hydrogen bonding capacity of the pyridine nitrogen atom was shown to be related to the strength of stacking interactions between pyridine rings [35–38]. The systems with stacking interactions between two pyridine molecules, which are hydrogen bonded to water molecules, also contain two water/pyridine parallel alignment interactions (Fig. 9.5). Therefore, in this system water/pyridine parallel alignment interactions should influence pyridine/pyridine-stacking interactions, in addition to water/pyridine hydrogen bonding.

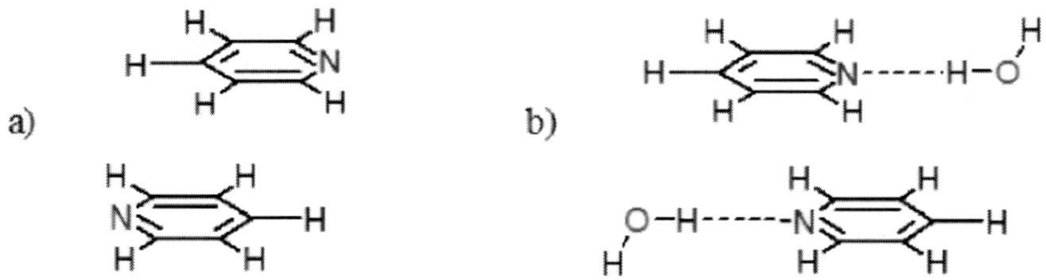

FIGURE 9.5 Stacking interaction between pyridine molecules (a) and between pyridine molecules hydrogen bonded with water (b).

Crystal structures archived in Cambridge Structural Database (November 2010 release, version 5.32) were screened for stacking interactions between pyridine molecules [39]. A total of 66 contacts with interplanar angle smaller than 10° and distance between centroids of pyridines below 4.6 Å [22] were found [39]. Geometrical parameters used for the analysis of the stacking interactions of pyridines are presented in Fig. 9.6. The analysis of torsion angle T showed the preference for values from 170° to 180° (head-to-tail orientation), since 56 of 66 interactions found possess this orientation. This is in agreement with previous highly accurate quantum chemical calculations [40–43]. The structures found by CSD search were screened for hydrogen bonds. Out of 66 stacking pyridine-pyridine interactions, 44 (or 67%) of them were of pyridines with hydrogen bonds. Stacking interactions of pyridine molecules with hydrogen bonds show preference for lower normal values R (3.4–3.5 Å) than stacking interactions of pyridines without hydrogen bonds (3.6–3.7 Å) (Fig. 9.7).

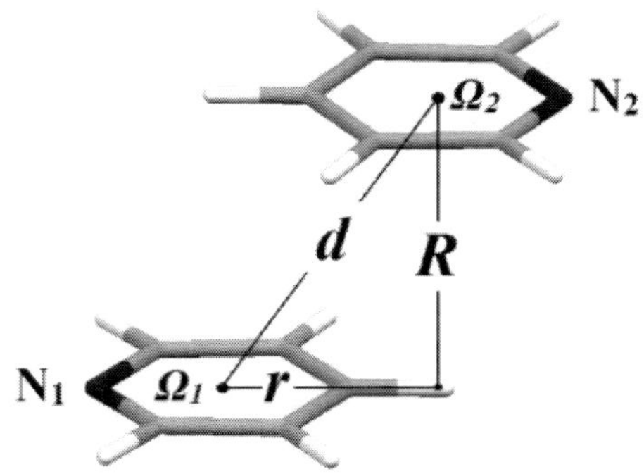

FIGURE 9.6 Geometrical parameters used to describe parallel interaction between two pyridine molecules; the distance between centroids (Ω_1 and Ω_2) of the pyridine rings is d; the normal distance R is the distance between the average planes of interacting rings; horizontal displacement (offset) r is the distance from the centroid of one ring (Ω_1) to the projection of the centroid of other ring (Ω_2) to the plane of the first ring; N_1-Ω_1-Ω_2-N_2 angle is torsion angle T.

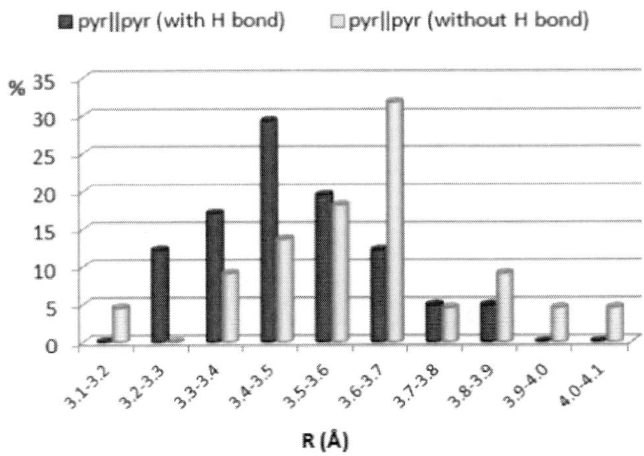

FIGURE 9.7 Distribution of values for normal distances R for stacking interactions between pyridine molecules with and without hydrogen bonds.

In order to compare geometries and energies of systems with and without water/pyridine hydrogen bonds and parallel alignment interactions, high level *ab initio* calculations were performed on pyridine||pyridine (Fig. 9.8) and HOH···pyr||pyr···HOH (Fig. 9.9) model systems [39], where "···" denotes hydrogen bonding and "||" denotes parallel orientation. MP2 method with BSSE correction and 6–311++G** basis set were used in Gaussian03 program [29]. The center of one pyridine molecule (Ω) was shifted along the N···Ω···C line, in the range of the offset values from −2.5 Å to 2.5 Å (Figs. 9.10 and 9.11). The negative offset values denote the geometries in which the nitrogen atoms are above the pyridine rings (Fig. 9.10 and 9.11), while the positive offset values correspond to geometries with the nitrogen atoms outside the pyridine rings (Figs. 9.8b, and 9.9b). The normal distances R were changed for every offset value. The hydrogen bonded water-pyridine dimer with H···N distance of 2.0 Å was used for the construction of HOH···pyr||pyr···HOH model systems (Fig. 9.9). While changing the pyridine-pyridine offset and normal distance values in these systems, the water-pyridine contacts with hydrogen bonding were kept rigid.

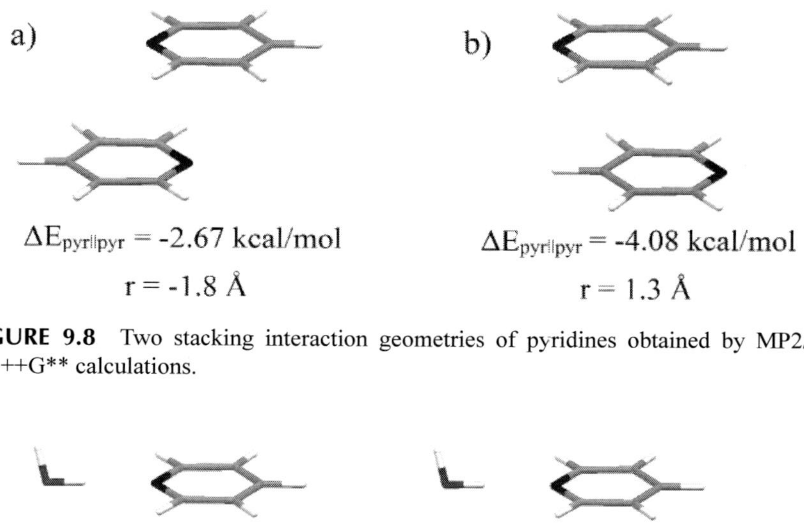

FIGURE 9.8 Two stacking interaction geometries of pyridines obtained by MP2/6–311++G** calculations.

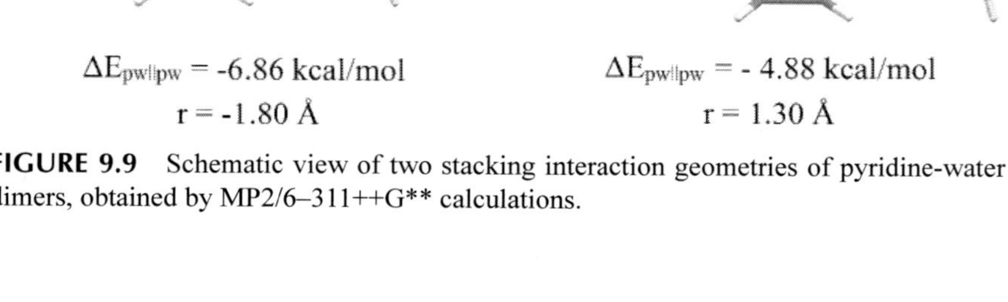

FIGURE 9.9 Schematic view of two stacking interaction geometries of pyridine-water dimers, obtained by MP2/6–311++G** calculations.

The geometry with the lowest energy in pyridine||pyridine system was obtained at positive offset value r = 1.3 Å (Fig. 9.8b), with energy of $\Delta E_{pyr||pyr}$ = −4.08 kcal/mol. This energy is very similar to estimated CCSD(T)/CBS energy (–3.99 kcal/mol). The most stable geometry with negative offset value (r = −1.8 Å, Fig. 9.8a) has the energy of $\Delta E_{pyr||pyr}$ = −2.67 kcal/mol.

In the investigated HOH···pyr||pyr···HOH model systems (Fig 9.9), each pyridine molecule simultaneously forms three interactions – stacking interaction with other pyridine molecule ($\Delta E_{pyr||pyr}$), classical hydrogen bond with one water molecule ($\Delta E_{pyr \cdots HOH}$) and parallel alignment interaction with the other water molecule ($\Delta E_{pyr||HOH}$) [18, 19]. The total interaction energy of the system was calculated using the expression: $\Delta E_t = E_{HOH \cdots pyr||pyr \cdots HOH} - 2E_{pyr} - 2E_{HOH}$. The energy of stacking interaction between two pyridine-water dimers was calculated using the expression: $\Delta E_{pw||pw} = \Delta E_t - 2\,\Delta E_{pyr \cdots HOH}$. There are two minima on potential curve (Fig. 9.10), one for negative offset ($\Delta E_{pw||pw}$ = −6.86 kcal/mol, r = −1.8 Å, Fig. 9.10) and one for positive offset ($\Delta E_{pw||pw}$ = −4.88 kcal/mol, r = 1.3 Å Fig. 9.9). Opposite to the system without hydrogen bonding, the geometry with N atom of one ring above the other ring is more stable due to two parallel-alignment water-pyridine interactions, which are substantially strong ($2\Delta E_{pyr||HOH}$ = −2.98 kcal/mol) when the nitrogen atoms are located above the rings and water molecules are close to the second pyridine molecule (Fig. 9.9a). The contribution of these parallel-alignment water-pyridine interactions in geometry with r = −1.8 Å (Fig. 9.9a) is even larger than the contribution of the stacking interaction between two pyridine molecules ($\Delta E_{pyr||pyr}$ = −2.67 kcal/mol).

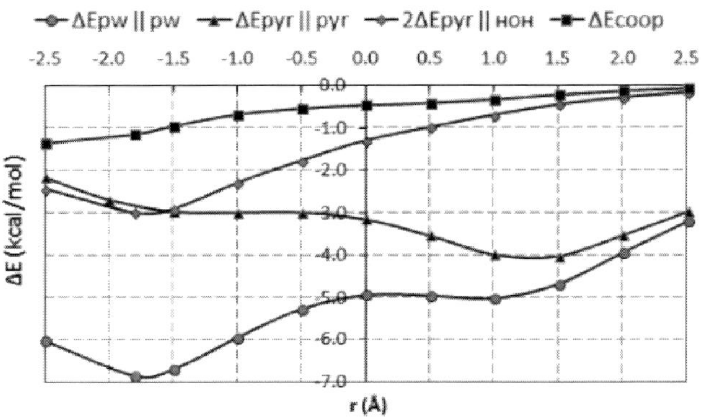

FIGURE 9.10 Energies of interaction between two pyridine–water (pw) dimers ($\Delta E_{pw||pw}$), two pyridine molecules ($\Delta E_{pyr||pyr}$), parallel-alignment water–pyridine interactions ($2\Delta E_{pyr||HOH}$), and the cooperative effect (ΔE_{coop}), as functions of offset values.

The results of *ab initio* calculations are in good agreement with the results of CSD search (Fig. 9.7). The calculated normal distances R are shorter for stacking interactions of pyridine molecules with hydrogen bonds than for pyridine molecules without hydrogen bonds (Fig. 9. 11). The results show that the involvement of pyridine nitrogen atom in hydrogen bonding significantly influences pyridine-stacking interactions, which are also substantially influenced by water/pyridine parallel alignment interactions.

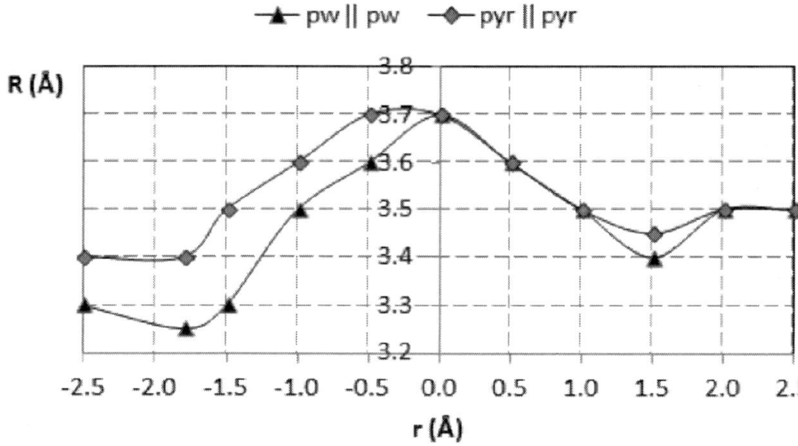

FIGURE 9.11 The distributions of normal distances (R) for parallel pyridine-pyridine dimers with and without hydrogen bonds, as functions of offset values.

The cooperative effect for HOH···pyr||pyr···HOH stacking interaction was also calculated (ΔE_{coop}, Fig. 9.10), using the expression: $\Delta E_{coop} = \Delta E_{pw||pw} - (\Delta E_{pyr||pyr} + 2\Delta E_{pyr||HOH} + \Delta E_{HOH/HOH})$, where $\Delta E_{HOH/HOH}$ is the energy of interaction between water molecules. The cooperativity varies from −0.05 to −1.35 kcal/mol (Fig. 9.10).

The difference in the interaction energies for the most stable pyridine–pyridine and HOH···pyr||pyr···HOH system is 2.78 kcal/mol (Fig. 9.10). The difference in correlation energies of these systems is only 1.38 kcal/mol (Table 9.3), meaning that the difference in interaction energy is not only caused by increased dispersion in HOH···pyr||pyr···HOH system, but also by increase of electrostatics.

TABLE 9.3 Calculated Stacking Interactions for Pyridine Dimer (pyr||pyr, Fig. 9.8) and Pyridine/water Dimers (pw||pw, Fig. 9.9) at HF/6–311++G** and HF/6–311++G** Levels and Correlation Energy in these Systems; $\Delta E_{CORR} = \Delta E_{MP2} - \Delta E_{HF}$

	r	ΔE_{HF}	ΔE_{MP2}	ΔE_{CORR}
pw ll pw	−1.8	2.06	−6.86	−8.92
pyr ll pyr		5.04	−2.67	−7.71
pw ll pw	+1.3	2.78	−4.88	−7.76
pyr ll pyr		3.46	−4.08	−7.54

9.3 PARALLEL INTERACTIONS OF AROMATIC RINGS AT LARGE HORIZONTAL DISPLACEMENTS

9.3.1 BENZENE/BENZENE PARALLEL INTERACTIONS AT LARGE HORIZONTAL DISPLACEMENTS

Cambridge Structural Database (November 2010 release, version 5.32) was searched for parallel benzene-benzene contacts [20]. Two benzene molecules were considered parallel if the angle between the mean planes of the rings was less than 10°. It was considered that the interaction of these molecules occurs if d < 6.0 Å and R < 4.0 Å (Fig. 9.12). Using these criteria, a total of 1824 parallel contacts were found within this search.

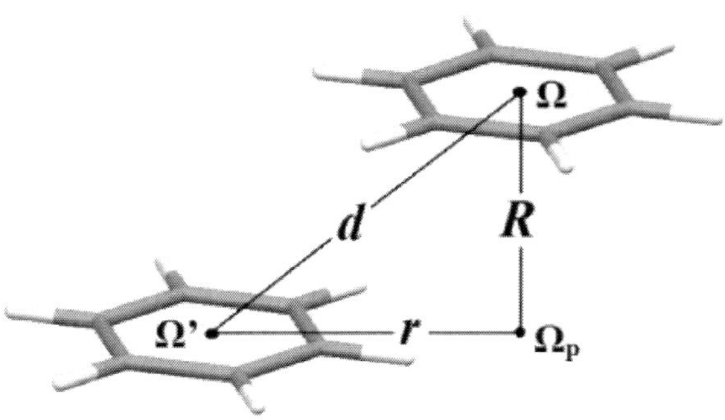

FIGURE 9.12 Geometrical parameters used for the description of parallel interactions between two benzene molecules; d is the distance between the centroids (Ω and Ω') of two benzene molecules; R is the normal distance between the planes of interacting rings; r is horizontal displacement (offset), that is, the distance from the centroid of one ring (Ω') to the projection of the centroid of other ring to the plane of the first ring (Ω_p).

Statistical analysis of the contacts shows the significant number of contacts at large horizontal displacements. The distribution of the offset values r (Fig. 9.13a) reveals that the most contacts (1056 of 1824 contacts, or 58%) have offset values in the range 4.5–5.5 Å, where two benzene molecules almost do not overlap (Fig. 9.14). A substantial number of interactions possess normal distances in range 3.0–3.6 Å, which is typical for stacking interactions [22–27]; however, the normal distances for contacts with offsets larger than 4.5 Å are predominantly below 3.0 Å (Fig. 9.13b).

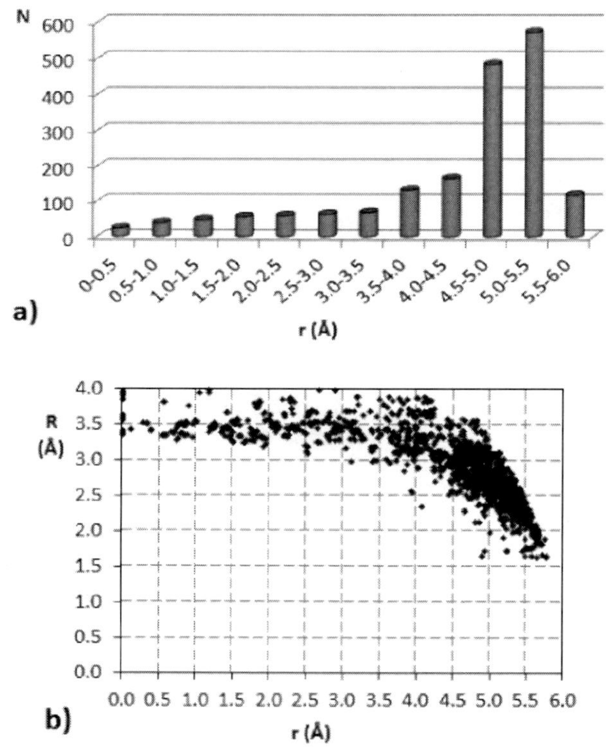

FIGURE 9.13 Geometrical characteristics of mutually parallel benzene-benzene contacts: (a) distribution of offset values (r) and (b) correlation of normal distances (R) with offset values (r).

DFT calculations on two mutually parallel benzene molecules were performed in order to determine interaction energies at large horizontal displacements [20]. Energies for three different orientations (Fig. 9.14) at different values of normal distances R for various offset values r in range 0.0–6.0 Å were calculated by using the dispersion corrected B2PLYP functional [44, 45] without BSSE correction and def2-TZVP basis set [46] implemented in ORCA program (version 2.8) [47].

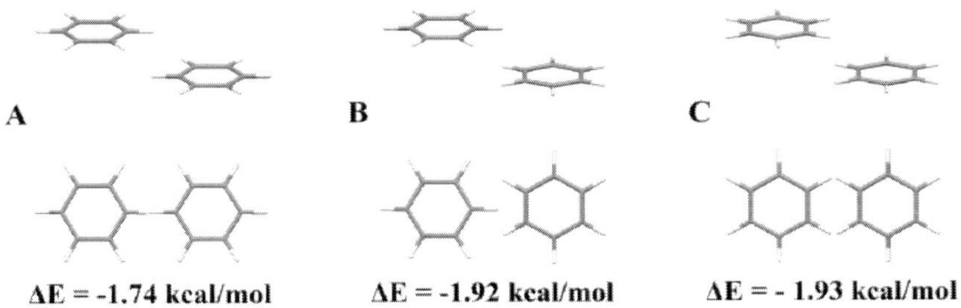

FIGURE 9.14 Two views of three parallel benzene-benzene orientations (A, B and C) used for B2PLYP-D/def2-TZVP calculations; geometries shown are at offset values of 5.0 Å

The calculated energies for face-to-face geometries are around −1.6 kcal/mol, while the strongest interactions are around −2.8 kcal/mol in the offset range 1.5–2.0 Å (Fig. 9.15a). At large offsets (4.0–5.0 Å) interaction is still substantially strong, with energy of about −2.0 kcal/mol (Fig. 9.15a), which is 71% of the strongest interaction energy between two benzene molecules. CCSD(T) energy at basis set limit at r = 5.0 Å for orientation B is −1.98 kcal/mol, which is very similar to the value obtained at B2PLYP-D/def2-TZVP level (–1.92 kcal/mol), indicating the high accuracy of the method and basis set used.

In all model systems, the calculated normal distances R were decreasing by increasing the offset values r (Fig. 9.15b), which is in good agreement with crystal data (Fig. 9.13b). The origin of the attraction at large offset values was estimated by comparing B3LYP [48–50] and B3LYP-D [45] energies at offsets of 1.5 Å and 5.0 Å. It was found that dispersion is reduced by approximately 50% at large offset, but repulsion is even more reduced, which results in substantial attraction.

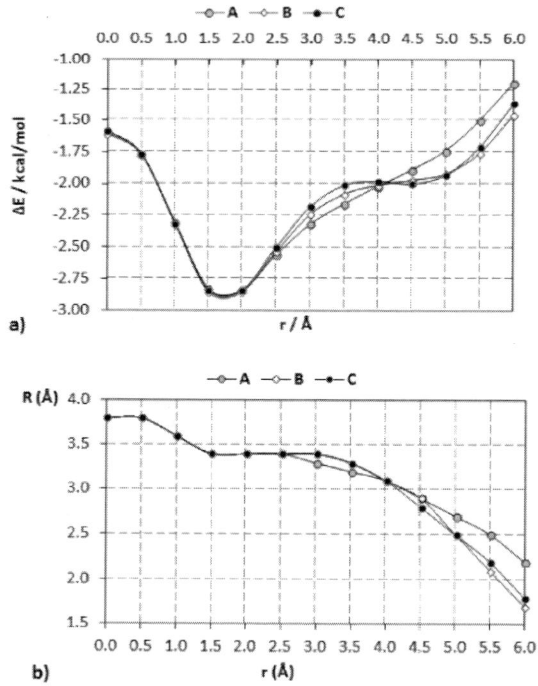

FIGURE 9.15 Calculated interaction energies (ΔE) for A, B and C (Fig. 9.14) orientations of benzene molecules (a) and the optimal normal distances R (b) plotted as functions of offset value r.

It is interesting that by optimizing the orientation C at B2PLYP-D level, the new minimum at benzene dimer potential curve was found. The optimized structure (Fig. 9.16) has the energy of −2.01 kcal/mol. Offset value of this minimum is r = 4.53 Å, normal distance is R = 2.82 Å, and angle between planes of benzene rings is 0.19°.

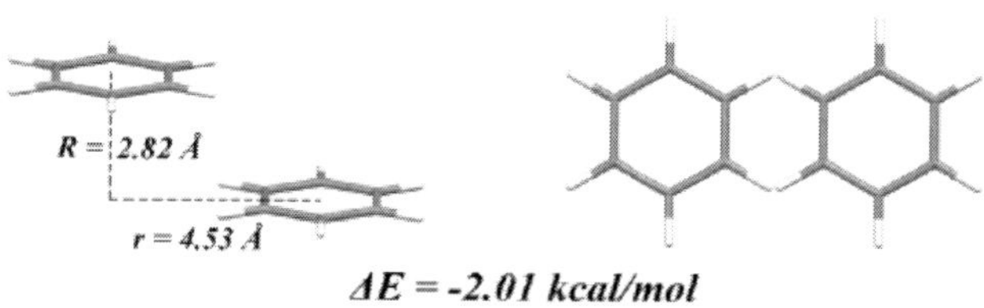

FIGURE 9.16 Two views of the new minimum of benzene dimer at large horizontal displacement.

The results of analysis of CSD data show that the most frequent geometries of two parallel benzene molecules are at offsets in range 4.5–5.5 Å, while calculations show that the strongest interactions are in range 1.5–2.0 Å. This disagreement is the consequence of the substantial interaction energy at large offsets and environment in crystal structures, since mutually parallel benzene molecules at large offsets have possibilities to form additional simultaneous interactions with surrounding molecules. The example is the crystal structure CENNUE – syncarpurea benzene solvate (Fig. 9.17), in which parallel benzene molecules at large offsets form additional CH/π interactions with other benzene molecules and with syncarpurea molecules, and also CH/O interactions with O-atoms of syncarpurea molecules.

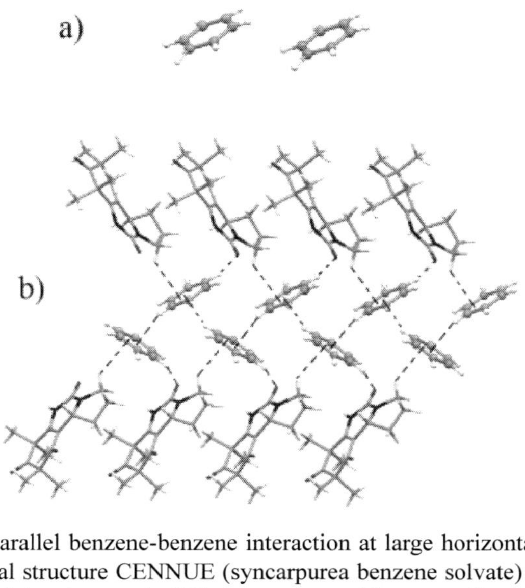

FIGURE 9.17 Parallel benzene-benzene interaction at large horizontal displacement (r = 5.10 Å) in crystal structure CENNUE (syncarpurea benzene solvate) (a); both benzene molecules form additional aromatic CH/π interactions with surrounding benzene molecules, CH/π interactions with syncarpurea molecules and CH/O interactions with O-atoms of syncarpurea molecules (b).

Interactions between mutually parallel aromatic molecules at large horizontal displacements are also significant in biomolecules. Preliminary Protein Data Bank search showed 1533 interactions between parallel phenylalanine aromatic phenyl residues, with 795 of these contacts (52%) at offset values larger than 4.0 Å [20], indicating very frequent occurrence of interactions between parallel phenyl rings at large horizontal displacements.

9.3.2 PYRIDINE-PYRIDINE INTERACTIONS AT LARGE HORIZONTAL DISPLACEMENTS

Cambridge Structural Database (November 2011 release, version 5.33) search was performed in order to study parallel pyridine-pyridine interactions [21]. It was considered that two pyridine rings are forming parallel interaction if the angle between the mean planes of the rings was less than 10°, the distance between centers of the rings was less than 6.0 Å and the normal distance between the ring planes R was less than 4.0 Å (Fig. 9.18). Since it was shown that hydrogen bonding significantly influences pyridine-pyridine stacking [35–39], the contacts containing hydrogen bonds between pyridine molecules and X-H species (X is O, N, S, F or Cl) with X···N distance shorter than 4.0 Å and X-H···N angle larger than 110° were excluded in the analysis of the data.

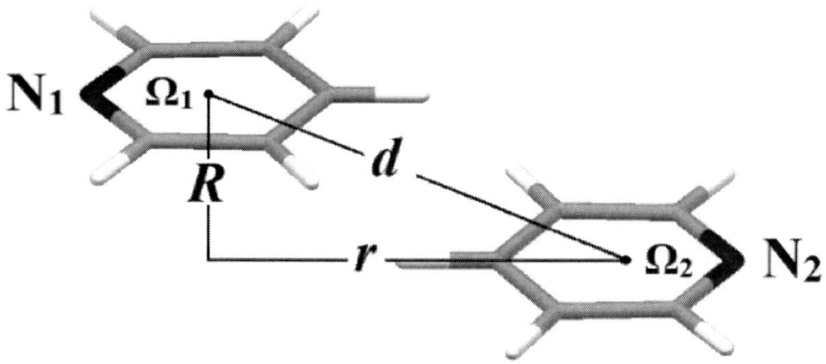

FIGURE 9.18 Geometrical parameters used to describe parallel interaction between two pyridine molecules; the distance between the centroids of pyridine rings rings is d; the normal distance R is the distance between the average planes of interacting rings; horizontal displacement (offset) r is the distance from the centroid of one ring (Ω_2) to the projection of the centroid of other ring (Ω_1) to the plane of the first ring; torsion angle T is N_1-Ω_1-Ω_2-N_2 angle.

The described CSD search gave 166 parallel pyridine-pyridine contacts without hydrogen bonds. The geometrical parameters used for statistical analysis of the contacts are shown in Fig. 9.19. The analysis shows very large share of contacts with large offset values of 4.0–6.0 Å (121 contacts, or 74%, Fig. 9.19b), similar to the data for parallel benzene/benzene contacts [20] shown in Chapter 2.1. At these large offsets two pyridine molecules almost do not overlap (Fig. 9.20). The most of the contacts have normal distances R in the range of 3.0–4.0 Å, which is typical for stacking interactions [22–27], but for offsets larger than 4.0 Å the normal distances can be below 3.0 Å. There is the clear preference for torsion angles T from 170° to 180°, which is head-to-tail orientation, found as the most stable orientation by previous quantum chemical calculations [35, 40–43].

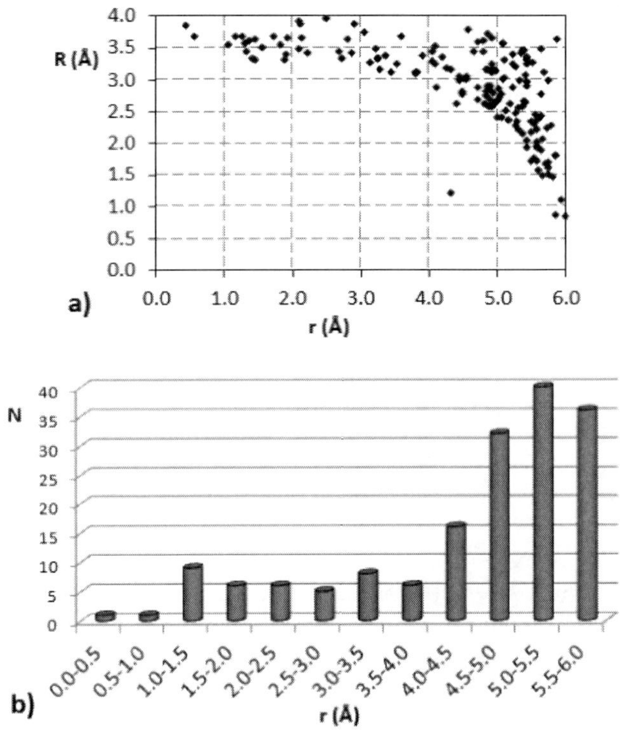

FIGURE 9.19 Geometrical characterization of parallel pyridine-pyridine interactions: (a) correlation of normal distances (R) with offset values (r); (b) distribution of offset values (r).

Calculations of interaction energies of parallel pyridine-pyridine interactions were done on three different head-to-tail pyridine-pyridine orientations (Fig. 9.20) [21], using B2PLYP-D2 [44, 45] functional without BSSE correction and def2-TZVP basis set incorporated in ORCA program (version 2.8). Monomer geometries were kept rigid, while for various offset values (r) normal distances (R) were varied. Offset values from −6.0 to +6.0 Å were investigated, where positive (+) offset values for App and Bpp orientations indicate that nitrogen atom of one ring moves away from the other pyridine ring, and negative (–) offset values mean that nitrogen atom of one ring overlaps with the other ring at small offset values (Fig. 9.20). Geometries with negative and positive offset values of Cpp orientation are equivalent.

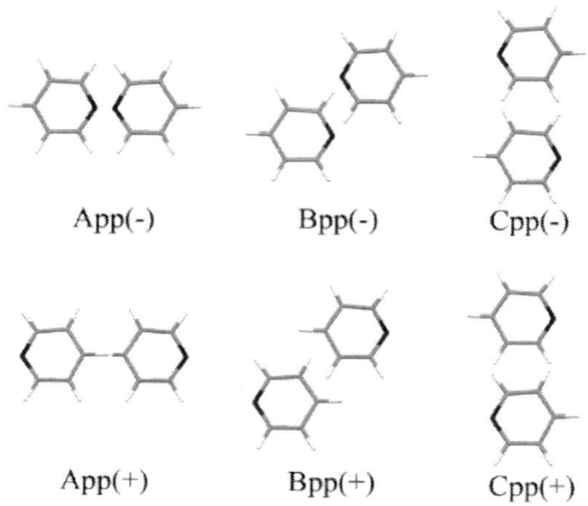

FIGURE 9.20 The top view of the parallel pyridine-pyridine orientations (App, Bpp, Cpp) used for calculations; geometries with positive and negative offset values of 5.0 Å are presented; geometries Cpp(+) and Cpp(–) are equivalent

Interaction energies for positive offsets in pyridine-pyridine dimers are similar for all three orientations (Fig. 9.21a). The calculated energies at minima of the potential curve at offset values of 1.5 Å for App(+), Bpp(+) and Cpp(+) orientations are −3.89, −4.09 and −4.12 kcal/mol, respectively. Interaction energies are substantially strong for all orientations at large positive offsets (r = 4.0–5.0 Å). The most stable orientation is Cpp(+), with energies of −2.46 and −2.04 kcal/mol for offsets of 4.0 and 5.0 Å, respectively. Orientations App(+) and Bpp(+) are energetically very similar, with interaction energies of about −2.1 and −1.7 kcal/mol for r = 4.0 Å and r = 5.0 Å, respectively. Overall, interaction energies at large positive horizontal displacements are smaller than the energy of face-to-face geometry (–2.91 kcal/mol) and are 50–60% of the energy of most stable geometries.

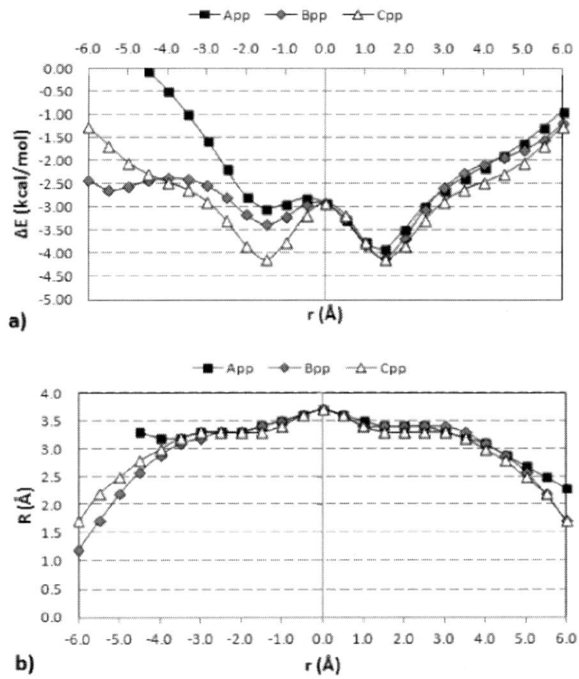

FIGURE 9.21 Calculated interaction energies (ΔE) (a) and the optimal normal distances (R) (b) for three different pyridine-pyridine orientations (Fig. 9.20) plotted as the functions of offset values (r).

Orientations with negative offset values are strongly influenced by the interactions of nitrogen atoms. Orientation Cpp(–) is equivalent to orientation Cpp(+) and therefore has the same values of interaction energies; it is the most stable orientation for negative offset values ($\Delta E = -4.12$ kcal/mol for $r = 1.5$ Å). The other two orientations, App(–) and Bpp(–), have smaller energies at minima, −3.02 kcal/mol and −3.37 kcal/mol, respectively. Interaction represented by App(–) orientation is the weakest due to unfavorable interactions of N atom with the center of the ring and with the N atom of the other molecule of pyridine; it becomes repulsive for offset values more negative than −4.5 Å. Repulsive N-N interactions in this orientation also cause high values of normal distances ($R > 3.2$ Å, Fig. 9.21b). In the Bpp(–) orientation, the influence of N atom on interaction energy is less significant. At offset value of $r = 5.5$ Å the curve has the new minimum, due to two additional attractive CH···N interactions (Fig. 9.20, Bpp(–) orientation). These additional interactions also cause significant decrease in the values of normal distances (Fig. 9.21b).

Similar as in benzene-benzene dimer, the origin of substantial interaction energy at large horizontal displacements was estimated by comparing DFT energies at offsets of 1.5 Å and 5.0 Å without and with dispersion correction, namely B3LYP [48–50] and B3LYP-D2 [45] energies. Calculated B3LYP energies were repulsive at both offsets, indicating the importance of dispersion component of interaction. The dispersion is significantly reduced at large horizontal displacement, but repulsion was also reduced, resulting in substantial attraction.

Although the most stable geometries for parallel pyridine-pyridine interactions are in offset range of 1.5–2.0 Å, in the crystal structures the preferred offset values are above 4.5 Å. This is the consequence of surroundings in the crystal structures and relatively strong interactions at these offsets. Similar to benzene-benzene dimers, parallel pyridine-pyridine dimers at large offsets can form multiple additional interactions with molecules from the environment. In crystal structure KINLIC (Fig. 9.22) both pyridines involved in parallel interaction at large offset form additional simultaneous CH/π interactions with other molecules.

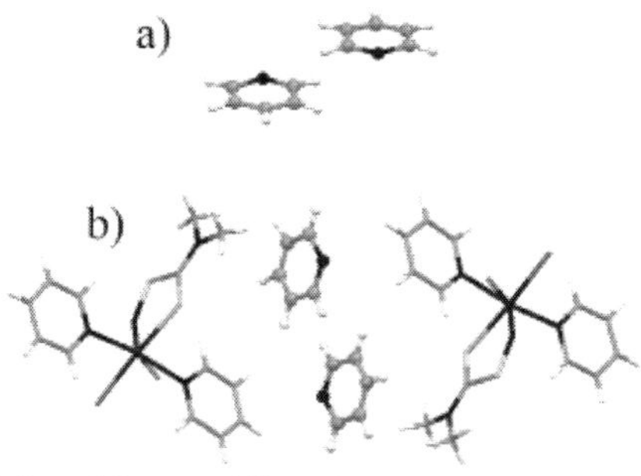

FIGURE 9.22 Parallel pyridine-pyridine interaction at large horizontal displacement (r = 4.29 Å) in crystal structure KINLIC (dibromo-(dimethylamino(thiocarbonyl)thiamin, S)-dipyridyl-titanium(IV) pyridine solvate) (a); both pyridine molecules also form two additional CH/π interactions at both sides of the rings with surrounding molecules (b).

9.3.3 BENZENE/PYRIDINE INTERACTIONS AT LARGE HORIZONTAL DISPLACEMENTS

By performing the search of Cambridge Structural Database (November 2011 release, version 5.33), using the same criteria as those for parallel pyridine/

pyridine contacts (the angle between the mean planes of the rings less than 10°, the distance between centers of the rings shorter than 6.0 Å and the normal distance between the ring planes R shorter than 4.0 Å, Fig. 9.18), no parallel benzene-pyridine contacts were found [21]. Therefore, the analysis of benzene/pyridine parallel interactions was limited to theoretical calculations.

Calculations of interaction energies of parallel benzene-pyridine interactions were done on three different benzene-pyridine orientations (Fig. 9.23) at B2PLYP-D2/def2-TZVP level in ORCA program (version 2.8) [47]. Monomer geometries were kept rigid, while for various offset values (r) normal distances (R) were varied. Offset values from −6.0 to +6.0 Å were investigated, where positive (+) offset values Abp and Bbp orientations indicate that nitrogen atom of pyridine ring moves away from benzene ring, and negative (–) offset values mean that nitrogen atom of pyridine ring overlaps with benzene ring at small offset values (Fig. 9.23). Geometries with negative and positive offset values of Cbp orientation are equivalent.

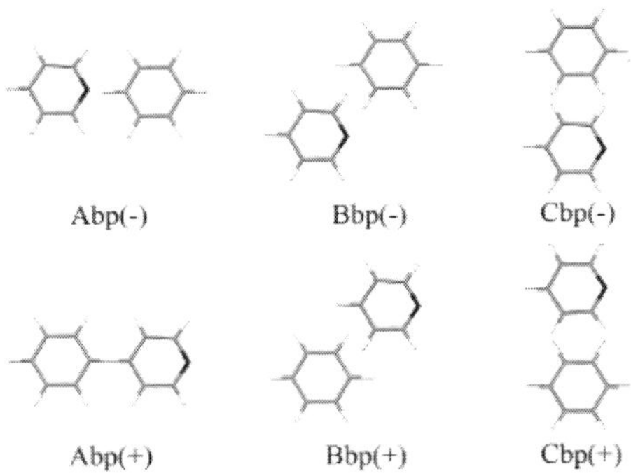

FIGURE 9.23 The top view of the parallel benzene-pyridine orientations (Abp, Bbp, Cbp) used for calculations; geometries with positive and negative offset values of 5.0 Å are presented; geometries Cbp(+) and Cbp(–) are equivalent.

The calculated interaction energies at minima of the potential curves for benzene-pyridine Abp(+), Bbp(+) and Cbp(+) orientations (Fig. 9.24) are very similar: −3.47, −3.54 and −3.38 kcal/mol. Interactions at large positive offsets (4.0–5.0 Å) are substantially strong. The strongest interaction energy for offset of 4.0 Å is found for Abp(+) orientation (ΔE = −2.20 kcal/mol), while at 5.0 Å the Cbp(+) orientation has the strongest interaction (ΔE = −1.95 kcal/mol). The interaction energies at large offsets are similar to the one calculated for face-to-face geometry (ΔE = −2.22 kcal/mol) and are 55–62% of the most stable energy.

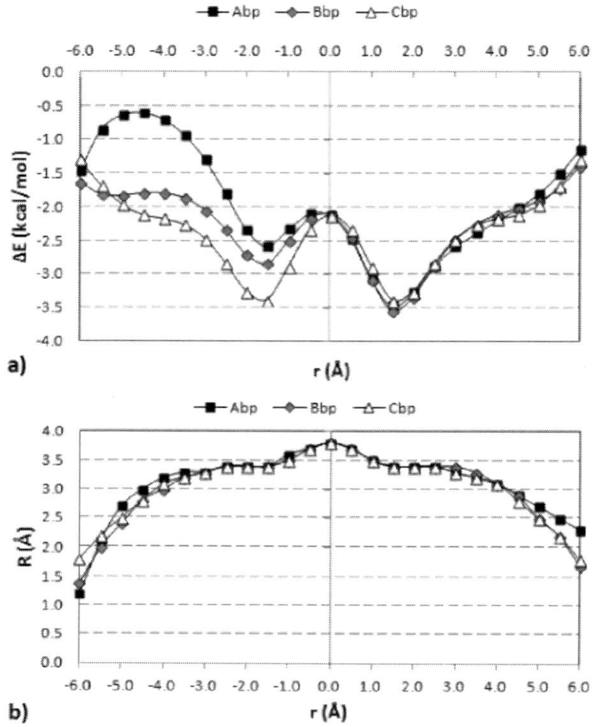

FIGURE 9.24 Calculated interaction energies (ΔE) (a) and the optimal normal distances (R) (b) for three different benzene-pyridine orientations (Fig. 9.23) plotted as the functions of offset values (r)

At negative offset values, the influence of N atom on interaction energies is significant, similar to that in pyridine-pyridine dimer. Similarly to Cpp(+) and Cpp(–) (Figure 21), Cbp(+) and Cbp(–) orientations (Fig. 9.23) are equivalent, and therefore interaction energy at minimum for Cbp(–) is −3.38 kcal/mol (Fig. 9.24a). The interaction within this orientation is stronger than interactions within Abp(–) and Bbp(–) orientations (–2.58 kcal/mol and −2.83 kcal/mol, respectively). In Abp(–) orientation there is a maximum on the curve at r = −4.5 Å as a consequence of the repulsive interactions of N atom with the C atom of benzene ring. However, the interaction becomes more favorable at more negative offset values as a consequence of attractive CH···N interactions (Fig. 9.23, orientation Abp(–)). In accord with the change from repulsive to attractive interactions, the R-value is reduced (Fig. 9.24b). At large negative offsets the orientation Bbp(–) is less stabilized than corresponding Bpp(–) orientation (Fig. 9.21a) because of only one attractive CH···N interaction (Fig. 9.23).

9.3.4 COMPARISON OF BENZENE/BENZENE, PYRIDINE/PYRIDINE AND BENZENE/PYRIDINE PARALLEL INTERACTIONS AT LARGE HORIZONTAL DISPLACEMENTS

By comparing the results of CSD searches for parallel benzene-benzene and pyridine-pyridine interaction, it can be concluded that both types of dimers show the preference for large horizontal displacements (Fig. 9.13a and b). However, the distribution of offset values for pyridine-pyridine contacts (Fig. 9.19b) shows another peak at offset of 1.0–1.5 Å, as a consequence of strong interaction (–4.12 kcal/mol) [21]. This peak does not exist at the offset distribution for benzene-benzene contacts (Fig. 9.13a), since interaction in parallel benzene dimer is significantly weaker (–2.85 kcal/mol) [20].

At large offset values pyridine-pyridine dimers can be stabilized by additional CH···N interactions (Fig. 9.20, Bpp(–) orientation). DFT calculations show that the minimum at potential curve for Bpp(–) orientation is at offset value of 5.5 Å, with energy of −2.62 kcal/mol. However, this minimum is only about 0.5 kcal/mol more stable than the geometries with proximal CH bonds of the two-pyridine molecules [App(+), Bpp(+), Cpp(+) and Cpp(–)]. Also, this additional interaction was found in only 30% of the contacts in the crystal structures with offsets in range 4.0–6.0 Å, and therefore is not the main reason for the large number of interactions at large offsets. In most of the contacts at large offsets the CH groups of two pyridine molecules are in close proximity.

Interaction energies for pyridine-pyridine, benzene-pyridine and benzene-benzene dimers (Table 9.4) show that pyridine-pyridine dimer is the most stable and benzene-benzene dimer is the least stable. In face-to-face (r = 0.0 Å) and minimum parallel-displaced (r = 1.5 Å) geometries, the pyridine dimer is the most stabilized system, due to the dipole moment in the pyridine molecule [42]. For that reason, the differences in stabilization in three types of dimers are significant for face-to-face and minimum configurations. However, at large offset values, the stabilization in the three dimers is very similar; the differences in energies are lower than 0.25 kcal/mol, and the dipole moment in pyridine molecule has no significant influence. Although the interaction at large horizontal displacement is strongest in pyridine-pyridine dimer, the smallest portion of the strongest interaction energy is preserved (50–60%, Table 9.4). In benzene-benzene dimer, the interactions at large offset values are weaker than in pyridine-pyridine and benzene-pyridine dimers, but their energy is the largest portion of the minimum energy (68–71%, Table 9.4).

TABLE 9.4 The Calculated Interaction Energies (in kcal/mol) at Four Offset Values (r) for Benzene-benzene (Fig. 9.14) Pyridine-pyridine (Fig. 9.20), Benzene-pyridine (Fig. 9.23) Dimers

	r = 0.0 Å	r = 1.5 Å	r = 4.0 Å		r = 5.0 Å	
	ΔE	ΔE	ΔE	%[a]	ΔE	%[a]
benzene-benzene[b]	–1.61	–2.85	–2.02	71	–1.93	68
pyridine-pyridine[c]	–2.91	–4.12	–2.46	60	–2.04	50
benzene-pyridine	–2.11	–3.54[d]	–2.20[e]	62	–1.95[f]	55

[a]Percentage of the strongest interaction energy;
[b]C orientation;
[c]Cpp(+) orientation;
[d]Bbp(+) orientation;
[e]Abp(+) orientation;
[f]Cbp(+) orientation.

9.4 CONCLUSION

Novel parallel stacking interactions of aromatic rings presented in this review show that aromatic rings have capability to form parallel interactions with substantial energy at very large parallel displacements (offsets), out of the ring, even beyond the C-H bond region, where two rings almost do not overlap. It is know that aromatic molecules can form various noncovalent interactions; however, the interactions at large offsets were neglected until recently. Recognizing these interactions in nanosystems can help to understand and predict properties of the systems.

KEYWORDS

- **aromatic rings**
- **benzene/pyridine interaction**
- **C-H interactions**
- **non-covalent interactions**
- **stacking displacements**
- **water/benzene interaction**

REFERENCES

1. An, B.-K.; Gierschner, J., Park, S. J. (2012). π-Conjugated Cyanostilbene Derivatives: A Unique Self-Assembly Motif for Molecular Nanostructures with Enhanced Emission and Transport. *Acc. Chem. Res.* 45, 544–554.
2. Ajayaghosh, A., Praveen, V. K. (2007). π-Organogels of Self-Assembled p-Phenylenevinylenes: Soft Materials with Distinct Size, Shape, and Functions. *Acc. Chem. Res.* 40, 644–656.
3. Chen, Y., Zhang, B., Liu, G., Zhuang, X. D., Kang, E. T. (2012). Graphene and its derivatives: switching ON and OFF. *Chem. Soc. Rev.* 41, 4688–4707.
4. Salonen, L. M., Ellermann, M., Diederich, F. (2011). Aromatic Rings in Chemical and Biological Recognition: Energetics and Structures. *Angew. Chem. Int. Ed.* 50, 4808–4842
5. Garcia, B., Garcia-Tojal, J., Ruiz, R., Gil-Garcia, R., Ibeas, S., Donnadieu, B., Leal, J. M. (2008). Interaction of the DNA bases and their mononucleotides with pyridine-2-carbaldehyde thiosemicarbazonecopper(II) complexes. Structure of the cytosine derivative. *J. Inorg. Biochem.* 102, 1892–1900.
6. Carroll, W. R., Pellechia, P., Shimizu, K. D. (2008). A Rigid Molecular Balance for Measuring Face-to-Face Arene−Arene Interactions. *Org. Lett.* 10, 3547–3550.
7. Mukhopadhyay, U., Choquesillo-Lazarte, D., Niclos-Gutierrez, J., Bernal, I. (2004). A critical look on the nature of the intramolecular interligand π, π-stacking interaction in mixed-ligand copper(II) complexes of aromatic side-chain amino acidates and α, α′-diimines. *CrystEngComm*, 6, 627–632.
8. Mosae, S. P., Suresh, E., Subramanian, P. S. (2009). Single stranded helical supramolecular architecture with a left handed helical water chain in ternary copper(II) tryptophan/diamine complexes. *Polyhedron*, 28, 245–252.
9. Mitchell, J. B. O., Nandi, C. L., McDonald, I. K., Thornton, J. M., Price, S. L. (1994). Amino/Aromatic Interactions in Proteins: Is the Evidence Stacked Against Hydrogen Bonding? *J. Mol. Biol.* 239, 315–331.
10. Wang, X., Sarycheva, O. V., Koivisto, B. D., McKie, A. H., Hof, F. A. (2008). Terphenyl Scaffold for π-Stacked Guanidinium Recognition Elements. *Org. Lett.* 10, 297–300.
11. Tomić, Z. D., Sredojević, D. N., Zarić, S. D. Stacking Interactions between Chelate and Phenyl Rings in SquarePlanar Transition Metal Complexes. *Cryst. Growth Des.* (2006). 6, 29–31.
12. Sredojević, D. N., Bogdanović, G. A., Tomić, Z. D., Zarić, S. D. (2007). Stacking vs. CH–π interactions between chelate and aryl rings in crystal structures of squareplanar transition metal complexes. *CrystEngComm,* 9, 793–798.
13. Tomić, Z. D., Novaković, S. B., Zarić, S. D. (2004). Intermolecular Interactions between Chelate Rings and Phenyl Rings in SquarePlanar Copper(II) Complexes. *Eur. J. Inorg. Chem.* 2215–2218.
14. Sredojević, D. N., Tomić, Z. D., Zarić, S. D. (2010). Evidence of Chelate−Chelate Stacking interactions in Crystal Structures of Transition-Metal Complexes. *Cryst. Growth Des.* 10, 3901–3908.

15. Sredojević, D. N., Vojislavljević, D. Z., Tomić, Z. D., Zarić, S. D. (2012). Parallel stacking interactions in squareplanar transition-metal complexes containing fused chelate and C6-aromatic rings. *Acta Crystallogr. B,* 68, 261–265.
16. Grimme, S. (2008). Do Special Noncovalent π–π Stacking Interactions Really Exist? *Angew. Chem. Int. Ed.* 47, 3430–3434.
17. Bloom, J. W. G., Wheeler, S. E. (2011). Taking the Aromaticity out of Aromatic Interactions. *Angew. Chem., Int. Ed.* 50, 7847–7849.
18. Ostojić, B. D., Janjić, G. V., Zarić, S. D. (2008). Parallel alignment of water and aryl rings – crystallographic and theoretical evidence for the interaction. *Chem. Commun.* 6546–6548.
19. Janjić, G. V., Veljković, D. Ž.; Zarić, S. D. (2011). Water/Aromatic Parallel Alignment Interactions. Significant Interactions at Large Horizontal Displacements. *Cryst. Growth Des.* 11, 2680–2683.
20. Ninković, D. B., Janjić, G. V., Veljković, D. Ž.; Sredojević, D. N., Zarić, S. D. (2011). What Are the Preferred Horizontal Displacements in Parallel Aromatic–Aromatic Interactions? Significant Interactions at Large Displacements. *ChemPhysChem,* 12, 3511–3514.
21. Ninković, D. B., Andrić, J. M., Zarić, S. D. (2013). Parallel Interactions at Large Horizontal Displacement in Pyridine–Pyridine and Benzene–Pyridine Dimers, *ChemPhysChem* 14, 237–243.
22. Janiak, C. J. (2000). A critical account on π–π stacking in metal complexes with aromatic nitrogen-containing ligands. *Chem. Soc., Dalton Trans.* 3885–3896.
23. Janjić, G., Andrić, J., Kapor, A., Bugarčić, Z. D., Zarić, S. D. (2010). Classification of stacking interaction geometries of terpyridyl squareplanar complexes in crystal structures. *CrystEngComm* 12, 3773–3779.
24. Sinnokrot, M. O., Sherrill, C. D. (2006). High-Accuracy Quantum Mechanical Studies of π–π Interactions in Benzene Dimers. *J. Phys. Chem. A* 110, 10656–10668.
25. Podeszwa, R., Bukowski, R., Szalewicz, K. (2006). Potential Energy Surface for the Benzene Dimer and Perturbational Analysis of π–π Interactions. *J. Phys. Chem. A* 110, 10345–10354.
26. Pitonak, M., Neogrady, P., Rezac, J., Jurecka, P., Urban, M., Hobza, P. (2008). Benzene Dimer: High-Level Wave Function and Density Functional Theory Calculations. *J. Chem. Theory Comput.* 4, 1829–1834.
27. Janowski, T., Pulay, P. (2007). High accuracy benchmark calculations on the benzene dimer potential energy surface. *Chem. Phys. Lett.* 447, 27–32.
28. Boys, S. F., Bernardi, F. (1970). The calculation of small molecular interactions by the differences of separate total energies. Some procedures with reduced errors. *Mol. Phys.* 19, 553–566.
29. *Gaussian 03,* Revision, C.02; Gaussian, Inc., Wallingford, CT, 2004.
30. Helgaker, T., Klopper, W., Koch, H., Noga, J. (1997). Basis-set convergence of correlated calculations on water. *J. Chem. Phys.* 106, 9639–9646.
31. Veljković, D. Ž.; Janjić, G. V., Zarić, S. D. (2011). Are C-H···O interactions linear? The case of aromatic CH donors. *CrystEngComm*, 13, 5005–5010.
32. Steiner, T. (2002). Hydrogen bonds from water molecules to aromatic acceptors in very high-resolution protein crystal structures. *Biophys. Chem.* 95, 195–201.

33. Malone, J. F., Murray, C. M., Charlton, M. H., Docherty, R., Lavery, A. J. (1997). X-H···π (phenyl) interactions: Theoretical and crystallographic observations. *J. Chem. Soc., Faraday Trans.* 93, 3429–3436.
34. Braga, D., Grepioni, F., Tedesco, E. (1998). X−H–π (X = O, N, C) Hydrogen Bonds in Organometallic Crystals, *Organometallics* 17, 2669–2672.
35. Mignon, P., Loverix, S., Proft, F. D., Geerlings, P. (2004). Influence of Stacking on Hydrogen Bonding: Quantum Chemical Study on Pyridine−Benzene Model Complexes. *J. Phys. Chem. A* 108, 6038–6044.
36. Ebrahimi, A., Habibi, M., Neyband, R. S., Gholipour, A. R. (2009). Cooperativity of -stacking and hydrogen bonding interactions and substituent effects on X-ben||pyr···H–F complexes. *Phys. Chem. Chem. Phys.* 11, 11424–11431.
37. Mignon, P., Loverix, S., Geerlings, P. (2005). Interplay between π–π interactions and the H-bonding ability of aromatic nitrogen bases. *Chem. Phys. Lett.* 401, 40–46.
38. Choudhury, R. R., Chitra, R. (2010). Stacking interaction between homostacks of simple aromatics and the factors influencing these interactions. *CrystEngComm* 12, 2113–2121.
39. Ninković, D. B., Janjić, G. V., Zarić, S. D. (2012). Crystallographic and ab Initio Study of Pyridine Stacking Interactions. Local Nature of Hydrogen Bond Effect in Stacking Interactions. *Cryst. Growth Des.* 12, 1060–1063.
40. Mishra, B. K., Sathyamurthy, N. (2005). π−π Interaction in Pyridine. *J. Phys. Chem. A* 109, 6–8.
41. Mishra, B. K., Arey, J. S., Sathyamurthy, N. (2010). Stacking and Spreading Interaction in N-Heteroaromatic Systems. *J. Phys. Chem. A* 114, 9606–9616.
42. Hohenstein, E. G., Sherrill, C. D. (2009). Effects of Heteroatoms on Aromatic π−π Interactions: Benzene−Pyridine and Pyridine Dimer. *J. Phys. Chem. A* 113, 878–886.
43. Piacenza, M., Grimme, S. (2005). Van der Waals Interactions in Aromatic Systems: Structure and Energetics of Dimers and Trimers of Pyridine. *ChemPhysChem* 6, 1554–1558.
44. Grimme, S. (2006). Semiempirical hybrid density functional with perturbative second-order correlation. *J. Chem. Phys.* 124, 034108/1–034108/16.
45. Grimme, S. (2006). Semiempirical GGA-type density functional constructed with a long-range dispersion correction. *J. Comput. Chem.* 27, 1787–1799.
46. Weigend, F., Ahlrichs, R. (2005). Balanced basis sets of split valence, triple zeta valence and quadruple zeta valence quality for H to Rn: Design and assessment of accuracy. *Phys. Chem. Chem. Phys.* 7, 3297–3305.
47. *ORCA*, version 2.8; University of Bonn, Bonn, Germany.
48. Becke, A. D. (1993). Density-functional thermochemistry. III. The role of exact exchange. *J. Chem. Phys.* 98, 5648–5652.
49. Becke, A. D. (1988). Density-functional exchange-energy approximation with correct asymptotic behavior. *Phys. Rev. A* 38, 3098–3100.
50. Lee, C., Yang, W., Parr, R. G. (1988). Development of the Colle-Salvetti correlation-energy formula into a functional of the electron density. *Phys. Rev. B* 37, 785–789.

CHAPTER 10

NANOCHIPS FOR MASS SPECTROMETRY AND APPLICATIONS IN BIOMEDICAL RESEARCH

ALINA D. ZAMFIR

CONTENTS

ABSTRACT

Development and implementation in bioanalytical sciences of micro/nanofluidics is attracting a high interest due to the advantages exhibited by these systems among which the ability to manipulate and analyze minute amounts of sample, increased throughput, reproducibility and sensitivity as well as reduced costs. Recently, by introducing mass spectrometry (MS) in combination with micro/nanofluidics the potential and the applicability of the chip systems in life sciences increased significantly. This chapter is dedicated to the technical and methodological developments as well as the first implementation in glycan analysis for biomedical research of several micro and nanofluidics for electrospray ionization mass spectrometry. The first parts of the chapter document in details the research conducted for: a) construction and optimization of two different sheathless interfaces for capillary electrophoresis coupling to mass spectrometry via electrospray ionization; b) coupling of fully automated silicon nanochip-based electrospray systems to three different mass spectrometers (quadrupole time-of-flight (QTOF), Fourier transform ion cyclotron resonance (FTICR), and high capacity ion trap (HCT)); c) coupling of thin polymer microsprayer chips to two different mass spectrometers (QTOF and FTICR MS); d) optimization of all coupled systems for functioning on-line in MS mode for screening, tandem MS (MS/MS) and multistage MS (MS^n) fragmentation by collision-induced dissociation (CID). The last part presents the state-of-the-art related to the most relevant applications of these analytical platforms, which were developed and introduced in glycomics for biomedical and clinical applications: screening, sequencing and structural analysis of *O*-glycopeptides expressed in the urine of patients suffering from Schindler's disease *vs.* age-matched healthy controls; mapping, sequencing, structural analysis and postulation of ganglioside/glycolipid biomarkers in healthy and diseased central nervous system.

The highlighted applications demonstrate not only the feasibility of the novel micro/nanofluidics-MS methods for clinical investigations but also their superiority in terms of analysis speed and sensitivity as well as data accuracy and reliability for *de novo* identification of valuable biomarkers.

10.1 MICRO AND NANOCHIPS FOR ELECTROSPRAY MASS SPECTROMETRY

10.1.1 ELECTROSPRAY IONIZATION MASS SPECTROMETRY

Mass spectrometry is an instrumental approach that allows for the determination of the molecular mass, therefore it is often called "the smallest scale in the world." The evolution of mass spectrometry has been marked by an ever-increasing demand for its application to problems of major difficulty such as biomolecule analysis, and the explosion of computer science. Though the new developments in the technology have created a complex and sophisticated array of instruments, the basic components of all mass spectrometers are the same: the ion source, the mass analyzer and the ion detector. The ion source ionizes the molecule of interest, then the mass analyzer differentiates the ions according to their mass-to-charge ratio (*m/z*) and finally, a detector measures the current of the ionic beam. Each of these elements exists in many forms and is combined to produce a wide variety of mass spectrometers with specialized characteristics.

Among all ionization techniques, electrospray (ESI) is one of the most fascinating as it is able to generate, at atmospheric pressure, ions directly from solution, which makes it applicable to a large class of nonvolatile substrates.

Initial experiments carried out by the physicist John Zeleny in 1917 [1] preceded the first description by Dole et al. in 1968 of the electrospray principle, including *the charge residue model* (CRM) which has survived as a main explanation for the controversial ESI process [2].

However, the well-defined breakthrough of ESI as a general ionization method came in 1988 when John B. Fenn presented his experiments on identification of polypeptides and proteins of 40 kDa molecular weight. Fenn showed that a molecular-weight accuracy of 0.01% could be obtained by applying a signal-averaging method to the multiple ions formed in the ESI process. The findings were based on experiments started in 1984 [3] in Fenn's laboratory at Yale, when electrospray and mass spectrometry were successfully combined for the first time. Fenn used his knowledge of free-jet expansion to improve Dole's method with a counterflow of gas for desolvation, eliminating resolvation of formed macromolecular ions. This discovery was closely followed by results from a Russian research group [4].

In ESI, basically, the liquid containing the analyte of interest is pumped through a metal capillary, which has an open end with a sharply pointed tip (Fig. 10.1).

The tip is attached to a voltage supply and its end faces a counter electrode plate. As the voltage is increased, the liquid becomes charged and due to charge-repulsion effect, it expands out of the capillary tip forming the so-called *Taylor* cone. Since all droplets contain the same electrical charge, at the very end of the cone, they emerge into a fine spray called ESI plume [5]. Depending on the polarity of the applied electric field, the charges may be positive or negative. The droplets are usually less than 10 micrometers across and contain both solvent and analyte molecules. The charged droplets move across the electric field existing between capillary and counter electrode and, under a curtain gas flow, the solvent molecules evaporate from the droplet. According to Dole's CRM, as the droplet size decreases while the total charge on the droplet is constant, the charge surface density increases until the droplet's surface tension is exceeded by the repulsive electric forces. At this critical point, the droplet explodes into smaller, still highly charged droplets. This process, called *Rayleigh explosion*, repeats itself until the analyte molecule is stripped of all solvent molecules, and is left as a multiply charged ion.

FIGURE 10.1 Electrospray ionization process (courtesy of New Objectives Inc.).

The number of charges retained by an analyte depends on such factors as the composition and pH of the electro sprayed solvent as well as the chemical nature of the sample. For small molecules (<2000 Daltons) ESI typically generates singly, doubly or triply charged ions, while for large molecules (>2000 Daltons) the ESI process typically generates a series of multiply charged species and the resultant ESI mass spectrum contains multiple peaks corresponding to the different charge states.

This feature brings complexity to the interpretation of the ESI mass spectra but concomitantly, as a first advantage, it adds to the information and can be used to improve the accuracy of the molecular-weight determination. The method of deducing this way the molecular weight was described in *the multiple charge theory* described by Fenn [6]. The theory showed that different charge states could be interpreted as independent measurements of molecular weight and that an averaging method based on the solution of simultaneous equations could provide accurate molecular weight estimations for large molecules. The complex charge pattern can simply be deconvoluted and the mass of the uncharged protein is determined to dramatically higher accuracy than if the interpretation of data was based on a single ion. The second advantage of multiple charging is the formation of ions with reduced *m/z* ratio measurable with good resolution by almost any type of analyzer with which ESI has been interfaced: magnetic sector, single or triple quadrupole, time-of-flight, quadrupole ion trap, Fourier-transform ion cyclotron resonance or hybrid quadrupole time-of-flight analyzer. All these make ESI the method of choice for large biopolymers and molecular aggregates or complexes that only have weak noncovalent interactions, such as protein-protein, enzyme-substrate or protein-ligand complexes.

10.1.2 MICRO- AND NANOFLUIDICS

At the beginning of the 90′ the continuous refinement of the electrospray as ion source in MS culminated with the low-flow (micro and nanoESI) systems, which provide sensitivities at subpicomolar level [7, 8]. However, in bioanalysis the major challenge for nanoESI MS was the high heterogeneity of the biological samples in terms of number of components and the diversity of their structure. Therefore, the combination of powerful liquid separation techniques with high sensitive MS detection modes started to attract a great interest due to the foreseen possibility to separate and directly identify the molecules in an on-line MS experiment [9–11]. The advantage of ESI MS to form ions directly from solution has established the technique as a convenient mass detector for high performance liquid chromatography (HPLC) or capillary electrophoresis (HPCE) [12–15]. While past attempts to couple LC or CE with mass spectrometry resulted in limited success, ESI has made online HPLC- and CE-MS possible, adding a new dimension to the capabilities of these techniques for biomolecule characterization. In particular, the high separation efficiency, sensitivity and selectivity offered by the capillary electrophoresis made the CE ESI MS coupling a method of choice in complex mixture analysis [13].

In the postgenome era, MS develops continuously as one of the most powerful analytical technique for structural elucidation of molecules originating from biological matrices. As shown before, potentials of MS for high-sensitive structural bioanalyses increased significantly after the introduction of ESI and MALDI methods from one side and the possibility to sequence complex ionic species by highly efficient dissociation techniques based on multiple stage MS (MS^n) on the other. In particular, in proteomics, glycomics and glycoproteomics, nanoESI MS^n in the positive as well as in the negative ion mode was shown to be capable in sequencing minute amounts of biological material thus providing straightforward information on various structural elements [16–19].

On the other hand, miniaturized analytical instrumentation is attracting growing interest in chemical, biochemical and structural analysis. Nowadays, massive effort is invested in MS interfacing to micro/nano-fluidic-based systems as front-end technologies for ESI [20–23]. The generic term *microfluidics* refers to all analytical tools where fluids can be driven in microstructured channels and/or narrow capillaries. Hence, in terms of MS interfacing, *micro/nano-fluidics* are: (i) stand-alone specialized devices/instruments like capillary electrophoresis, micro and nano LC, etc.; (ii) integrated mono-or polyfunctional micro/nanosystems such as silicon, glass, or polymer chips; (iii) complex devices combination of automated sample delivery and chip-based ionization such as automated chip-based robots for sample MS infusion by ESI.

10.1.2.1 CAPILLARY ELECTROPHORESIS/ELECTROSPRAY IONIZATION MASS SPECTROMETRY

Capillary electrophoresis (CE) is an instrumental evolution of traditional slab gel electrophoretic techniques and is based on differences in solute velocity in an electric field [24–26]. In capillary electrophoresis the electromigration of the analytes is taking place in narrow-bore capillaries, which includes CE in the category of microfluidic devices. The narrow capillaries allow the application of a high electric field up to 0.6 kV/cm thus enhancing a very high efficiency of the separation.

In Fig. 10.2, the scheme of a capillary electrophoresis setup with UV detection is depicted. The assembly consists of a fused silica separation capillary of 20–100 μm i.d. 20–100 cm length, two buffer vials A and B and another one containing the solution of analyte, a high voltage power supply (delivering up to 30–40 kV and 200–250 μA) and a detector which can be of various types: UV detector, electrochemical detector, laser induced fluorescence (LIF) or MS.

The two ends of the capillary are immersed in the vials containing electrolyte together with two electrodes. A high voltage source delivers the potential difference between the electrodes necessary for the separation. The UV detector is placed at 5–10 cm distance from the outlet capillary.

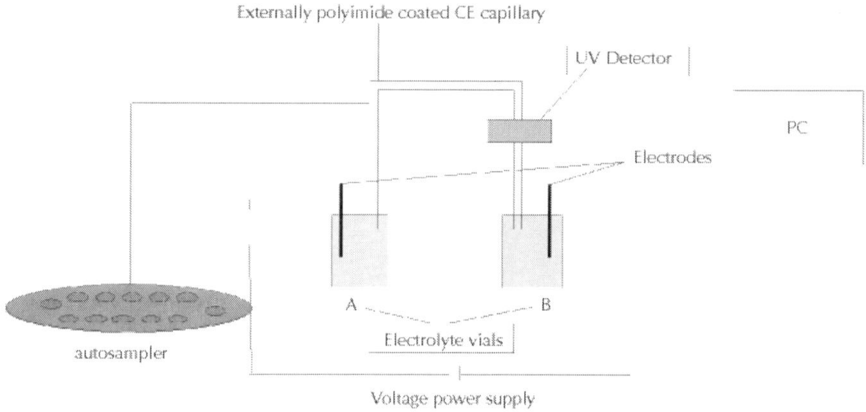

FIGURE 10.2 Basic CE setup [27].

The sample, usually dissolved in electrolyte is injected into the capillary by either (i) hydrodynamic injection using pressure or vacuum application (ex. 0.5 psi) while the injection end of the capillary is inserted in the vial containing the analyte solution, or (ii) electrokinetic injection induced by application of potential difference. The dispersion processes limits the amount injected in the capillary. A practical limit of injection plug length is less than 1–2% of the total capillary length meaning nl or pl volumes.

Different types of capillaries are used for the CE separation: glass, Teflon, polymer capillaries etc. However, fused silica capillaries are the best option, as they meet all requirements claimed by the CE technique: chemically and physically resistant, transparent to UV radiation, able to dissipate Joule heat, narrow internal diameters [12, 27]. The silica capillaries are externally "coated" with a layer of polyimide and internally may be uncoated, or coated with polymers suppressing the adsorption of the analyte to the capillary walls.

CE separation is usually carried out at constant potential in direct polarity (injection at anode and detection at cathode) or reverse polarity (injection at cathode and detection at anode). However, constant current mode can also be performed, especially when the temperature control of the capillary is not efficient. Not commonly used, gradients or steps in the voltage may be useful in simultaneous analysis of compounds having very different electrophoretic mobility [28, 29].

In principle, CE separates the species according to their migration velocity under the influence of a high electric field [28, 29]. The difference in solute velocity is given by the different electrophoretic mobility as expressed by the Eq. (1):

$$\mathbf{v}=\mu_e\mathbf{E} \tag{1}$$

where **E** is the electric field vector and $\mu_{e\text{-}}$ the electrophoretic mobility, constant for a certain ion and medium and determined by the electric force balanced by the frictional one (Eqs. (1.3) and (1.4)):

$$\mu_e=q/6\pi\eta r \tag{2}$$

$$\mu_a=\mu_e+\mu_{eof} \tag{3}$$

where: q-ion charge, η-viscosity, r-ion radius, $\mu_{a\text{-}}$ apparent electrophoretic mobility and $\mu_{eof\text{-}}$ mobility of the electro osmotic flow.

The above relations show that the electrophoretic mobility of an analyte is depending on several factors: charge of the analyte, pH of the solution, viscosity, temperature inside the capillary, *m/z* ratio, the applied electric field, dimensions of the capillary. Therefore, the choice of the electrolyte is a key step in performing an efficient separation, imposing the optimization of all its parameters: pH, ionic strength, chemical composition, and concentration.

An essential parameter of the CE separation is the *electro osmotic flow* (EOF), a consequence of the surface charge on the inner capillary wall. Under certain solution conditions (pH > 3.0) fused silica surface possess an excess of negative charges due to the ionization of silanol group. The counter ions, which balance the surface charge, forming the diffuse double layer at the capillary wall, create a potential difference [27]. When applying a voltage across the capillary, the positive ions of the electrical double layer are attracted toward the cathode. Due to solvation, the ionic movement drags the bulk flow solution creating the electro osmotic flow under the electric field. $\mathbf{F}_{eof}$, the force of the EOF, is one order of magnitude higher than $\mathbf{F}_e$ the electrophoretic force; therefore EOF causes the movement of all species regardless the charge in the same direction (Fig. 10.3).

The separation occurs under the action of different $\mathbf{F}_e$ [27] so that if the mixture contains positive and negative ions as well as neutral species, (Fig. 10.3) in forward or direct polarity the first will elute the positive ions, followed by the molecules, which did not undergo ionization. The negative ions, dragged by $\mathbf{F}_{eof}$ will also migrate toward the cathode but will elute later. In reverse polarity, the EOF is oriented against the desired direction of ion motion. It may be suppressed by careful reconsideration of solution parameters (pH < 3.0). In this case the $\mathbf{F}_e$ becomes concomitantly the drift and separation force-giving rise to high separation efficiency and resolution.

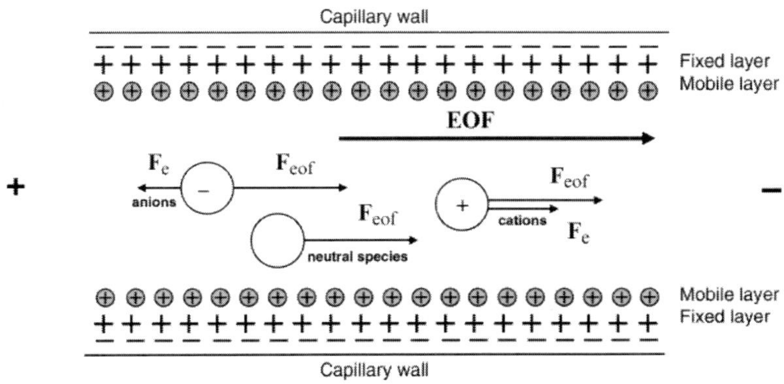

FIGURE 10.3 Capillary electrophoresis mechanism [27].

10.1.2.2 ON-LINE CE ESI MS COUPLING

The utility of CE towards biomolecule screening can be greatly enhanced by mass spectrometric (MS) detection and identification. By combined CE MS technique, not only molecular masses of the separated components can be directly measured with good accuracy but also specific fragment ions may be generated by tandem MS (MS/MS) from individual components in a complex mixture, to deduce the molecular structure.

Direct coupling of CE to MS has been introduced for over 10 years and nowadays the most attractive method for coupling capillary separation technique to MS is via ESI interface [30–32]. The on-line CE ESI MS coupling requires however, an interface fulfilling the requirements related to an efficient transfer of the sample from the CE capillary into MS without affecting the CE separation efficiency. Numerous parameters are influencing the CE ESI MS analysis and are to be taken into account for optimization [33]: i) the choice of an electrolyte compatible with both the ionic species formation/separation by CE and electrospray process; ii) the interface design and configuration; iii) range of the applied CE and ESI potentials; iv) performance of the formed CE MS electric circuit; v) fine positioning of the sprayer with respect to the MS sampling orifice; vi) the general solution and instrumental parameters such as buffer and sample concentration, pH, injection time and pressure, capillary temperature, etc.

The most widely used interface for CE ESI MS setup is the sheath liquid interface which was conceived in various configurations but basically in all of them a sheath liquid, serves to ensure the electrical contact between the CE capillary and the ESI sprayer. In principle, the make-up liquid picks up the analyte eluting from the CE capillary in a solvent appropriate for ESI and the whole resulting mixture is then sprayed into the mass spectrometer.

In all sheath flow interfaces, the make-up liquid mixes the CE buffer and the sample, therefore, the major drawback of this interface is the reduced sensitivity consequence of sample dilution. To overcome the disadvantages of the sheath flow interfaces, *Olivares* et al. [32] proposed a novel CE ESI MS setup, which eliminated the addition of a make-up liquid and used instead a stainless steel capillary sheath for electrical contact. This type allows for a sheathless CE ESI MS design spraying the sample directly from the CE microspray tip into the mass spectrometer. The effects of the use of this kind of microsprayers include higher analyte concentration, lower spraying potential, closer positioning of the sprayer to the orifice of the mass spectrometer and significant improvement of ion transfer into MS [33]. Moreover, ionization and desolvation of the generated droplets are also improved contributing to the significant gain in sensitivity. Hence, the best alternative for the sheath flow design is the high sensitivity sheathless interface. The most effective sheathless design is based on coating the sprayer tip with a conductive layer to provide the electrical contact needed for both CE and ESI. Very popular sheathless interfaces are those based on applying either noble metals [34], or carbon [35] on the spraying tip, which may be either the CE capillary itself, etched as a microsprayer, or a spraying needle butted to the CE column.

Despite the great number of CE ESI interface types, so far, no general consensus has been reached regarding the best design and no approach generally valid for the analysis of complex biomolecule mixtures has been reported.

Most of the on-line CE ESI MS couplings are, however, constructed and completely functional for peptide and protein characterization but only a few of them have been introduced for complex carbohydrate system analysis [33].

10.1.2.3 CHIP-BASED ELECTROSPRAY IONIZATION

The actual trends in analytical science are toward the high-throughput measurements based on the nanotechnology achievements in automatization and miniaturization of devices. In bioscience integrated, fully automated nanosystems functioning on the basis of the *"lab-on-a-chip"* principle have been demonstrated to provide one of the most rapid, sensitive and accurate analysis [20, 36].

The potential of the modern chip-based ESI systems considerably broadened the area of MS applicability in life sciences. The option for miniaturized, integrated devices for sample infusion into MS is driven by several technical, analytical and economical advantages such as [37]: 1) simplification of the laborious chemical and biochemical strategies required currently for MS research; 2) high throughput nanoanalysis/identification of biomolecules; 3) elimination of the time-consuming optimization procedures; 4) increase of the sensitivity by drastically reduction of the sample and reagent consumption, sample handling and potential sample loss; 5) high reproducibility of the experiments; 6) potential to discover novel biologically relevant structures due to increased ionization efficiency; 7) high signal-to-noise ratio; 8) reduced *in-source* fragmentation; 9) flexibility and broad area of applicability; 10) low cost of analysis and chip production; 11) possibility for unattended high-throughput experiments reducing the man power and man intervention; 12) possibility to perform several stages of sample preparation in a single integrated unit followed by direct MS structural analysis; 13) elimination of possible cross-contamination and carry-overs; 14) flexibility for different configurations, upgrading and modifications; 15) minimal infrastructure requirements for optimal functioning; 16) reduction of the ion source size facilitating manipulation and efficient ion transfer by precise positioning towards the MS sampling orifice.

Two types of chip-based devices are currently being investigated and used for ESI MS studies. The first category is represented by the out-of plane devices, where hundreds of nozzle-like nanospray emitters are integrated onto a single silicon substrate from which electrospray is established perpendicular to the substrate [38]. These devices are particularly well suited to high-throughput sample delivery to ESI MS [20, 39] and have the potential to completely replace flow-injection analysis assays. Moreover, the technical quality of the nanosprayers obtained by silicon microtechnology is so high, and the experiments so reproducible, that such devices have been found in some instances to give more robust and quantitative analyses than LC- or CE MS [20] and to be able to suppress the need for LC or CE separation prior to MS analysis.

Due to very efficient ionization properties, silicon-based nanochip ESI MS preferentially forms multiply charged ions, and the *in-source* fragmentation of labile groups attached to the main biomolecular framework is minimized. These features further enhance the significantly increased ionization efficiency and sensitivity of the analysis.

The second category consists of planar or thin microchips, made from glass [40] or polymer material [41, 42], embedding a microchannel at the end of which electrospray is to be generated in-plane, on the edge of the microchip. In recent years, the progress in polymer-based microsprayer systems was promoted by development of simpler methods for accurate plastic replications and ease to create lower-cost disposable chips. Though clearly amenable to automation for high-throughput analysis, these designs are best suited to the integration of other analytical functions prior to sample delivery to the MS system, such as sample cleanup, analyte separation by CE, and chemical tagging [43, 44]. Moreover, in comparison with glass nanospray capillaries, these thin microsprayers were found to provide superior stability of the spray with time, improved signal-to-noise ratios at various flow rates and high flexibility and adaptability to different ion source configurations, with the additional advantage of cheaper production costs compared to silicon technologies.

Several approaches for interfacing MS to polymer chips providing flow rates, which include them in categories from nano- to microsprayers, were reported [20, 45, 46]. Different MS instruments such as single and triple quadrupole MS ion trap and ultra high resolution Fourier-transform ion cyclotron resonance mass spectrometry (FTICR MS) were adapted to polymer-based chip ESI and contributed significant benefits for various studies. However, the general areas of implementation and applications of chip technologies either silicon-based or polymer microchips have by far been primarily genomics, drug discovery, and proteomics. Despite the high potential and performance exhibited by chip based MS methodologies, the glycomics field has not significantly benefited from the chip technology. Optimization of ESI MS and chip ESI MS for operation in the necessary negative ion mode was considered a challenging task, and application to carbohydrate analysis limited by their high structural diversity and microheterogeneity [20]. Moreover, in addition to the low ionization efficiency that ultimately leads to decreased sensitivity, each class of carbohydrates requires particular and defined conditions to promote chip ionization and detection by MS. These conditions are depending on the type of the labile attachments on the sugar chain, the ionizability of the functional groups, the hydrophilic and/or hydrophobic nature, branching of the sugar chains, etc.

Modern achievements in robotics are currently also intensively introduced in the MS field, tending to substitute the existing classical ESI which requires either manual loading or pumping of the sample liquid through the electrospray capillary. Automated, robotized and programmed systems for sample delivery into MS are significantly increasing the analysis throughput and efficiency, allowing minimization of sample volumes and handling.

Though of maximum efficiency and intensively used for proteomic analysis and genomic surveys, the latest developments in robotized chip-based ESI MS have been only to a limited extent introduced in glycomics. This can be rationalized also by the special requirements, described above, for ionization/detection of long saccharide chains and saccharides having labile attachments, which are more difficult to be fulfilled by high throughput experiments. Due to the high heterogeneity of the carbohydrate mixtures originating from biological sources, automated chip ESI MS analysis has often to be preceded by laborious chromatographic or electrophoretic methods in conjunction with other biochemical tools. Additionally, as mentioned before, for efficient ionization and/or fragmentation in tandem MS, different instrumental conditions are required by each particular carbohydrate category therefore the optimization stage is rather laborious. All these attributes of carbohydrate mass spectrometry, made this biomolecule category less amenable to chip-based and high-throughput methods [20]. However, in the second part of this chapter it will be demonstrated that adequate chip MS and automated chip ESI tandem MS strategies may lead to successful implementation of this technology in glycomics for biomedical research.

10.1.3 INTERFACING MICRO/NANOFLUIDICS SYSTEMS TO MASS SPECTROMETRY

10.1.3.1 QUADRUPOLE TIME-OF-FLIGHT (QTOF) MASS SPECTROMETER

QTOF MS instrument incorporates a high performance quadrupole analyzer equipped with a prefilter assembly to protect the main analyzer (time-of-flight) from contaminating deposits, and an orthogonal acceleration time-of-flight (TOF) analyzer. A hexapole collision cell between the two MS analyzers is used to induce fragmentation by collision-induced dissociation (CID) to assist the structural investigations [12, 20]. Ions emerging from the second mass analyzer are detected by a microchannel plate (MCP) detector and ion counting system. A post acceleration photomultiplier detector situated after the orthogonal acceleration cell, is used to detect the beam passing through the first stage of the instrument for tuning and optimization.

In QTOF MS, the ionic pathway is the following: the Z-spray source leads the ions/ionized droplets perpendicular to the spray direction through the counerelectrode (sampling cone) lets them pass the heater and further, via turbulent motion into the RF-only hexapole ion guide. The advantage of this set-up is that nonionized species do not enter the analyzer region. In the MS mode the quadrupole acts as a wide band-pass filter to transmit a wide mass range (RF-only mode) through the hexapole collision cell into the pusher region of the TOF analyzer, whereas in the MS/MS mode the quadrupole operates in the normal resolving mode and is used to select precursor ions. The precursor ions are then accelerated into the RF-only collision cell, where CID occurs using Ar as collision gas. The ionic beam of the fragment ions is then focused into pusher by acceleration, focus steer and tube lenses. The pusher is pulsing a section of the beam towards a reflectron, which reflects the ions back to the detector. As the ions are traveling from the pusher to the detector they are separated in mass according to their flight times, with ions of the highest *m/z* ratio arriving latter. The pusher may operate at repetition frequencies of up to 20 kHz resulting in a full spectrum being recorded by the detector every 50 μs. Each spectrum is summed in the histogram memory of a time to digital converter (TDC) until the histogram spectrum is transferred into a host PC running a software system, which controls the instrument, acquires and processes the data.

Unlike scanning instruments, the TOF performs parallel detection of all masses within a spectrum at high sensitivity, resolution and acquisition rates. The latter characteristic is of very high importance when the MS is functioning coupled to high performance CE (HPCE) or other fast chromatographic separation techniques [12] since each spectrum is representative of the sample composition at that point of time, irrespective of how rapidly the sample composition is changing.

10.1.3.2 COUPLING OF HPCE TO QTOF MASS SPECTROMETER

The most accessible approach for ESI MS characterization of CE separated components is the off-line collection of fractions. Fraction collection can be performed using different techniques such calculating the time window when a compound has migrated to the end of the capillary or using a pre-run to obtain a CE profile and estimate the migration velocities [12, 47]. Unlike on-line coupling, off-line method provides higher flexibility toward system optimization since the CE instrument and the mass spectrometer can be adjusted independently and optimized separately. Additionally, postseparation treatment of the samples, prior to MS analysis, like concentration by solvent evaporation, modification of buffer composition, dialysis, centrifugation etc. are possible [47]. However, using the CE instrument as a fraction collector lowers the eluted sample concentrations by the method principle itself, which imposes the collection of a few nanoliters into tens of microliter volumes of electrolyte. Lack of sensitivity is therefore the specific drawback of this approach often limiting the extension of its applicability toward minute amount of samples deriving from biological matrices. Anyway, off-line CE ESI MS method may be successfully used as a prerequisite for the direct on-line coupling [47].

Due to its sensitivity and high reproducibility, on-line CE ESI MS coupling gained in popularity as the most convenient approach to interfacing CE and MS instruments [33]. It allows minimum sample handling, which significantly reduces potential sample loss and uses the mass spectrometer as an extremely selective and sensitive detector acquiring either the total (TIC) or extracted (XIC) ion chromatograms. However, a number of already widely known problems are inherently associated even with this method [47]. From the separation point of view, a fundamental and general concern is that the best suited CE buffers are usually incompatible with the electrospray process being nonvolatile and causing unstable behavior with the ESI source. Therefore, the compatibility of the CE electrolyte as a spraying agent is one of the major difficulties encountered when interfacing either off-line or on-line the CE to ESI MS. From this perspective, carbohydrates either oligosaccharides or glycoconjugates provide even more limited number of options because of the restrictive required conditions for ion formation, separation and detection [12, 48]. The necessity of high sensitivity, high-speed MS acquisition, especially for on-line tandem MS, and optimization of the coupling for detecting carbohydrate ions in the negative mode [49] represent the second class of requirements for a successful and efficient on-line CE MS glycoscreening.

Sheathless design provides the best sensitivity, therefore, for biological applications when the amount of available analyte is limited, sheathless configuration is practically the only possible option. This type of design allows for two variants and both of them have been optimized, tested and implemented in carbohydrate research with biological and clinical relevance.

The first interface was based on a one-piece CE column with its terminus *in-house* modeled as a microsprayer [48]. The emitter at the CE column terminus is a critical part because it has to have simultaneously microsprayer properties and to act as an electrode as well. This means that it should have a smooth and well-defined tip, which also maintains the electrical contact across the CE buffer and produces a fine spray under mild electrospray voltage conditions. In addition, it should be resistant to manipulations, durable, and suitable to certain type of analysis and not to introduce dead volumes because this dramatically deteriorates the CE separation efficiency.

The second interface, of improved sensitivity and ionization efficiency, consists of two pieces, the CE column being butted to a nanosprayer needle [50]. In this case special attention had to be paid to the butted region and system assembling into the *in-house* made holder, a device responsible also for delivering the high-voltage.

10.1.3.2.1 SHEATHLESS ON-LINE CE/MICROESI QTOF MS INTERFACE

In this setup the ESI emitters were manufactured at the CE capillary terminus so that, the interface consists of single separation column [48]. CE capillaries were sharpened at one end by locally heating in a flame of corresponding melting temperature. Meanwhile, the capillary tubing was manually pulled apart. In order to obtain a very well defined tip orifice, the tiny long wire obtained by pulling under flame has been removed with a tapered ceramic cutter under visual inspection with a stereomicroscope. After this operation, the measured outer diameter of the tip was around 100 μm. The orifice size has been calculated from $d_2=d_1d_{02}/d_{01}$ (where d_{01} and d_1 are the initial and final outer diameters, respectively and d_{02} the initial inner diameter) where it was assumed that by pulling under flame the inner and outer diameters decreased proportionally. Thus, from a tip of 90 μm o.d. an orifice diameter of about 15 μm resulted. Although a capillary tip with even smaller inner diameter would give a better sensitivity, practical aspects such the ease of production, use and manipulation as well as the possibility of capillary blockages determined the choice of the tip diameter.

To further reduce the outer diameter of the tapered tips and to smooth the edge, the tips were etched in hydrofluoric acid (HF) a procedure that turned out to be a crucial step in manufacturing the CE microsprayer tips. Measurements performed using a tapered capillary tip without HF etching showed that the obtained sharp tip did not generate a stable spray. After etching, a good quality steady spray under mild ESI voltages was achieved at the low electro osmotic flows of about 50 nL/min.

The next stage of CE ESI tip producing was the metallic deposition. Although in recent years several metallization techniques have been proposed, lack of metallic deposition durability was one of the major drawback in almost all of designs. In this study all CE tips were coated with copper. The coating process was performed by smearing the surface of the tip with a thin liquid layer of copper suspension in dimethylether. Copper was deposited as a thick layer on a length of 5–6 cm from the capillary tapered end and as a very thin layer in the near vicinity of the tip.

The stability of the electrical contact at CE terminus is another crucial factor for successful on–line CE ESI MS analysis. The stability of the CE current was tested in the running buffers and the CE current, which indicates the electrical performances of the circuit, was found stable in both positive and negative ion mode for CE voltages of 15–30 kV.

Several advantages of this design for sheathless CE ESI MS applications in glycoscreening are to be mentioned [12, 33, 48]. First, a one-piece separation column provides a better separation and a higher reproducibility since, in comparison with the butted one, there is no the misalignment danger which could require, during the same experiment, several interventions for finding appropriate connection points between the tip and original column. Next, copper deposition proved to be very resistant and stable in inevitable contact with aggressive media such as highly alkaline buffers used for sugar CE separation in direct polarity. Finally, although the copper coat is not everlasting, the electrical contact could be maintained under ESI voltage around 30 h without need for supplementary Cu deposition. Once the electrical contact degraded, the copper coat from the surface of the tip could be several times readjusted retaining the original emitter.

In order to realize the electrical contact to the ESI power supply, the shaped CE column was introduced into a metallic body, home made stainless steel device for clenching the capillary. The whole system was screwed onto the ESI high voltage plate of the QTOF MS (Fig. 10.4). Exchange between the conventional source and home-built CE MS interface did not claim for any definitive dismantling or special mechanical modifications [48].

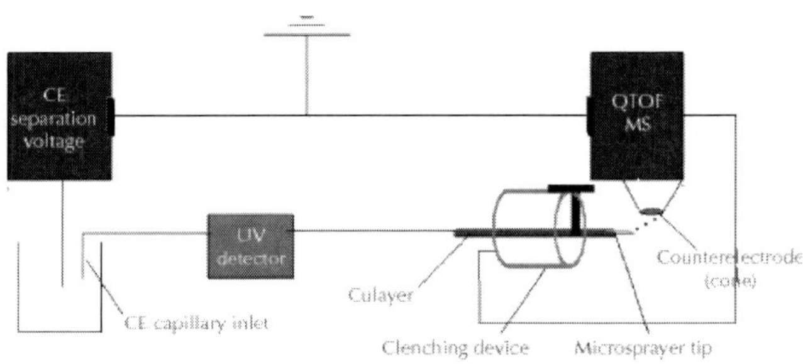

FIGURE 10.4 Electrical scheme of the sheathless on-line CE microESI QTOF MS [48].

The tapered tip extended 1–2 cm over the holder and the position of CE emitter was adjusted in the vicinity of the entrance hole of the counter electrode sampling cone, at a distance less than 5 mm, by the source assembly which can be manipulated in *x*, *y* and *z* direction via micrometer screws, although fine positioning of the microsprayer tip turned out not to be a critical parameter.

10.1.3.2.2 SHEATHLESS ON-LINE CE NANOESI QTOF MS INTERFACE

The highest sensitivity achieved by using the interface described above was 20 pmol/μL. The low amount of glycans resulting after the usual sample extraction/preparation protocols, the mild ESI source parameters and the high ionization efficiency necessary for proper ionization of some carbohydrate categories such as glycosaminoglycans, required the development of a novel nanosprayer design to address all these issues. In this new setup [50–52], the CE column was butted to a commercial (New Objectives Company) nanosprayer needle by using a home-made joint. The resulting two-pieces-column is incorporated into the stainless steel clenching device, designed and constructed to allow the application of the ESI voltage onto the needle. The whole set up was mounted by a set of screws directly onto the ESI high voltage QTOF plate. The electrical scheme of the CE nanoESI QTOF MS setup is depicted in Fig. 10.5.

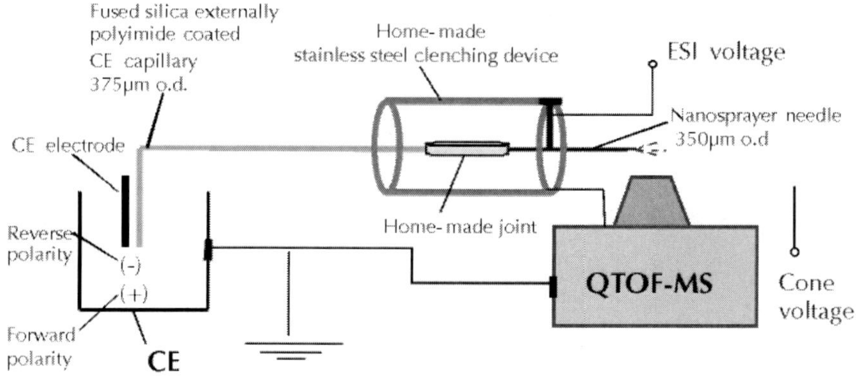

FIGURE 10.5 Electrical scheme of the sheathless on-line CE nanoESI QTOF MS [50].

The spray could be initiated at values of 6–900 V applied to the nanoESI needle and 12–30 V for the sampling cone potential without the need of nebulizer gas. Leakages in the butted region or broadening of the total ion current (TIC) peaks were not observed proving that no misalignment of the connection or gap between the CE column and the nanosprayer needle were created.

The sheathless CE nanosprayer system was optimized for glycoscreening in both forward and reverse polarity and for two different buffer systems ammonium acetate/ammonia at high pH values (strongly alkaline for forward polarity) and formic acid/ammonia low pH values (strongly acidic for reverse polarity).

10.1.3.3 COUPLING OF THE FULLY AUTOMATED NANOCHIP-BASED ELECTROSPRAY IONIZATION TO QTOF MASS SPECTROMETER

Fully automated nanochip-based electrospray, NanoMate™ 100 incorporating ESI Chip technology (Advion BioSciences, Ithaca, USA), was coupled to the QTOF MS and for the first time optimized for glycomics surveys [51]. The robot and the incorporated silicon chip are presented in Fig. 10.6.

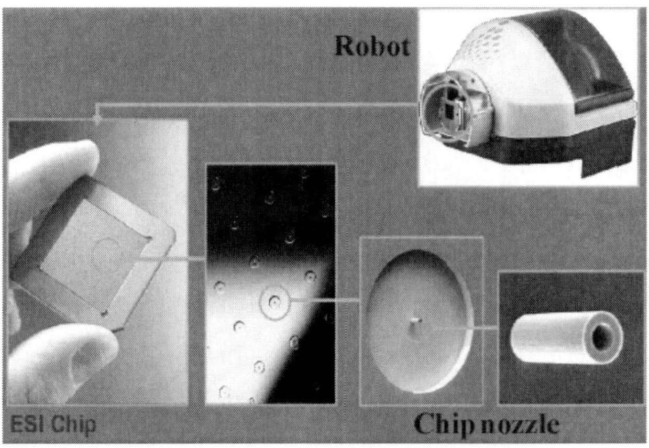

FIGURE 10.6 NanoMate™ robot (courtesy of Advion BioSciences).

NanoMate™ 100, the world's first fully automated nanoelectrospray system, is a robotic device that provides an automated nanoelectrospray ion source for mass spectrometers. The system is capable of infusing samples at low flow rates (50–100 nL/min) in an automated fashion. In addition to the robot itself, the key component of the system is the ESI nanochip. The robot holds a 96-well sample plate and a 96-pipette tip tray. Automated sample analysis is achieved by loading a disposable, conductive pipette tip on a moveable sampling probe, aspirating sample via a syringe pump, and moving the sampling probe to engage against the back of the ESI nanochip.

Nanoelectrospray is initiated by applying a head pressure and voltage to the sample in the pipette tip. Each nozzle and tip is used only once in order to eliminate carryover and contamination typical to conventional autosamplers. The nanochip is an array of nanoelectrospray nozzles of 10 μm internal diameter etched in a planar silicon chip. The chip is fabricated from a monolithic silicon substrate using deep reactive ion etching (DRIE) and other standard microfabrication techniques.

The inert coating on the surface allows a variety of acidic and organic compositions and concentrations to be used to promote ionization without degradation of the nozzle. As visible in Fig. 10.6 a channel extends from the nozzle through the nanochip. A unique feature of the chip is the incorporation of the ESI ground potential into the spray nozzle.

Conventional electrospray devices define the electric field by the potential difference between the spray device (fluid potential) and the mass spectrometer inlet. In the nanochip ESI, the electric field around the nozzle tip is formed from the potential difference between the conductive silicon substrate and the voltage applied to the fluid via the conductive pipette tip. The distance is only a few microns, so the field is dense, and the distance is not variable. The distance that defines the electric field is about 1000 times smaller than the distance of the nozzle to the mass spectrometer. Therefore, the mass spectrometer position and voltage though crucial for efficient ion transfer into analyzer, do not play any role in forming the chip electrospray, thus essentially decoupling the ESI process from the inlet of the mass spectrometer.

The robot was mounted to the QTOF mass spectrometer via a special bracket allowing rough adjustment of the robot position with respect to the sampling cone [51]. The fine positioning was driven by the ChipSoft software under Windows^{XP}, which controls the robot. The position of the electrospray chip was adjusted with respect to the sampling cone potential to give raise to an optimal transfer of the ionic species into the mass spectrometer. In order to prevent any contamination, for some experiments glass coated microtiter plates were used. 2–5 μL aliquots of the working sample solutions were loaded onto the 96-well plate. The robot was in most cases programmed to aspirate the whole volume of sample, followed by 2 μL of air into the pipette tip and then deliver the sample to the inlet side of the nanochip. Electrospray was initiated by applying voltages within 1.45 kV to 1.9 kV and a head pressure of 0.3 to 1.3 p.s.i. The value of these parameters turned out to be critical for generating and maintaining a long-lasting, stable spray of carbohydrate solutions, regardless the solvent. Following sample infusion and MS analysis, the pipette tip was ejected and a new tip and nozzle were used for each sample, thus preventing any cross-contamination or carry-over. In these studies, the whole coupled assembly was for the first time optimized to function in the negative ion mode which is the ionization mode the best suited for the investigation of complex *O*-glycan systems. By optimizing the NanoMate-ESI-QTOF MS assembly for sugar ionization and sequence requirements a solid methodology with major advantages in comparison with the capillary-based nanoESI was developed.

The analytical applications demonstrated the feasibility of the fully automated nanochip ESI QTOF for high performance glycoscreening, sequencing and identification. This, has shown its potential to discover novel carbohydrate variants of potential diagnostic value in complex biological mixtures, due to increased sensitivity, reproducibility and ionization efficiency and the ability to generate a sustained and constant electrospray.

10.1.3.4 INTERFACING A THIN CHIP MICROSPRAYER SYSTEM TO QTOF MS

QTOF mass spectrometer was for the first time interfaced [37] to a disposable polymer microchip with integrated microchannels and electrodes conceived by DiagnoSwiss (Lausanne, Switzerland). The chip was microfabricated by semiconductor techniques including photolitography. Basically, a photoresist is patterned on a 75 μm thick, copper-coated polyimide foil through a printed slide acting as a mask. The photoresist is developed, and chemical etching is afterwards used to remove the deprotected copper where microchannels are to be patterned. The polyimide layer is plasma-etched to the desired depth. The final microchannels are 120 μm wide, 45 μm deep (nearly "half moon" cross section), with gold-coated microelectrodes placed at the bottom of the microchannel. A 35 μm polyethylene/polyethylene terephthalate is laminated to close the channels (Fig. 10.7).

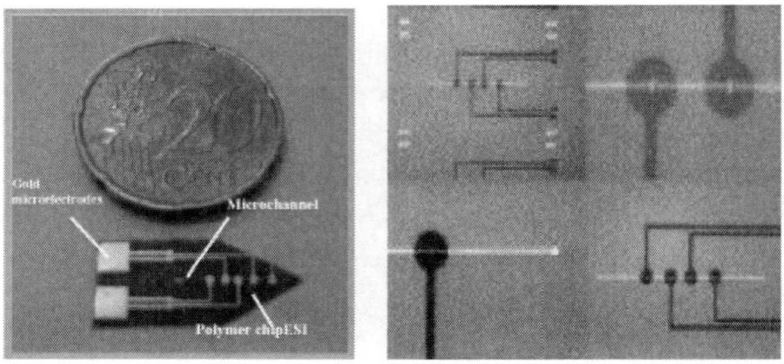

FIGURE 10.7 Plasma etched disposable polyimide microchip for electrospray with integrated 100 micrometer gold electrodes (courtesy of DiagnoSwiss).

For MS coupling one end of each channel was manually cut in a tip shape, which was visually inspected with a stereomicroscope. This way, the outlet of the microchannel is located on the edge of the chip, providing a in-plane electrospray. For sample dispensing, a reservoir was pasted over the inlet of the microchannel.

The whole chip/reservoir assembly was mounted to the QTOF MS [37]. In order to realize the electrical contact to the ESI power supply, the ESI QTOF sampler was removed and the chip system was directly connected to the ESI high voltage plate, which is a fixed part of QTOF conventional Micromass ESI source. Exchange between the original source and chip system interface did not claim for any definitive dismantling or special mechanical modifications to either of the original assembly and no further modifications on the TOF/MS analyzer were necessary. The position of chip emitter was adjusted in the vicinity of the entrance hole of the QTOF MS counter electrode (sampling cone) by the source assembly, manipulated in x, y and z direction via micrometer screws. The microsprayer tip was placed at a distance of about 5 mm. The electrical contact was ensured by a conductive wire with one terminal connected to the chip electrode and the other fixed on the ESI high voltage plate. The spray could be initiated at values of 2–3 kV, in the negative ion mode, applied to the nanoESI plate and 80–100 V applied to the sampling cone without the need of nebulizer gas.

In Fig. 10.8, a photograph of the QTOF source assembly with mounted polymer chip is presented.

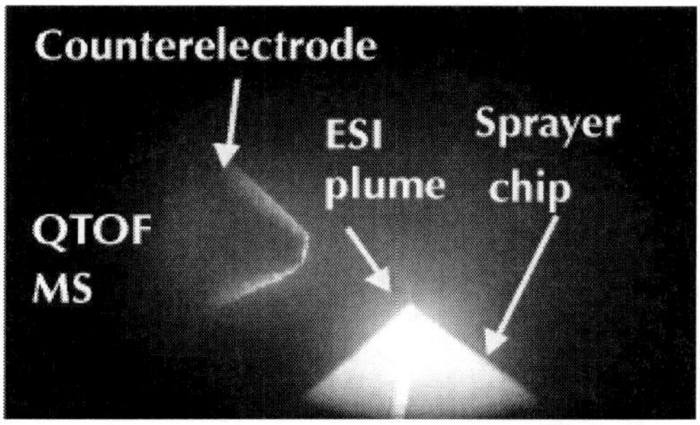

FIGURE 10.8 Photograph of the polymer-chip functioning coupled to the QTOF MS [37].

The photograph has been taken immediately after application of the negative voltages on both chip and counter electrode [37]. The ESI plume is clearly visible demonstrating the instant initiation of the electrospray.

10.1.3.5 FOURIER TRANSFORM ION CYCLOTRON RESONANCE (FTICR) MASS SPECTROMETER

FTICR MS has matured to become an indispensable tool in bioanalytical studies for analysis of complex mixtures, such as those encountered in glycomics [52, 53]. The unique features of the FTICR MS in comparison to all other MS methods are the ultra-high resolution exceeding 10^6 and the mass-determination accuracy very often below 1 ppm. Additionally, FTICR MS provides the advantage of several ion fragmentation techniques based on precursor dissociation such as collision induced-dissociation (sustained off-resonance SORI CID), infrared laser multiphoton (IRMPD), or electron capture dissociation (ECD) as well as the possibility to perform multiple stage MS (MS^n). For the analysis of native glycoconjugate mixtures, nanoESI FTICR MS and multiple stage MS in the negative ion mode was shown to be most substantial for screening and sequencing of complex carbohydrate mixtures originating from biological sources [54].

The most important part of the FTICR MS instrument is the analyzer cell, which resides in a strong, homogeneous magnetic field. The analyzer cell can take on different geometries but generally consists of a front and back trapping electrode, two opposite excitation electrodes and two opposite detection electrodes. The analyzer cell is in fact a low pressure (10^{-10} mbar) Penning trap in which ions can be stored for extended periods of time. The timescale of the experiment is another particular feature of FTICR MS, therefore it may be used to study slow (and fast) ion-molecule reactions, slow conformational changes in biomolecules, the dissociation of very large molecules with a large number of degrees of freedom, and many more processes that require both gas-phase ions and time to complete.

The role of the analyzer cell is to determine the mass-to-charge ratio of the ions stored in it. Each ion moving in a spatially uniform magnetic field will describe a circular, so-called cyclotron, motion as a result of the *Lorenz* force and the centrifugal force operating on it in opposite directions. The angular frequency of this motion is given by:

$$\omega_c = qB_0/m \tag{4}$$

where ω_c is the unperturbed cyclotron frequency and is solely depending on the magnetic field induction B_0 and the mass-to-charge ratio m/q. Modern superconducting magnets with a field strength ranging between 7 and 15 T usually drift only several ppm per year, so the cyclotron frequency can be an extremely accurate measure of m/q ratio.

The ions are exposed to an oscillating electric field that produces a net outward electric force on the ions for a limited period of time. This oscillating field is created by applying an RF potential on the two-excitation electrodes and is referred to as the excitation pulse. The ions will only experience a net continuous outward force if the frequency of the oscillating electric field is resonant with the cyclotron frequency of the ions. To ensure excitation of all ions trapped in the ICR cell, an RF pulse comprising multiple frequencies is employed such that all ions of different m/q ratios are exposed to a net outward electric force for the same amount of time. The radius after excitation is shown to be independent of m/q as long as the magnitude of the excitation signal is constant with frequency.

After excitation, the radius of the ion cloud increases and all ions with the same mass-to-charge ratio move coherently in a circular orbit. This coherently moving ensemble of charges at a radius close to the cell electrodes will induce an oscillating differential image current in two the opposite detection electrodes. This image current is then amplified and digitized yielding a time domain signal or transient containing signal contributions from all excited ions in the cell.

On the FTICR instrument, CID is performed by exciting an isolated ion to a higher cyclotron radius (and, therefore, to higher kinetic energy) in the presence of an increased background pressure of a neutral gas. Collisions occur as a result of the reduced path length and increased ion velocity; this leads to a transfer of energy to the two collision partners with mostly kinetic energy transfer to the neutral and conversion to internal energy in the ion. This internal energy is rapidly redistributed about the ion's structure and, if it locally exceeds the energy required for dissociation, the ion breaks apart.

There are multiple strategies [52, 53] for increasing the kinetic energy of the ions but the one used in most of the studies is the sustained off-resonance ion (SORI) irradiation/excitation.

SORI is a soft excitation technique enabling the operator to focus on the lowest energy fragmentation pathways. In SORI, the precursor ion is subjected to dipolar radiation at a frequency slightly offset from its cyclotron radius. This results in the ion alternately increasing and decreasing in radius and kinetic energy over the course of the SORI excitation so that collisions deposit lower internal energies per collision (typically 0.3 V), but many more collisions occur, hundreds per second, typically. As the internal energy accumulates, assuming that cooling mechanisms such as infrared radiative cooling are slower than the internal energy build-up, it is rapidly randomized throughout the ion and the lowest dissociation pathways are sampled. The product ions formed in this way have cyclotron frequencies separated far enough from the SORI frequency so that their continued excitation is minimal and any subsequent collisions serve only to cool their residual kinetic energy.

10.1.3.6 COUPLING OF THE FULLY AUTOMATED NANOCHIP-BASED ELECTROSPRAY IONIZATION TO FTICR MASS SPECTROMETER

Introduction of the fully automated chip-based nanoelectrospray in combination with QTOF-tandem MS for the first time in glycomics demonstrated the major advantages of this approach for structural investigation of complex carbohydrate systems. In this context, the NanoMate robot has been [55] coupled to the FTICR MS at 9.4 T and optimized in the negative ion mode to combine in one system automated sample delivery on the chip along with maximal sensitivity, ultra-high resolution, accurate mass determination and efficient tandem MS.

A specially designed interface conceived by Bruker Daltonik (Bremen, Germany) was constructed in order to obtain a viable coupling of the Nanomate system to the Bruker Apex II Apollo ion source. The coupling interface consisted of a prototype-mounting bracket [55].

Initiation of the electrospray and efficient transfer of the ionic species into MS have been accomplished by a fine adjustment of the NanoMate nanochip position with respect to the FTICR ion transfer capillary. The Apex II metal-coated glass capillary was set to create a slightly attractive potential for the ESI generated negative ions and the capillary exit voltage was set values minimizing the *in-source* ion fragmentation. The ESI generated ions were accumulated in the hexapole located after the second skimmer of the ion source and then transferred into the ion cyclotron resonance cell. The generated ions were trapped by *Sidekick* trapping.

For SORI CID experiments the precursor ions were isolated by application of a broadband excitation pulse to eliminate all ions except those of interest. The robot was programmed to aspirate the sample solution and to submit it to (–) nanochip ESI FTICR MS experiments. The generation of the chip electrospray required an initial high backpressure, which, was slightly decreased to lower values during acquisition.

The fine-tuning on the *x*-, *y*- and *z*-axis of the chip nozzle toward the MS inlet represented a critical step in the optimization of the ionic species transfer into the FTICR MS inlet. In addition, a significant increase of the spray stability and intensity of the MS signal was achieved by applying a low potential of 50 V to the ion transfer capillary of the FTICR MS instrument. Under these well-defined conditions a constant and stable spray, significant increase of sensitivity (1 pmol/μL) and high intensity of the MS signal over long signal acquisition time were obtained.

Another major benefit of the NanoMate FTICR MS system was the high ionization efficiency, with formation of multiply charged glycoconjugate ions, which in conjunction with the high sensitivity, resolution and mass accuracy allowed for the identification of minor glycoforms previously undetectable by any other MS-related technique.

The capability of the automated chip ESI FTICR MS approach for complete structural elucidation by SORI CID MS^2 was also evaluated and proved to provide accurate and structural-informative fragmentation data [55].

10.1.3.7 INTERFACING A THIN CHIP MICROSPRAYER SYSTEM TO FTICR MS

The Apex II FTICR mass spectrometer equipped with a 9.4 T superconducting actively shielded magnet and the Bruker Apex II Apollo ESI ion source was interfaced to the disposable thin polymer microchip. For the first time such a system was optimized in the negative ion mode, to detect carbohydrate ions [56].

For sample loading and FTICR interfacing, the chip was sandwiched in a home-made chip holder with an integrated reservoir (Fig. 10.9).

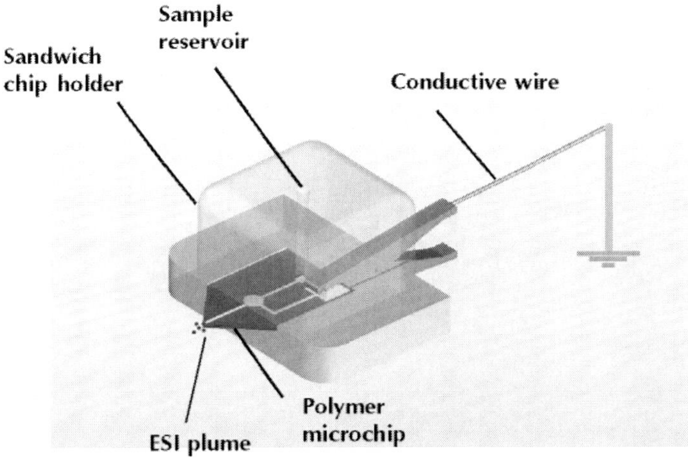

FIGURE 10.9 Schematic of the polymer chip incorporated into the sandwich holder for coupling with the FTICR MS [56].

The chip was positioned into the holder with the microchannel in contact with the reservoir and the front part extruding a few mm. The chip was grounded via a conductive wire connected to the terminal electrode. The chip was coupled to the Bruker Apex II ion source by an in-laboratory constructed mounting system [56]. The interface consists of a metal plate mounted to the Apollo ion source by two 90° brackets. The plate and the 90° brackets featured two slots for the screws, thus providing *x*- and *z*-axis movement and some *y*-axis variability.

The chip holder was attached to the metal plate and carefully positioned to point toward the entrance orifice of the Apex II capillary. Mounting of the chip interface system did not require any significant dismantling of, or irreversible mechanical modifications to any of the original Apollo source components. Moreover, the exchange between the original source and chip interface is quick and simple.

The Apex II metal-coated glass transfer capillary was kept in the range 1500–2500 V while the chip was grounded. This potential difference facilitated the ESI driven by the electro osmotic flow. The generated ions were accumulated for 0.05 to 0.3 s in the hexapole located after the second skimmer of the ion source, and then transferred into the ion cyclotron resonance cell.

The in-house made mounting system provided a robust and viable interfacing of the polymer chip to the FTICR MS instrument. The positioning/alignment of the polymer chip on the *x*-, *y*- and *z*-axes turned out to be a crucial step in the initiation and long-term maintenance of the electrospray, in particular with respect to the direct-spray configuration of the Bruker Apex II instrument. However, under optimized conditions, the polymer chipESI FTICR MS system provided a high ionization yield, an extremely stable and long-lasting spray, high sensitivity, and minimization of in-source fragmentation of labile moieties such as sialic acid residues.

All these advantages along with the high mass accuracy detection provided by the FTICR instrument made this technique a real option for achieving improved and detailed structural characterization of oligosaccharides and glycoconjugates.

10.1.3.8 HIGH CAPACITY ION TRAP (HCT) MASS SPECTROMETER

High capacity ion trap (HCT) mass spectrometer is currently one of the most efficient types of ion trap instruments. Released by Bruker Daltonics company a few year ago, HCT provides outstanding ion trap performance in terms of sensitivity, speed and mass accuracy. HCT exhibits an up to 15 fold higher ion storage capacity than the regular trapping instruments, which contributes to the dramatic increase in sensitivity [53]. The instrument is actually the fastest and most sensitive ion trap mass spectrometer. The high ion capacity, dynamic range, speed and multistage fragmentation (CID up to MS^{11}) make this instrument ideal for high throughput glycomics and proteomics as well as for quantitative analyses.

The HCT ultra PTM (posttranslational modifications) instrument is equipped with electron transfer dissociation (ETD) source using fluoranthene as the reagent. In contrast to CID (previously the only available fragmentation technique on ion traps) ETD induces specific $N\text{-}C_{\alpha}$ bond cleavages of peptide backbone with the preservation of the posttranslational modification [54–55] and consequently with generation of ions that are diagnostic for the modification site(s). Together, ETD and CID as well as alternating ETD/CAD feasible on HCT MS, may significantly increase the sequence coverage and give added confidence to protein, glycoprotein and glycopeptide identification [57].

10.1.3.9 COUPLING OF THE FULLY AUTOMATED NANOCHIP-BASED ELECTROSPRAY IONIZATION TO HCT MASS SPECTROMETER

In view of the advantages of HCT MS with CID and ETD MS^n, it was conceived the first combination of fully automated nanochip ESI with HCT MS [58] to yield a platform on which high throughput glycomics to be feasible.

The mass spectrometer employed these studies was a HCT Ultra PTM Discovery from Bruker Daltonics (Bremen, Germany). The HCT MS is interfaced to a PC running the Compass integrated software package, which includes the Hystar and Esquire modules for instrument controlling and chromatogram/spectrum acquisition as well as Data Analysis portal for storing the ion chromatograms and processing the MS data. The robot was coupled to the HCT Ultra mass spectrometer [58] via an in-laboratory made mounting system. For NanoMate interfacing, the conventional Bruker electrospray ion source was detached and all HCT instrumental settings and electrical connections were readapted to functioning in the MS-decoupled ESI regime of the NanoMate system. The robot was set up on three O-xyz adjustable supports and connected to the HCT MS nebulizer nitrogen supply pipeline. The position of the electrospray chip was adjusted with respect to the HCT counter electrode to ensure an optimal transfer of the ionic species to the mass spectrometer.

NanoMate-HCT MS system demonstrated a high reliability and versatility as it could be successfully applied to a broad class of biomolecules, which required different instrumental conditions for ionization, detection, screening in MS and sequencing by CID and ETD multistage MS. As described in the next sections, NanoMate-HCT coupling was optimized for compositional and fragmentation analysis in positive and negative ion mode of biomolecules such as peptides and proteins [57], glycolipids/gangliosides [58–60], *N*- and *O*-glycans [61, 62] and small molecules as well [63].

10.2 APPLICATIONS OF MICRO/NANOFLUIDICS FOR ELECTROSPRAY TO STRUCTURAL ANALYSIS OF GLYCANS IN BIOMEDICAL RESEARCH

Carbohydrates represent a class of biopolymers with high degree of structural complexity. They are polyhydroxylated aldehydic and ketonic compounds classified as monosaccharides, oligosaccharides and polysaccharides according to the size of the molecules and related to the number of monomeric units connected by glycosidic bonds. Carbohydrates are present either as oligosaccharides or as glycoconjugates in which the oligosaccharide chain is covalently linked to an aglycon, frequently another biopolymer such as a protein and/or a lipid. Carbohydrates occur ubiquitously in nature displayed on macromolecules and the surface of cells being involved in basic biological functions, such as antigen recognition machinery, cellular adhesion of bacteria and viruses, and protein folding, stability and trafficking [64]. Particular structures were found biomarkers of severe diseases and others to play an essential role in fertilization and embryogenesis. Due to the large number of saccharide building blocks and variety of linkages between them, this biopolymer category has also a high potential to carry information.

The large discrepancy between the extreme diversity of the glycoforms found in nature, their high biological importance and mostly an infime quantity available from biological sources, employed lately massive work in development of sensitive and specific methods for compositional mapping of heterogeneous glycan mixtures and the structural elucidation of their single components. The complete structural analysis of carbohydrates includes: (a) molecular weight determination; (b) identification of number and type of saccharide components; (c) determination of sequence and patterns of branching; (d) determination of glycosidic attachment sites and their anomeric configuration; (e) identification of the type and conformation of glycozyl ring; (f) determination of their secondary structure. In the last years, ESI MS demonstrated its potential for structural elucidation of carbohydrates being able to provide information related to (a)–(f) determinants, which significantly increased in amount and precision after introduction of nanoESI MS and tandem MS.

In the following parts of this chapter it will be demonstrated the contribution of micro and nanofluidics/ESI MS to elucidating complex issues raised in biomedical research, in particular those related to the structural identification of glycoconjugates with potential biomarker value. Throughout this part, glycan-related fragment ions were assigned according to the nomenclature [65] introduced by Domon and Costello (Fig. 10.10).

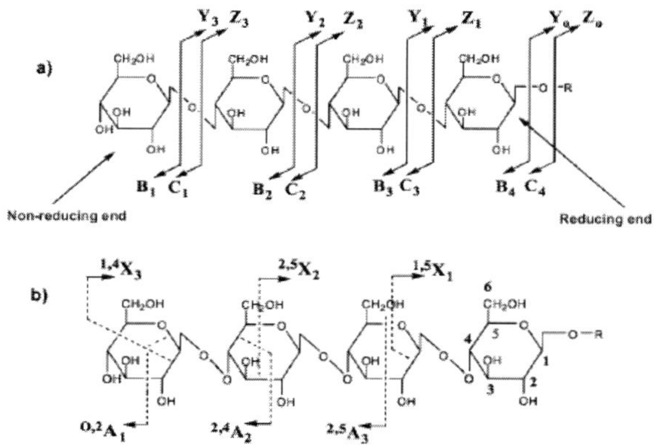

FIGURE 10.10 Types of fragment ions in tandem mass spectra of linear polysaccharides and glycoconjugates and their assignment according to the nomenclature [65]. (a) ions produced by cleavage of glycosidic linkages; (b) ring cleavage ions.

10.2.1 O-GLYCOPEPTIDES FROM URINE OF PATIENTS SUFFERING FROM SCHINDLER DISEASE

Schindler disease is a recently recognized autosomal recessive disorder caused by the deficient activity of α-*N*-acetylgalactosaminidase (NAGA), a lysosomal hydrolase previously known as α-galactosidase B.

Clinically, the disease is rather heterogeneous with three different phenotypes identified to date. The most severe form is the type I, an infantile-onset *neuroaxonal dystrophy*. It has been described [66] in three related German infants: two siblings born from consanguineous parents and a distant cousin. All three children were born after a normal pregnancy, labor and delivery. The sibs are currently alive, in the state, which is described below, while their cousin died unexpectedly at 18 months of life from apnea during a seizure with prolonged convulsion.

The clinical course experienced by the siblings was characterized by three stages: (i) apparently normal development in the first 9 to 12 months; (ii) a period of developmental delay followed by rapid regression starting with the second year of life (with the younger brother deteriorating faster); (iii) progressive neurological impairment resulting by 3 to 4 years of age in cortical blindness, deafness, spasticity, myoclonus, decorticate posturing and profound psychomotor retardation and little, if any, contact to the environment. At 4 and 5 years of age, respectively the affected brothers had developmental skills at the newborn level, did not have anymore voluntary movements, any contact to the environment and response to the stimuli. They did not appear to see or hear, were incontinent and dependent on tube feeding. Both brothers survived episodes of pneumonias due to the diligent nursing effort of their parents but remained to date in this vegetative state.

A milder form, *Schindler* disease type II (also called *Kanzaki* disease) is an adult-onset disorder characterized by *angiokeratoma corporis diffusum* and mild intellectual impairment [67]. To date three affected adults: one Japanese and two Spanish sibs, have been identified and are alive.

Schindler disease type III, described [68] very recently in two Dutch sibs and one unrelated French child is an intermediate and variable form with manifestations ranging from seizures and psychomotor retardation in infancy to a milder autism, with speech and language delay and marked behavioral difficulties in early childhood.

In all types of this rare inherited lysosomal storage disease, the severe enzymatic defect (enzyme residual activity ranging from 0.5% to 2% in plasma, lymphoblasts and fibroblasts) leads to an abnormal accumulation of sialylated and asialo-glycopeptides and oligosaccharides with α-*N*-acetylgalactosaminyl residues (mucine type of *O*-glycosylation) in various tissues and body fluids.

In human urine, complex carbohydrates are catabolic products excreted either as free oligosaccharides or linked to peptides, and their structures and amounts are known to vary under different physiological and pathological conditions. In all three types of *Schindler* disease, the deficient NAGA was found to cause glycopeptiduria and the concentration of *O*-glycans in urine was estimated to be 100 times higher than in healthy controls. For this reason, screening, structural characterization and complete identification of *O*-GalNAc glycosylated aminoacids and peptides extracted from patients' urine is of major diagnostic importance.

The developed arsenal of micro and nanofluidics/mass spectrometry methods presented has been employed *de novo* for the analysis of *O*-glycosylated peptides in the urine of the two German siblings diagnosed with *Schindler* disease type I. The complex mixtures of *O*-glycosylated peptides were extracted, purified, separated and prefractionated as described in details in Ref. [69, 37, 55, 56].

CE in off-line conjunction with negative ion mode nanoESI QTOF CID MS/MS was first developed for assessing the glycopeptide mixture heterogeneity and identification of the components. In the collected CE fractions 11 structural elements typical for *O*-glycosylation of proteins, like expression of core 1 and 2 type *O*-glycans with different numbers of *N*-acetyllactosaminyl repeats and different degrees of sialylation, could be directly detected and identified by optimized MS and CID MS/MS experiments.

A significant extension of the sensitivity limit for detection of minor components in this mixture was achieved by a novel analytical approach based on sheathless on-line forward polarity CE negative ESI QTOF MS and MS/MS [70]. The method revealed structures elongated by fucosylation and/or extended chains with higher degree of sialylation not detectable before and not known to be present in the mixture. Detailed structural information upon the separated species was obtained by data-dependent analysis carried out in the high speed automated "on-the-fly" MS-MS/MS mode switching which was for the first time introduced as fragmentation method in glycomics [70].

For development of a more efficient protocol based on CE separation and mass spectrometric screening of glycopeptide expression in the patient urine, the QTOF MS was coupled to the sheathless nanoESI interface in-laboratory designed for such purposes. The system was optimized for operating in the negative ion mode (MS) and reverse CE polarity [71]. So far, the on-line reverse polarity CE (–) ESI-QTOF MS (^{RP}CE (–) nano-ESI QTOF MS) was carried out under low buffer pH, low concentration and with coated capillaries to suppress the EOF, and pressure assistance to reduce the diffusion processes and the analysis time. However, a major drawback of the pressure assisted sheathless ^{RP}CE (–) nanoESI QTOF MS is the considerable decrease of separation efficiency and resolution. Therefore, such an approach was not considered beneficial toward the separation/detection of all components in these complex mixtures. For this reason, the development of a sheathless ^{RP}CE (–) nano-ESI QTOF MS method, based solely on the migration of components in electrical field without assistance of pressure and coating of the capillaries has been implemented by total reconsideration of the solution and CE instrument parameters and operation mode [71].

The spectra derived from the most prominent detected TIC-peaks (Fig. 10.11) clearly indicated that the mixture is dominated by the Ser-, Thr- and Thr-Pro-linked tetrasaccharide bearing two sialic acid moieties, hexasaccharide bearing two sialic acids and monosialo trisaccharides (Fig. 10.12). These results are in agreement with the data obtained by all previous experiments.

A detailed inspection of data, revealed that a larger number of minor components, doubly and triply charged ions corresponding to molecular masses up to 4000 Da, previously not detectable in this complex mixture were detected.

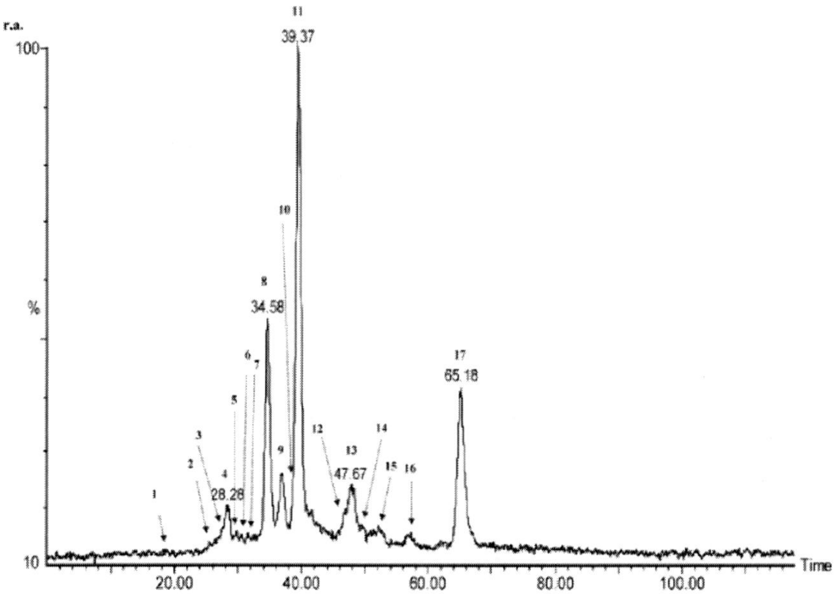

FIGURE 10.11 ^{RP}CE (–) nanoESI QTOF TIC MS of the fraction of *O*-glycosylated sialylated peptides from urine of the patient suffering from Schindler's disease. c = 0.75 mg/mL buffer (5 pmol injected); CE voltage, −25 kV; CE buffer, 0.1 mM methanol/water (6:4%v/v) formic acid, pH 2.8; CE capillary length, 130 cm [71].

The low ionic intensity exhibited by these components can be rationalized by their low abundance in this mixture showing a high dynamic range proportions. Nevertheless, the potential of this approach to separate and detect with high sensitivity even less abundant components, previously not accessible due to overlapping of isobaric structures and/or low content in the original biological material, is of major importance for progress in detailed identification of all structures related to this disease. The assignment of some of these species according to their molecular ions was conducted under the hypothesis that modification of glycopeptides by sulfation and acetylation could be present.

Thus at this point four ions could be attributed to the: (SO_3)Neu5AcGal-GalNAc-Ser and -Thr linked, respectively; (SO_3)Neu5AcGal$_2$GlcNAcGal-NAc-Ser/Thr and (SO_3)Neu5Ac$_2$Gal$_2$ GlcNAcGalNAc-Ser/Thr.

To detect and identify all *O*-glycoforms present in patients' urine, new accurate methods for MS mapping and sequencing were required. Therefore, the performance of the NanoMate system to provide long-lasting electrospray signal, rendering reliable conditions for high sensitivity, was of a particular usefulness for detection and sequencing of minor glycopeptide components, previously not accessible for fragmentation from such complex mixtures. The mixture heterogeneity was assessed by NanoMate robot in direct coupling first with QTOF MS and CID MS/MS via direct infusion (Figs. 10.13 and 10.14 a, b) [51] and subsequently in off-line coupling with CE [72].

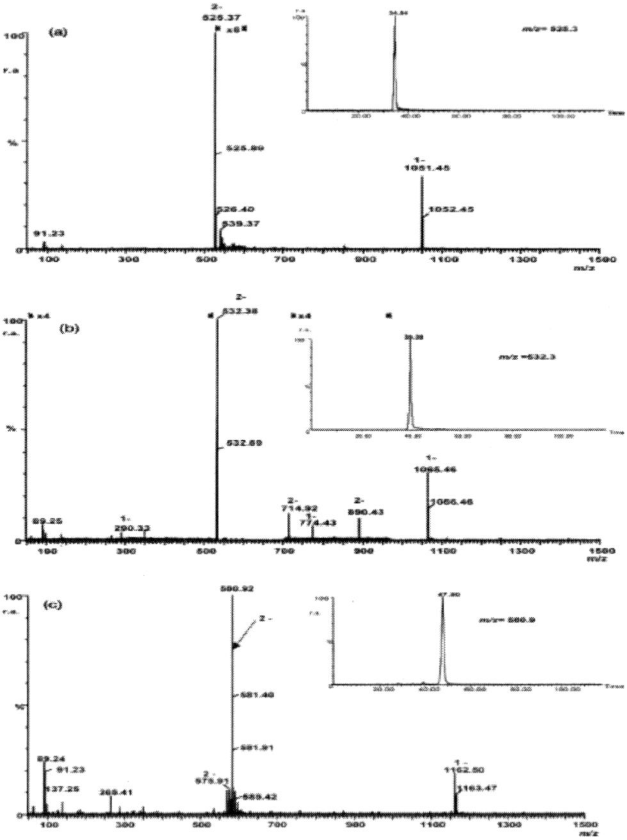

FIGURE 10.12 (a) ^{RP}CE (–) nanoESI QTOF MS by combining across the extracted ion chromatogram (XIC) of the ion at *m/z* 525.3 corresponding to $Neu5Ac_2HexHexNAc$-Ser. Inset: XIC of the ion at *m/z* 525.3 processed from the TIC-MS in Fig. 11; (b) ^{RP}CE (–) nano-ESI QTOF MS obtained by combining across the XIC of the ion at m/*z* 532.3 corresponding to $Neu5Ac_2HexHexNAc$-Thr. Inset: XIC of the ion at *m/z* 532.3 processed from the TIC MS in Fig. 10.11; (c) ^{RP}CE (–) nanoESI QTOF MS obtained by combining across the XIC of the ion at *m/z* 580.9 corresponding to $Neu5Ac_2HexHexNAc$- Thr-Pro. Inset: XIC of the ion at *m/z* 580.9 processed from the TIC MS in Fig. 11 [71].

Fully automated chip electrospray (NanoMate robot) was also coupled to FTICR mass spectrometry. The system was applied to high-performance glycoscreening and sequencing of *O*-glycopeptides from urine of *Schindler* disease patients [55]. NanoMate/FTICR MS screening provided a spectrum of extremely high complexity (Fig. 15) and, besides the already known species, revealed a high number of doubly and triply charged ions detected as lower abundant components within a relative narrow *m/z* ranges, like 700–780, 820–870, and 1050–1100.

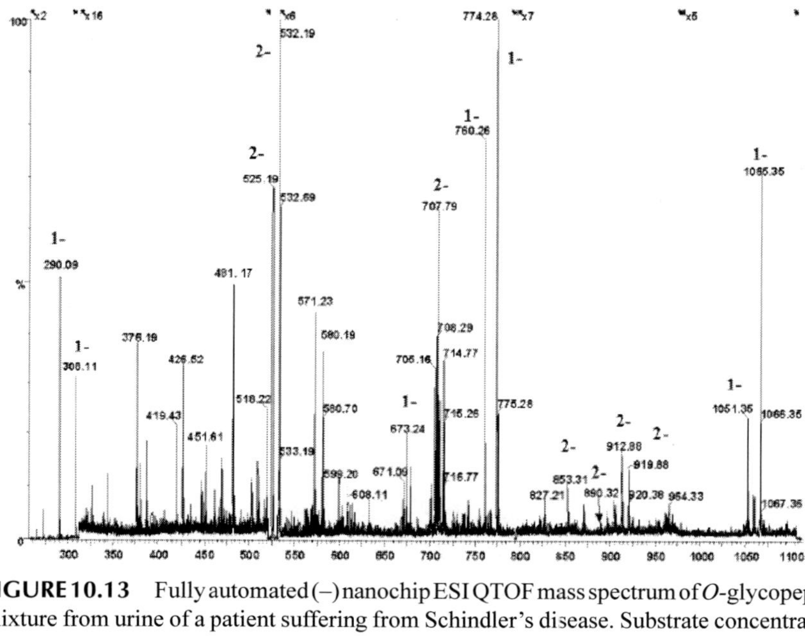

FIGURE 10.13 Fully automated (–) nanochip ESI QTOF mass spectrum of *O*-glycopeptide mixture from urine of a patient suffering from Schindler's disease. Substrate concentration, 3 pmol/μL in MeOH. Sampling cone potential, 30 V [51].

An unambiguous structural assignment at a mass accuracy below 10 ppm could be achieved for the structures Ser- and Thr- linked hexasaccharide and the nonasaccharides $Neu5Ac_3Hex_2HexNAc_4$-Ser/H_2O and $Neu5Ac_3Hex_2$ $HexNAc_4$-Thr/H_2O. Four new components, not identified so far by any other method, were detected by this method and assigned with a mass accuracy well below 10 ppm. Interestingly, the method disclosed the presence of two previously unknown undecasaccarides bearing three sialic acid moieties detected as triply dehydrated sodiated counterpart ions of $Neu5Ac_3Hex_3HexNAc_5$-Ser and $Neu5Ac_3Hex_3HexNAc_5$-Thr.

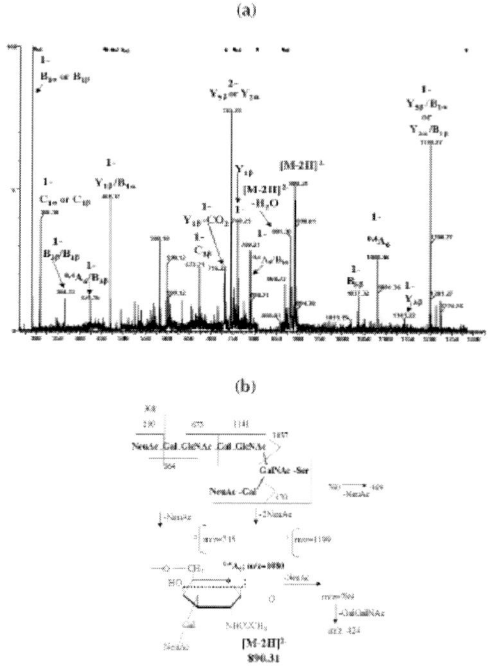

FIGURE 10.14 Fragmentation spectrum (a) and scheme (b) obtained by fully automated nanochip ESI QTOF CID MS/MS derived by using as the precursor ion $NeuAc_2Gal_3GlcNAc_2GalNAc$-Ser detected as a doubly charged ion at *m/z* 890.32. Collision energy range (25–40) eV; Sampling cone potential 30 V [51].

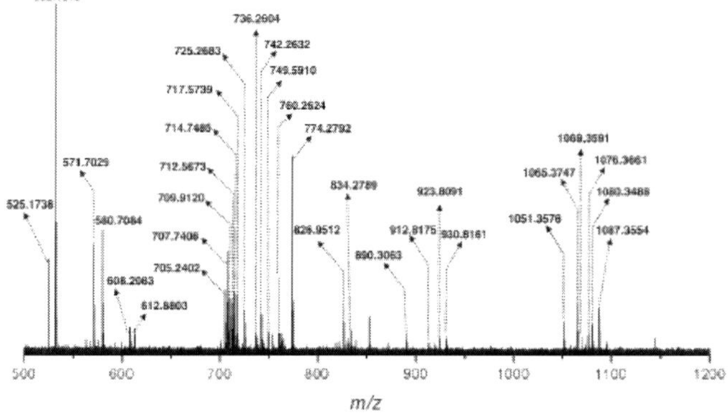

FIGURE 10.15 Fully automated (–) nanochip ESI FTICR mass spectrum of the fraction from urine of a patient suffering from Schindler's disease. Sample concentration: 5 pmol/µL in MeOH [55].

For comparative study, a mixture of glycopeptides extracted from urine of a age-matched healthy control person was subjected to compositional and structural analysis by (–) NanoMate/FTICR MS (Fig. 10.16) and SORI CID MS/MS (55), thin chip polymer microspray FTICR MS [56] and thin chip polymer microspray /QTOF MS and MS/MS (Fig. 10.17) [37] at the same ionization/detection and sequencing conditions as those employed for the mixtures originating from patient urine.

The mixture was found to contain a reduced number of species, having as the dominant components structures of shorter chains and lower degree of sialylation (maximum 2) such as: monosialo Ser, Thr-, and Thr-Pro-linked trisaccharide and Thr-Pro linked disialo tetrasaccharides, monosialo Ser- and Thr linked pentasaccharides of lower abundance, disialo Ser- and Thr- linked hexasaccharides and Thr-Pro linked disialo hexasaccharide much less abundant. The structure of the latter components were identified by NanoMate / FTICR SORI CID MS/MS experiment, which showed that the molecule composition is identical to the hexasaccharide mono- and dipeptide found in the patient urine.

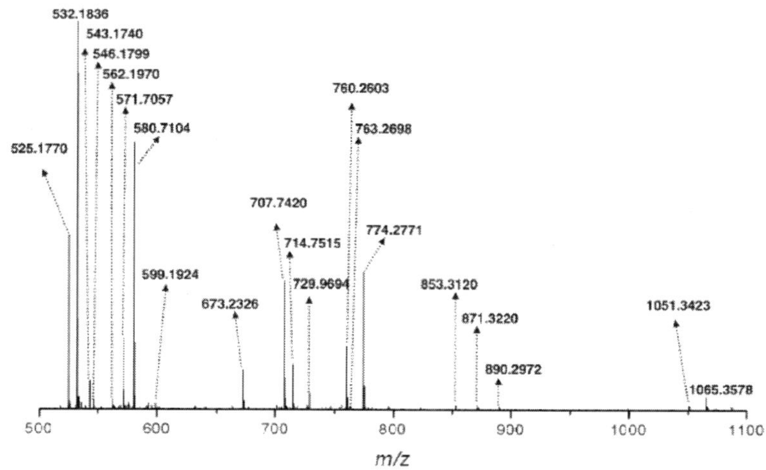

FIGURE 10.16 Fully automated (–) nanochip ESI FTICR mass spectrum of the mixture from urine of a healthy control person. Sample concentration: 7 pmol/μL in MeOH. Number of scans: 61 [55].

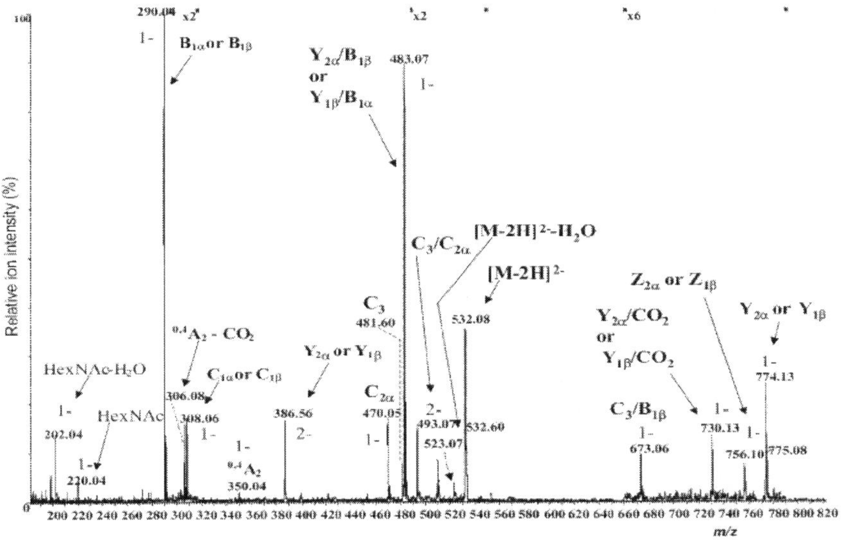

FIGURE 10.17 Thin polymer microchip ESI QTOF CID MS/MS of the $NeuAc_2HexHexNAc$-Ser doubly charged ion at *m/z* 532.08. ESI voltage 2.8 kV; sampling cone potential 100 V; solvent: MeOH; average sample concentration 5 pmol/μL; collision energy 40 eV; signal acquisition 30 scans; sample consumption 1.23 pmols [37].

Octasaccharides were barely represented in the spectrum and only at low abundance Ser-linked disialo octasaccharide was detected as a doubly charged ion. Longer chains and/or species of higher sialylation degree were not found in the urine of the healthy infant.

10.2.2 HUMAN BRAIN GANGLIOSIDES

Gangliosides, sialylated glycosphingolipids (GSLs), consist of sialylated (mono- to poly) oligosaccharide chain of variable length attached to the ceramide portion of different composition with respect to types of sphingoid base and fatty acid residues.

Designation and structure of the gangliosides according to Ref. [73] is the following:

LacCer, Galβ4Glcβ1Cer; Gg_3Cer, GalNAcβ4Galβ4Glcβ1Cer; Gg_4Cer, Galβ3GalNAcβ4Galβ4Glcβ1Cer; nLc_4Cer, Galβ4GlcNAcβ3Galβ4Glcβ1Cer; **GD3**, II^{3}-α-$(Neu5Ac)_{2}$-LacCer; **GT3**, II^{3}-α-$(Neu5Ac)_{3}$-LacCer; **GM2**, II^{3}-α-Neu5Ac-Gg_3Cer; **GD2**, II^{3}-α-$(Neu5Ac)_{2}$-Gg_3Cer; **GM1a**, II^{3}-α-Neu5Ac-Gg_4Cer; **GM1b**, IV^{3}-α-Neu5Ac-Gg_4Cer; **GD1a**, IV^{3}-α-Neu5Ac, II^{3}-

α-Neu5Ac-Gg_4Cer; **GD1b**, II^{3-}α-$(Neu5Ac)_{2-}$$Gg_4$Cer; **GT1b**, IV^{3-}α-Neu5Ac, II^{3-}α-$(Neu5Ac)_{2-}$$Gg_4$Cer; **GQ1b**, IV^{3-}α-$(Neu5Ac)_{2}$, II^{3-}α-$(Neu5Ac)_{2-}$$Gg_4$Cer; **3'-nLM1** or **nLM1**, IV^{3-}α-Neu5Ac-nLc_4Cer; **nLD1**, disialo-nLc_4Cer (IV^{3-}α-$(Neu5Ac)_{2-}$$nLc_4$Cer.

This variability of molecular constitution gives rise to a high number of species classified into oligosaccharide series according to the major oligosaccharide core structure. In Svenerholm system of abbreviations from above the fact that we are dealing with gangliosides is indicated by the letter G, the number of sialic acid residues is stated by M for mono-, D for di-, T for tri-, and Q for tetrasialoglycosphingolipids. A number is then assigned to the individual compound, which referred initially to its migration order in a certain chromatographic system. The ceramide portion is embedded in the outer leaflet of the plasma membrane, while a hydrophilic oligosaccharide chain protrudes into the extracellular environment. Gangliosides are enriched in the microdomains, functional membrane units, participating in cell-to-cell recognition/communication and cell signaling, modulating or triggering various biological events. The central nervous system (CNS) contains the highest content of gangliosides: neuronal membranes contain at least several times higher concentrations of gangliosides then the extraneural cell types, highlighting their special role in the CNS [74].

Ganglioside composition is species- and cell type-specific and changes specifically during brain development, maturation, aging and disease or neurodegeneration. For this reason, gangliosides are considered valuable tissue stage- and/or diagnostic markers and even potential therapeutic agents [75].

In human brain, the brain region-specific differences in ganglioside composition and quantity as well as in their distribution and cell surface expression have been demonstrated primarily by thin-layer chromatographic (TLC), immunochemical and immunohistochemical methods [76, 77]. These observations were based only on comparison concerning the major species due to detection limitations of the used methods. The region-specific differences most probably reflect the chemical basis of a high complexity of brain organization and the functional specialization of regions. This important investigation issue is still far from systematic characterization. As an example, cerebellum a highly specialized part of the brain showed some characteristic differences in composition of major ganglioside species in comparison to the cerebrum. Moreover, function, behavior and even survival of the cerebellar neurons strongly depend on the cellular expression of certain, even less abundant, ganglioside species.

Detailed and unambiguous compositional mapping and structural elucidation of individual ganglioside components are therefore of crucial necessity for systematic characterization of the brain region-specific ganglioside compositions in health and disease [49]. Such a study is of major importance for correlating the composition and structure specificity with the functional specialization of the particular region [78] and pathological state respectively [58, 60, 79]. Efficient separation and detailed MS structural characterization of gangliosides from biological sources are basic prerequisites for the further developed research strategies tending to elucidate specific function of each particular structure and to use it, accordingly, as therapeutic agents in treatment of diseases and/or as specific diagnostic markers.

10.2.2.1 ANALYSIS OF GANGLIOSIDES FROM NORMAL TISSUES

Healthy central nervous system (CNS) contains the highest amount of gangliosides: neuronal membranes hold at least several times higher concentrations of gangliosides/glycosphingolipids than the extraneural cell types, highlighting their special role at the CNS level [80]. Mapping of the gangliosides expressed in different regions of normal human brain using classical approaches based on TLC, immunochemical and immunohistochemical methods offered a low amount of information because of the detection limitations of these methods and their low throughput.

As a part of the efforts upon the implementation of the fully automated chip-based mass spectrometry in the field of the complex carbohydrate analysis, a general methodology for screening of glycosphingolipids under optimized conditions in terms of ionization, sensitivity, automated sequencing, speed of analysis and limitation of sample consumption, was probed [81].

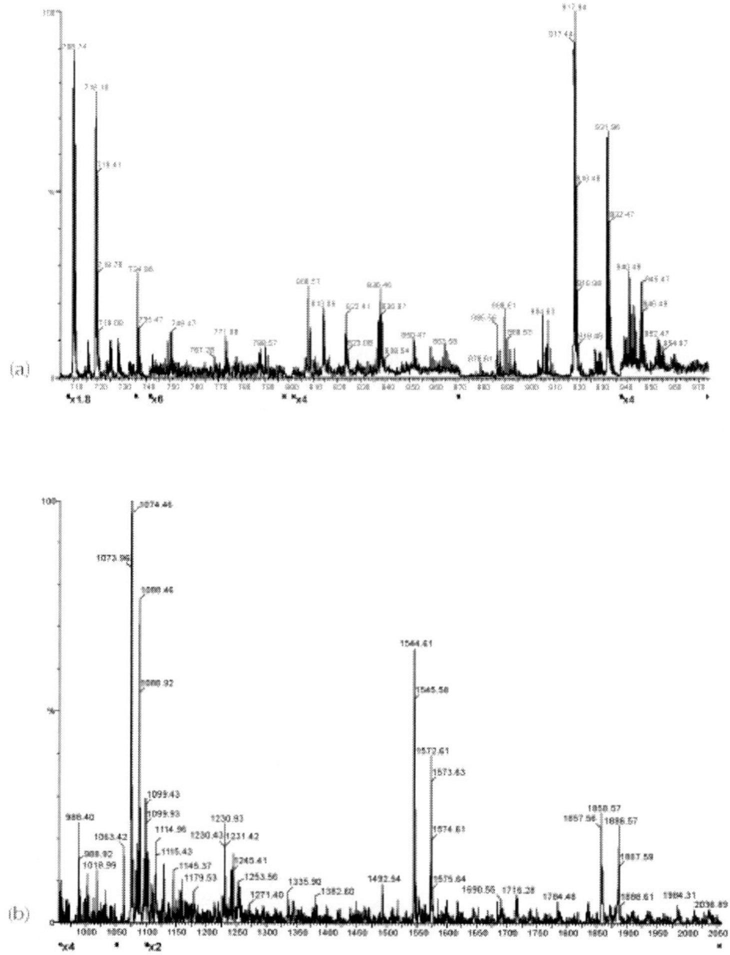

FIGURE 10.18 Fully automated (–) nanochip ESI QTOF MS of the ganglioside mixture from gray matter of normal human cerebellum. Sample concentration 2–3 pmol/μL in MeOH; acquisition time 3 min; sampling cone potential 45–135 V. a) *m/z* (700–980). b) *m/z* (980–2050) [81].

Automated nanochip ESI QTOF MS (Fig. 10.18a, b; Table 10.1) and CID MS/MS was optimized in the negative ion mode for characterization of a complex ganglioside mixture from normal human cerebellar tissue to demonstrate its general feasibility for ganglioside analysis [81], and its advantages in comparison with capillary-based ESI MS and MS/MS [39].

TABLE 10.1 Composition of Single Components in the Ganglioside Mixture from Gray Matter of Normal Human Cerebellum as Detected by a Fully Automated (–) Nanochip ESI QTOF MS [81]

Type of Molecular Ion	*m/z* (monoisotopic)		Assigned structure
	Detected	Calculated	
$(M+2Na-4H)^{2-}$	611.40	611.35	GM3 (d18:1/18:0)
$(M-H)^{-}$	1179.57	1179.74	
$(M-H)^{-}$	1382.60	1382.82	GM2 (d18:1/18:0)
$(M-2H)^{2-}$	734.96	734.91	GD3 (d18:1/18:0)
$(M+Na-2H)^{-}$	1492.78	1492.81	
$(M-2H)^{2-}$	748.99	748.93	GD3 (d18:1/20:0)
$(M-H)^{-}$	1518.51	1518.85	GM1, nLM1 and /or LM1 (d18:0/16:0)
$(M-2H)^{2-}$	771.98	771.93	GM1, nLM1 and /or LM1
$(M-H)^{-}$	1544.61	1544.85	(d18:1/18:0)
$(M-2H)^{2-}$	786.00	785.92	GM1, nLM1 and /or LM1
$(M-H)^{-}$	1572.61	1572.85	(d18:1/20:0)
$(M-2H)^{2-}$	836.46	836.45	GD2 (d18:1/18:0)
$(M-2H)^{2-}$	850.47	850.47	GD2 (d18:1/20:0)
$(M-2H)^{2-}$	917.44	917.48	GD1, nLD1 and /or LD1 (d18:1/18:0)
$(M+Na-3H)^{2-}$	928.45	928.47	
$(M-H)^{-}$	1835.62	1835.96	
$(M+Na-2H)^{-}$	1857.56	1857.95	
$(M-2H)^{2-}$	926.44	926.48	GD1, nLD1 and /or LD1 (t18:0/18:0)
$(M-2H)^{2-}$	924.44	924.49	GD1, nLD1 and /or LD1 (d18:1/19:0)
$(M-2H)^{2-}$	931.46	931.49	GD1, nLD1 and /or LD1 (d18:1/20:0)
$(M+Na-3H)^{2-}$	942.44	942.48	
$(M-H)^{-}$	1885.60	1885.98	
$(M-2H)^{2-}$	940.49	940.50	GD1, nLD1 and /or LD1 (t18:0/20:0)
$(M-2H)^{2-}$	938.44	938.50	GD1, nLD1 and /or LD1 (d18:1/21:0)

TABLE 10.1 *(Continued)*

Type of Molecular Ion	*m/z* (monoisotopic)		Assigned structure
	Detected	**Calculated**	
$(M\text{-}2H)^{2-}$	945.47	945.51	GD1, nLD1 and /or LD1 (d18:1/22:0)
$(M\text{-}2H)^{2-}$	954.46	954.51	GD1, nLD1 and /or LD1 (t18:0/22:0)
$(M\text{-}2H)^{2-}$	952.47	952.52	GD1, nLD1 and /or LD1 (d18:1/23:0)
$(M\text{-}2H)^{2-}$	958.46	958.52	GD1, nLD1 and /or LD1 (d18:1/24:1)
$(M\text{-}2H)^{2-}$	966.44	966.53	GD1, nLD1 and /or LD1 (d18:1/25:0) or (d20:1/23:0)
$(M\text{-}2H)^{2-}$	988.40	988.49	Fuc-GD1 (d18:1/18:2)
$(M\text{-}2H)^{2-}$	990.40	990.51	Fuc-GD1 (d18:1/18:0)
$(M\text{-}2H)^{2-}$	999.41	999.51	Fuc-GD1 (t18:0/18:0)
$(M\text{-}2H)^{2-}$	1002.41	1002.51	Fuc-GD1 (d18:1/20:2)
$(M\text{-}2H)^{2-}$	1004.42	1004.52	Fuc-GD1 (d18:1/20:0)
$(M\text{-}2H)^{2-}$	1013.44	1013.53	Fuc-GD1 (t18:0/20:0)
$(M\text{-}2H)^{2-}$	1018.99	1019.02	GalNAc-GD1 (d18:1/18:0)
$(M\text{-}2H)^{2-}$	1032.93	1033.03	GalNAc-GD1 (d18:1/20:0)
$(M\text{-}3H)^{3-}$	708.39	708.35	GT1 (d18:1/18:0)
$(M\text{-}2H)^{2-}$	1062.96	1063.03	
$(M+Na\text{-}3H)^{2-}$	1073.92	1074.02	
$(M+2Na\text{-}4H)^{2-}$	1084.93	1085.01	
$(M\text{-}3H)^{3-}$	714.41	714.35	GT1 (t18:0/18:0)
$(M+Na\text{-}3H)^{2-}$	1082.92	1083.02	
$(M\text{-}3H)^{3-}$	717.75	717.69	GT1 (d18:1/20:0)
$(M\text{-}2H)^{2-}$	1076.97	1077.04	
$(M+Na\text{-}3H)^{2-}$	1087.95	1088.03	
$(M+2Na\text{-}4H)^{2-}$	1098.92	1099.02	
$(M\text{-}3H)^{3-}$	723.75	723.70	GT1 (t18:0/20:0)
$(M+Na\text{-}3H)^{2-}$	1096.93	1097.04	
$(M+Na\text{-}3H)^{2-}$	1094.95	1095.04	GT1 (d18:1/21:0)

TABLE 10.1 *(Continued)*

Type of Molecular Ion	m/z (monoisotopic)		Assigned structure
	Detected	Calculated	
$(M\text{-}3H)^{3-}$	727.11	727.04	GT1 (d18:1/22:0)
$(M\text{+}Na\text{-}3H)^{2-}$	1101.92	1102.05	
$(M\text{+}Na\text{-}3H)^{2-}$	1108.92	1109.06	GT1 (d18:1/23:0)
$(M\text{+}Na\text{-}3H)^{2-}$	1114.96	1115.06	GT1 (d18:1/24:1)
$(M\text{-}3H)^{3-}$	722.39	722.35	*O*-Ac-GT1 (d18:1/18:0)
$(M\text{-}3H)^{3-}$	731.74	731.70	*O*-Ac-GT1 (d18:1/20:0)
$(M\text{-}2H)^{2-}$	1128.95	1129.05	Fuc-GT1 (d18:1/17:0)
$(M\text{-}2H)^{2-}$	1144.89	1145.06	Fuc-GT1 (t18:0/18:0)
$(M\text{-}2H)^{2-}$	1159.89	1159.08	Fuc-GT1 (t18:0/20:0)
$(M\text{-}3H)^{3-}$	805.40	805.38	GQ1 (d18:1/18:0)
$(M\text{+}Na\text{-}4H)^{3-}$	812.73	812.71	
$(M\text{+}2Na\text{-}4H)^{2-}$	1230.43	1230.56	
$(M\text{+}3Na\text{-}5H)^{2-}$	1241.43	1241.55	
$(M\text{-}3H)^{3-}$	814.74	814.72	GQ1 (d18:1/20:0)
$(M\text{+}Na\text{-}4H)^{3-}$	822.07	822.05	
$(M\text{+}2Na\text{-}4H)^{2-}$	1244.42	1244.57	
$(M\text{-}3H)^{3-}$	819.38	819.38	*O*-Ac-GQ1 (d18:1/18:0)
$(M\text{+}Na\text{-}4H)^{3-}$	826.73	826.71	

The sample investigated in this study was a native mixture of gangliosides extracted from the gray matter of a normal adult human (20 years of age) cerebellum, without pathological signs according to morphoanatomical and histopathological examination, originating from a healthy subject who died in a traffic accident.

The automated nanochip ESI QTOF MS approach optimized for ganglioside analysis provided a new insight into the structural diversity of ganglioside expression in human cerebellar gray matter and complex molecular architecture of the species. It was found that, in comparison with capillary-based ESI MS, a higher sensitivity and closer representation upon the mixture composition could be achieved.

By nanochip ESI MS screening, 44 glycoforms expressing high heterogeneity in the ceramide motifs, as well as biologically relevant peripheral modifications such as *O*-acetylation and fucosylation have been identified. By combining the fully automated chip ESI MS infusion with automatic selection and fragmentation of the precursor ion, a complete set of structural data and sequence ions could be obtained for polysialylated single ganglioside species GT1(d18:1/18:0) in a native mixture of high complexity, within short analysis time and with drastically reduced sample consumption.

To test the feasibility and advantages of the thin chip polymer-based microsprayer system in combination with QTOF MS/MS concerning information that could be provided by both MS and CID MS/MS, as well as to define the corresponding appropriate conditions for the GSL molecular class detection and structural characterization, a rather structurally complex polysialylated ganglioside fraction, GT1, was chosen as the testing sample [37].

The analyzed GT1 ganglioside fraction, showing migration properties of GT1b species in high performance thin-layer chromatography (HPTLC), was isolated from the total native ganglioside mixture purified from a normal adult human cerebrum (45 years of age). By this approach a reproducible compositional mapping of eight molecular components in the GT1 fraction mixture was obtained from both triply and doubly charged formed molecular ions related to gangliosides containing a number of lipid variants. Furthermore, the high sequencing efficiency of the microchip ESI QTOF MS/MS (Fig. 10.19) resulted in information-rich fragmentation pattern. This feature was of particular importance for elucidating the presence of structural isomers or isobars as distinct species as in many cases these species play a particular physiological and/or pathological role and therefore they might have a specific diagnostic relevance.

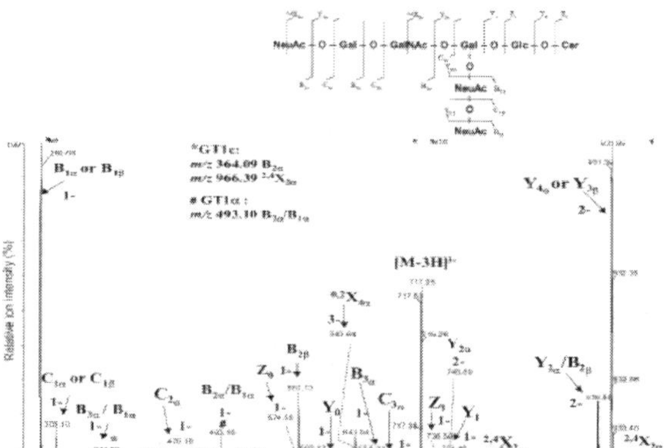

FIGURE 10.19 Thin polymer microchip ESI QTOF CID MS/MS of the triply charged ion at m/*z* 717.50 corresponding to GT1 (d18:1/20:0). ESI voltage 3 kV. Sampling cone potential 100 V. Collision energy (40–70) eV. Inset: GT1 (d18:1/20:0) fragmentation pathway by CID [37].

10.2.2.2 ANALYSIS OF GANGLIOSIDE EXPRESSION AND STRUCTURE IN PATHOLOGICAL TISSUES

10.2.2.2.1 PRIMARY BRAIN TUMORS

Tumorigenesis/malignant transformation is accompanied by aberrant cell surface composition, particularly due to irregularities in glycoconjugate glycosylation pathways. Various glycozyl epitopes constitute tumor-associated antigens [82]. Some of them promote invasion and metastases, while some other suppress tumor progression [82].

Gangliosides are among the molecules bearing characteristic glycozyl epitopes causing such effects. Glycosphingolipid-dependent cross-talk between glycozynapses interfacing tumor cells with their host cells has been even recognized as a basis to define tumor malignancy [83]. Structural elucidation of individual ganglioside components in normal human brain as well as their spatial-temporal distribution was an essential requirement for investigation of primary brain tumors gangliosides. Specific changes of ganglioside pattern in brain tumors vs. normal brain, correlating with tumor histopathological origin, malignancy grade, invasiveness and progression have been observed [84]. A decrease in the regular ganglioside profile and an increase in the structures detected only in small amounts in normal brain tissue was found in primary brain tumors [85, 86], demonstrating a direct correlation between ganglioside composition and histological type and grade of the tumors and an option to use this feature as biochemical marker in early histopathological diagnosis, grading and prognosis of tumors.

Glycoantigens and lipoantigens have been recognized as relevant and potentially valuable diagnostic and prognostic markers and tumor molecular targets for development/production of specific antitumor drugs, such as GSL-based vaccines, but their investigation in this regard has been neglected comparing to proteins [87].

In the last years several biophysical methods have been developed for the investigation of ganglioside expression in severe brain tumors. Ganglioside profiling, their quantification and correlation to histomorphology and grading of human gliomas has been studied [88] using a newly developed microbore HPLC method. The use of infrared (IR) spectroscopy as an adjunct to histopathology in detecting and diagnosing human brain tumors was also demonstrated [89]. In another study [90] ganglioside expression in human glioblastoma was determined by confocal microscopy of immunostained brain sections using antiganglioside monoclonal antibodies. However, a large number of low abundant tumor-associated species could not be detected by these conventional analytical methods. Systematic studies of ganglioside composition in human brain tumors are still restricted to several major components and many less abundant species with possible biomarker values could not be structurally characterized. This emphasized the need for detailed and systematic screening and structural characterization of brain tumor glycoconjugate composition, which could adequately be achieved only combining up-to-date, ultra-sensitive, high-resolution methodological approaches of detection and sequencing of biomolecules, such as advanced MS methods based on chip ESI sometimes complemented by immunochemical and chromatographic techniques.

The first nanochip-based ESI MS method for ganglioside analysis from human brain malignant alterations was introduced in 2007 [91]. The ganglioside composition and structure were characterized for human brain gliosarcoma obtained during surgical procedure, using the combination of NanoMate robot and QTOF MS.

Five microliter aliquots of the ganglioside mixture working sample solutions were loaded and submitted for MS screening in negative ion mode detection (Fig. 10.20). By chip ESI QTOF MS more than 25 species dominated by GD3 and a high abundance of O-acetylated GD3 species could be observed.

High intensity ions corresponding to GM3 and GD2 species carrying different ceramides were present as well. Several considerably abundant ions related to GM2, GM1, and/or their isomers nLM1 and LM1, as well as to GD1 species characterized by heterogeneity in composition of their ceramide moieties, were found.

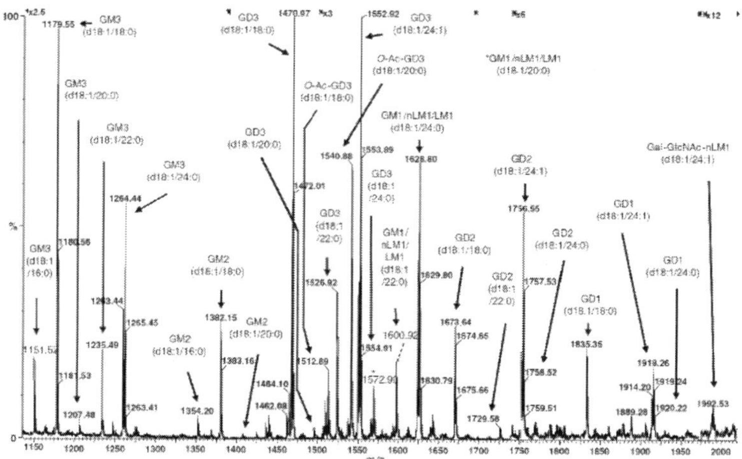

FIGURE 10.20 (–) Nanochip ESI QTOF MS of the native gliosarcoma ganglioside mixture. ESI voltage, 1.60 kV; sampling cone, 80 V; acquisition, 2 min; average sample consumption, 0.5 pmol [91].

To provide a consistent structural identification, in the same study [91] several detected species were subjected to fine analysis by tandem MS. Sequencing data defined the composition and detailed structure of several gliosarcoma-associated species among which GD3 (d18:1/24:1) OAc-GD3 (d18:1/20:0) GD2 (d18:1/18:0), GM1a (d18:1/18:0), GM1b, nLM1 or LM1 (d18:1/18:0). A particular attention was paid to *O*-Ac-GD3 molecule because this ganglioside could by itself be responsible for the protection of tumor cells from apoptosis. Its sequencing pattern offered the structural support to postulate a novel *O*-Ac-GD3 isomer, *O*-acetylated at the inner Neu5Ac-residue.

Two years later, the research continued with the investigation of ganglioside composition and structure in human brain hemangioma, a benign tumor, using advanced mass spectrometry methods based on NanoMate HCT and CID MS^n [70]. The obtained mass spectrum revealed 29 different ganglioside species dominated by mono- and disialylated structures. Two acetylated species, *O*-Ac-GM4 (d18:0/29:0) and *O*-Ac-GD2 (d18:1/23:0), the last one correlated with the reduced malignancy grade of the cerebral tumor were discovered. For fine structural analysis of the unusual, hemangioma-associated GT1, CID MS^2 at variable RF signal amplitudes within 0.6–1.0 V was applied. Five different fragment ions supported a structure of GT1c-type bearing (d18:0/20:0) ceramide. To confirm this assignment, the ion corresponding to GD1b (d18:0/20:0) was submitted to CID MS^2 under identical conditions. It was found that GD1b structure has the same lipid constitution as the previously sequenced GT1 however, an oligosaccharide core lacking one Neu5Ac residue. The NanoMate-based system developed and optimized for determination of ganglioside expression and structure in human brain hemangioma was able to detect an elevated number of species and, most importantly, to correlate the presence of *O*-Ac-GD2 with the low malignancy grade of the investigated cerebral tumor.

In 2012, a strategy combining HPTLC, laser densitometry and fully automated nanochip-based nanoelectrospray performed on a NanoMate robot coupled to QTOF MS was developed for mapping and structural identification of gangliosides extracted and purified from human angioblastic meningioma [92]. While HPTLC pattern indicated only six fractions migrating as GM3, GM2, GM1, GD3, GD1a (nLD1, LD1), GD1b, and possibly GD2, due to the high sensitivity, mass accuracy and ability to ionize minor species in complex mixtures, nanochip ESI QTOF MS was able to discover 34 distinct components of which two asialo, four GM4, nine GM3, two GM2, two GD3, nine GM1 and six GD1 differing in their ceramide compositions. All structures presented long-chain bases with 18 carbon atoms, while the length of the fatty acid was found to vary from C11 to C25. MS screening results indicated also that the diversity of the expressed GM1structures is higher than expected in view of the low proportions evidenced by densitometric quantification. Simultaneous fragmentation analysis of meningioma-associated GM1 (d18:1/24:1) and GM1 (d18:1/24:2) by MS/MS using CID confirmed the structure of the ceramide moieties and provide data on the glycan core, which document for the first time that both GM1a and GM1b isomers are expressed in meningioma tissue.

10.2.1.2.2 BRAIN METASTASES

In 2011 the research started to be oriented towards the first optimization and application of chip-based nanoelectrospray (NanoMate robot) MS for the investigation of gangliosides in secondary brain tumors [60]. A native ganglioside mixture extracted and purified from brain metastasis of lung adenocarcinoma (male patient 73-yr-old) was screened by NanoMate robot coupled to a quadrupole time-of-flight MS *vs.* a native ganglioside mixture from an age matched healthy brain tissue (subject deceased in a traffic accident), sampled and analyzed under identical conditions [60]. This comparative assay highlighted a considerable difference in the number and type of ganglioside components expressed in brain metastasis *vs.* healthy brain tissue.

Healthy cerebellar tissue was found to contain a higher variety of structures differing in their sialylation degree, from short, monosialylated (GM) to large, polysialylated carbohydrate chains (GH) and also ganglioside chains modified by *O*-acetyl (*O*-Ac) and fucosyl (Fuc) attachments. GM1 (d18:1/18:0) or (d18:0/18:1), GM1 (d18:1/20:0) or (d18:0/20:1) and Fuc-GM1(18:1/18:0) were detected as abundant singly charged ions at *m/z* 1544.92, 1572.92 and 1690.95 respectively. Beside these species, highly abundant doubly charged ions at *m/z* 917.44 and 931.45 assigned to disialylated GD1 components, with (d18:1/18:0) or (d18:0/18:1) and (d18:1/20:0) or (d18:0/20:1), respectively, were identified. Healthy brain sample is dominated by mono-, di- and trisialylated structures. 28 distinct *m/z* signals correspond to 44 possible GM-type species, 44 *m/z* signals correspond to 63 possible GD- type species, and 32 *m/z* signals are attributable to 59 GT- type species. Most of these structures have a tetrasaccharide sugar core and exhibit high heterogeneity in their ceramide composition. Additionally, 6 possible tetrasialylated structures (GQ) and only one asialo species (GA) could be detected. Notable is the presence of a hexasialylated GH2 species having (d18:0/24:1) or (d18:1/24:0) Cer constitution. This species was detected as $(M-3H)^{3-}$ at *m/z* 968.34, and was not found in the pathological brain sample. 18 possible GG species modified by fucosylation as well as 30 possible *O*-acetylated GG variants were also identified. Most of the fucosylated components are of GM1 and GD1-type with different fatty acid and/or sphingoid base compositions in the Cer moiety. Unlike fucosylation, *O*-acetylation was found for a higher variety of glycoforms such as GM3, GM1, GD3, GD2, GD1, GT3, GT2, GT1 and GQ1, which differ not only in oligosaccharide chain composition but also in their sialylation status. Interestingly, 4 possible di-*O*-Ac GG variants of GT2, GM1 and GM3 were detected as well [60].

In contrast to the healthy cerebellar tissue, the ganglioside mixture extracted from brain metastasis of lung adenocarcinoma exhibited mostly species of short oligosaccharide chains and reduced overall sialic acid content. More than a half, from the total of 59 different ions detected and corresponding to 125 possible structures in brain metastatic tissue, represented monosialylated species of GM1, GM2, GM3 and GM4-type. Besides the large number of monosialylated components, 8 asialo species of GA1 and GA2-type bearing ceramides of variable constitution were discovered. GD1, GD2 and GD3 as well as GT1, GT2 and GT3 with short carbohydrate chains, expressing different ceramide portions were also identified in the mixture. Ganglioside components modified by Fuc or *O*-Ac could also be detected, but in a different pattern than in healthy brain; most *O*-acetylated gangliosides are monosialo species of GM3, as well as short GT3- and GT2- type, while fucosylated components are represented by monosialo species of GM3 and GM4 structure, di- and trisialylated GD1 and GT3 exhibiting high heterogeneity in their ceramide motifs.

The most abundant singly charged ions at *m/z* 1150.17, 1168.01, 1515.29 and 1627.41 were assigned to GA2 (d18:0/22:0) or GM3 (d18:1/16:1); GM3 (t18:0/16:0); sodiated GD3 (d18:1/18:0) or (d18:0/18:1) or GM1 (d18:1/16:1) or (d18:0/16:2) or GM1 (d18:2/16:0) or *O*-Ac-GD3 (d18:0/18:0) and sodiated GD3 (d18:1/26:0) or (d18:0/26:1) or GM1 (d18:0/24:2) or (d18:1/24:1) or (d18:2/24:0).

MS data indicated the presence in the metastatic tissue of several unusual monosialylated species modified by fucosylation or *O*-acetylation such as Fuc-GM4, Fuc-GM3, di-*O*-Ac-GM3, *O*-Ac-GM3 (Table 2). These species were previously reported as fetal brain-associated, developmentally regulated antigens, which are only minor components of the normal brain [58].

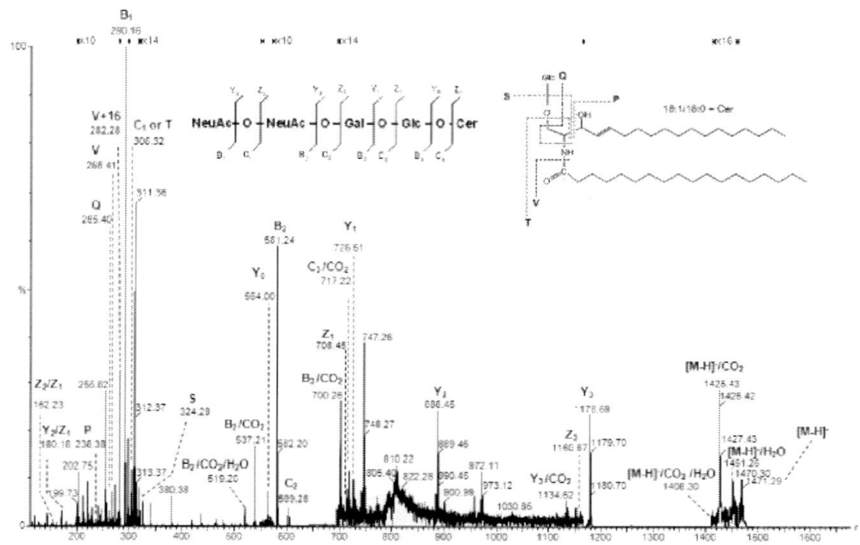

FIGURE 10.21 (–) Nanochip ESI QTOF CID MS/MS of the singly charged ion at *m/z* 1471.29 corresponding to GD3 (d18:1/18:0) from brain metastasis of lung adenocarcinoma. Acquisition time 1 min. Insets: fragmentation schemes of the oligosaccharide core and ceramide moiety [60].

GD3 (d18:1/18:0) was reported to enhance tumor cell proliferation, invasion and metastasis in a variety of brain tumor cells, especially in glioma and neuroblastoma [91]. GD3 influence tumor angiogenesis and metastasis by stimulating VEGF release from tumor cells, hence its structural characterization is of high biological importance.

By tandem MS using CID, the oligosaccharide core of the brain metastasis-associated GD3 (d18:1/18:0) species was structurally elucidated (Fig. 10.21).

TABLE 10.2 Ganglioside and asialo-ganglioside species from brain metastasis of lung adenocarcinoma detected by (–) nanochip ESI QTOF MS analysis of complex native ganglioside mixture [60]

m/z (monoisotopic) theoretical	*m/z* (monoisotopic) experimental	Mass accuracy (ppm)	Molecular ion	Proposed structure
875.19	874.91	33	$(M-H)^-$	LacCer(d18:1/17:0)
933.31	932.99	35	$(M-H)^-$	LacCer(d18:0/21:0)

TABLE 10.2 *(Continued)*

m/z (monoisotopic) theoretical	*m/z* (monoisotopic) experimental	Mass accuracy (ppm)	Molecular ion	Proposed structure
947.34	947.19	16	$(M-H)^-$	LacCer(d18:0/22:0)
949.22	949.24	21	$(M+2Na-3H)^-$	LacCer(d18:0/19:0)
964.24	963.90	35	$(M-H)^-$	GM4(d18:0/14:0)
982.19	981.94	25	$(M+Na-2H)^-$	GM4(d18:1/14:1)
984.21	983.87	34	$(M+Na-2H)^-$	GM4(d18:1/14:0) or GM4(d18:0/14:1)
1122.48	1122.23	22	$(M-H)^-$	GA2(d18:0/20:0)
1138.44	1138.15	25	$(M-H)^-$	Fuc-GM4(d18:0/16:0)
1138.48		29		GA2(t18:0/20:0)
1150.49	1150.17	28	$(M-H)^-$	GA2(d18:1/21:0)
1150.40		20	$(M-H)^-$	GM3(d18:1/16:1)
1168.42	1168.01	35	$(M-H)^-$	GM3(t18:0/16:0)
1178.46	1178.14	27	$(M-H)^-$	GM3(d18:1/18:1)
1179.74	1180.10	30	$(M-H)^-$	GM3(d18:1/18:0)
1182.49	1182.21	24	$(M-H)^-$	GM3(d18:0/18:0)
1184.37	1184.08	24	$(M-H)^-$	O-Ac-GA1(d18:1/10:0)
1194.50	1194.15	29	$(M-H)^-$	GM3(d18:1/19:0) or GM3(d18:0/19:1)
1206.51	1206.33	15	$(M-H)^-$	GM3(d18:1/20:1)
1206.64		26	$(M-H)^-$	GA2(d18:0/26:0)
1222.51	1222.19	26	$(M-H)^-$	O-Ac-GM3(d18:1/18:0)
1222.55		29	$(M-H)^-$	GM3(d18:0/21:1) or GM3(d18:1/21:0)
1234.56	1234.22	27	$(M-H)^-$	GM3(d18:1/22:1)
1234.52		24		O-Ac- GM3(d18:1/19:1)
1248.55	1248.18	33	$(M-H)^-$	O-Ac-GM3(d18:1/20:1)
1248.59		30		GM3(d18:1/23:1)
1248.59	1249.02	34	$(M-H)^-$	GM3(d18:1/23:0)
1260.60	1260.33	21	$(M-H)^-$	GM3(d18:1/24:2)

TABLE 10.2 *(Continued)*

m/z (monoisotopic) theoretical	*m/z* (monoisotopic) experimental	Mass accuracy (ppm)	Molecular ion	Proposed structure
1262.62	1262.35	21	$(M-H)^-$	GM3(d18:1/24:1)
1264.63	1264.19	35	$(M-H)^-$	GM3(d18:1/24:0)
1276.61	1277.01	31	$(M-H)^-$	O-Ac-GM3(d18:1/22:0)
1276.64		29		GM3(d20:1/23:1)
1276.69		25	$(M-H)^{-(}$-$H_2O)$	GM3(d18:0/26:0)
1278.66	1278.21	35	$(M-H)^-$	GM3(d20:1/23:0)
1278.61		31	$(M-H)^-$	O-Ac-GM3(d18:1/22:0) or O-Ac-GM3(d18:0/22:1)
1288.67	1289. 04	29	$(M-H)^-$	GM3(d18:1/26:2) or GM3(d18:2/26:1)
1288.67		29	$(M-H)^-$	GM3(d20:1/24:2)
1292.68	1292.23	35	$(M-H)^-$	GM3(d18:1/26:0) or GM3(d18:0/26:1)
1296.54	1296.24	23	$(M-H)^-$	Fuc-GM3(d18:1/16:1)
1296.59		27	$(M-H)^-$	O-Ac-GA1(d18:1/18:0)
1296.61		28	$(M-H)^-$	GA1(d18:0/21:0) or GA1(d18:0/21:0)
1405.65	1405.21	31	$(M+Na-2H)^-$	GM2(d18:1/18:0)
1420.68	1420.80	8	$(M-H)^-$	O-Ac-GM2(d18:2/18:2)
1435.59	1435.21	26	$(M+Na-2H)^-$	GD3(d18:1/14:1) or GD3(d18:0/14:2) or GD3(d18:2/14:0)
1441.66	1441.19	33	$(M-H)^-$	GD3(d18:1/16:1) or GD3(d18:0/16:2) or GD3(d18:2/16:0)
1471.73	1471.28	31	$(M-H)^-$	GD3(d18:1/18:0)
1493.71	1493.23	32	$(M+Na-2H)^-$	GD3(d18:1/18:0)

TABLE 10.2 *(Continued)*

m/z (monoisotopic) theoretical	*m/z* (monoisotopic) experimental	Mass accuracy (ppm)	Molecular ion	Proposed structure
1515.69	1515.29	26	$(M+2Na-3H)^-$	GD3(d18:1/18:0) or GD3(d18:0/18:1)
1515.74		30	$(M-H)^-$	GM1(d18:1/16:1) or GM1(d18:0/16:2) or GM1(d18:2/16:0)
1515.78		32	$(M-H)^-$	O-Ac-GD3(d18:0/18:0)
1515.71	1516.01	20	$(M+Na-2H)^-$	GD3(d18:1/20:2) or GD3(d18:0/20:3) or GD3(d18:2/20:1)
1515.75		17	$(M-H)^-$	GM1(d18:2/16:0) or GM1(d18:1/16:1) GM1(d18:0/16:2
1527.83	1528. 16	22	$(M-H)^-$	GD3(d18:0/22:0)
1541.79	1542. 19	26	$(M-H)^-$	GM1(d18:1/18:2) or GM1(d18:2/18:1) or GM1(d18:0/18:3)
1569.78	1570.29	32	$(M+2Na--3H)^-$	GD3(d18:1/22:0) or GD3(d18:0/22:1)
1569.83		29	$(M-H)^-$	GM1(d18:1/20:1) or GM1(d18:0/20:2) or GM1(d18:2/20:0)
1569.77		33	$(M+Na--2H)^-$	GD3(d18:0/24:2) or GD3(d18:1/24:1) or GD3(d18:2/24:0)
1597.88	1598.09	13	$(M-H)^-$	GM1(d18:0/22:2) or GM1(d18:1/22:1) or GM1(d18:2/22:0)
1611.77	1612.17	25	$(M+2Na-3H)^-$	GM1(d18:1/20:2)
1625.89	1625.40	30	$(M+2Na-3H)^-$	GD3(d18:1/26:1) or GD3(d18:0/26:2) or GD3(d18:2/26:0)
1624.92		30	$(M-H)^-$	GM1(d18:1/24:2)

TABLE 10.2 *(Continued)*

m/z (monoisotopic) theoretical	*m/z* (monoisotopic) experimental	Mass accuracy (ppm)	Molecular ion	Proposed structure
1627.90	1627.41	30	(M+2Na-3H)$^-$	GD3(d18:0/26:1) or GD3(d18:1/26:0)
1626.93		29	(M-H)$^-$	GM1(d18:0/24:2) or GM1(d18:1/24:1) or GM1(d18:2/24:0)
1629.92	1629.42	31	(M-H)$^-$	GM1(d18:0/24:1) or GM1(d18:1/24:0)
1628.94		29	(M-H)$^-$	di-O-Ac-GM1(d18:1/18:0)
1659.79	1660.18	23	(M+3Na-4H)$^-$	GM1(d18:1/22:3) or GM1(d18:0/22:4) or GM1(d18:2/22:2)
1674.87	1675.23	21	(M+Na-2H)$^-$ (-H_2O)	GD2 (d18:1/18:2)
1748.97	1749.39	24	(M+Na-2H)$^-$	GD2 (d18:1/22:1)
1766.97	1767.28	18	(M-H)$^-$ (-H_2O)	GT3 (d18:1/20:1)
1785.07	1785.37	17	(M-H)$^-$	O-Ac-GD2(d18:1/23:0) or O-Ac-GD2(d18:0/23:1)
1833.81	1833.28	29	(M-H)$^-$	GT3(d18:0/23:0)
1833.07		11	(M-H)$^-$	O-Ac-GT3 (d18:0/20:0)
1861.12	1861.24	6	(M-H)$^-$	O-Ac-GT3--lactone(d18:0/22:0)
1861.12		6	(M-H) (-H_2O)	O-Ac-GT3(d18:0/22:0)
1879.09	1879.39	16	(M+Na--2H)$^-$	O-Ac-GT3 (d18:2/22:1)
1879.10		15	(M-H)$^-$(--H_2O)	Fuc-GT3(d18:0/17:0)
1879.99		32	(M-H)$^-$	GT2(d18:1/12:1) or GT2(d18:2/12:0)
1909.16	1909.03	7	(M-H)$^-$	GD1 (d18:1/22:0)
1960.21	1959.84	19	(M-H)$^-$(-2H_2O)	GT2(d18:0/20:0)
1960.12		14	(M-H)$^-$	GT2(d18:1/18:3) or GT2(d18:2/18:2)

TABLE 10.2 *(Continued)*

m/z (monoisotopic) theoretical	m/z (monoisotopic) experimental	Mass accuracy (ppm)	Molecular ion	Proposed structure
1990.17	1989.78	20	$(M+Na-2H)^-$	GT2(d18:0/18:0)
1990.19		21	$(M-H)^-$	GT2(d18:0/20:3) or GT2(d18:1/20:2) or GT2(d18:2/20:1)
1990.19	1990.83	32	$(M-H)^-$	GT2(d18:1/20:1) or GT2(d18:0/20:2) or GT2(d18:2/20:0)
2005.20	2005.63	21	$(M-H)^-$	Fuc-GD1(d18:1/20:2)
2006.19		28		O-Ac-GT2(d18:1/18:1)
2048.23	2048.80	28	$(M-H)^-$	di-O-Ac-GT2(d18:0/18:0)
2048.10		34	$(M-H)^-(-H_2O)$	GT1(d18:2/14:2) or GT1(d18:3/14:1)

At the same time, a number of Cer-derived fragment ions allowed also the postulation of the lipid moiety composition. Optimized MS/MS conditions enabled also the structural assessment of Fuc-GM1 (d18:1/18:0) detected in healthy brain. It was found that the identified Fuc-GM1 is an atypical isomer bearing the labile Fuc residue at the inner Gal molecule together with one Neu5Ac attached at the same monosaccharide.

From the methodological point of view it is noteworthy to mention that nanochip ESI QTOF MS and CID MS/MS were able to provide compositional and structural characterization of native ganglioside mixtures from secondary brain tumors with a remarkable analysis pace and sensitivity. In view of the flow rate delivered by the nanochip ESI, which under the applied conditions was around 100 nL/min, 1 min acquisition time at a sample concentration of only 2.5 pmol/μL corresponds to 250 fmol biological extract consumption. Thus a MS screening followed by CID MS/MS required only 500 fmols of material. For all these reasons, the bioanalytical platform demonstrated here for determination of glycolipid molecular markers in brain tumors has real perspectives of development into a routine, ultrafast and sensitive method applicable to other types of cancer and molecular markers.

10.3 CONCLUSIONS AND PERSPECTIVES

In the postgenome era, the trends in all analytical sciences are toward miniaturization of devices and high-throughput experiments. In bioscience, integrated, fully automated nanosystems have been demonstrated to provide one of the most rapid, sensitive and accurate analysis. Mass spectrometry has the potential to revolutionize the nanobioanalytics and help in understanding many essential biological phenomena and events.

Due to the above two observations, the implementation of the modern micro and nanofluidic devices and nanochip-based technology is the purpose of the current research in the field and massive efforts are presently invested for the routinely introduction of the "lab-on-a-chip" principle in MS. The high potential of these systems to discover novel structures of biological importance makes them ideal for identification of unknown, minor components in complex biological mixtures. Furthermore, the capability of structural elucidation of biomolecules possibly indicative of pathological states gives this method clear perspectives for use in clinical diagnostics.

Consequently, biological mass spectrometry with nanofluidics for electrospray, though at its very beginning, is a relevant example of the nanotechnology/biology successful merging, which substantiate Stanley Fields optimism: *"Because the technology provides the tools and biology the problems, the two should enjoy a happy marriage."*

ACKNOWLEDGMENTS

This work was supported by the Romanian National Authority for Scientific Research, ANCS-UEFISCDI, projects PN-II-PCCA-2011-142, PCE-ID-2011-0047, RU-TE-2011-0008 and by the EU Commission, project FP7 Marie Curie-PIRSES-"MS-Life"-2010-269-256. The permissions for figure reproduction granted by the American Chemical Society, Royal Society of Chemistry, Wiley and Oxford University Press are gratefully acknowledged.

KEYWORDS

- **brain tumors**
- **electrospray spectroscopy**
- **Fourier transform ion cyclotron resonance (FTICR)**
- **high capacity ion trap (HCT)**
- **mass spectroscopy**
- **silicon nanocips**

REFERENCES

1. Zeleny, J. (1917). Instability of Electrified Liquid Surfaces. *Phys. Rev.* 10, 1–6.
2. Dole, M., Mack, L., Hines, R., Mobley, R., Ferguson, L., Alice, M. (1968). Molecular Beams of Macroions. *J. Chem. Phys*. 49, 2240–2250.
3. Yamashita, M., Fenn, J. B. (1984). *J. Phys. Chem.* Electrospray Ion Source. Another Variation on the free-jet theme. 88, 4451–4459.
4. Aleksandrov, M. L., Gall, L. N., Krasnov, V. N., Nikolaev, V., Pavlenko, V. A., Shkurov, V. A. (1984). Extraction of Ions from Solutions at Atmospheric Pressure – a Method for Mass Spectrometric Analysis of Bioorganic Substances. *Dokl. Akad. Nauk SSSR* 277, 379–383.
5. Masselon, C., Pasa-Tolić, L., Tolić, N., Anderson, G. A., Bogdanov, B., Vilkov, A. N., Shen, Y., Zhao, R., Qian, W. J., Lipton, M. S. (2005). Targeted Comparative Proteomics by Liquid Chromatography-Tandem Fourier Ion Cyclotron Resonance Mass Spectrometry. *Anal. Chem*. 77, 400–406.
6. Fenn, J. B. (2003). Electrospray Wings for Molecular Elephants *Angew. Chem. Int. Ed. Engl*. 42, 3871–3894.
7. Wilm, M., Mann, M. (1996). Analytical Properties of the Nanoelectrospray Ion Source. *Anal. Chem*. 68, 1–8.
8. Hanisch, F. G., Jovanović, M., Peter-Katalinić, J. (2001). Glycoprotein Identification and Localization of *O*-Glycozylation Sites by Mass Spectrometric Analysis of Deglycozylated/Alkylaminylated Peptide Fragments. *Anal. Biochem.* 290, 47–59.
9. Smith, R. D., Olivares, J. A., Nguyen, N. T., Udseth, H. R. (1988). Capillary Zone Electrophoresis-Mass Spectrometry Using an Electrospray Ionization Interface. *Anal. Chem.* 60, 436–441.
10. Stalmach, A; Albalat, A; Mullen, W; Mischak, H. (2013). Recent Advances in Capillary Electrophoresis Coupled to Mass Spectrometry for Clinical Proteomic Applications. *Electrophoresis* 34, 1452–1464.
11. Kriger, M. S., Cook, K. D., Ramsey, R. S. (1995). Durable Gold-Coated Fused Silica Capillaries for Use in Electrospray Mass Spectrometry. *Anal. Chem*. 67, 385–389.
12. Zamfir, A. D., Peter-Katalinić, J. (2004). Capillary Electrophoresis-Mass Spectrometry for Glycoscreening in Biomedical Research. *Electrophoresis* 25, 1949–1963.

13. Ramsey, R. S., McLuckey, S. A. (1995). Capillary Electrophoresis/Electrospray Ionization Ion Trap Mass Spectrometry Using a Sheathless Interface. *J. Microcol. Sep.* 7, 461–469.
14. Chang, Y. Z., Her, G. R. (2000). Sheathless Capillary Electrophoresis/Electrospray Mass Spectrometry Using a Carbon-Coated Fused-Silica Capillary. *Anal Chem.* 72, 626–630.
15. Thorsen, T., Maerkl, S. J., Quake, S. R. (2002). Microfluidic Large-Scale Integration. *Science* 298, 580–584.
16. Groisman, A., Enzelberger, M., Quake, S. R. (2003). Microfluidic Memory and Control Devices. *Science* 300, 955–958.
17. Raju, B., Ramesh, M., Srinivas, R., Chandrasekhar, S., Kiranmai, N., Sarma, V. U. (2011). Differentiation of Positional Isomers of Hybrid Peptides Containing Repeats of β-Nucleoside Derived Amino Acid (β-Nda-) and L-Amino Acids by Positive and Negative Ion Electrospray Ionization Tandem Mass Spectrometry (ESI-MS^n) *J. Am. Soc. Mass Spectrom.* 22, 703–717.
18. Wang, H., Wong, C. H., Chin, A., Taguchi, A., Taylor, A., Hanash, S., Sekiya, S., Takahashi, H., Murase, M., Kajihara, S., Iwamoto, S., Tanaka, K. (2011). Integrated Mass Spectrometry-Based Analysis of Plasma Glycoproteins and Their Glycan Modifications. *Nat. Protoc.* 6, 253–269.
19. Zehl, M., Pittenauer, E., Jirovetz, L., Bandhari, P., Singh, B., Kaul, V. K., Rizzi, A., Allmaier, G. (2007). Multistage and Tandem Mass Spectrometry of Glycozylated Triterpenoid Saponins Isolated from Bacopa monnieri: Comparison of the Information Content Provided by Different Techniques. *Anal. Chem.* 79, 8214–8221.
20. Zamfir, A. D., Bindila, L., Lion, N., Allen, M., Girault, H. H., Peter-Katalinic, J. (2005). Chip Electrospray Mass Spectrometry for Carbohydrate Analysis. *Electrophoresis* 26, 3650–3673.
21. Dethy, J. M., Ackermann, B. L., Delatour, C., Henion, J. D., Schultz, G. A. (2003). Demonstration of direct bioanalysis of drugs in plasma using nanoelectrospray infusion from a silicon chip coupled with tandem mass spectrometry. *Anal. Chem.* 75, 805–811.
22. Kim, S., Rodgers, R. P., Blakney, G. T., Hendrickson, C. L., Marshall, A. G. (2009). Automated Electrospray Ionization FT-ICR Mass Spectrometry for Petroleum Analysis. *J. Am. Soc. Mass Spectrom.* 20, 263–268.
23. Sainiemi, L., Nissilä, T., Kostiainen, R., Ketola, R. A., Franssila, S. A Microfabricated Silicon Platform with 60, Microfluidic Chips for Rapid Mass Spectrometric Analysis. *Lab Chip.* (2011). 11, 3011–3014.
24. Klepárník, K. (2013). Recent Advances in the Combination of Capillary Electrophoresis with Mass Spectrometry: From Element to Single-Cell Analysis. *Electrophoresis* 34, 70–85.
25. Kohler, I., Schappler, J., Rudaz, S. (2013). Microextraction Techniques Combined with Capillary Electrophoresis in Bioanalysis. *Anal. Bioanal. Chem.* 405,125–141.
26. Ramautar, R., Somsen, G. W., de Jong, G. J. (2013). CE-MS for Metabolomics: Developments and Applications in the Period 2010–2012. *Electrophoresis* 34, 86–98.
27. Zamfir, A. D., Flangea, C., Serb, A., Zagrean, A.-M.; Rizzi, A., Sisu, E. (2013). Separation and Identification of Glycoforms by Capillary Electrophoresis with Electrospray Ionization Mass Spectrometric Detection. *Meth. Mol. Biol.* 951, 145–169.

28. Bao, J., Krylov, S. N. (2012). Volatile Kinetic Capillary Electrophoresis for Studies of Protein-Small Molecule Interactions. *Anal. Chem.* 84, 6944–69447.
29. Kubáň, P., Kobrin, E. G., Kaljurand, M. (2012). Capillary Electrophoresis-A New Tool for Ionic Analysis of Exhaled Breath Condensate. *J. Chromatogr. A* 1267, 239–245.
30. Bonvin, G., Schappler, J., Rudaz, S. (2012). Capillary Electrophoresis-Electrospray Ionization-Mass Spectrometry Interfaces: Fundamental Concepts and Technical Developments. *J. Chromatogr. A* 1267, 17–31.
31. Hirayama, A., Tomita, M., Soga, T. (2012). Sheathless Capillary Electrophoresis-Mass Spectrometry with A High-Sensitivity Porous Sprayer for Cationic Metabolome Analysis. *Analyst* 137, 5026–5033.
32. Olivares, J. A., Nguyen, N. T., Yonker, C. R., Smith, R. D. (1987). On-Line Mass Spectrometric Detection for Capillary Zone Electrophoresis. *Anal. Chem.* 59, 1230–1232.
33. Zamfir, A. D. (2007). Recent Advances in Sheathless Interfacing of Capillary Electrophoresis and Electrospray Ionization Mass Spectrometry. *J. Chromatogr. A* 1159, 2–13.
34. McComb, M. E., Perreault, H. Design of a Sheathless Capillary Eletrophoresis-Mass Spectrometry Probe for Operation with a Z-Spray Ionization Source. *Electrophoresis* (2000). 21, 1354–1362.
35. Nilsson, S., Svedberg, T. M., Pettersson, J., Bjorefors, T. F., Markides, K., Nyholm, L. (2001). Evaluations of the Stability of Sheathless Electrospray Ionization Mass Spectrometry Emitters Using Electrochemical Techniques. *Anal. Chem.* 73, 4607–4616.
36. Chen, Y. R., Her, G. R. (2003). A Simple Method for Fabrication of Silver-Coated Sheathless Electrospray Emitters. *Rapid Commun. Mass Spectrom.* 17, 437–441.
37. Zamfir, A. D., Lion, N., Vukelic, Ž.; Bindila, L., Rossier, J.,.Girault, H., Peter-Katalinić, J. (2005). Thin Chip Microsprayer System Coupled to Quadrupole Time-Of-Flight Mass Spectrometer for Glycoconjugate Analysis. *Lab. Chip* 5, 298–307.
38. Lion, N., Gellon, J. O., Girault, H. H. (2004). Flow-Rate Characterization of Microfabricated Polymer Microspray Emitters. *Rapid Commun. Mass Spectrom.* 18, 1614–1620.
39. Flangea, C., Serb, A., Sisu, E., Zamfir, A. D. (2011). Chip-Based Mass Spectrometry of Brain Gangliosides. *Biochim. Biophys. Acta (Molec & Cell Biol. of Lipids)* 1811, 513–535.
40. Deng, Y., Henion, J., Li, J., Thibault, P., Wang, C., Harrison, D. J. (2001). Chip-Based Capillary Electrophoresis/Mass Spectrometry Determination of Carnitines in Human Urine. *Anal. Chem.* 73, 639–646.
41. Roussel, C., Dayon, L., Lion, N., Rohner, T. C., Josserand, J., Rossier, J. S., Jensen, H., Girault, H. H. (2004). Generation of Mass Tags by the Inherent Electrochemistry of Electrospray for Protein Mass Spectrometry. *J. Am. Soc. Mass Spectrom.* 15, 1767–1779.
42. Rossier, J. S., Youhnovski, N., Lion, N., Damoc, E., Reymond, F., Girault, H. H., Przybylski, M. (2003). Thin-Chip Microspray System for High-Performance Fourier-Transform Ion-Cyclotron Resonance Mass Spectrometry of Biopolymers. *Angew. Chem. Int. Ed. Engl.* 42, 53–58.
43. Ramos-Payán, M. D., Jensen, H., Petersen, N. J., Hansen, S. H., Pedersen-Bjergaard, S. (2012). Liquid-Phase Microextraction in a Microfluidic-Chip–High Enrichment

and Sample Clean-up From Small Sample Volumes Based on Three-Phase Extraction. *Anal. Chim. Acta* 20, 735, 46–53.

44. Petersen, N. J., Pedersen, J. S., Poulsen, N. N., Jensen, H., Skonberg, C., Hansen, S. H., Pedersen-Bjergaard, S. (2012). On-Chip Electromembrane Extraction for Monitoring Drug Metabolism in Real Time by Electrospray Ionization Mass Spectrometry. *Analyst*. 137, 3321–3327.
45. Hua, Y., Jemere, A. B., Dragoljic, J., Harrison, D. J. (2013). Multiplexed Electrokinetic Sample Fractionation, Preconcentration and Elution for Proteomics. *Lab Chip*. 13, 2651–2659.
46. Vázquez, M., Paull, B. (2010). Review on Recent and Advanced Applications of Monoliths and Related Porous Polymer Gels in Micro-Fluidic Devices. *Anal. Chim. Acta* 668, 100–113.
47. Zamfir, A. D., Flangea, C., Altmann, F., Rizzi, A. M. (2011). Glycozylation Analysis of Proteins, Proteoglycans And Glycolipids by CE-MS. *Adv. Chromatogr.* 49, 135–186.
48. Zamfir, A. D., Dinca, N., Sisu, E., Peter-Katalinić, J. (2006). Copper-Coated Microsprayer Interface for On-Line Sheathless Capillary Electrophoresis Electrospray Mass Spectrometry of Carbohydrates. *J. Sep. Science* 29, 414–422.
49. Sisu, E., Flangea, C., Serb, A., Rizzi, A., Zamfir, A. D. (2011). High-Performance Separation Techniques Hyphenated to Mass Spectrometry for Ganglioside Analysis. *Electrophoresis* 32, 1591–1609.
50. Zamfir, A. D., Seidler, D., Schonherr, E., Kresse, H., Peter-Katalinić, J. (2004). On-Line Sheathless Capillary Electrophoresis/Nanoelectrospray Ionization-Tandem Mass Spectrometry for the Analysis of Glycosaminoglycan Oligosaccharides. *Electrophoresis* 25, 2010–2016.
51. Zamfir, A. D., Vakhrushev, S., Sterling, A., Niebel, H., Allen, M., Peter-Katalinić, J. (2004). Fully Automated Chip-Based Mass Spectrometry for Complex Carbohydrate System Analysis. *Anal. Chem.* 76, 2046–2054.
52. Fukui, K., Takahashi, K. (2012). Infrared Multiple Photon Dissociation Spectroscopy and Computational Studies of *O*-Glycozylated Peptides. *Anal. Chem.* 84, 2188–2194.
53. Jmeian, Y., Hammad, L. A., Mechref, Y. (2012). Fast and Efficient Online Release of *N*-Glycans from Glycoproteins Facilitating Liquid Chromatography-Tandem Mass Spectrometry Glycomic Profiling. *Anal. Chem.* 84, 8790–8796.
54. Froesch, M., Bindila, L., Zamfir, A. D., Peter-Katalinić, J. (2003). Sialylation Analysis of *O*-Glycozylated Sialylated Peptides from Urine of Patients Suffering From Schindler's Disease by Fourier Transform Ion Cyclotron Resonance Mass Spectrometry and Sustained off-Resonance Irradiation Collision-Induced Dissociation. *Rapid Commun. Mass Spectrom.* 17, 2822–2832.
55. Froesch, M., Bindila, L., Baykut, G., Allen, M., Peter-Katalinić, J., Zamfir, A. D. (2004). Coupling of Fully Automated Chip Electrospray to Fourier Transform Ion Cyclotron Resonance Mass Spectrometry for High-Performance Glycoscreening and Sequencing. *Rapid Commun. Mass Spectrom.* 18, 3084–3092.
56. Bindila, L., Froesch, M., Lion, N., Vukelic, Ž., Rossier, J., Girault, H., Peter-Katalinić, J., Zamfir, A. D. (2004). A Thin Chip Microsprayer System Coupled to Fourier Transform Ion Cyclotron Resonance Mass Spectrometry for Glycopeptide Screening. *Rapid Commun. Mass Spectrom.* 18, 2913–2920.

57. Flangea, C., Schiopu, C., Capitan, F., Mosoarca, C., Manea, M., Sisu, E., Zamfir, A. D. (2013). Fully Automated Chip-Based Nanoelectrospray Combined with Electron Transfer Dissociation for High Throughput Top-Down Proteomics. *Cent. Eur. J. Chem.* 11, 25–34.
58. Almeida, R., Mosoarca, C., Chirita, M., Udrescu, V., Dinca, N., Vukelić Ž.; Allen, M., Zamfir, A. D. (2008). Coupling of Fully Automated Chip-Based Electrospray Ionization to High Capacity Ion Trap Mass Spectrometer for Ganglioside Analysis. *Anal. Biochem.* 378, 43–52.
59. Serb, A. F., Sisu, E., Vukelić, Z., Zamfir, A. D. (2012). Profiling and Sequencing of Gangliosides from Human Caudate Nucleus by Chip-Nanoelectrospray Mass Spectrometry. *J. Mass Spectrom.* 47, 1561–70.
60. Zamfir, A. D., Serb, A., Vukelić, Ž.; Flangea, C., Schiopu, C., Fabris, D., Capitan, F., Sisu, E. (2011). Assessment of the Molecular Expression and Structure of Gangliosides in Brain Metastasis of Lung Adenocarcinoma by an Advanced Approach Based on Fully Automated Chip-Nanoelectrospray Mass Spectrometry. *J. Am. Soc. Mass Spectrom.* 22, 2145–2159.
61. Flangea, C., Schiopu, C., Sisu, E., Serb, A., Przybylski, M., Seidler, D. G., Zamfir, A. D. (2009). Determination of Sulfation Pattern in Brain Glycosaminoglycans by Chip-Based Electrospray Ionization Ion Trap Mass Spectrometry. *Anal. Bioanal. Chem.* 395, 2489–2498.
62. Flangea, C., Sisu, E., Seidler, D. G., Zamfir, A. D. (2012). Analysis of Oversulfation in Biglycan Chondroitin/Dermatan Sulfate Oligosaccharides by Chip-Based Nanoelectrospray Ionization Multistage Mass Spectrometry. *Anal. Biochem.* 420, 155–162.
63. Stefanut, M., Cata, A., Pop, R., Mosoarca, C., Zamfir, A. D. (2011). Anthocyanins HPLC-DAD and MS Characterization, Total Phenolic and Antioxidant Activity of Some Berries Extracts. *Anal. Lett.* 44, 2843–2855.
64. Guthals, A., Bandeira, N. (2012). Peptide Identification by Tandem Mass Spectrometry with Alternate Fragmentation Modes. *Mol. Cell. Proteomics.* 11, 550–557.
65. Domon, B., Costello, C. E. (1988). A Systematic Nomenclature for Carbohydrate Fragmentations in FAB-MS/MS Spectra of Glycoconjugates. *Glycoconj. J.* 5, 397–409.
66. Wang, A. M., Schindler, D., Desnick, R. J. (1990). Schindler Disease: The Molecular Lesion in the Alpha-N-Acetylgalactosaminidase Gene that Causes an Infantile Neuroaxonal Dystrophy. *J. Clin. Invest.* 86, 1752–1756.
67. Desnick, R. J., Wang, A. M. (1990). Schindler Disease: An Inherited Neuroaxonal Dystrophy Due to Alpha-N-Acetylgalactosaminidase Deficiency. *J. Inherit. Metab. Dis.* 13, 549–559.
68. Sakuraba, H., Matsuzawa, F., Aikawa, S., Doi, H., Kotani, M., Nakada, H., Fukushige, T., Kanzaki, T. (2004). Structural and Immunocytochemical Studies on Alpha-*N*-Acetylgalactosaminidase Deficiency (Schindler/Kanzaki Disease). *J. Hum. Genet.* 49, 1–8.
69. Vakhrushev, S. Y., Zamfir, A. D., Peter-Katalinić, J. (2004). An Cross-Ring Cleavage as a General Diagnostic Tool for Glycan Assignment in Glycoconjugate Mixtures. *J. Am. Soc. Mass Spectrom.* 15, 1863–1868.

70. Zamfir, A. D., Peter-Katalinić, J. (2001). 22Glycoscreening by Sheathless on-Line Capillary Electrophoresis/Electrospray Quadrupole Time-of-Flight Tandem Mass Spectrometry. *Electrophoresis*, 2448–2457.
71. Bindila, L., Peter-Katalinić, J., Zamfir, A. D. (2005). Sheathless Reverse Polarity Capillary Electrophoresis/Electrospray Mass Spectrometry for the Analysis of Underivatized Glycans. *Electrophoresis* 26, 1488–1499.
72. Bindila, L., Almeida, R., Sterling, A., Allen, M., Peter-Katalinić, J. Zamfir, A. D. (2004). Off-Line Capillary Electrophoresis/Fully Automated Chip-Based Electrospray Ionization Quadrupole Time-of-Flight Mass Spectrometry and Tandem Mass Spectrometry for Glycoconjugate Analysis. *J. Mass Spectrom.* 39, 1190–1201.
73. Svennerholm, L. (1994). Designation and Schematic Structure of Gangliosides and Allied Glycosphingolipids. *Prog. Brain Res.* 101, XI-XIV.
74. Fuller, M. (2010). Sphingolipids: The Nexus Between Gaucher Disease and Insulin Resistance. *Lipids Health Dis.* 11, 9–113.
75. Salminen, A., Kaarniranta, K. (2009). Siglec Receptors and Hiding Plaques in Alzheimer's Disease. *J. Mol. Med.* 87, 697–701.
76. Kasperzyk, J. L., El-Abbadi, M. M., Hauser, E. C., d'Azzo, A., Platt, F. M., Seyfried, T. N. (2004). *N*-Butyldeoxygalactonojirimycin Reduces Neonatal Brain Ganglioside Content in a Mouse Model of GM1 Gangliosidosis. *J. Neurochem.* 89, 645–653.
77. Kusunoki, S., Kaida, K., Ueda, M. (2008). Antibodies Against Gangliosides and Ganglioside Complexes in Guillain-Barré Syndrome: New Aspects of Research. *Biochim. Biophys. Acta* 1780, 441–444.
78. Serb, A., Schiopu, C., Flangea, C., Vukelić, Ž.; Sisu, E., Zagrean, L., Zamfir, A. D. (2009). High-Throughput Analysis of Gangliosides in Defined Regions of Fetal Brain by Fully Automated Chip-Based Nanoelectrospray Ionization Multistage Mass Spectrometry. *Eur. J. Mass Spectrom.* 15, 541–553.
79. Schiopu, C., Serb, A., Capitan, F., Flangea, C., Sisu, E., Vukelic, Z., Przybylski, M., Zamfir, A. D. (2009). Determination of Ganglioside Composition and Structure in Human Brain Hemangioma by Chip-Based Nanoelectrospray Ionization Tandem Mass Spectrometry. *Anal. Bioanal. Chem.* 395, 2465–2477.
80. Futerman, A. H., van Meer, G. (2004). The Cell Biology of Lysosomal Storage Disorders. *Nat. Rev. Mol. Cell. Biol.* 5, 554–565.
81. Zamfir, A. D., Vukelic, Z., Bindila, L., Almeida, R., Sterling, A., Allen, M., Peter-Katalinić, J. (2004). Fully Automated Chip-Based Nanoelectrospray Tandem Mass Spectrometry of Gangliosides from Human Cerebellum. *J. American Soc. Mass Spectrom.* 15, 1649–1657.
82. Copp, A. J., Greene, N. D. (2010). Genetics and Development of Neural Tube Defects. *J. Pathol.* 220, 217–230.
83. Hakomori, S., Handa, K. (2002). Glycosphingolipid-Dependent Cross-Talk Between Glycozynapses Interfacing Tumor Cells with Their Host Cells: Essential Basis to Define Tumor Malignancy. *FEBS Lett.* 531, 88–92.
84. Becker, R., Rohlfs, J., Jennemann, R., Wiegandt, H., Mennel, H. D., Bauer, B. L. (2000). Glycosphingolipid Component Profiles or Human Gliomas–Correlation to Survival Time and Histopathological Malignancy Grading. *Clin. Neuropathol.* 19, 119–125.

85. Omran, O. M., Saqr, H. E., Yates, A. J. (2006). Molecular Mechanisms of GD3-Induced Apoptosis in U-1242 MG Glioma Cells. *Neurochem Res.* 31, 1171–1180.
86. Yin, J., Miyazaki, K., Shaner, R. L., Merrill Jr, A. H., Kannagi, R. (2010). Altered Sphingolipid Metabolism Induced by Tumor Hypoxia – New Vistas in Glycolipid Tumor Markers. *FEBS Lett.* 584, 1872–1878.
87. Wakabayashi, M., Okada, T., Kozutsumi, Y., Matsuzaki, K. (2005). GM1 Ganglioside-Mediated Accumulation of Amyloid Beta-Protein on Cell Membranes. *Biochem. Biophys. Res. Commun.* 328, 1019–1023.
88. Wagener, R., Rohn, G., Schilinger, G., Schroder, R., Kobbe, B., Ernestus, R. I. (1999). Ganglioside Profiles In Human Gliomas: Quantification by Microbore High Performance Liquid Chromatography and Correlation to Histomorphology and Grading. *Acta Neurochir.* 141, 1331–1345.
89. Steiner, G., Shaw, A., Choo-Smith, L. P., Abuid, M. H., Schackert, G., Sobottka, S., Steller, W., Salzer, R., Mantsch, H. H. (2003). Distinguishing and Grading Human Gliomas by IR Spectroscopy. *Biopolymers* 72, 464–471.
90. Hedberg, K. M., Dellheden, B., Wikstrand, C. J., Fredman, P. (2000). Monoclonal Anti-GD3 Antibodies Selectively Inhibit the Proliferation of Human Malignant Glioma Cells in Vitro. *Glycoconj. J.* 17, 717–726.
91. Vukelić, Ž.; Kalanj-Bognar, S., Froesch, M., Bindila, L., Radić, B., Allen, M., Peter-Katalinić, J., Zamfir, A. D. (2007). Human Gliosarcoma-Associated Ganglioside Composition Is Complex and Highly Distinctive as Evidenced by High-Performance Mass Spectrometric Determination and Structural Characterization. *Glycobiology* 17, 504–515.
92. Schiopu, C., Vukelić, Ž.; Capitan, F., Sisu, E., Zamfir, A. D. (2012). Chip-Nanoelectrospray Quadrupole Time-of-Flight Tandem Mass Spectrometry of Meningioma Gangliosides: A Preliminary Study. *Electrophoresis* 33, 1778–1786.

CHAPTER 11

THE "HOW-TO" GUIDE TO COMPUTATIONAL CRYSTALLOGRAPHY

EMRE S. TASCI, ALESSANDRO STROPPA, DOMENICO DI SANTE, GIANLUCA GIOVANNETTI, SILVIA PICOZZI, and J. MANUEL PEREZ-MATO

CONTENTS

ABSTRACT

Symmetry analysis is a powerful tool for studying many properties of crystalline materials. Here we review the basic theory of group theory as applied to material science and we show how we can combine computational crystallography as well as density functional theory for studying the ferroelectric properties of materials.

11.1 INTRODUCTION

A material showing a spontaneous electric polarization that can be reversed by the application of an external electric field is said to be ferroelectric. Ferroelectricity is an important topic in condensed-matter physics with applications in memory devices [1]. The Rochelle salt is the first ferroelectric crystal based on organic molecules [2], discovered in 1920. The lightness, flexibility and nontoxicity of organic ferroelectrics make them promising materials in the emerging field of organic electronics [3].

The purpose of the present chapter is to describe how computational crystallography combined with density functional theory calculations represents a useful tool for studying ferroelectricity in crystalline compounds. Our approach is based on the description of the main physical concepts, in an intuitive and qualitative way rather than using a formal mathematical approach. The outline is as follows: we will give an introduction to density functional theory and to modern theory of polarization; then we will proceed to a brief overview of group theory and the basic computational crystallography tools; finally we will discuss case studies such as organic ferroelectrics in order to provide a step-by-step procedure for the analysis of the ferroelectric polarization by using symmetry mode analysis.

11.1.1 DENSITY FUNCTIONAL THEORY

The investigation of the microscopic properties of matter is a very complicated task. One has to solve the Schrödinger equation of the system,

$$H\Psi = E\Psi \tag{1}$$

The systems in which we are interested in are ions, say M, with charge Z_j and mass M_j, and electrons, say N. In the nonrelativistic approximation, the Hamiltonian appearing in Eq. (1) reads:

$$H=\sum_{i=1}^{N}\left(-\frac{1}{2}\nabla_{\vec{r}_i}^2\right)+\sum_{j=1}^{M}\left(-\frac{1}{2M_j}\nabla_{\vec{R}_j}^2\right)-\sum_{i=1}^{N}\sum_{j=1}^{M}\frac{Z_j}{\left|\vec{r}_i-\vec{R}_j\right|}+\sum_{i<j}^{N}\frac{1}{r_{ij}}+\sum_{i<j}^{M}\frac{Z_iZ_j}{R_{ij}} \tag{2}$$

where atomic units have been used ($e^2=m_e=\hbar=1$). This equation is in practice impossible to solve and thus one needs some reasonable approximations. The first approximation, is the separation of the ionic degrees of freedom from the electronic ones. This is the so called Born-Oppenheimer (or adiabatic) approximation. Ions are much heavier then electrons and then the electrons follow adiabatically the ions' movements, remaining close to their instantaneous ground state (GS). In this way the ions act as external potential source for the electronic problem, which therefore depends parametrically on the ionic positions,

$$\left(H\{\vec{R}\}-E\{\vec{R}\}\right)\Psi\{\vec{r};\vec{R}\}=0 \tag{3}$$

$$H\{\vec{R}\}=\sum_{i=1}^{N}\left(-\frac{1}{2}\nabla_{\vec{r}_i}^2\right)-\sum_{i=1}^{N}\sum_{j=1}^{M}\frac{Z_j}{\left|\vec{r}_i-\vec{R}_j\right|}+\sum_{i<j}^{N}\frac{1}{r_{ij}} \tag{4}$$

Where $\vec{r},\vec{R}$ refer to the collection of the electronic and ionic degrees of freedom, respectively. Once the electronic problem is solved (ideally for each ionic position), the dynamics of the ions can be studied using the electronic energy $E\{\vec{R}\}$ as *effective* potential. Usually ions are heavy enough to be treated as classical particles, which is therefore a further reasonable approximation. However, Eqs. (3) and (4) do not seem to be easier to solve then Eq. (2) due to the presence of the two-body term due to electron-electron interaction which couples all the electronic degrees of freedom. Beside the Schrödinger equation, a new formalism has been developed: it is the Density Functional Theory (DFT) [4, 5] and it overcomes the many-body difficulties providing an alternative exact new formalism, which is much easier to handle.

The original idea, which started the development of DFT, is due to Hohenberg and Kohn (HK) [6]. They showed that the external potential acting on the electrons is uniquely determined (up to a trivial additive constant) by the electron GS density,

$$n(\vec{r})=\langle\Psi|\hat{\mathrm{n}}|\Psi\rangle=\int d\vec{r}_2\cdots d\vec{r}_n\left|\Psi(\vec{r},\vec{r}_2,\cdots,\vec{r}_n)\right|^2 \tag{5}$$

where Ψ [1] is the GS of the system and $\hat{\mathrm{n}}(\vec{r})$ is the density operator,

$$\hat{n}(\vec{r}) = \sum_{i}^{N} \delta(\vec{r} - \vec{r}_i) \tag{6}$$

Since $n(\vec{r})$ also determines the number of electrons N, and since V_{ext} and N fix the Hamiltonian of the system, it turns out that the electron density completely determines the system. The energy can be written:

$$E[n] = F_{HK}[n] + \int V_{ext}(\vec{r})n(\vec{r})d\vec{r} \tag{7}$$

with

$$F_{HK}[n] = \langle \Psi_{GS}[n] | T + V_{ee} | \Psi_{GS}[n] \rangle \tag{8}$$

where T and V_{ee} are the kinetic energy and the electron-electron interaction operators, respectively; and $\Psi_{GS}[n]$ is the GS of the system. Note that $F_{HK}[n]$ does not depend on the external potential and therefore it is a universal functional. Using the variational principle HK proved a second theorem which states that the GS density of the system is the one that minimizes E[n], and the minimum of E[n] is the GS energy E_0. The importance of these two results is clear: the only quantity that is needed is the electron density, no matter how many electrons are present in the system! The exact form of the functional F[n] is not known, however, its existence justifies the large work that is still in progress to improve the approximations available nowadays. One year after the publication of the HK paper Kohn and Sham (KS) invented an indirect method to solve the problem [7]. The idea is to write the energy functional as an "easy" part plus a "difficult" part.

$$F[n] = T_0[n] + E_H[n] + E_{XC}[n] \tag{9}$$

where $T_0[n]$ is the GS kinetic energy of an auxiliary noninteracting system whose density is the same as the one of the real system, $E_H[n]$ is the repulsive electrostatic energy of the classical charge distribution $n(\vec{r})$ and $E_{xc}[n]$ is the exchange-correlation (xc) energy defined through Eq. (9). Minimizing the total energy E[n] under the constraints of orthonormality for the one-particle orbitals of the auxiliary system, $\int \psi_i^*(\vec{r})\psi_j(\vec{r})d\vec{r} = \delta_{ij}$, one finds a set of one-particle Schrödinger equations:

$$\left[-\frac{1}{2}\nabla^2 + V_{KS}(\vec{r})\right]\Psi_i(\vec{r}) = \varepsilon_i \Psi_i(\vec{r}) \tag{10}$$

[1]At this stage, the spin variable is not included for simplicity.

where the KS potential is

$$V_{KS}(\vec{r}) = V_{ext}(\vec{r}) + \int \frac{n(\vec{r}')}{|\vec{r}-\vec{r}'|} d\vec{r}' + V_{XC}(\vec{r}); \qquad V_{XC} = \frac{\partial E_{XC}[n]}{\partial n(\vec{r})} \tag{11}$$

and

$$n(\vec{r}) = \sum_{i=1}^{N} f(\varepsilon_i - \varepsilon_F) |\Psi_i(\vec{r})|^2 \tag{12}$$

with f(x) being the Fermi-Dirac distribution and ε_F the Fermi energy fixed by the condition

$$\int n(\vec{r}) d\vec{r} = N \tag{13}$$

These are the famous KS equations, they must be solved self-consistently because V_{KS} is a functional of the orbitals itself. When self-consistency is achieved, the electronic energy of the system is:

$$E = \sum_{i=1}^{N} f(\varepsilon_i - \varepsilon_F)\varepsilon_i - \frac{1}{2}\int \frac{n(\vec{r})n(\vec{r}')}{|\vec{r}-\vec{r}'|} d\vec{r} + E_{XC}[n] - \int V_{XC}(\vec{r}) n(\vec{r}) d\vec{r} + E^{ion} \tag{14}$$

where E^{ion} is the ionic electrostatic repulsion term. This would be the exact electronic GS energy of the system if we knew $E_{xc}[n]$. Unfortunately the exact form of the xc energy is not (yet) known.

11.1.2 LOCAL DENSITY APPROXIMATION

The most widely used approximation for the xc energy is the so called *Local Density Approximation*: the dependence on the density is of the form

$$E_{xc}^{LDA}[n] = \int n(\vec{r}) \varepsilon_{xc}(n(\vec{r})) d\vec{r} \tag{15}$$

and $\varepsilon_{xc}(n)$ is taken to be the exchange and correlation energy per particle of a uniform electron gas whose density is $n(\vec{r})$. This has been accurately calculated using Monte Carlo simulations [8] and parameterized in order to be displayed in an analytic form [9].

By construction this approximation yields exact results if the density of the system is uniform, and should not be very accurate for those systems whose density is highly nonhomogeneous, as for example for atoms and molecules.

However, it turns out to work better than expected for a wide range of materials. In molecules, for example, the LDA usually overestimates the binding energies, but it yields in general good results for equilibrium distances and vibrational frequencies.

Within this approximation, the exchange and correlation part of the KS potential is

$$V_{xc}^{LDA}(\vec{r}) = \frac{\delta E_{xc}^{LDA}[n]}{\delta n(\vec{r})} = \varepsilon_{xc}(n(\vec{r})) + n(\vec{r})\frac{\partial \varepsilon_{xc}(n)}{n} \quad (16)$$

11.1.3 INFINITE (PERIODIC) SYSTEMS

Now we have a complete procedure to calculate the GS density and energy of the system. Given the Hamiltonian of the system one has to find the eigenvalues and the eigenvectors of the KS equations (Eq. (10)). The number of eigenvectors needed is proportional to the number of atoms in the system. It is clear that only systems containing a finite number of atoms can be studied in practice. However, if the system is a periodic solid, then the KS potential is a periodic function of the lattice[2],

$$V_{KS}(\vec{r} + \vec{R}) = V_{KS}(\vec{r}) \quad (17)$$

for each Bravais lattice vector $\vec{R}$. Thus, we can use the Bloch theorem to write the general form of the solution of Eq. (10),

$$\Psi_i(\vec{r}) = \Psi_{nk}(\vec{r}) = e^{i\vec{k}\cdot\vec{r}} u_{nk}(\vec{r}) \quad (18)$$

where $\vec{k}$ is a vector in the first Brillouin zone (BZ) of the reciprocal lattice, $u_{nk}(\vec{r})$ is a function with the periodicity of the Bravais lattice,

$$u_{nk}(\vec{r} + \vec{R}) = u_{nk}(\vec{r}) \quad (19)$$

and n is a degeneracy index which accounts for the band number [10]. Placing Eq. (18) in Eq. (10) we have:

$$\left[\left(-i\nabla + \vec{k}\right)^2 + V_H(\vec{r}) + V_{xc}(\vec{r}) + V_{ext}(\vec{r};\vec{k})\right] u_{nk}(\vec{r}) = \varepsilon_{nk} u_{nk}(\vec{r}) \quad (20)$$

[2]Again, for the sake of simplicity, we omit the spin variable in the following.

To calculate the electronic properties of the system, such as the total energy and the ionic forces, integrations in the whole BZ are needed. As a consequence one in principle needs an infinite number of solutions of Eq. (20), that is, for each $\vec{k}$ point of the BZ, a number of Eigen states which depends on the number of electrons in the unit cell. Of course, in practice only a finite number of solutions can be found. However, one can check the quality of the approximation by systematically increasing the number of $\vec{k}$ points. If these points are badly chosen, the convergence is very slow. The right way to produce the points had been firstly invented by Baldereschi [11], and then improved by Chadi and Cohen [12] and Monkhorst and Pack (MP) [13]. If the quantity to be integrated is smooth, then a small number of special $\vec{k}$ points is needed. In the case of metals some additional care is needed. The integrals now do not have to be evaluated in the whole BZ anymore, but only inside the Fermi surface. This is the same thing of doing integrals in the whole BZ cutting the integrands at the Fermi surface and keeping them zero outside. This fact results in a discontinuity of the function to be integrated, which therefore is not smooth at all. For this reason a larger number of $\vec{k}$ points is usually needed for metals. To speed up the convergence with respect the $\vec{k}$ points sampling, a trick has been invented by Fu & Ho [14]: one artificially smoothens the functions near the Fermi energy and substitutes the cut across the Fermi surface with an appropriate Fermi-Dirac like behavior. This fact corresponds to the introduction of a fictitious electronic temperature which introduces an error in the integrals that depends on the electronic temperature and on the smearing function used [15]; however, it can be easily controlled.

11.1.4 THE MODERN THEORY OF POLARIZATION

The macroscopic electric polarization of materials plays a fundamental role in the phenomenological description of dielectrics. In this section, we will briefly discuss some aspects of the quantum theory of polarization of crystalline solids and the role assumed in this theory by the geometric Berry phase [16]; for more details on the formalism we refer to the original works of King-Smith and Vanderbilt [17] and of Resta [18].

Consider a crystal of volume $V = N\Omega$, formed by an arbitrary large number N of identical unit cells of volume Ω. The average electric polarization of the crystal, for example, the electric dipole per unit volume, is related to the electronic charge density $n(\vec{r})$ by the expression

$$\vec{P} = \vec{P}_{ion} + \vec{P}_{el} = \frac{1}{N\Omega}\left[\sum_j z_j e\vec{R}_j - e\int_{N\Omega} d\vec{r}\, n(\vec{r})\vec{r}\right] \qquad (21)$$

where e is the absolute value of the electronic charge and $\vec{R}_j$ are the positions within the crystal of the nuclei of charge $z_j e$. The average polarization in Eq. (21) is also called macroscopic polarization, or simply polarization, of the crystal. The polarization vector $\vec{P}$ as defined in Eq. (21) depends on the details of the unit cell chosen to build up the crystal, whereas infinitesimal changes of polarization are independent of how the crystal has been assembled and are thus bulk properties. For these reasons, all the physical effects related to changes of polarization can be evaluated unambiguously and compared with experimental measurements. What could be measured experimentally are changes of polarization from a crystal structure with inversion symmetry and a polar one, in which that symmetry has been broken by polar distortions. Let λ be a continuous parameter varying from 0 to 1 that denotes the normalized amplitude to the polar distortion between the two structures. For any assigned value of λ, let $\Psi_n(\vec{k},\vec{r},\lambda)$ indicate the Kohn-Sham one-electron orbitals, and we will now focus on the change of electronic polarization as λ varies. From the knowledge of the parameter-dependent orbitals $\Psi_n(\vec{k},\vec{r},\lambda)$, we can express the electronic contribution to the average crystal polarization in the form

$$\begin{aligned}\vec{P}_{el}(\lambda) &= -\frac{e}{N\Omega}\int_{N\Omega} d\vec{r}\, n(\vec{r})\vec{r} \\ &= -\frac{2e}{N\Omega}\sum_{n\vec{k}}\left\langle \Psi_n(\vec{k},\vec{r},\lambda)\middle|\vec{r}\middle|\Psi_n(\vec{k},\vec{r},\lambda)\right\rangle\end{aligned} \tag{22}$$

where the factor 2 takes into account the spin degeneracy and the sum is over all occupied bands of the semiconductor or insulator under study. Indicating the periodic part of the Bloch functions with $u_n(\vec{k},\vec{r},\lambda)$, Eq. (22) can be written as

$$\vec{P}_{el}(\lambda) = -\frac{2e}{N\Omega}\sum_{n\vec{k}}\left\langle u_n(\vec{k},\vec{r},\lambda)\middle|\vec{r}\middle|u_n(\vec{k},\vec{r},\lambda)\right\rangle \tag{23}$$

As we noted before, only changes in polarization have real physical meaning, so we are interested in variations of $\vec{P}_{el}(\lambda)$ with respect to λ:

$$\frac{\partial\vec{P}_{el}(\lambda)}{\partial\lambda} = -\frac{2e}{N\Omega}\sum_{n\vec{k}} 2\,\mathrm{Re}\left\langle u_n(\vec{k},\vec{r},\lambda)\middle|\vec{r}\middle|\frac{\partial}{\partial\lambda}u_n(\vec{k},\vec{r},\lambda)\right\rangle \tag{24}$$

where Re stands for the real part. After some manipulations, equation Eq. (23) can be recast in the usual form

$$\frac{\partial \vec{P}_{el}(\lambda)}{\partial \lambda} = -\frac{2e}{N\Omega}\sum_{n\vec{k}} 2\,\mathrm{Im}\left\langle \nabla_{\vec{k}} u_n\left(\vec{k},\vec{r},\lambda\right) \middle| \frac{\partial}{\partial \lambda} u_n\left(\vec{k},\vec{r},\lambda\right)\right\rangle \tag{25}$$

the total change ΔP_{el} in polarization is obtained by integrating Eq. (24) in $d\lambda$ within the range $0 \le \lambda \le 1$ and in the Brillouin zone. In practice, integration over the three-dimensional Brillouin zone is carried out performing integrations over one variable, say for example, k_z, once a number of special points are chosen for the other two variables k_x and k_y. From these assumptions, the final form that one can get for the n-th band contribution, after a little of algebra, is

$$\Delta P_{el} = -\frac{e}{\pi ab}\gamma_n(C) = -\frac{e}{\pi ab}\mathrm{Im}\oint_C \left\langle u_n\left(k_z,\vec{r},\lambda\right) \middle| \nabla_{k_z,\lambda} u_n\left(k_z,\vec{r},\lambda\right)\right\rangle d\vec{l} \tag{26}$$

where $\gamma_n(C)$ is the Berry phase of the cell-periodic wave functions moving along the circuit C identified by the rectangle $-\pi / c \le k_z \le \pi / c$ and $0 \le \lambda \le 1$ in the (k_z, λ) space. From a physical-mathematical point of view, the Berry phase is the phase acquired by a quantum system described by a parameter-dependent Hamiltonian moving along a circuit C on a given adiabatic parameter-dependent surface.

A quantity strongly related to the macroscopic polarization is the *Born Effective Charge* tensor defined as

$$Z^*_{i,\alpha,\beta} = \frac{\Omega}{|e|}\frac{\partial P_a}{\partial u_{i,\beta}} \tag{27}$$

For example, the ratio between the changes in the α-th component of polarization due to an infinitesimal displacement u of the i-th atom in the β direction. The knowledge of the Born Effective Charge tensor could be useful in the determination of atoms which play an active role in ferroelectric transitions, because for these atoms the Z^* charge is often much larger than the nominal one.

11.1.5 BRIEF OVERVIEW OF GROUP THEORY

Nature has its own rules, and wherever we look, we observe a pattern, be it a leaf, a microorganism or a crystal. Finding the similarities, classifying and associating the symmetries not only simplifies the problem but gives us powerful tools to tackle problems. Specifically, group theory appears as a fundamental tool in the study of ferroelectric materials. In this section, we shall briefly recall some basic definitions of the group theory.

11.1.6 SYMMETRY OPERATIONS, WYCKOFF POSITIONS, SPACE GROUPS

In its most general definition, a symmetry operation is a transformation that when applied on a system, leaves the system unchanged. They are mathematically represented by matrices. For example, in 3D a 3x3 matrix is used for rotations, reflections and inversions while a 3x1 matrix represent the translational information. A transformation can be interpreted in two ways: one either assumes that the coordinate system is affected, thus the physical system do not move (passive transformation); or it can be seen as the displacement of the physical system while the coordinate system is fixed (active transformation). The symmetry operations throughout this text will be assumed as active.

As an example of the matrix representation, consider one of the 3-fold symmetry operators belonging to the space group $F\bar{4}3c$ – its rotational and translational components are given as:

$$R = \begin{pmatrix} 0 & 0 & 1 \\ 1 & 0 & 0 \\ 0 & 1 & 0 \end{pmatrix}; t = \begin{pmatrix} 0 \\ \frac{1}{2} \\ \frac{1}{2} \end{pmatrix} \tag{28}$$

It should be emphasized that the specific values for the rotational and translational matrices depend on the choice of the unit cell and its origin; they will be represented with other sets of values when the setting is changed, that is, a different unit cell is used. These relations between the settings (thus the unit cells) are defined by a transformation *(P, p)*. One of the conventional choices of these settings has been designated as the "standard setting" and the transformations *(P, p)* to and from this standard setting are tabulated for the most commonly used alternative settings.

Other than relating different settings of the same structure to each other, a transformation matrix is also used to "transform" a group into another space group's setting when there exists a group-subgroup relation.

For 3D, the symmetry operations act on positions of atoms, so the position vectors are written as vectors with 3 components:

$$\vec{x} = \begin{pmatrix} x_0 \\ y_0 \\ z_0 \end{pmatrix} \tag{29}$$

and the action of a symmetry operation on a position can now be calculated as:

$$\vec{x}' = (R \mid t)\vec{x} = R\vec{x} + t = \begin{pmatrix} 0 & 0 & 1 \\ 1 & 0 & 0 \\ 0 & 1 & 0 \end{pmatrix} \times \begin{pmatrix} x_0 \\ y_0 \\ z_0 \end{pmatrix} + \begin{pmatrix} 0 \\ \frac{1}{2} \\ \frac{1}{2} \end{pmatrix} = \begin{pmatrix} z_0 \\ x_0 \\ y_0 \end{pmatrix} + \begin{pmatrix} 0 \\ \frac{1}{2} \\ \frac{1}{2} \end{pmatrix} = \begin{pmatrix} z_0 \\ x_0 + \frac{1}{2} \\ y_0 + \frac{1}{2} \end{pmatrix} \tag{30}$$

if the position vector $\vec{x}$ is represented as a row-vector (i.e., if given as the transpose of the column-vector representation), then the symmetry operator acts from the right and its transpose is taken into account:

$$\begin{aligned} \vec{x}_{row} = (x_0, y_0, z_0) \rightarrow \vec{x}'_{row} &= \vec{x}_{row} R^T + t^T \\ &= (x_0, y_0, z_0) \times (0, \tfrac{1}{2}, \tfrac{1}{2}) \\ &= (z_0, x_0 + \tfrac{1}{2}, y_0 + \tfrac{1}{2}) \end{aligned}$$

These matrices can also be given in *(abc)* or in *(xyz)* notations:

- The *(abc)*notation is used primarily for denoting transformation matrices (P, p). The first three parameters separated by comma (functions of a, b, c) give the change of the unit cell basis vectors, while the three values after the semicolon holds the translational information of the origin. Thus, a transformation matrix such as:

  $$-\tfrac{2}{3}a - \tfrac{1}{3}b + \tfrac{2}{3}c, -b, 2a + b; \tfrac{1}{6}, \tfrac{1}{3}, \tfrac{1}{3}$$

  represents the transformation (P, p):

  $$P = \begin{pmatrix} -\frac{2}{3} & 0 & 2 \\ -\frac{1}{3} & -1 & 1 \\ \frac{2}{3} & 0 & 0 \end{pmatrix}; p = \begin{pmatrix} \frac{1}{6} \\ \frac{1}{3} \\ \frac{1}{3} \end{pmatrix}$$

  (notice that the coefficients are read in columns).
- The *(xyz)* notation is used primarily for denoting symmetry operations. In contrast to the *(abc)* notation, the coefficients are placed into the matrix in rows, for example, the operation in Eq. (28) is written in *(xyz)* form as:

  $$z, x + \frac{1}{2}, y + \frac{1}{2}$$

It must be emphasized that, even though from a mathematical point of view a transformation matrix and a symmetry operation can be treated under the same heading, crystallography strictly distinguishes the two.

A symmetry operation indicates that the new position obtained using the corresponding symmetry operator will have the same properties (same surroundings, same species, etc.) as those of the originating site. In a similar way, the set of such equivalent positions derived by applying the symmetry operators of the group belong to the same Wyckoff Position. Furthermore, if the number of such positions is equal to the number of the symmetry operators, then these positions are called *general positions*, while if the number of such positions are less than the total number of the symmetry operators of the group (meaning that some generated positions overlap with the existing ones, for example, application of a reflection operation on a point lying on the mirror plane), then these types of positions are called *special positions*[3].

A group is a set of objects (in our case, the symmetry operations) that obey to the following basic axioms:

1. Closure: For any two elements of the group, their consecutive action is also an element of the set: $a, b \in G \rightarrow a \otimes b = c \in G$. Here, $\otimes$ is a binary operation (group law) defined on the group elements;
2. Associativity: For any three elements of the group, $a, b, c \in G$, the following relation holds: $(a \otimes b) \otimes c = a \otimes (b \otimes c)$ as long as the order is preserved;
3. Identity element: There must exist an element E such that: $a \otimes E = E \otimes a = a$.
4. Inverse element: For any element a of the group, there must be an element b in the group such that: $a \otimes b = b \otimes a = E$.

   (from here on, the group operation symbol '$\otimes$' will be suppressed, hence an operation such as '$a \otimes b$' will be denoted simply as 'ab')

A space group is a special group that contains rotation, inversion, reflection operators as well as translation operators and the combination of them. In 3D, there are 230 such space groups. Their elements and basic properties are tabulated in the International Tables of Crystallography, Volume A [19] and also available from the GENPOS [20] and WYCKPOS [21] tools (along with other tools) on the Bilbao Crystallographic Server *(22–24)*. Screenshots from these tools for space group $P4_2$ are presented in Figs. 11.1 and 11.2. One can also use the designated interface (EXPLORE SYMMETRY [25]) to display space

[3]In a more general definition, Wyckoff Positions are the set of positions that are invariant under the application of same symmetry operators (i.e., points having common *site-symmetry groups*)

group information coded using the Computational Crystallography Toolbox (cctbx) [26].

General Positions of the Group 77 ($P4_2$)

1	x,y,z	$\begin{pmatrix} 1 & 0 & 0 & 0 \\ 0 & 1 & 0 & 0 \\ 0 & 0 & 1 & 0 \end{pmatrix}$	1
2	-x,-y,z	$\begin{pmatrix} -1 & 0 & 0 & 0 \\ 0 & -1 & 0 & 0 \\ 0 & 0 & 1 & 0 \end{pmatrix}$	2 0,0,z
3	-y,x,z+1/2	$\begin{pmatrix} 0 & -1 & 0 & 0 \\ 1 & 0 & 0 & 0 \\ 0 & 0 & 1 & 1/2 \end{pmatrix}$	4^+ (0,0,1/2) 0,0,z
4	y,-x,z+1/2	$\begin{pmatrix} 0 & 1 & 0 & 0 \\ -1 & 0 & 0 & 0 \\ 0 & 0 & 1 & 1/2 \end{pmatrix}$	4^- (0,0,1/2) 0,0,z

FIGURE 11.1 General positions of the space group $P4_2$ as listed by the GENPOS tool.

11.1.7 Q&A (QUESTION AND ANSWER): WHAT DOES "CENTRIC," "ACENTRIC," "CENTROSYMMETRIC," "NON-CENTROSYMMETRIC," "POLAR" SPACE GROUPS MEAN?

A centric (also referred to as "centrosymmetric") space group is a group that contains inversion symmetry among its symmetry operator elements. Suppose that the inversion center is located at the origin of the unit cell: then, for every position (x_0, y_0, z_0) in the unit cell complying the symmetry relations of the space group there exists a position $(-x_0, -y_0, -z_0)$ related to the original one by the inversion symmetry. Having a center imposes a restriction on the arbitrary translation of the sites. Note that the inversion center is not limited to the origin – the role can also be assumed by another point. If the inversion center is located at (x_1, y_1, z_1), then the coordinates of the two points related by inversion symmetry are (x_0, y_0, z_0) and $(2x_1 - x_0, 2y_1 - y_0, 2z_1 - z_0)$. The space groups that do not contain an inversion symmetry operation are called "noncentrosymmetryic" (or "acentric").

Although all of the centric/centrosymmetric groups are nonpolar, *not all of the nonpolar groups are centric!* For a counterexample, consider the symmetry operations and Wyckoff positions of the space group P222: it has only 2-fold rotations (in addition to the identity operation) and no inversion operation yet its Wyckoff positions are fixed and do not allow an arbitrary translation in any direction.

To easily check if a space group is polar, one can directly refer to its associated point group: if it's one of the 10 polar point groups (which are: 1, 2, m, mm2, 4, 4 mm, 3, 3 m, 6, 6 mm) then the space group is also polar.

Space groups can be referred by their Hermann-Maugin symbols as well as their index number in the ITA (from 1 to 230[4]). With respect to their lattice types, the space groups are grouped as in Table 11.1.

Wyckoff Positions of Group 77 ($P4_2$)

Multiplicity	Wyckoff letter	Site symmetry	Coordinates
4	d	1	(x,y,z) (-x,-y,z) (-y,x,z+1/2) (y,-x,z+1/2)
2	c	2..	(0,1/2,z) (1/2,0,z+1/2)
2	b	2..	(1/2,1/2,z) (1/2,1/2,z+1/2)
2	a	2..	(0,0,z) (0,0,z+1/2)

FIGURE 11.2 Wyckoff positions of the space group $P4_2$ as listed by the WYCKPOS tool.

TABLE 11.1 Lattice Types with Respect to Space Group Numbers.

ITA#	Lattice Type	Lattice Properties
1–2	Triclinic	$a \neq b \neq c; \alpha \neq \beta \neq \gamma$
3–15	Monoclinic	$a \neq b \neq c; \alpha = \gamma = 90°, \beta \neq 90°$
16–74	Orthorhombic	$a \neq b \neq c; \alpha = \beta = \gamma = 90°$
75–142	Tetragonal	$a = b \neq c; \alpha = \beta = \gamma = 90°$
143–167	Trigonal	$a = b \neq c; \alpha = \beta = 60°, \gamma = 90°$
168–194	Hexagonal	$a = b \neq c; \alpha = \beta = 60°, \gamma = 90°$
195–230	Cubic	$a = b = c; \alpha = \beta = \gamma = 90°$

[4]You can access to the complete list from http://www.cryst.ehu.es/cgi-bin/cryst/programs/nph-table .

11.1.8 CASE STUDY: SYMMETRY OPERATORS AND GENERAL POSITIONS OF P4$_2$ (#77)

Other than the three basic unit cell translations $(1|100),(1|010),(1|001)$ where 1 in the rotational part represents the 3x3 unit matrix, the P4$_2$ has four symmetry operations $\{1,(4^+|00\frac{1}{2}),2,(4^-|00\frac{1}{2})\}$ which have already been listed with their matrix representations in Fig 11.1.

In the most general case, applying these four operators on an arbitrary point located at $A(x_0,y_0,z_0)$ will generate four points as demonstrated in Fig. 11.3(a).

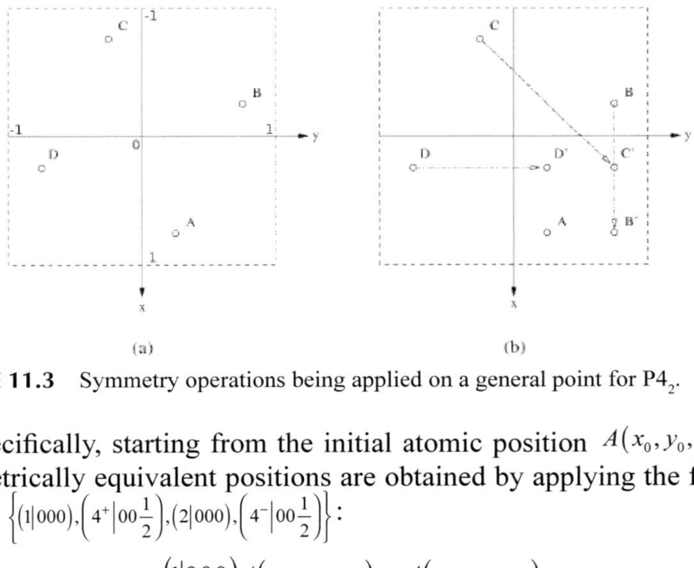

FIGURE 11.3 Symmetry operations being applied on a general point for P4$_2$.

Specifically, starting from the initial atomic position $A(x_0,y_0,z_0)$, the symmetrically equivalent positions are obtained by applying the four operators $\left\{(1|000),\left(4^+|00\frac{1}{2}\right),(2|000),\left(4^-|00\frac{1}{2}\right)\right\}$:

$$(1|000)A(x_0,y_0,z_0)=A(x_0,y_0,z_0)$$

$$\left(4^+|0\ \frac{1}{2}\right)A(x_0,y_0,z_0)=B\left(-y_0,x_0,z_0+\frac{1}{2}\right)$$

$$(2|000)A(x_0,y_0,z_0)=C(-x_0,-y_0,z_0)$$

$$\left(4^-|0\ \frac{1}{2}\right)A(x_0,y_0,z_0)=D\left(y_0,-x_0,z_0+\frac{1}{2}\right)$$

(notice the different z-components of the positions: while A and C points lie on the xy-plane, B and D points lie above that plane).

Since we are dealing with space groups, unit cell translations can be used to find the symmetrically equivalent positions in a single unit cell:

$$(1|100)B = B'$$
$$(1|100)(1|010)C = C'$$
$$(1|010)D = D'$$

hence obtaining the elements of the general 4d Wyckoff orbit for the space group $P4_2$ as indicated in Fig. 11.3 (b). It is important to keep in mind that each unit cell is identical, that is, the basic translations are used to populate the remaining unit cells likewise.

With the specification of symmetry operations, it is obvious that the information of only 1 site for the whole orbit is sufficient. This site is properly called the representative and the minimal unit cell volume sufficient to define the whole unit cell by means of the application of the symmetry operators is called the asymmetrical unit.

11.1.9 Q&A: HOW COME THE COORDINATES OF THE ATOMS ARE BETWEEN 0 AND 1?

The coordinates of the atoms are usually represented in the so-called "direct" (also called "rational," "fractional") coordinates where the Cartesian components of the coordinates are divided by the a, b, c parameters of the unit cell $(0 \le \{x, y, z\} < 1)$. So, for example, an atom that is located in the middle of any component of the unit cell directions (basis vectors) will have 1/2 corresponding to that direction; $\left(\frac{1}{2}, \frac{1}{2}, \frac{1}{2}\right)$ always designating the center of the unit cell; (x0z) a point on the B faces; $(x_0, y_0, z_0 + 1)$ being the corresponding atom in the c-wise upper cell of the reference unit cell, etc.

It should be stressed that even though in all the 3 directions the unit cell extends from 0 to 1, it doesn't necessarily mean neither $a = b = c$ nor $\alpha = \beta = \gamma$!

11.1.10 GROUP-SUBGROUP RELATIONS AND COSET DECOMPOSITION

Let G and H be two space groups satisfying the group axioms. If G contains all the elements of H, then G is said to be a supergroup of H, and in a similar manner, H is called as a subgroup of G. This relation is represented by the symbol: G>H.

A group and a subgroup may be connected via different paths and using different transformation matrices. An example to group-subgroup relations is given in Fig. 11.4. One can visualize these relations via the SUBGROUP-GRAPH tool [27].

If there is no intermediate group between a group G and its subgroup H, then the subgroup H is said to be a maximal subgroup of G, and equivalently, G is called a minimal supergroup of H. Thus, any group-subgroup relation G>H can be written as a consecutive chain of minimal supergroups:

$$G > H \rightarrow G > Z_n > \cdots > Z_1 > H \tag{31}$$

We will use this deduction when we are searching for the high symmetry phase of a given low symmetry structure.

Since the supergroup contains all the symmetry operators of the sub-groups, it is possible to write the supergroup in terms of the low symmetry group and the missing elements in the subgroup. This process is the so-called coset decomposition.

The procedures for calculating the coset decomposition of G in terms of H are:

1. Take the subgroup H as the first coset.
2. Pick an element g_i from G that is not already present in any of the cosets obtained so far and apply this to H (from left or right leading to left/right coset decomposition).
3. Go to step #2 until all the elements of G are contained in the cosets.

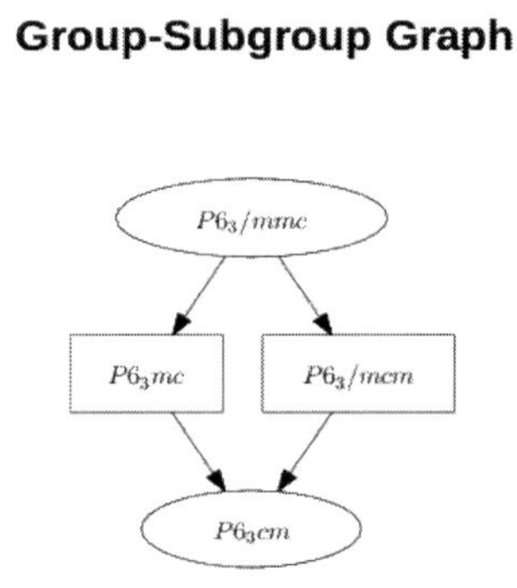

FIGURE 11.4 Group-subgroup relation between the space-groups $P6_3/mmc$ (#194) and $P6_3cm$ (#185) obtained via the SUBGROUPGRAPH tool.

11.1.11 CASE STUDY: COSET DECOMPOSITION OF P4/MMM (#123) WITH RESPECT TO P4 (#75)

G:P4/mmm and H:P4 are related by the transformation matrix (a, b, c; 0, 0, 0). Their symmetry operators are:

- (G) P4/mmm: $\{1, 4_z^+, 2_z, 4_z^-, 2_y, 2_x, 2_{xx}, 2_{x\bar{x}}, \bar{1}, m_z, -4_z^+, -4_z^-, m_y, m_x, m_{xx}, m_{x\bar{x}}\}$
- (H) P4: $\{1, 4_z^+, 2_z, 4_z^-\}$

(The list of the operators, as well as their matrix representations and geometric interpretations can be obtained via the GENPOS tool of the Bilbao Crystallographic Server).

The coset decomposition starts with taking the subgroup P4 as the first coset (i.e., picking the identity operator as the first operator):

$$1\{1, 4_z^+, 2_z, 4_z^-\} = \{1, 4_z^+, 2_z, 4_z^-\}$$

For the second coset, we can't pick 4_z^+, 2_z or 4_z^- since they are already included in the 1st coset, so we must pick another operator, say 2_y:

$$2_y\{1, 4_z^+, 2_z, 4_z^-\} = \{2_y, 2_{xx}, 2_x, 2_{x\bar{x}}\}$$

The third coset is constructed by picking any of the not yet obtained ones (i.e., one of $\{\bar{1}, m_z, -4_z^+, -4_z^-, m_y, m_x, m_{xx}, m_{x\bar{x}}\}$). Let's pick $\bar{1}$, then the 3rd coset is constructed as:

$$\bar{1}\{1, 4_z^+, 2_z, 4_z^-\} = \{\bar{1}_y, -4_z^+, m_z, -4_z^-\}$$

Pick any of the remaining four operators to construct the final coset:

$$m_y\{1, 4_z^+, 2_z, 4_z^-\} = \{m_y, m_{x\bar{x}}, m_x, m_{xx}\}$$

Thus, the coset decomposition of P4/mmm (#123) with respect to P4 (#75) is:

$$\{1, 4_z^+, 2_z, 4_z^-\}, \{2_y, 2_{xx}, 2_x, 2_{x\bar{x}}\}, \{\bar{1}_y, -4_z^+, m_z, -4_z^-\}, \{m_y, m_{x\bar{x}}, m_x, m_{xx}\}$$

The COSETS [28] tool of the Bilbao Crystallographic Server can be used to automatically construct the cosets of a group-subgroup pair (the results are given in the (xyz) format).

11.2 PHASE TRANSITIONS

Even though 1st order phase transitions in general can not be connected with a group-subgroup relation, all 2nd order phase transitions and many 1st order phase transitions can be decomposed into a hierarchy of supergroup, intermediate subgroups and the destination subgroup as explained in previous section.

The involved space groups and their symmetries can be derived by employing the pseudosymmetry concept where the displacements required to reach the low symmetry structure compatible with a higher symmetry are calculated and considered. In this way, starting from a low symmetry "root" structure, an ascending path towards high symmetries can be considered, and the transformation matrices relating the different phases can be computed.

The intermediate symmetries do not necessarily are involved during the real transition but they are still useful to refine and classify the transition mechanism. An interesting case, concerning the "existence" of such a phase of $YMnO_3$ has been studied in detail by Orobengoa et al. [29].

The search of the unknown parent structure is a powerful application of the group theory that enables us to predict potential (displacive) ferroelectrics. In this section, we'll see how to construct the group-subgroup trees along with their interrelations by using the PSEUDO tool [30]; symmetry mode analysis of the involved distortion modes is also possible via the AMPLIMODES tool [29, 31] of the Bilbao Crystallographic Server.

11.2.1 SYMMETRY AND BREAKING OF IT

Consider for example a system with a reflection symmetry as shown in Fig. 11.5. Since we know that it possesses a mirror plane m_y, whose effect is to change the sign of the y component of the atomic position, in order to define a given system that contains two atoms in the unit cell, the specification of only 1 of the sites (e.g., the A site) is sufficient to describe the whole system. The space group for this system is designated as Pm and it contains two operations: the identity E, and the reflection m. The "P" denotes the primitive unit cell, meaning that only translations along the basis vectors $\{(1|100),(1|010),(1|001)\}$ are present. So, the other site in the same unit cell C site is obtained from site A by first applying the reflection operator, therefore reaching to the B site and finally applying the translation operator along the y-direction.

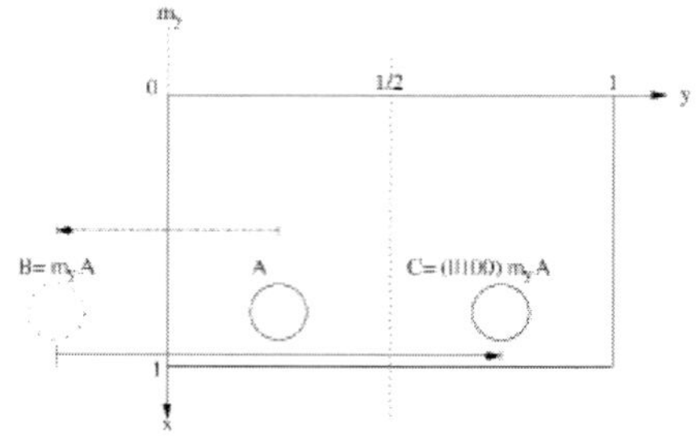

FIGURE 11.5 A system defined in Pm symmetry.

Assuming the unit cell dimensions are given by the $a,b,c,\alpha,\beta,\gamma$ parameters, with α: being the angle between $\vec{b}$ and $\vec{c}$; β: between $\vec{a}$ and $\vec{c}$; γ: between $\vec{a}$ and $\vec{b}$, this system can be described as:

6
a b c α β γ
A 1 2c $x_0 y_0 z_0$

where "A" is the symbol of the species of the atom occupying that position; "1" is the site index (to differ in case there are more than 1 of the same species); "2c" is the corresponding Wyckoff position with the number "2" designating that this position yields two symmetrically equivalent positions in the unit cell for this setting; and 6 corresponds to the ITA number for the space group Pm.

If we neglect the symmetries (i.e., describing the system in the P1 symmetry where the only symmetry element is the identity operation), then we would have to write each of the atomic site position explicitly. This time the same system would be referred to as:

1
a b c α β γ
A 1 1a $x_0 y_0 z_0$
A 2 1a $x_0 - y_0 + 1\ z_0$

11.2.1.1 Q&A WHAT ARE THE MOST COMMON STRUCTURE DATA FORMATS?

There are different formats to represent structural information. The format used in the example above is used in the Bilbao Crystallographic Server (BCS). For a bulk structure, one has to specify the six unit cell parameters

$a, b, c, \alpha, \beta, \gamma$ and the coordinates of the occupied sites. The lattice parameters can be represented using the 3 basis vectors $\vec{a}, \vec{b}, \vec{c}$ (their xyz projections) as well.

As mentioned above, to avoid to write every atomic site position, one can use the coordinates of the atoms representative of the orbits and use symmetry operations to generate all the other atomic positions.

The most commonly used and widely accepted format for bulk material is the International Union of Crystallography's (IUCr) CIF format [32]. Most of the ab-initio programs accept and output the structure information in P1 symmetry. STRCONVERT [33], a tool in the Bilbao Crystallographic Server converts between various different structures formats such as BCS, CIF, VASP [34] and VESTA [35], also allowing editing and symmetry detection using the integrated FINDSYM [36] program.

11.2.1.2 Q&A MY STRUCTURE LOOKS DIFFERENT AFTER CONVERSION—WHY IS THAT?

There are more than one way to describe the same structure. Most of the commonly used settings are classified by the IUCr and for each unit cell, a "standard setting" has been chosen in case many choices exist. One can convert the data from a nonstandard setting into the corresponding standard setting via the SETSTRU [37] tool of the Bilbao Crystallographic Server.

Now, suppose that the atoms have been slightly distorted from their ideal positions as in Fig. 11.6 the atom previously located at position A goes to A' and B to B' such that they are not related anymore by the reflection operation, thus the reflection symmetry is said to be lost and the symmetry is broken.

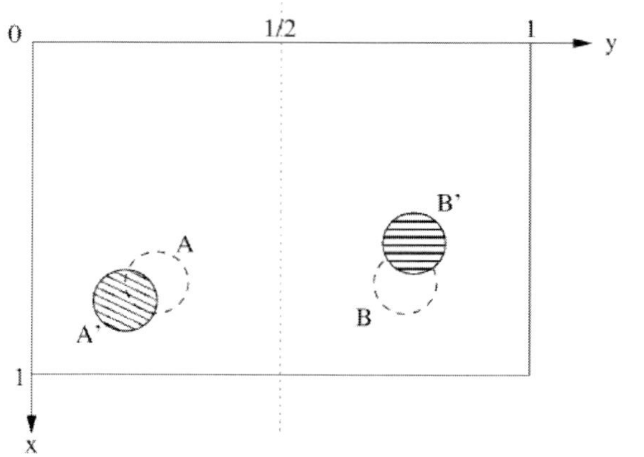

FIGURE 11.6 Distortion induced structure previously in Pm setting, now in P1.

11.2.2 PSEUDOSYMMETRY

If the magnitudes of the displacements that cause the breaking of symmetry are sufficiently small, the low symmetry structure can be analyzed in order to find a pseudosymmetric higher symmetry space group.

For all 2nd order and many of the first order phase transitions, the low symmetry structure's space group is related to that of the high symmetry parent's space group by means of a group-subgroup relation. If the parent phase is not known, then one can look for higher symmetries by a step by step procedure using the minimal supergroups approach (see Eq. (31)). This description of the parent phase in terms of the distorted phase is nothing but the coset decomposition process explained in subsection "Group-subgroup relations and coset decomposition."

Coming back to our previous example, we want to check the compatibility with the Pm of our distorted P1 structure. We use the coset decomposition:

$$Pm = P1 + mP1$$

assuming that, the missing operator is the reflection operator. Applying the reflection operator on the distorted A' and B' sites, we derive the A'' and B'' points respectively, as shown in Fig. 11.7. If the sites were related by mirror symmetry, then the points generated by applying the symmetry operation would give overlapping sites (A' with B'' and B' with A''). Due to the presence of the distortion, they appear very close to each other. By assuming a possible relation between these sites (this assumption is checked by introducing a tolerance in the relative distance), we calculate the ideal site position via averaging the associated sites so that this ideal site will be symmetrical under the operation.

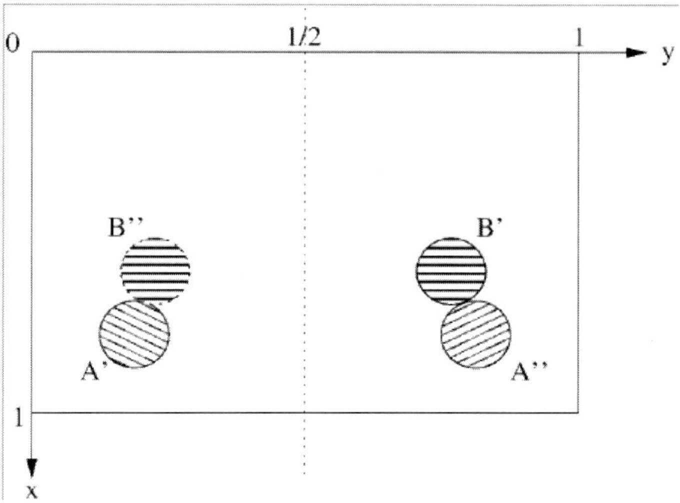

FIGURE 11.7 Searching for a possible symmetry in the distorted system by applying sought symmetry operator.

Thus we arrive at points C' and C'' that will be the idealized sites for (A' & B'') and (B' & A''), respectively, as shown in Fig. 11.8.

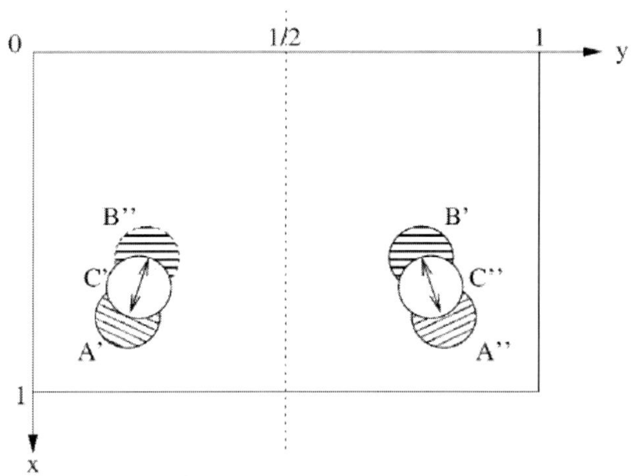

FIGURE 11.8 The idealized points that will comply with the reflection symmetry.

Even though for a simple system with few atomic sites and low symmetry this procedure is simple and straight-forward, for more complex systems the calculations become easily complicated and we need to use computer tools. The PSEUDO [38] tool of the Bilbao Crystallographic Server performs the analysis of pseudosymmetry for low symmetry structures by applying the procedures briefly summarized here. Further details can be found in Capillas et al. [30].

11.2.2.1 CASE STUDY: HIGH SYMMETRY PHASE OF BATIO$_3$ P4MM (#99)

Consider the well-known tetragonal P4 mm phase of the $BaTiO_3$ perovskite at the room temperature as observed by Buttner et al. [39]:

99
3.9998 3.9998 4.0180 90.0 90.0 90.0
4
Ba 1 1a 0.000 0.000 0.000
Ti 1 1b 0.500 0.500 0.482
O 1 1b 0.500 0.500 0.016
O 2 2c 0.500 0.000 0.515

The P4 mm is a noncentrosymmetric space group, that is, a general point (x_0, y_0, z_0) doesn't necessarily have a complementary at ($-x_0$, $-y_0$, $-z_0$).

P4 mm has a centrosymmetric minimal supergroup, P4/mmm (#123) and we have to find the new atomic positions which satisfy the symmetry operations of the higher space group symmetry. We can input the structure data of the P4 mm phase into PSEUDO, which returns the necessary shifts given in Fig. 11.9. The columns under the $u_{\{xyz\}}$ headings show the deviations from the ideal positions in rational coordinates while the $|u|$ column gives the displacement in Å.

1# Supergroup *P4/mmm* (123): a,b,c ; 0,0,0 and index 2

Displacements:

Atom	Idealized Coordinates	u_x	u_y	u_z	$\|u\|$
Ba1	(0.0000, 0.0000, 0.0000)	0.000000	0.000000	0.000000	0.0000
Ti1	(0.5000, 0.5000, 0.5000)	0.000000	0.000000	-0.018000	0.0723
O1	(0.5000, 0.5000, 0.0000)	0.000000	0.000000	0.016000	0.0643
O2	(0.5000, 0.0000, 0.5000)	0.000000	0.000000	0.015000	0.0603

FIGURE 11.9 Displacements necessary to comply with the P4/mmm supergroup settings.

The structure in the idealized coordinates with respect to the high symmetry phase (P4/mmm) is also presented among the results of PSEUDO as:

123
3.9998 3.9998 4.0180 90.00 90.00 90.00
4
Ba 1 1a 0.000000 0.000000 0.000000
Ti 1 1d 0.500000 0.500000 0.500000
O 1 1c 0.500000 0.500000 0.000000
O 2 2e 0.500000 0.000000 0.500000

where these new positions are gained by subtracting the deviations given in Fig. 11.9 from the low symmetry structure. These high symmetry positions which have been derived purely from group theory have been observed in real experiments as well [39]. The two systems are visualized side by side in Fig. 11.10. Notice the small displacements of the O atoms at the B faces (z>0.5 in P4 mm; z=0.5 in P4/mmm) and the Ti atom in the inside (z<0.5 in P4 mm; z=0.5 in P4/mmm) so the total polarization at the P4/mmm is zero. As the temperature is increased, the structure goes into the cubic phase $Pm\overline{3}m$ (#221) [39]. With PSEUDO, we are searching for the parent phase that is possibly stable at higher temperatures.

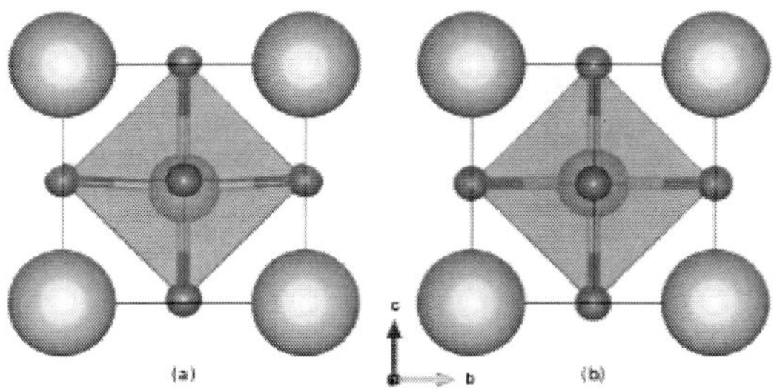

FIGURE 11.10 $BaTiO_3$ (a) Low symmetry phase (P4 mm); (b) High symmetry phase (P4/mmm) (Figures obtained using the VESTA program [40]).

11.2.3 SYMMETRY MODE ANALYSIS

Once the high and low symmetry structures and the transformation matrix that describes the symmetry break of the phase transition are known, it is useful to analyze the nature of the transition by the so-called *symmetry mode analysis*. Much like the decomposition of a vector into its -preferable- orthogonal components, the atomic sites distortions that causes the breaking of the symmetry can also be described using independent symmetry modes. To illustrate this, consider a CO_2 molecule as shown in Fig. 11.11. The movements of this linear molecule along in its symmetry axis can be fully described -in analogy with a three masses (2×*M*; 1×*m*) and two springs system- by the combination of the two modes indicated with the black and white arrows in the figure: In the 1st mode represented by the white arrows, the O atoms move in the opposite direction with the same magnitude while the central C atom doesn't move; the 2nd mode is just the translation of the molecule all-together.

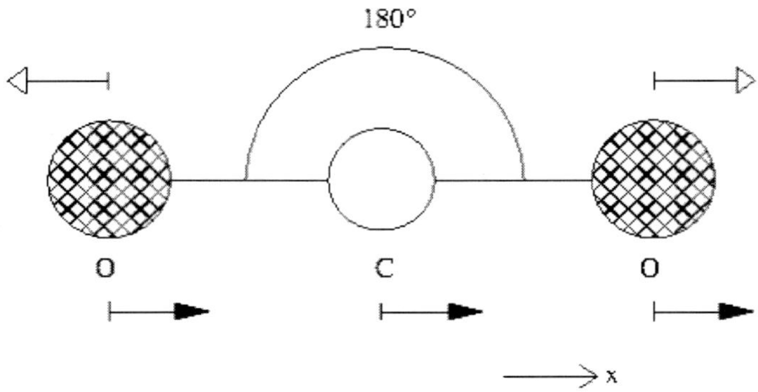

FIGURE 11.11 Linear CO_2 molecule with the symmetry modes indicated.

These two modes are orthogonal to each other as can be verified by constructing the displacement vectors of the atoms in each mode and multiplying them:

$$m_1 = [-1 \quad 0 \quad 1]; m_2 = [1 \quad 1 \quad 1]$$

$$m_1 \cdot m_2^T = 0$$

where the vectors are defined as the displacements of the O, C, O atoms from their equilibrium positions in the *x* positive direction. Inspecting the system, it can be easily seen that there are two compatible symmetry operations (in 1-dimension): the identity operator *E*, and the inversion operator *I*.

For more complex systems and having freedom in more than 1 dimension, the situation is more complicated as there will be D×N displacements where D is the dimensionality and

N is the number of atoms in the system. Thus, decomposing the displacements in terms of the symmetry modes greatly simplifies the transformations and makes it easier to visualize and understand the phase transition. The total distortion is the combined result of contributions from different symmetry modes. For an extensive discussion of the derivation and analysis of the symmetry modes, please refer to Wherrett's book [41]. Fortunately, this task of listing the modes compatible for a given group-subgroup pair has been implemented in computer programs as ISOTROPY [42]) and the SYMMODES [43, 44] tool of the Bilbao Crystallographic Server serve for this purpose. Figure 11.12 shows the screenshot from the SYMMODES tool for G: C2/c (#15), H: Cc (#9), transformation matrix: a, b, c;0,0,0 and Wyckoff positions: 8f, 4e, 4b.

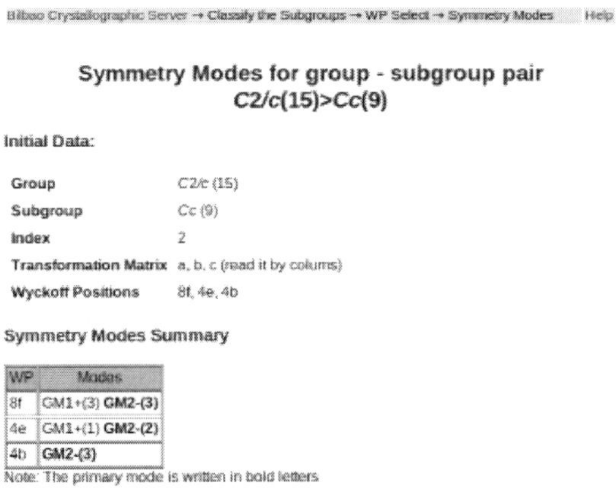
Bilbao Crystallographic Server → Classify the Subgroups → WP Select → Symmetry Modes Help

Symmetry Modes for group - subgroup pair
C2/c(15)>Cc(9)

Initial Data:

Group	C2/c (15)
Subgroup	Cc (9)
Index	2
Transformation Matrix	a, b, c (read it by colums)
Wyckoff Positions	8f, 4e, 4b

Symmetry Modes Summary

WP	Modes
8f	GM1+(3) **GM2-(3)**
4e	GM1+(1) **GM2-(2)**
4b	**GM2-(3)**

Note: The primary mode is written in bold letters

FIGURE 11.12 SYMMODES results for C2/c → Cc with a, b, c;0,0,0 and Wyckoff positions: 8f, 4e, 4b.

The GM2– and the GM1+ modes correspond to translations. In order to relate the symmetry modes to the actual atomic displacements, it is necessary to use another tool called AMPLIMODES [29, 31, 45] of the Bilbao Crystallographic Server and ISODISTORT [46]. These tools are used to calculate the contribution (i.e., the amplitudes) of each of the symmetry modes.

We now give an example of the analysis for the case study of the 1-cyclobutene-1,2-dicarboxylic acid (CBDC, $C_6H_6O_4$).

11.2.3.1 CASE STUDY: COMPLETE STUDY OF CBDC ($C_6H_6O_4$)

The 1-cyclobutene-1,2-dicarboxylic acid, CBDC in short has been studied by Bellus et al. [47], with Cambridge Structural Database Id: CBUDCX where they have reported a polar Cc structure along with the occupied positions.

11.2.3.1.1. RECOVERY OF THE ACTUAL SYMMETRY FROM P1

This reported structure then has been relaxed using ab-initio techniques [48]. Almost all the ab-initio calculation software packages present the resulting system in the P1 space group's setting, so after the calculations, it is important to recover the actual symmetry for the system. Thus, to recover the real symmetry, one can employ the STRCONVERT [33] tool of the Bilbao Crystallographic Server, which, among other utilities such as conversion and structure manipulation also offers the symmetry analysis with an interface to FINDSYM [36] application. Using STRCONVERT, the following structure data (belonging to the Cc space group as expected) is obtained from the reported P1 result of the ab-initio calculations:

```
# Space Group
9
# Lattice parameters
5.4812 13.4236 8.5216 90.000 100.520 90.000
16
# [atom type] [number] [WP] [x] [y] [z]
C 1 4a 0.142200 0.235000 0.716930
C 2 4a 0.934680 0.242470 0.820530
C 3 4a 0.749140 0.053880 0.848520
C 4 4a 0.189990 0.042520 0.627060
C 5 4a 0.901730 0.131800 0.794200
C 6 4a 0.082890 0.125510 0.704010
H 1 4a 0.329920 0.252290 0.781220
H 2 4a 0.776290 0.289530 0.769280
H 3 4a 0.971430 0.953310 0.719690
H 4 4a 0.489050 0.973720 0.474990
H 5 4a 0.106900 0.277320 0.604880
H 6 4a 0.999020 0.265700 0.944750
O 1 4a 0.583500 0.084860 0.931390
O 2 4a 0.354930 0.057810 0.546320
O 3 4a 0.779770 0.963680 0.818660
O 4 4a 0.108540 0.951640 0.647500
```

11.2.3.1.2 FINDING THE HIGH SYMMETRY PARENT PHASE

The next step is to investigate if there is a pseudosymmetry to a higher centrosymmetric supergroup. As mentioned previously, for searching a high symmetry parent phase, PSEUDO can be used. In Fig. 11.13, possible parent structures with the maximum atomic displacements needed in order to displace the atoms of low Cc structure to reach the high symmetry setting are shown, as obtained from PSEUDO. The maximum displacement is 0.4131 Å, which is a reasonable value. The transformation matrix connecting the subgroup to the supergroup is also reported.

Summary search results

Pseudosymmetry search among minimal supergroups.

Case #	Supergroup G	Index i	Index i_k	(P,p)	Tr. Matrix	Δ_{max}	u_{max}
1	*Pc* (007)	2	2	**2a,2b,c** ; 0,0,0	[2 0 0] [0] [0 2 0] [0] [0 0 1] [0]	>tol	-
2	*Cm* (008)	2	2	**a,b,2c** : 0,0,0	[1 0 0] [0] [0 1 0] [0] [0 0 2] [0]	>tol	-
3	*Cm* (008)	2	2	**a,b,a+2c** : 0,1/4,0	[1 0 1] [0] [0 1 0] [1/4] [0 0 2] [0]	>tol	-
4	*C2/c* (015)	2	1	**a,b,c** ; 0,0,0	[1 0 0] [0] [0 1 0] [0] [0 0 1] [0]	0.8262	0.4131
5	*Fdd2* (043)	2	1	**-c,b,1/2a+1/2c** ; 1/,1/8,1/	[0 0 1/2] [0] [0 1 0] [1/8] [-1 0 1/2] [0]	>tol	-
6	*Fdd2* (043)	2	1	**-b,a,1/2b+1/2c** ; 1/8,1/,1/	[0 1 0] [1/8] [-1 0 1/2] [0] [0 0 1/2] [0]	>tol	-

FIGURE 11.13 Considerations of possible minimal supergroups for CBDC by Pseudo.

PSEUDO also report the distortions of the atoms from their idealized positions in the high symmetry C2/c phase and these are shown in Fig. 11.14.

4# Supergroup C2/c (015): a,b,c ; 0,0,0 and index 2

Displacements:

Atom	Idealized Coordinates	u_x	u_y	u_z	|u|
O1	(0.6143, 0.0713, 0.9425)	-0.030784	0.013525	-0.011141	0.2542
O1_2	(0.3857, 0.0713, 0.5575)	-0.030784	-0.013525	-0.011141	0.2542
O5	(0.8356, 0.9577, 0.8356)	-0.055844	0.006020	-0.016916	0.3239
O5_2	(0.1644, 0.9577, 0.6644)	-0.055844	-0.006020	-0.016916	0.3239
C1	(0.1038, 0.2387, 0.6982)	0.038441	-0.003735	0.018734	0.2452
C1_2	(0.8962, 0.2387, 0.8018)	0.038441	0.003735	0.018734	0.2452
C9	(0.9094, 0.1287, 0.7951)	-0.007689	0.003145	-0.000891	0.0592
C9_2	(0.0906, 0.1287, 0.7049)	-0.007689	-0.003145	-0.000891	0.0592
C17	(0.7796, 0.0482, 0.8607)	-0.030434	0.005680	-0.012206	0.1953
C17_2	(0.2204, 0.0482, 0.6393)	-0.030434	-0.005680	-0.012206	0.1953
H1	(0.2768, 0.2709, 0.7560)	0.053106	-0.018620	0.025254	0.4131
H1_2	(0.7232, 0.2709, 0.7440)	0.053106	0.018620	0.025254	0.4131
H5	(0.0539, 0.2715, 0.5801)	0.052961	0.005810	0.024819	0.3357
H5_2	(0.9461, 0.2715, 0.9199)	0.052961	-0.005810	0.024819	0.3357
H17	(0.0000, 0.9533, 0.7500)	-0.028569	0.000000	-0.030306	0.2765
H21	(0.5000, 0.0000, 0.5000)	-0.010949	-0.026280	-0.025006	0.4108

NOTE: u_x, u_y and u_z are given in relative units. \|u\| is the absolute displacement given in Å

FIGURE 11.14 The displacements of the atomic sites in the Cc phase in order to reach the high symmetry C2/c phase, as calculated by Pseudo.

Finally, Pseudo lists the idealized symmetrized structure C2/c, with the atomic positions both in the subgroup and supergroup (parent phase) settings. In the C2/c setting the result is:

15
5.4812 13.4236 8.5216 90.00 100.52 90.00
9
O 1 8f 0.614285 0.071335 0.942535
O 2 8f 0.835615 0.957660 0.835580
C 1 8f 0.103760 0.238735 0.698200
C 2 8f 0.909420 0.128655 0.795095
C 3 8f 0.779575 0.048200 0.860730
H 1 8f 0.276815 0.270910 0.755970
H 2 8f 0.053940 0.271510 0.580065
H 3 4e 0.000000 0.953310 0.750000
H 4 4b 0.500000 0.000000 0.500000

11.2.3.1.3 ANALYSIS OF THE DISTORTIONS

Now that we have both the structural data of the polar and centro-symmetric phases and the transformation matrix connecting them, we can analyze the distortions in terms of the symmetry modes (if the transformation matrix is not known then the STRUCTURE RELATIONS [49] tool of the Bilbao Crystallographic Server can be used to derive the relating transformation).

By using AMPLIMODES, the amplitudes of the symmetry modes associated with the Wyckoff positions of the parent phase are calculated. For the CBDC, these positions are 8f, 4e and 4b and the related modes had already been summarized in Fig. 11.15.

Symmetry Modes Summary

Atoms	WP	Modes
C9 O5 H1 H5 C1 O1 C17	8f	GM1+(3) **GM2-(3)**
H17	4e	GM1+(1) **GM2-(2)**
H21	4b	**GM2-(3)**

Note: The primary mode is written in bold letters

Summary of Amplitudes

K-vector	Irrep	Direction	Isotropy Subgroup	Dimension	Amplitude (Å)
(0,0,0)	GM1+	(a)	C2/c (15)	22	0.0003
(0,0,0)	GM2-	(a)	Cc (9)	26	1.6458

Global distortion: 1.6458 Å

FIGURE 11.15 The amplitudes of the symmetry modes calculated by Amplimodes for CBDC C2/c → Cc.

In addition to the overall amplitudes, the form of the modes is accessible by clicking on the 'Detailed information' link located at the end of the AMPLIMODES results page. The visualization of each mode's effect can be obtained by following the link to STRCONVERT's tools. For the CBDC, the GM2- mode's resulting displacements are visualized using STRCONVERT's internal Jmol [50] routine and presented in Fig. 11.16 (ISODISTORT also offers an on-the-fly visualization of the symmetry modes).

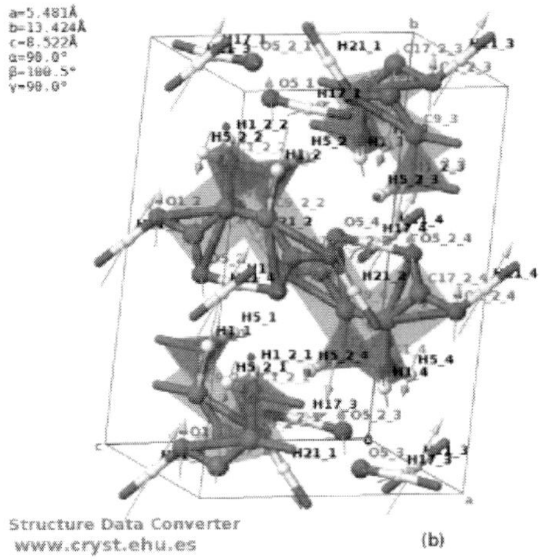

FIGURE 11.16 The GM2- mode's effect for CBDC.

More details and detailed discussions can be found in Refs. [48, 51].

11.2.3.2 CASE STUDY: PHMDA ($C_9H_8O_2$)

2-phenylmalondialdehyde's low temperature structure has been observed in 1977 by Semmingsen using X-ray diffraction and it was refined in the $Pna2_1$ (#33) space group, with a suggestion of a possible pseudosymmetry with respect to a higher symmetry, Pnab (#60) [52].

Using the experimental data, the structure is relaxed with ab-initio methods and the symmetry of the resulting structure (reported in P1 space group settings by the ab-initio calculation package) is recovered using the FINDSYM program implemented in STRCONVERT tool of the Bilbao Crystallographic Server, confirming the $Pna2_1$ symmetry.

To analyze the pseudosymmetry with respect to a higher symmetry space groups' settings, Pseudo is employed. Among the possible minimal supergroups of Pna2_1, only three of them are calculated to be compatible with the default maximum allowed displacement value of 2 Å as listed in Table 11.2.

TABLE 11.2 Compatible Minimal Supergroups of the Pna2_1PhMDA Molecule

Supergroup	(P, p)	u_{max}
Ama2 (#40)	a, b, c;0,1/4,0	0.9867 Å
Pbcn (#60)	c, b, –a;0,1/4,0	0.2483 Å
Pnma (#62)	a, –c, b;0,0,0	0.8698 Å

It should be noted that different notations used for the high symmetry phase which have been mentioned in the original experimental paper as Pnab and reported by PSEUDO as Pbcn, although they both refer to the space group that is designated as #60 in ITA. This is due to the use of different setting for the same space group symmetry. Bilbao Crystallographic Server's tools use by default the standard setting, and it is strongly recommended to work in this setting, when using the Bilbao Server tools. There is a tool designated to convert a given structure from one setting to another: SETSTRU [37].

Coming back to our example, Pseudo reports the high symmetry phase structure data (in Pbcn setting) as:

```
60
5.5530 17.1600 7.6830 90.00 90.00 90.00
12
O 1  8d 0.348800 0.544610 0.073225
C 1  8d 0.176530 0.514460 0.162420
C 5  4c 0.000000 0.558090 0.250000
C 13 4c 0.000000 0.644520 0.250000
C 17 8d 0.191260 0.685840 0.325465
C 21 8d 0.189465 0.767070 0.327510
C 25 4c 0.000000 0.807630 0.250000
H 1  4a 0.500000 0.500000 0.000000
H 5  8d 0.164365 0.450350 0.170115
H 13 8d 0.342120 0.654435 0.383255
H 17 8d 0.336490 0.798320 0.390655
H 21 4c 0.000000 0.871170 0.250000
```

which is related to the low-symmetry Pna2_1 phase with the transformation c, b, –a; 0,1/4,0. Even though PSEUDO, by default, provides this idealized high symmetry structure transformed to the low symmetry settings via the transformation matrix, the transformation of any structure can be performed by the TRANSTRU [53] tool in general. If necessary, the high symmetry structure described in the default Pbcn setting can be converted to the Pnab setting via SETSTRU by selecting the initial and final settings as shown in the snapshot Fig. 11.17.

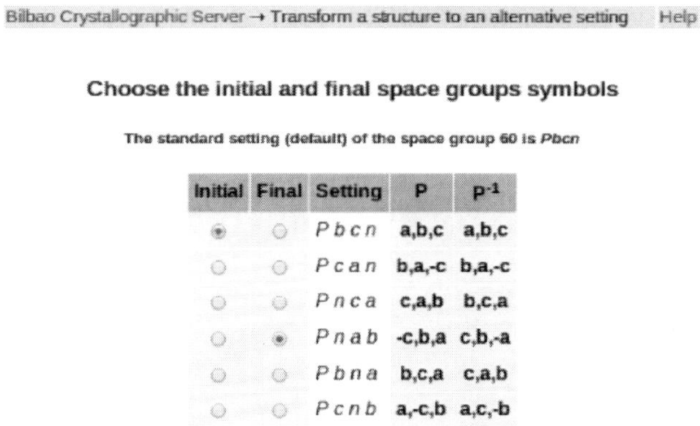

FIGURE 11.17 Selecting the initial and final settings in SETSTRU.

In the results page, SETSTRU gives a comprehensive information on the relation of the atoms between the two settings.

Now that we have both the parent and the low symmetry phases and the transformation relating them, we can proceed to use the AMPLIMODES tool to analyze the distortions in terms of the symmetry modes, which yields the results summarized in the snapshot Fig. 11.18 and the GM3-contribution is visualized in Fig. 11.19. The screenshot taken from the ISODISTORT program is also included in Fig. 11.20 where the unit cells of the parent and distorted phases are shown explicitly; note the different orientations of the coordinate system and how they are related by the transformation matrix.

Symmetry Modes Summary

Atoms	WP	Modes
H5 H13 H17 C1 O1 C17 C21	8*d*	GM1+(3) **GM3-(3)**
C5 C25 C13 H21	4*c*	GM1+(1) **GM3-(2)**
H1	4*a*	**GM3-(3)**

Note: The primary mode is written in bold letters

Summary of Amplitudes

K-vector	Irrep	Direction	Isotropy Subgroup	Dimension	Amplitude (Å)
(0,0,0)	GM1+	(a)	Pbcn (60)	25	0.0000
(0,0,0)	GM3-	(a)	Pna2_1 (33)	32	0.9503

Global distortion: 0.9503 Å

FIGURE 11.18 AMPLIMODES' symmetry modes analysis for PhMDAPbcn → $Pna2_1$.

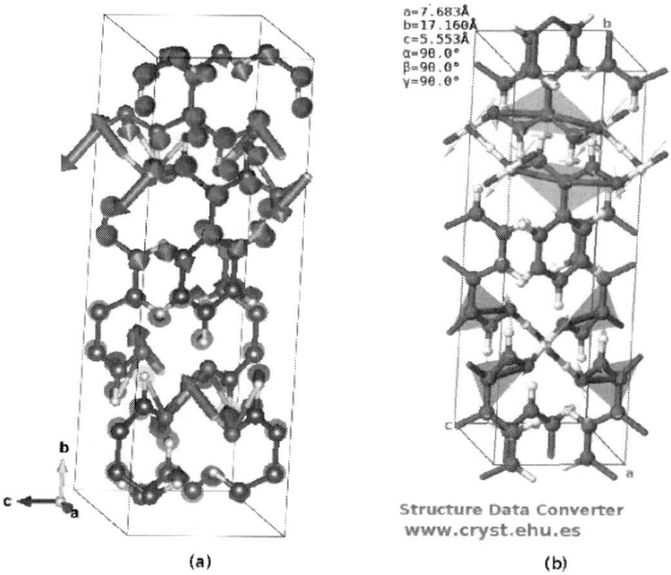

FIGURE 11.19 The visualization of the symmetry mode GM3- contribution for PhMDA as rendered by: (a) VESTA (b) Jmol.

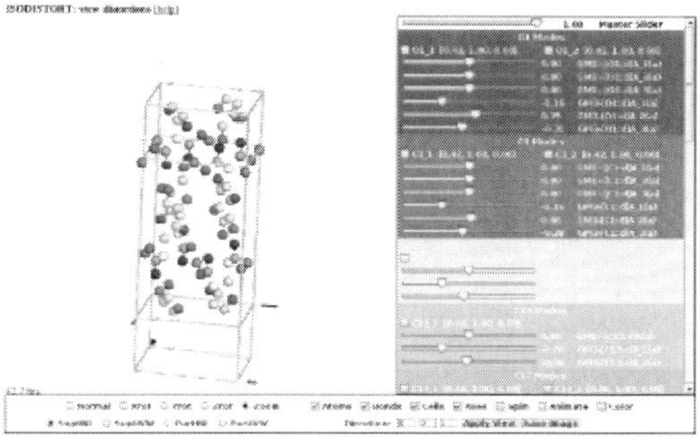

FIGURE 11.20 Screenshot of ISODISTORT's visualization of the symmetry modes analysis of PhMDA.

11.3 CONCLUSION

In this chapter we tried to give a brief overview of the group theory and its application to the solid state physics via the main computational crystallographic tools as they are implemented in the Bilbao Crystallographic Server (http://www.cryst.ehu.es). Moreover, we have tried to outline the main physical concepts underlying the symmetry mode analysis, explaining their utilization in studying ferroelectric materials focusing on the organic ferroelectrics CBDC and PhMDA as case studies from the recent literature (for additional ferroelectric case studies, the reader can refer to Refs. [54–57]). Further advanced applications of the symmetry mode analysis to complex materials can be found in Refs. [58–68]; symmetry analysis related to hybrid improper Ferro electricity can be found in Refs. [69–75].

Group theory and application of symmetry is a powerful tool in solid state calculations since they both simplify and reduce the calculations, automatically validating the process on the run. Last but not least, it gives a clear understanding of the behavior of the materials involved in phase transitions. Thanks to modern computers, most of the analyzes mentioned in this chapter can be easily performed using a standard PC.

ACKNOWLEDGMENTS

A.S. would like to thank Prof. A. Sayede for the kind invitation at Université d'Artois, Faculté des Sciences Jean Perrin where this work was finalized (June 2013). A.S. greatly appreciate discussions with Prof. J.M. Rondinelli and Prof. C.J. Fennie during the visiting stay at Drexel and Cornell University, respectively (April 2013). E.S.T. acknowledges financial support from TUBITAK through the 2232 fellowship program.

KEYWORDS

- **density functional theory**
- **extended periodic systems**
- **group theory in crystals**
- **many-body nanosystems**
- **phase transition in nanostructures**
- **pseudosymmetry**

REFERENCES

1. Lines, M. E., Glass, A. M. (1977). *Principles and Applications of Ferroelectrics and Related Materials*; Oxford Univ. Press: New York.
2. Valasek, J. (1921). Piezoelectric and Allied Phenomena in Rochelle Salt. *Phys. Rev.* 17, 475–481.
3. Horiuchi, S., Tokura, Y. (2008). Organic ferroelectrics. *Nat. Mater.* 7, 357–366.
4. Parr, R., Yang, W. *Density Functional Theory*, Springer-Verlag, 1990.
5. Dreizler, R., Gross, E., *Density Functional Theory*, Springer-Verlag, 1990.
6. Hohenberg, P., Kohn, W. (1964). Inhomogeneous Electron Gas. *Phys. Rev.* 136, B864-B871.
7. Kohn, W., Sham, L. J. (1965). Self-Consistent Equations Including Exchange and Correlation Effects. *Physical Review* 140, 1133–1138.
8. Ceperly, D., Alder, B. (1980). Ground State of the Electron Gas by a Stochastic Method. *Phys. Rev. Lett.* 45, 566–569.
9. Perdew, J., Zunger, A. (1981). Self-interaction correction to density-functional approximations for many-electron systems. *Phys. Rev. B* 23, 5048–5079.
10. Ashcroft, N., Mermin, N. *Solid State Physics*, Holt, Rinehart and Wiston, 1976.
11. Baldereschi, A. (1973). Mean-Value Point in the Brillouin Zone. *Phys. Rev. B* 7, 5212–5215.
12. Chadi, D. J., Cohen, M. (1973). Special Points in the Brillouin Zone. *Phys. Rev. B* 8, 5747–5753.
13. Monkhorst, H., Pack, J. (1976). Special points for Brillouin-zone integrations. *Phys. Rev. B* 13, 5188–5192.
14. Fu, C., Ho, K. (1983). First-principles calculation of the equilibrium ground-state properties of transition metals: Applications to Nb and Mo. *Phys. Rev. B* 28, 5480–5486.
15. Methfessel, M., Paxton, A. (1989). High-precision sampling for Brillouin-zone integration in metals. *Phys. Rev. B* 40, 3616–3621.

16. Berry, M. (1984). Quantal Phase Factors Accompanying Adiabatic Changes. *Proc. R. Soc. Lond. A* 392, 45–57.
17. King-Smith, R., Vanderbilt, D. (1993). Theory of polarization of crystalline solids. *Phys. Rev. B* 47, 1651–1654.
18. Resta, R. (1994). Macroscopic polarization in crystalline dielectrics: the geometric phase approach. *Rev. Mod. Phys.* 66, 899–915.
19. Paufler, P., Ed. *International Tables for Crystallography, 5th edition*, IUCr, 2006.
20. Bilbao Crystallographic Server, "GENPOS: Generators and General Positions." http://www.cryst.ehu.es/cryst/get_gen.html (accessed Jul 20, 2013).
21. Bilbao Crystallographic Server, "WYCKPOS: Wyckoff Positions." http://www.cryst.ehu.es/cryst/get_wp.html (accessed Jul 20, 2013).
22. Aroyo, M. I., Perez-Mato, J. M., Capillas, C., Kroumova, E., Ivantchev, S., Madariaga, G., Kirov, A., Wondratschek, H. (2006). Bilbao Crystallographic Server: I. Databases and crystallographic computing programs. *Zeitschrift fur Kristallographie* 221, 15–27.
23. Aroyo, M. I., Kirov, A., Capillas, C., Perez-Mato, J. M., Wondratschek, H. (2006). Bilbao Crystallographic Server. II. Representations of crystallographic point groups and space groups. *Acta Crystallographica Section A* 62, 115–128. DOI: 10.1107/S0108767305040286
24. Aroyo, M. I., Perez-Mato, J. M., Orobengoa, D., Tasci, E., de la Flor, G., Kirov, A. (2011). Crystallography online: Bilbao crystallographic server. *Bulg. Chem. Commun.* 43, 183–197.
25. Computational Crystallography Toolbox, "cctbx – Explore symmetry." http://cci.lbl.gov/cctbx/explore_symmetry.html (accessed Jul 20, 2013).
26. Grosse-Kunstleve, R. W., Sauter, N. K., Moriarty, N. W., Adams, P. D. (2002). The Computational Crystallography Toolbox: crystallographic algorithms in a reusable software framework. *Journal of Applied Crystallography* 35, 126–136.
27. Bilbao Crystallographic Server, "SUBGROUPGRAPH: Group-Subgroup Lattice and Chains of Maximal Subgroups." http://www.cryst.ehu.es/cryst/subgroupgraph.html (accessed Jul 20, 2013).
28. Bilbao Crystallographic Server, "COSETS: Coset Decomposition." http://www.cryst.ehu.es/cryst/cosets.html (accessed Jul 20, 2013).
29. Orobengoa, D., Capillas, C., Aroyo, M. I., Perez-Mato, J. M. (2009). AMPLIMODES: symmetry-mode analysis on the Bilbao Crystallographic Server. *Journal of Applied Crystallography* 42, 820–833.
30. Capillas, C., Tasci, E. S., de la Flor, G., Orobengoa, D., Perez-Mato, J. M., Aroyo, M. I. (2011). A new computer tool at the Bilbao Crystallographic Server to detect and characterize pseudosymmetry. *Z Kristallogr* 226, 186–196.
31. Perez-Mato, J. M., Orobengoa, D., Aroyo, M. I. (2010). Mode crystallography of distorted structures. *Acta Crystallographica Section A* 66, 558–590.
32. Hall, S. R., Allen, F. H., Brown, I. D. (1991). The crystallographic information file (CIF): a new standard archive file for crystallography. *Acta Crystallographica Section A* 47, 655–685.
33. Bilbao Crystallographic Server, "STRCONVERT: Structure Data Converter & Editor." http://www.cryst.ehu.es/cgi-bin/cryst/programs/mcif2vesta/index.php (accessed Jul 20, 2013).

34. Hafner, J., Kresse, G., Vogtenhuber, D., Marsman, M., "VASP: Vienna Ab initio Simulation Package." http://www.vasp.at/ (accessed Jul 20, 2013).
35. Momma, K., Izumi, F., "VESTA: Visualization for Electronic and Structural Analysis." http://jp-minerals.org/vesta/en/ (accessed Jul 20, 2013).
36. Stokes, H. T., Campbell, B. J., Hatch, D. M., "FINDSYM: Program for Identifying the Space Group Symmetry of a Crystal." http://stokes.byu.edu/iso/findsym.html (accessed Jul 20, 2013).
37. Bilbao Crystallographic Server, "SETSTRU: Transform a structure to an alternative setting." http://www.cryst.ehu.es/cryst/setstru.html (accessed Jul 20, 2013).
38. Bilbao Crystallographic Server, "PSEUDO: Pseudosymmetry Search." http://www.cryst.ehu.es/cryst/pseudosymmetry.html (accessed Jul 20, 2013).
39. Buttner, R. H., Maslen, E. N. (1992). Structural parameters and electron difference density in BaTiO3. *Acta Crystallographica Section B* 48, 764–769.
40. Momma, K., Izumi, F. (2011). VESTA 3, for three-dimensional visualization of crystal, volumetric and morphology data. *Journal of Applied Crystallography* 44, 1272–1276.
41. Wherrett, B. *Group Theory for Atoms, Molecules and Solids*, Prentice Hall International, Incorporated, 1986.
42. Stokes, H., Hatch, D., Campbell, B., "ISOTROPY," http://stokes.byu.edu/iso/isotropy.php (accessed Jul 20, 2013).
43. Bilbao Crystallographic Server, "SYMMODES: Symmetry Modes." http://www.cryst.ehu.es/cryst/symmodes.html (accessed Jul 20, 2013).
44. Capillas, C., Kroumova, E., Aroyo, M. I., Perez-Mato, J. M., Stokes, H. T., Hatch, D. M. (2003). SYMMODES: a software package for group-theoretical analysis of structural phase transitions. *Journal of Applied Crystallography* 36, 953–954. DOI: 10.1107/S0021889803003212
45. Bilbao Crystallographic Server, "AMPLIMODES: Symmetry mode analysis." http://www.cryst.ehu.es/cryst/amplimodes.html (accessed Jul 20, 2013).
46. Campbell, B. J., Stokes, H. T., Tanner, D. E., Hatch, D. M. (2006). ISODISPLACE: a web-based tool for exploring structural distortions. *Journal of Applied Crystallography* 39, 607–614.
47. Bellus, D., Mez, H-.C.; Rihs, G. (1974). Synthesis and reactivity of compounds with cyclobutane rings. Part III. Cyclobut-1-ene-1,2-dicarboxylic acid. X-Ray crystal structure and exceptional stereoselectivity in its Diels-Alder reaction with cyclopentadiene. *J. Chem. Soc. Perkin Trans.* 2, 884–890.
48. Stroppa, A., D. Sante, D., Horiuchi, S., Tokura, Y., Vanderbilt, D., Picozzi, S. (2011). Polar distortions in hydrogen-bonded organic ferroelectrics. *Phys. Rev. B* 84, 014101.
49. Bilbao Crystallographic Server, "Structure Relations." http://www.cryst.ehu.es/cryst/rel.html (accessed Jul 20, 2013).
50. Jmol, "Jmol: an open-source Java viewer for chemical structures in 3D." http://www.jmol.org/ (accessed Jul 20, 2013).
51. Horiuchi, S., Kumai, R., Tokura, Y. (2011). Hydrogen-Bonding Molecular Chains for High-Temperature Ferroelectricity. *Adv. Mater.* 23, 2098–2103.
52. Semmingsen, D. (1977). The Crystal Structure of Phenylmalondialdehyde at −162°C. *Acta Chem. Scand. B* 31, 114–118.

53. Bilbao Crystallographic Server, "TRANSTRU: Transform Structure." http://www.cryst.ehu.es/cryst/transtru.html (accessed Jul 20, 2013).
54. Fu, D-.W.; Cai, H-.L.; Yuanming Liu, Q. Y., Zhang, W., Zhang, Y., Chen, X-.Y.; Giovannetti, G., Capone, M., Li, J., Xiong, R-.G. (2013). Diisopropylammonium bromide is a high temperature molecular ferroelectric crystal. *Science* 339, 425–428.
55. Horiuchi, S., Tokunaga, Y., Giovannetti, G., Picozzi, S., Itoh, H., Shimano, R., Kumai, R., Tokura, Y. (2010). Above-room-temperature ferroelectricity in a single-component molecular crystal. *Nature* 463, 789–792.
56. Giovannetti, G., Kumar, S., Stroppa, A., V. D. Brink, J; Picozzi, S. (2009). Multiferroicity in TTF-CA organic molecular crystals predicted through ab-initio calculations. *Phys. Rev. Lett.* 103, 266401.
57. Giovannetti, G., Kumar, S., Pouget, J-.P.; Capone, M. (2012). Unraveling the polar state in TMTTF2-PF6 organic crystals. *Phys. Rev. B* 85, 205146.
58. Balachandran, P. V., Rondinelli, J. M. (2013). Interplay of octahedral rotations and breathing distortions in charge ordering perovskite oxides. arXiv e-prints [Online] http://arxiv.org/abs/1303.0903 (accessed Jul 20, 2013).
59. Wu, H., Yu, H., Yang, Z., Hou, X., Su, X., Pan, S., Poeppelmeier, K. R., Rondinelli, J. M. (2013). Designing a Deep-Ultraviolet Nonlinear Optical Material with a Large Second Harmonic Generation Response. *J. Am. Chem. Soc.* 135, 4215–4218.
60. Cammarata, A., Rondinelli, J. M. (2012). Spin-assisted covalent bond mechanism in 'charge-ordering' perovskite oxides. *Phys. Rev. B* 86, 195144.
61. Rondinelli, J. M., Coh, S. (2011). Large Isosymmetric Reorientation of Oxygen Octahedra Rotation Axes in Epitaxially Strained Perovskites. *Phys. Rev. Lett.* 106, 235502.
62. Rondinelli, J. M., Spaldin, N. A. (2010). Electron-lattice instabilities suppress cuprate-like electronic structures in $SrFeO_3/SrTiO_3$ superlattices. *Phys. Rev. B* 81, 085109.
63. Rondinelli, J. M., Spaldin, N. A. (2010). Substrate coherency driven octahedral rotations in perovskite oxide films. *Phys. Rev. B* 82, 113402.
64. Yamauchi, K. (2013). Theoretical Prediction of Multiferroicity in $SmBaMn_2O_6$. *J. Phys. Soc. Jpn.* 82, 043702.
65. Yamauchi, K., Picozzi, S. (2012). Orbital degrees of freedom as origin of magnetoelectric coupling in magnetite. *Phys. Rev. B* 85, 085131.
66. Yamauchi, K., Barone, P., Picozzi, S. (2011). Theoretical investigation of magnetoelectric effects in $Ba_2CoGe_2O_7$. *Phys. Rev. B* 84, 165137.
67. Yamauchi, K., Picozzi, S. (2010). Interplay between Charge Order, Ferroelectricity, and Ferroelasticity: Tungsten Bronze Structures as a Playground for Multiferroicity. *Phys. Rev. Lett.* 105, 107202.
68. Picozzi, S., Yamauchi, K., Sanyal, B., Sergienko, I. A., Dagotto, E. (2007). Dual Nature of Improper Ferroelectricity in a Magnetoelectric Multiferroic. *Phys. Rev. Lett.* 99, 227201.
69. Rondinelli, J. M., Fennie, C. J. (2012). Octahedral Rotation-Induced Ferroelectricity in Cation Ordered Perovskites. *Adv. Mat.* 24, 1961–1968.
70. Mulder, A. T., Benedek, N. A., Rondinelli, J. M., Fennie, C. J. (2013). Turning ABO_3 Antiferroelectrics into Ferroelectrics: Design Rules for Practical Rotation-Driven Ferroelectricity in Double Perovskites and $A_3B_2O_7$ Ruddlesden-Popper Compounds. Advanced Functional Materials [Online] http://dx.doi.org/10.1002/adfm.201300210 (accessed Jul 20, 2013).

71. Gou, G., Rondinelli, J. M. (2013). Predicted strain-induced isosymmetric ferri-to-ferroelectric transition with large piezoelectricity. arXiv e-prints [Online] arXiv:1304.4911 http://arxiv.org/abs/1304.4911 (accessed Jul 20, 2013).
72. Stroppa, A., Jain, P., Barone, P., Marsman, M., Perez-Mato, J. M., Cheetham, A. K., Kroto, H. W., Picozzi, S. (2011). Electric Control of Magnetization and Interplay between Orbital Ordering and Ferroelectricity in a Multiferroic Metal–Organic Framework. *Angew. Chem. Int. Ed.* 50, 5847–5850.
73. Stroppa, A., Barone, P., Jain, P., Perez-Mato, J. M., Picozzi, S. (2013). Hybrid Improper Ferroelectricity in a Multiferroic and Magnetoelectric Metal-Organic Framework. *Adv. Mater.* 25, 2284–2290.
74. Sante, D., Stroppa, D., Picozzi, A. S. (2012). Structural, Electronic and Ferroelectric properties of Croconic Acid Crystal: a DFT study. *Phys. Chem. Chem. Phys.* 14, 14673–14681.
75. Di Sante, D., Stroppa, A., Picozzi, S. (2013). J. Am. Chem. Soc., 135, 18126.

INDEX

D

G

H

M

O

P

Q

W

X

Z